S0-FQV-525

RAMAN, INFRARED, AND NEAR-INFRARED CHEMICAL IMAGING

Edited by

SLOBODAN ŠAŠIĆ
Pfizer
Groton, Connecticut, USA

YUKIHIRO OZAKI
Kwansei Gakuin University
Sanda, Japan

WILEY

A JOHN WILEY & SONS, INC., PUBLICATION

Published by John Wiley & Sons, Inc., Hoboken, New Jersey
Published simultaneously in Canada

For general information on our other products and services or for technical support, please contact our
Customer Care Department within the United States at (800) 762-2974, outside the United States at (317)
572-3993 or fax (317) 572-4002.

Wiley also publishes its books in a variety of electronic formats. Some content that appears in print may not be
available in electronic formats. For more information about Wiley products, visit our web site at
www.wiley.com.

Library of Congress Cataloging-in-Publication Data:

Raman, infrared, and near-infrared chemical imaging / edited by Slobodan
Šašić, Yukihiro Ozaki.
      p. cm.
   Includes index.
   ISBN 978-0-470-38204-2 (cloth)
  1. Raman spectroscopy. 2. Infrared imaging. 3. Imaging systems in
chemistry. I. Šašić, Slobodan. II. Ozaki, Y. (Yukihiro)
   QD96.R34R36 2011
   543′.57–dc22
                                        2010010781

Printed in the United States of America

10 9 8 7 6 5 4 3 2 1

CONTENTS

PREFACE

Vibrational spectroscopy-based chemical imaging is a comparatively new imaging approach that has become truly operational in the past 10 years. It applies to samples more frequently met in academic rather than industrial laboratories, and it is used for exploration rather than routine analysis. However, it is steadily improving, gaining recognition in various industries, and finding use for solving a variety of real-world problems. The key points of chemical imaging based on Raman, infrared, or near-infrared response/spectra are chemical specificity and richness of information that stems from the collection of full-range spectra. Such wealth of data is usually best handled by the linear algebra-based algorithms that have once been considered fairly advanced but today are much more of a commonplace among this chemical imaging community. Applications of linear algebra (known as chemometrics in this field) are possible due to the assumed linearity of the responses from the imaged sample. Thus, quite often, a chemical imaging application (in this context) is a nice example of usefulness of chemometrics for extracting the information that would normally be unattainable or ambiguous by simply following response at a wavenumber that is considered indicative of a sought component (which, in fact, is the most commonly followed approach in the broadly popular imaging techniques). Many cases depicted in this book necessitate chemometrics for obtaining meaningful images.

Hardware is another key element in the development of chemical imaging. The improvement in combining spectrometers with microscopes (which roughly stands for a chemical imaging instrument) has been tremendous in recent times and certainly hugely contributed to this type of chemical imaging to become more widely used. It may not be an exaggeration to say that at present it is the applications that are somewhat behind what technically can be obtained from the available instruments. In this book, particular attention is paid to hardware.

The same holds for software. There are a couple of commercial software to chose from, and users quite frequently individually employ programming languages (with Matlab being unquestionably predominant) to tackle these complex 3D (or 4D) data sets via routines that do not really involve much more than skillfully combining existing algorithms. Here, it is not so much about improving the computational approaches as many of them have been used quite broadly for diverse problems, but rather finding suitable applications in this imaging field to demonstrate the ability to extract reliable information buried somewhere among thousands of spectra with hundreds of data points.

This book tries to portray all facets of chemical imaging via vibrational spectra. It starts with introducing vibrational spectroscopy, addresses hardware in more and software in less details (due to frequent references to computational details in the applications), and then methodically lists applications in several fields of which the biomedical and pharmaceutical ones probably dominate as for the number and impact of the publications, followed by no less promising and important food and polymers. A glimpse into future is also taken by listing several cutting-edge experimental endeavors. Each of the chapters in the book aims at covering all the three vibrational spectroscopy mechanisms (Raman, IR, and NIR) with Raman being given some more attention in the closing chapters. While, in essence, producing a chemical image may not be a tough task and can be done easily in some cases, it takes an expert with substantial knowledge of spectroscopy and chemometrics to single-handedly tackle demanding cases and unravel useful information from the labyrinth of data intrinsic to such experiments. The editors believe that the authors of this book are such individuals, world-class scientists, and authorities in respective fields. We hope that the joint effort presented in this book is an influential source of information on what chemical imaging is, how and what for to use it, and where to look for additional information. We hope readers will enjoy reading it as much as the authors enjoyed writing.

February 2010

SLOBODAN ŠAŠIĆ AND YUKIHIRO OZAKI

CONTRIBUTORS

Nils Kristian Afseth, Nofima Mat As, Ås, Norway

Ulrike Böcker, Nofima Mat As, Ås, Norway

R. A. Crocombe, Thermo Fischer Scientific, Billerica, MA, USA

Janie Dubois, Malvern Instruments, Columbia, MD, USA

Carol R. Flach, Department of Chemistry, Newark College, Rutgers University, Newark, NJ, USA

Paul Geladi, Unit of Biomass Technology and Chemistry, Swedish University of Agricultural Sciences, Umeå, Sweden

Ad Gerich, Schering Plough, Oss, The Netherlands

Hans Grahn, Department of Neuroscience, Division of Behavioral Neuroscience, Karolinska Institute, Stockholm, Sweden

Hiro-o Hamaguchi, Department of Chemistry, School of Science, The University of Tokyo, Tokyo, Japan

Xiaoxia Han, State Key Laboratory of Supramolecular Structure and Materials, Jilin University, Changchun, China

Mohammad Kamal Hossain, Department of Chemistry, School of Science and Technology, Kwansei Gakuin University, Sanda, Japan

R. A. Hoult, PerkinElmer, Buckinghamshire, UK

Yu-San Huang, Department of Chemistry, School of Science, The University of Tokyo, Tokyo, Japan

Tamitake Itoh, National Institute of Advanced Industrial Science and Technology, Takamatsu, Kagawa, Japan

Jianhui Jiang, State Key Laboratory of Chemo/Biosensing and Chemometrics, Hunan University, Changsha, China

Olga Jilkina, Institute for Biodiagnostics, National Research Council of Canada, Winnipeg, Manitoba, Canada

Hideaki Kano, Department of Chemistry, School of Science, The University of Tokyo, Tokyo, Japan; PRESTO (Precursory Research for Embryonic Science and Technology), Japan Science and Technology Agency, Saitama, Japan

Yasutaka Kitahama, Department of Chemistry, School of Science and Technology, Kwansei Gakuin University, Sanda, Japan

Sergei G. Kazarian, Department of Chemical Engineering, Imperial College London, London, UK

Linda H. Kidder, Malvern Instruments, Columbia, MD, USA

Valery V. Kupriyanov, Institute for Biodiagnostics, National Research Council of Canada, Winnipeg, Manitoba, Canada

E. Neil Lewis, Malvern Instruments, Columbia, MD, USA

Gurjit S. Mandair, Department of Chemistry, University of Michigan, Ann Arbor, MI, USA

Marena Manley, Department of Food Science, Stellenbosch University, Stellenbosch, South Africa

Richard Mendelsohn, Department of Chemistry, Newark College, Rutgers University Newark, NJ, USA

David J. Moore, ISP Corporation, Wayne, NJ, USA

Michael D. Morris, Department of Chemistry, University of Michigan, Ann Arbor, MI, USA

Yasuaki Naito, Department of Chemistry, Gakushuin University, Tokyo, Japan

Matthew P. Nelson, ChemImage Corporation, Pittsburgh, PA, USA

Yukihiro Ozaki, Department of Chemistry, School of Science and Technology, Kwansei Gakuin University, Sanda, Japan

C. C. Pelletier, NASA-Jet Propulsion Laboratory, Gales Ferry, CT, USA

M. J. Pelletier, Pfizer, Groton, CT, USA

Slobodan Šašić, Analytical Development, Pfizer, Groton, CT, USA

Harumi Sato, Department of Chemistry, School of Science and Technology, Kwansei Gakuin University, Sanda, Japan

J. Sellors, PerkinElmer, Buckinghamshire, UK

Laurence Senak, ISP Corporation, Wayne, NJ, USA

R. Anthony Shaw, Institute for Biodiagnostics, National Research Council of Canada, Winnipeg, Manitoba, Canada

Rintaro Shimada, Department of Chemistry, School of Science, The University of Tokyo, Tokyo, Japan

Hideyuki Shinzawa, Research Institute of Instrumentation Frontier, Advanced Industrial Science and Technology (AIST), Chubu, Nagoya, Japan

Michael G. Sowa, Institute for Biodiagnostics, National Research Council of Canada, Winnipeg, Manitoba, Canada

Athiyanathil Sujith, National Institute of Technology Calicut, Calicut, Kerala, India

Patrick J. Treado, ChemImage Corporation, Pittsburgh, PA, USA

Patrick S. Wray, Department of Chemical Engineering, Imperial College London, London, UK

N. A. Wright, Applied Instrument Technologies, Hamilton Sundstrand, Pomona, CA, USA

Ru-Qin Yu, State Key Laboratory of Chemo/Biosensing and Chemometrics, Hunan University, Changsha, China

Lin Zhang, Analytical Development, Pfizer, Groton, CT, USA

1

SPECTROSCOPIC THEORY FOR CHEMICAL IMAGING

M. J. PELLETIER

Pfizer, Groton, CT, USA

C. C. PELLETIER[1]

NASA–Jet Propulsion Laboratory, Gales Ferry, CT, USA

1.1 INTRODUCTION

All images require some type of contrast to differentiate regions of interest in a field of view. The most common source of image contrast is variation in the intensity of reflected light. Contrast can, however, be based upon any measurable property of the sample that can be expressed as a function of location. Contrast is improved for measurements having a wider dynamic range and by measuring a larger number of variables for each pixel, as in color versus black-and-white photography. Contrast may also be enhanced with one or more of a wide range of techniques including digital image processing and structured illumination. This book will focus on chemical images generated using vibrational spectroscopic contrast. Such contrast is generated by quantifying one or more attribute(s) of an infrared absorption, infrared emission, or Raman scattering spectrum for each pixel. By providing a window into the spatial distribution of properties such as molecular composition, structure, state, and concentration, images based on vibrational spectroscopies open up a new way of seeing the world.

Imaging can be accomplished by measuring a property from the entire field of view simultaneously (global imaging) or by measuring a property from individual points in the field of view sequentially and combining the points to create the image (mapping). Since mapping requires a large number of measurements, each measurement must be relatively fast for mapping to be practical. For example, an image consisting of 640×480 pixels contains over 300,000 measurements and would take more than 3.5 days to acquire if each measurement required 1 s. Mapping speed can be increased by simultaneously measuring a property at multiple points in a subregion of the field of view and combining those subregions to create the image. The subregion may consist of a single column of measurement points (line imaging) or may contain multiple columns (mosaic imaging). In most cases, even global imaging requires multiple frames, each containing different spectroscopic information, to be collected sequentially and overlaid to form a single image. Sample changes during the course of sequential measurements can confound the interpretation of spectral images.

This chapter provides an introduction and theoretical background for vibrational spectroscopies, as used to produce chemical images. Infrared, Raman, and related spectra result from the interactions of electromagnetic radiation with molecular vibrations, so this chapter begins with a description of relevant aspects of molecular vibration, followed by a section on electromagnetic radiation and its interactions with matter. Next are three sections on infrared spectroscopies, divided by spectral region. After that, several different types of Raman spectroscopy that are used for chemical imaging are described. The final section briefly presents the use of Raman and infrared spectroscopies for creating large chemical images by remote sensing. Remote sensing is probably responsible for the majority of chemical images created because of its use in mapping the atmosphere, planets including Earth, and moons, and in astronomy.

---

[1] Retired

1.2   MOLECULAR VIBRATIONS

A chemical bond between two atoms can be modeled as a spring connecting two point masses. If the spring follows Hook's law, the force it applies between the two point masses will be proportional to the spring displacement from its lowest energy position. The system, called a harmonic oscillator, will have a single resonant vibrational frequency, v, given by

$$v = \frac{1}{2\pi c} \left(\frac{k}{\mu}\right)^{1/2} \qquad (1.1)$$

where c is the speed of light, k is the force constant, and μ is the reduced mass, $m_a m_b / (m_a + m_b)$.

Equation 1.1 describes the vibrational frequency of diatomic molecules reasonably well. Increasing the strength of the chemical bond increases the vibrational frequency. Increasing the atomic mass reduces the vibrational frequency.

The force applied by a chemical bond does not follow Hook's law exactly, though. Atoms have finite size and cannot occupy the same space. As a result, the repulsive force increases much more quickly than Hook's law would predict as the atoms get close together. As the atoms get further apart the chemical bond weakens, approaching zero strength at infinite separation, again violating Hook's law. Deviations from Hook's law are amplified by disparity between the molecular masses. Vibrating systems that do not follow Hook's law are called anharmonic, and the extent to which they deviate from an ideal harmonic oscillator is called anharmonicity. Anharmonicity has a relatively small role in most forms of Raman spectroscopy, a somewhat larger role in mid-infrared (mid-IR) spectroscopy, and is of primary importance in near-infrared (NIR) spectroscopy.

Vibrations in molecules containing more than two atoms are more complicated. The total number of different, or normal, vibrations (ignoring anharmonicity) in a molecule with n atoms is $3n - 5$ for a linear molecule and $3n - 6$ for a nonlinear molecule. For example, an anthracene molecule has 24 atoms and therefore 66 normal vibrations. Some of these vibrations have exactly the same frequency (called degenerate vibrations). Other vibrations produce no signal for a particular type of vibrational spectroscopy due to symmetry constraints. As a result of these spectral simplifications, even most large molecules have manageable vibrational spectra.

Oscillators sharing a common atom may exert forces on each other when they oscillate. If the oscillator frequencies are very different from each other, each oscillator remains fairly independent of the other. If the frequencies are similar, though, the oscillators can couple, essentially forming a new single oscillator with new frequencies. Consider the linear CO_2 molecule. Both carbon–oxygen bonds are identical. They couple to form a single oscillator having two different

vibrations. One vibration consists of each carbon–oxygen bond stretching in phase, resulting in a vibration where the carbon atom does not move. This in-phase vibration is an example of a symmetric vibration. The other vibration consists of the carbon–oxygen bonds stretching out of phase with each other, resulting in a vibration where the carbon atom moves and the oxygen atoms do not. This out-of-phase vibration is an example of an antisymmetric vibration. In general, the antisymmetric vibration tends to be at higher frequency and the symmetric vibration tends to be at lower frequency than the natural frequency of the uncoupled oscillators.

Groups of atoms in a molecule that are not vibrationally coupled to the rest of the molecule, to a first approximation, have about the same frequencies of vibration in any molecule. This makes it possible to associate a vibrational frequency with a particular chemical functional group, such as a carbonyl group or a phenyl ring, without considering the rest of the molecule. These general-purpose vibrational frequencies are called group frequencies. Tabulations of group frequencies are typically refined to include the small frequency shifts caused by properties of the rest of the molecule, such as a weakening of the oscillator bond strength due to electron density withdrawal by the rest of the molecule. Tabulations of characteristic frequencies also are specific to a type of vibrational spectroscopy, since vibrations that produce a strong signal with one type may produce little or no signal with a different type of vibrational spectroscopy.

Molecular vibrations are often classified into groups that are intuitively descriptive of the vibrational motion. An oscillation in bond length is called a "stretch." An oscillation in bond angle is called a "deformation" or "bend." More specialized descriptions include terms such as "wag," "rock," or "breathing mode." Another way to classify molecular vibrations is by their symmetry properties using group theory. It can be shown that vibrations having certain symmetry properties will theoretically produce exactly zero signal for some types of spectroscopy, but not for other types of spectroscopy. Rules derived from symmetry considerations that identify vibrations expected to produce no spectroscopic signal are called selection rules. A detailed explanation of the use of group theory in vibrational spectroscopy is given in Refs 1 and 2.

1.3   INTERACTIONS BETWEEN ELECTROMAGNETIC RADIATION AND MATTER

1.3.1   Electromagnetic Radiation

Electromagnetic radiation consists of electric and magnetic fields oscillating in phase with each other and perpendicular to both each other and the direction of propagation. Gamma rays, X-rays, ultraviolet (UV) radiation, visible light, NIR and mid-IR radiation, terahertz (far-infrared) radiation, microwaves, and radio waves are all forms of electromagnetic radiation, differing only in their decreasing frequencies of oscillation. The energy of electromagnetic radiation is quantized. The smallest unit of light is the photon, having an energy, E, given by $E = hv$, where v is the frequency of the electromagnetic radiation and h is Planck's constant.

Electromagnetic radiation can be thought of either as a particle (photon) or as a wave. We will use the representation that is most intuitive when describing phenomena involving electromagnetic radiation. For simplicity, we will use the term "light" as synonymous with electromagnetic radiation of any frequency, rather than just those frequencies that are visible to the human eye.

Light travels at 2.99792458×10^8 m/s in a vacuum. The speed of light can be used to convert time into distance, thereby providing depth resolution. Raman, mid-infrared, and near-infrared spectroscopies have all been used this way to make three-dimensional chemical images of objects in the atmosphere, such as clouds or discharge plumes.

Light of a particular frequency can be specified by its wavelength (the distance light travels during one oscillation cycle of the electric field), its wavenumber (the number of oscillating cycles per centimeter), or its energy (e.g., Joules per photon). For example, light having a frequency of 6.00×10^{14} Hz has a wavelength of 500 nm, a wavenumber of 20,000 cm^{-1}, and an energy of 3.98×10^{-19} J/photon, or 57.2 kcal/mol of photons.

Another important property of light is coherence. Coherence is a nonrandom relationship between photons. Coherence may be spatial (photon relationships based on photon location and/or direction) or temporal (relationships based on time when maxima in the oscillation fields of the photons occur). For example, a thermal light source is temporally incoherent because there is no mechanism coordinating the time that different photons are emitted. Lasers are temporally coherent because the process of stimulated emission causes the created photons to be in phase with the photons that stimulated the emission. Some spectroscopic processes such as coherent anti-Stokes Raman spectroscopy (CARS) or Raman gain spectroscopy rely on establishing temporal coherence between photons. Spectroscopic techniques such as FTIR (Fourier transform infrared spectroscopy) or OCT (optical coherence tomography) rely on establishing temporal coherence from nominally incoherent light sources.

1.3.2 Absorption and Emission of Light

A material having an internal process, such as molecular vibration, that is resonant with the frequency of incident light can be excited to a higher energy state by absorbing some of the light. The higher energy state usually relaxes back to the lowest energy state quickly by releasing heat and/or light.

The strength of optical absorption or emission can be used to determine analyte concentration. Beer's law [3] relates analyte concentration to the strength of optical absorption, regardless of whether the transition involves an electronically, vibrationally, or rotationally excited state:

$$A_\lambda = -\log T = a_\lambda bc \qquad (1.2)$$

where A_λ is the absorbance at wavelength λ, a_λ is the molar absorptivity at wavelength λ, b is the path length, c is the analyte concentration, and T is the transmittance, that is, ratio of transmitted intensity to incident intensity.

Emission intensity is also proportional to analyte concentration.

1.3.3 Refractive Index

Light slows down relative to its speed in a vacuum when traveling through matter. The ratio of the speed of light in a vacuum to that in a material is the refractive index of that material. Light incident on a planar interface between two transparent materials of different refractive indices is bent as a result of this speed change if the light is not perpendicular to the interface. The bending at this interface is described by Snell's law:

$$n_1 \sin \theta_1 = n_2 \sin \theta_2 \qquad (1.3)$$

where n_1 is the refractive index of the first material, θ_1 is the angle of light with respect to interface normal in the first material, n_2 is the refractive index of the second material, and θ_2 is the angle of light with respect to interface normal in the second material.

The refractive index of a material changes with the wavelength of the light, as well as with the temperature of the material.

Light is also reflected at an interface between two transparent materials having different refractive indices. The reflected intensity is given by the Fresnel equations [4]

$$\begin{aligned} R_\perp &= \frac{I_R}{I_i} = \left(\frac{n_1 \cos \theta_1 - n_2 \cos \theta_2}{n_1 \cos \theta_1 + n_2 \cos \theta_2} \right)^2, \\ R_{//} &= \frac{I_R}{I_i} = \left(\frac{n_2 \cos \theta_1 - n_1 \cos \theta_2}{n_1 \cos \theta_2 + n_2 \cos \theta_1} \right)^2 \end{aligned} \qquad (1.4)$$

where $R_\perp$ is the reflectance of light polarized perpendicular to the plane of incidence, $R_{//}$ is the reflectance of light polarized parallel to the plane of incidence, I_R is the intensity of reflected light, I_i is the intensity of incident light, n_1 is the refractive index of the first material, θ_1 is the angle of light with respect to interface normal in the first material, n_2 is the refractive index of the second material, and θ_2 is the angle of light with respect to interface normal in the second material.

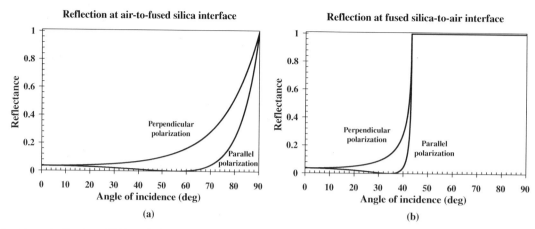

FIGURE 1.1   Fresnel reflection of 500 nm light at the interface between air and fused silica. (a) Light traveling from air into fused silica; (b) light traveling from fused silica into air.

Figure 1.1a and b shows the reflectivity of an interface between air ($n = 1.000$) and fused silica ($n = 1.462$) for both polarizations of 500 nm light. The reflectivity of the interface for light traveling into the higher refractive index material generally increases with angle of incidence, except for a reduction to zero at Brewster's angle θ_B ($\theta_B = \mathrm{Tan}^{-1}(n_2/n_1)$, or 55.6° in this example) for light polarized parallel to the plane of incidence. Light traveling from the higher index material experiences total internal reflection for incidence angles greater than the critical angle θ_c ($\theta_c = \mathrm{Sin}^{-1}(n_2/n_1)$, or 43.2° in this example). During total internal reflection, no energy is transmitted through the interface. An evanescent field does extend into the lower refractive index material, though. If a higher refractive index material, or an absorbing material, is placed in the evanescent field, energy can be transferred through the evanescent field to this material. The energy transfer process is called "attenuated total reflection" or ATR [5]. The evanescent field rapidly decays with distance from the interface:

$$d_p = \frac{\lambda}{2\pi n_1 (\sin^2 \phi - n_{21}^2)^{1/2}} \qquad (1.5)$$

where d_p is the penetration depth of the evanescent field, λ is the wavelength of light, n_1 is the refractive index of the ATR crystal, ϕ is the internal angle of incidence, and n_{21} is the refractive index ratio of the sample to the ATR crystal.

ATR has an important role in some spectroscopic imaging techniques discussed later in this book.

As the dimensions of the refractive index interface approach the wavelength of light, diffraction effects dominate, and the interaction between light and the refractive index interface, which we now call a particle, is best described as Mie scattering. Mie scattered light travels in all directions from the particle. Most of the Mie scattered light intensity travels in the forward direction for particles 5–10 optical wavelengths in size. The scattered intensity becomes less directed as the particle size becomes smaller.

Molecules can scatter light in two different ways. If scattering does not change the energy of the light, it is called elastic scattering or Rayleigh scattering. Inelastic scattering or Raman scattering changes the energy of the scattered light. Raman scattering is described in greater detail later in this chapter. The intensity of molecular scattering is proportional to the fourth power of the optical frequency. Rayleigh scattering of polarized light is strongest in directions perpendicular to the electric field of the light and goes to zero in the direction parallel to the electric field. The polarization dependence of Raman scattered light is more complex and is described later in this chapter.

Light can be scattered many times when it interacts with a material consisting of a dense collection of many particles. The optical path through such a sample is best represented by a distribution of paths whose median can be 10–100 times longer than an unscattered path through the material. A large number of scattering events also tend to depolarize the light. Light that is ultimately reflected from a material after multiple scattering events is called diffusely reflected light. Similarly, light that is ultimately transmitted by a material after multiple scattering events is called diffusely transmitted light.

Diffuse reflectance and diffuse transmission usually degrade images. Spectroscopic imaging systems are often designed to minimize the detection of diffusely scattered light emanating from a sample. One exception is the use of spatially localized diffuse reflectance for depth discrimination and depth profiling. The most probable light paths through a highly scattering material form a banana-shaped volume connecting the point where light enters the material to a point where light exits the material, as illustrated in Figure 1.2. The depth of the material probed by the light increases with increasing separation of the optical entrance

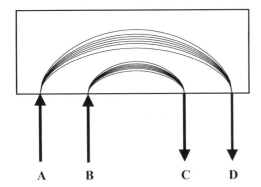

FIGURE 1.2 Depth discrimination in a highly scattering material by spatially resolved diffuse reflectance measurements. The most probable light paths connecting spatially separated excitation and collections points form a banana-shaped volume in the sample. Using points A and D for excitation and detection probes a greater depth than using points B and C for excitation and detection.

and exit points. This approach for depth discrimination and volume imaging has been extensively used in near-infrared spectroscopy [6–9]. Raman spectra have been collected this way through scattering media [10, 11], so perhaps Raman diffuse reflectance imaging is also possible.

1.3.4 Thermal Emission

All materials are continuously emitting radiation simply because they are at a temperature above absolute zero. If they are in equilibrium with their environment, they are also absorbing an equivalent amount of energy from the environment in order to maintain a constant temperature. A material that completely absorbs all frequencies of incident optical radiation, called an ideal blackbody source, has an emission spectrum given by [12]

$$H_\lambda(T) = \frac{2hc^2}{\lambda^5(e^{hc/\lambda kT}-1)}, \; H_v(T) = \frac{2hc^2v^3}{e^{hcv/kT}-1} \quad (1.6)$$

where $H_\lambda(T)$ is the spectral radiant energy density per nanometer, $H_v(T)$ is the spectral radiant energy density per wavenumber, λ is the wavelength of light, v is the wavenumber of light, T is the temperature in Kelvin, h is Planck's constant $(6.626 \times 10^{-34}\,\text{J s})$, c is the speed of light $(2.998 \times 10^8\,\text{m/s})$, and k is Boltzmann's constant $(1.3807 \times 10^{-23}\,\text{J/K})$.

Real materials do not totally absorb all frequencies of electromagnetic radiation. Their emission spectra consist of an ideal blackbody emission spectrum multiplied by their absorbance spectrum, where absorbance is expressed as the fraction of light absorbed. For example, an ideal blackbody has a fractional absorbance of 1, and a completely transparent object has a fractional absorbance of 0.

Thermal emission spectroscopy can determine the absorption spectrum of a material from its spontaneous emission of light. Laboratory samples are often heated to improve the quality of the data. Chemical imaging based on thermal emission spectroscopy is extensively used for remote sensing, which is described in more detail at the end of this chapter.

1.3.5 Fluorescence

Fluorescence is one process where an electronically excited state decays to a lower electronic state by emitting a photon. An energy level diagram describing fluorescence is shown in Figure 1.3. The excited state is usually the lowest vibrational level of the first excited singlet electronic state. The lower state is usually one of many vibrational levels in the electronic ground state. The emission bands to the vibrational levels in the ground state overlap spectrally giving a relatively broad fluorescence emission spectrum with few spectral features.

The lifetime of the fluorescence process is typically on the order of 1–10 ns. Kinetically competing processes that return a molecule from the electronic excited state to the ground state without the emission of a photon, called dark reactions, reduce the fluorescence lifetime. They also reduce the fluorescence quantum yield, defined as the number of fluorescence photons produced divided by the number of molecules in an excited state capable of producing fluorescence photons. The process of reducing the fluorescence quantum yield is called fluorescence quenching. Highly fluorescent molecules have quantum yields very close to 1.

Absorption of a photon is the most common mechanism for creating the excited state necessary for fluorescence emission. Other mechanisms include chemical excitation

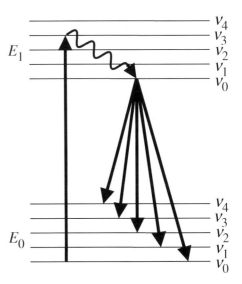

FIGURE 1.3 Energy level diagram illustrating fluorescence.

(chemiluminescence) or electron bombardment (cathodoluminescence). The absorption spectrum for the process of creating excited states for fluorescence emission is called the fluorescence excitation spectrum. The absorption spectrum of a material is the sum of the fluorescence excitation spectrum and the absorption spectrum of all processes that do not produce fluorescence. Materials having a single fluorescent species have the same emission spectrum at all excitation wavelengths, since fluorescence almost always occurs from the lowest vibrational level of the first electronic excited state, regardless of how that state got populated.

The fluorescence emission spectrum from impurities is often the combined spectra of many different chemical compounds. The observed fluorescence therefore has properties that differ from the fluorescence of a pure material. Excitation and emission spectra may be spectrally broader. The shape of the emission spectrum may change significantly with changing excitation wavelength, or during fluorescence quenching. The fluorescence decay rate may become very nonexponential due to different impurity fluorescence lifetimes. Since materials having multiple fluorescent species can have multiple excitation spectra, excitation–emission matrices are used to describe their fluorescence. Excitation–emission matrices are three-dimensional plots on axes of excitation wavelength, emission wavelength, and emission intensity.

Fluorescence imaging is a powerful and very popular chemical imaging technique, but it is outside the scope of this book. We include fluorescence here because it is often a serious nuisance that limits the capabilities of Raman chemical imaging. This limitation will be discussed in greater detail later in the book. In the context of Raman spectroscopy, the term fluorescence is often used generically to mean any process (often unknown) that produces a spectrally broad background intensity. Phosphorescence is one example of a nonfluorescence process that may be mistaken for fluorescence in a Raman measurement.

1.4 MID-INFRARED ABSORPTION SPECTROSCOPY

The mid-infrared spectral range includes wavelengths from about 2.5 to 25 μm. This corresponds to about 4000–400 cm^{-1} or 11–1.1 kcal/mol. Absorption in this spectral region is due to molecular vibrations that modulate the dipole moment of the molecule. The energy of these vibrations is small compared to the energy of a chemical bond. For example, the C–H bond energy of 98 kcal/mol is 12 times greater than the vibrational energy, 8.4 kcal/mol, of the C–H stretching vibration at 2950 cm^{-1}.

Table 1.1 lists some characteristic mid-infrared absorption frequencies of common functional groups. Much more extensive tables are given in Refs 13–15. Tables such as these provide a good starting point for estimating the spectral

TABLE 1.1 Mid-Infrared Characteristic Frequencies for Several Common Functional Groups

Vibration	Shift (cm^{-1})	Group
OH stretch (dilute solution)	3600–3700	–OH in alcohols and phenols
OH stretch (solids and liquids)	3250–3420	–OH in alcohols and phenols
NH$_2$ antisymmetric stretch (solids)	3340–3360	–NH$_2$ in primary amines
H-bonded OH stretch; very broad	2400–3100	–OH in carboxylic acids
=C–H stretch	3000–3100	Unsaturated hydrocarbons
C–H stretch	2850–2990	Aliphatic hydrocarbons
C≡N stretch	2200–2260	Nitriles
Overtones and combination bands	1650–2000	Substituted benzene rings
C=O stretch	1650–1870	Carbonyl compounds
C=O stretch	1740–1750	Esters
C=O stretch	1700–1720	Ketones
NH$_2$ deformation	1580–1650	Primary amines
Ring stretch, sharp peak	1590–1615	Benzene ring in aromatics
COO$^-$ antisymmetric stretch	1560–1610	Carboxylic acid salts
Antisymmetric CH$_3$ deformation	1440–1465	CH$_3$ in aliphatics
Symmetric CH$_3$ deformation	1370–1380	CH$_3$ in aliphatics
C–O stretch	1015–1200	Alcohols
Si–O–Si antisymmetric stretch	1000–1100	Siloxanes
C–Br stretch	500–650	C-Br in bromo compounds
COC bend	430–520	Ethers
CNC bend	400–510	Amines

Compiled from Ref. 15.

location of mid-infrared absorption bands that may be analytically useful for chemical imaging. They are also useful for assigning bands observed in a spectrum of a known material to chemical groups in the material.

Huge libraries of mid-infrared spectra are available that can provide the experimentally observed spectrum of most common materials, often eliminating the need to estimate spectra from characteristic frequency tables. These commercial libraries can be supplemented by custom libraries or small sets of experimental spectra collected from standards. When the spectrum of a desired material is not available from libraries, the spectra of several related materials from the library can be used as a highly specific characteristic frequency table to estimate the desired spectrum.

TABLE 1.2 Mid-Infrared Molar Absorptivities for Several Common Functional Groups

Band Descripton	Sample	Band Position (cm^{-1})	Molar Absorptivity (L/(mole cm))	Path Length for 1 AU in Neat Material (μm)
OH stretch	Water	3404	100	1.8
OH bend	Water	1643	22	8.3
CH stretch	Toluene	3025	53	20.2
CH bend	Toluene	728	302	3.5
Ring stretch	Toluene	1496	94	11.4
CH bend	Dichloromethane	1265	109	5.9
CH stretch	Dichloromethane	3054	8	83.1
CH stretch	Benzene, 25°C	3036	79	11.2
CH bend	Benzene, 25°C	673	397	2.2
Ring stretch	Benzene, 25°C	1478	102	8.7
CH stretch	Acetonitrile	2944	5.83	90.5
CH bend	Acetonitrile	1445	15.54	34.0

Compiled from Refs 16–20.

The intensity of mid-infrared absorption by a molecular vibration is proportional to the square of the change in dipole moment. Functional group absorptivities are not as generally useful as functional group frequencies, however, because dipole moments are much more sensitive to neighboring group effects. The absorptivities of molecular vibrations do follow Beer's law, however, so mid-infrared molar absorptivities are useful for measuring analyte concentrations. Table 1.2 lists reported molar absorptivities for vibrations in some common materials. The relatively strong mid-infrared absorption of these materials requires sample path lengths to be on the order of 10 μm or less to yield undistorted spectra.

Not all molecular vibrations absorb light. For example, the symmetric stretching vibration of carbon dioxide described earlier has the changing dipole moment of one C–O bond exactly cancelled out by the changing dipole moment of the other C–O bond. Since this vibration has zero change in its dipole moment, it cannot absorb infrared light. More generally, group theory can be used to identify vibrations having a symmetry that causes any change in the dipole moment of one chemical bond to be cancelled by a corresponding change in another chemical bond. Such vibrations do not absorb light, and are called symmetry forbidden.

Mid-infrared absorption chemical images can be created by measuring spectra of external light intensity *not* absorbed by the sample. Three different techniques to do this are based on measuring light intensity after transmission through the sample, after reflectance from the sample, and after ATR. All three techniques can produce images by mapping or by global imaging. ATR can be used in a different mode to measure mid-infrared depth profiles by changing the penetration depth of the evanescent wave [21]. This can be done by varying the angle of incidence at the point of total internal reflection or by using ATR elements having different refractive indices.

Mid-infrared depth profiles can also be created by measuring the light intensity absorbed by a material using photoacoustic spectroscopy [22]. Absorbed light produces a thermal wave that travels back to the surface of the sample. Some of the thermal wave energy couples into gas at the sample interface producing sound that is detected by a sensitive microphone. The penetration depth into the sample is determined by the modulation frequency of the mid-infrared light, which can be changed by changing the scan speed of a Fourier transform mid-infrared instrument. Sampling depths typically range from several to 100 μm.

Mid-infrared chemical images can be created from the spontaneous thermal emission spectra of objects as well, since an object's absorption spectrum can be deduced from its emission spectrum. Between −20 and +50 °C, typical of environmental temperatures on the Earth, the wavelength of maximum ideal blackbody intensity is between 9 and 11.5 μm. These emission wavelengths are not only in the center of the highly predictive mid-infrared fingerprint spectral region, but also in the atmospheric transmission window between 8 and 14 μm. This makes mid-infrared emission spectroscopy especially attractive for remote sensing. Laboratory applications of mid-infrared emission spectroscopy often involve sample heating, since sensitivity increases with increasing temperature difference between the sample and the detector.

1.5 FAR-INFRARED AND TERAHERTZ SPECTROSCOPY

The far-infrared, terahertz, and submillimeter spectral regions are all labels for approximately the same interval in the electromagnetic spectrum. This spectral interval includes wavelengths ranging from about 25 to 1000 μm. This corresponds to about 400–10 cm^{-1}, 12–0.3 THz, or 1.14–0.0286 kcal/mol. Room-temperature thermal energy,

kT, is in this spectral range at about $207\,\mathrm{cm}^{-1}$. The different names, and differing spectral limits for the region, have been associated with different bodies of experimental technique. Both "far-infrared" and "submillimeter" imaging and spectroscopy are copious in the astronomical and remote sensing literature. "Terahertz" has become more often associated with measurements in this spectral region using innovative new light sources and detection methods based on femtosecond lasers, quantum cascade lasers, or nonlinear optical techniques.

Light absorption in the far-infrared region requires dipole moment oscillation at lower frequencies than in the mid-infrared spectral region, implying harmonic oscillators with greater masses and/or weaker bond strengths. Intramolecular vibrations contributing to this spectral region include stretching of bonds involving heavy atoms, organic skeletal bending modes, torsional modes (restricted rotational motion about single bonds), and ring puckering of small-ring molecules. Intermolecular vibrations, between different molecules associated by hydrogen bonding or electrostatic interactions, also occur in this spectral region, as well as crystal lattice modes of polymers and inorganic solids. Pure rotational transitions of light, gas-phase molecules extend from the microwave region into the far-infrared region. Table 1.3 gives some typical molar absorptivities for some far-infrared absorption bands.

Blackbody excitation sources for spectroscopy are very weak in the far-infrared spectral region. While other light sources such as the HCN laser, quantum cascade lasers, and difference-frequency generation optics are available for making traditional transmission measurements in this spectral region, terahertz spectroscopy [27] has arisen as the primary far-infrared imaging technology. Terahertz spectroscopy uses unique light sources and detection methods that give it capabilities not available to traditional absorption spectroscopy. Briefly, pulses of terahertz radiation are generated by illuminating a biased photoconductive antenna with ultrashort pulses of near-infrared light from a titanium sapphire laser. The pulses are detected with a similar time-gated photoconductive antenna. The transit time, phase, and amplitude of the subpicosecond terahertz pulse are recorded after it interacts with the sample, making possible the calculation of distance and both the refractive index and absorption spectrum of the sample.

Terahertz instruments are usually operated in either an imaging mode or a spectroscopic mode, though combination 3D spectroscopic imaging devices have been reported [28]. The imaging mode measures the reflections from the sample that occur at interfaces between compositions of differing refractive index. The refractive indices of the materials in adjacent layers are determined from the intensity of the reflection using the Fresnel equations. The thickness of a layer is determined from the time between reflections from the interfaces at the start and end of the layer. Each pixel in a two-dimensional map can provide a depth profile of the

TABLE 1.3   Far-Infrared Molar Absorptivities for Some Common Materials

Sample	Spectral Position (cm^{-1})	Molar Absorptivity (L/(mole cm))	Path Length for 1 AU in Neat Material (mm)
Benzene	300	0.110	8.1
Benzene	185	180	0.005
Benzene	33[a]	0.24	3.7
Toluene	345	1.63	0.66
Toluene	33[a]	0.5	2.1
Dichloromethane	285	2.77	0.23
Methanol	34[a]	2.22	0.18
Water	198	9.35	0.019
Water	32[a]	1.82	0.10
Hexane	33[a]	0.09	15

Compiled from Refs 20 and 23–26.
[a] Slowly varying background absorption, not a peak location.

sample refractive index. A three-dimensional map of an entire pharmaceutical tablet can be collected in several minutes. The lateral spatial resolution is diffraction limited at hundreds of micrometers, while the axial resolution is determined by the instrument time resolution, which is usually on the order of $30\,\mu\mathrm{m}$.

The spectroscopic mode of a terahertz instrument can determine the absorption spectrum of a sample by Fourier transforming a terahertz pulse transmitted by the sample. The use of transmission rather than reflection eliminates spectral artifacts that would result from including pulse reflection in the Fourier transformation. Spectra of individual layers can be obtained, however, by using a windowed Fourier transform between reflected pulses when carrying out terahertz imaging. The quality of these spectra may be compromised, by refractive index heterogeneity in the layer.

Far-infrared emission spectroscopy is extensively used for making chemical images of deep space. This topic will be covered in greater detail in Section 1.8.

1.6   NEAR-INFRARED ABSORPTION SPECTROSCOPY

The near-infrared spectral range includes wavelengths from about 0.78 to $2.5\,\mu\mathrm{m}$. This corresponds to about $12{,}820$–$4000\,\mathrm{cm}^{-1}$ or 37–11 kcal/mol. The energies of these vibrations are greater than those of the mid-infrared fundamental vibrations, but still considerably less than the bond energies of the vibrating chemical bonds. For example, the C—H bond energy of 98 kcal/mol is 5.8 times greater than the $5870\,\mathrm{cm}^{-1}$ C—H stretching vibration's first overtone energy of 16.8 kcal/mol.

Absorption in this spectral region is due to overtone and combination bands of molecular vibrations that modulate the dipole moment of the molecule. Overtone bands result from the simultaneous absorption of two or more vibrational quanta by the same vibration. Combination bands result from the simultaneous absorption of two or more vibrational quanta by two or more different vibrations of the same symmetry and of the same functional group.

An overtone or combination band has an energy slightly below the sum of the individual energies of the related fundamental vibrations. For example, the fundamental C–H stretching vibrations for aliphatic hydrocarbons occur in the region of 2800–$3000\,cm^{-1}$. Their first, second, and third overtones occur in the spectral regions of 5555–5882, 8264–8696, and 10929–$11664\,cm^{-1}$, respectively [29]. This deviation from the sums of fundamental vibrational energies is due to the anharmonicity of the vibration. In fact, the very existence of overtone and combination bands requires vibrational anharmonicity, and the intensity of these absorption bands increases with increasing anharmonicity.

Spectral interpretation in the near-infrared spectral region is less developed than it is in the mid-infrared spectral region. The wide diversity of possible combination bands that are severely overlapped makes definitive band assignments difficult. An excellent compilation of band assignments for important molecules in the near-infrared spectral region is given in Ref. 29. Published band assignments for model molecules, along with estimates of band locations by summing the energies of the fundamental Vibrations involved, provide a good starting point for estimating the spectral location of near-infrared absorption bands that may be analytically useful for chemical imaging. Libraries of near-infrared spectra are also available.

Overtone and combination bands tend to be weaker by a factor of 10–100 for each additional fundamental vibrational quantum contributing to their absorption. For example, a first overtone absorption band (two vibrational quanta) will be 10–100 times stronger than the corresponding second overtone absorption band (three vibrational quanta),

but 1/10 to 1/100 as strong as fundamental absorption bands in the mid-infrared region. If combination and overtone bands involving different numbers of vibrational quanta are present in the same spectral region, the intensities of those composed of the fewest vibrational quanta will tend to dominate those composed of more vibrational quanta. Stretching vibrations involving a hydrogen atom, mainly C–H, O–H, and N–H vibrations, have the highest fundamental vibrational energies, so their overtones and combination bands tend to dominate the near-infrared spectra of organic molecules. The large mass disparity between hydrogen and the atom to which it is bonded increases the anharmonicity of the vibration, further enhancing the dominance of these vibrations in near-infrared spectroscopy.

Near-infrared spectra are strongly affected by hydrogen bonding. Hydrogen bonding not only changes the bond strength of hydrogen-bond donors such as O–H and N–H bonds, but also reduces their anharmonicity, and therefore, the intensities of their near-infrared absorption bands. Overtone and combination bands involving O–H and N–H groups are more intense when those groups are not hydrogen bonded than when they are.

Temperature also has an important impact on near-infrared spectra. Anharmonicity of absorbing groups tends to increase with temperature. Temperature also affects hydrogen bonding, which, in turn, affects near-infrared spectra. As a result, band intensities and frequencies may change significantly with small changes in temperature.

Tables of near-infrared functional group absorptivities are mainly useful for semiquantitative estimations rather than for quantitative analysis due to the dependence of near-infrared absorbance on the sample microenvironment. The absorptivities of molecular vibrations do follow Beer's law, however, so near-infrared molar absorptivities are useful for measuring analyte concentrations. Table 1.4 lists reported molar absorptivities for vibrations in some common materials.

TABLE 1.4   Near-Infrared Molar Absorptivities for Several Common Functional Groups

Band Descripton		Sample	Band Position (cm^{-1})	Molar Absorptivity (L/(mol cm))	Pathlength for 1 AU in Neat Material (mm)
NH	$v + \delta$	1-Propaneamine	4942	0.98	0.84
NH	$2v$	1-Propaneamine	6550	0.58	1.4
NH	$3v$	1-Propaneamine	9560	0.05	16.3
aliphatic OH	$v + \delta$	1,3-Propanediol	4782	0.69	1.1
aliphatic OH	$2v$	1,3-Propanediol	6750	0.16	4.4
aromatic OH	$2v$	2,5-Dichlorophenol[a]	6901	2.23	0.44
CH	$v + \delta$	n-Hexane	4334	3.43	0.38
CH	$2v$	n-Hexane	5800	0.62	2.1
CH	$3v$	n-Hexane	8396	0.11	11.8
CH	$2v + \delta$	n-Hexane	7182	0.09	13.9
OH	$v + \delta$	Water	5173	1.07	0.17
OH	$2v$	Water	6886	0.26	0.70

Compiled from Refs 23 and 30.
[a] Slowly varying background absorption, not a peak location.

Overall, near-infrared absorptivities are much weaker than those in the mid-infrared spectral region. Neat materials often require sample path lengths on the order of millimeters to get undistorted spectra of two-quantum overtone and combination bands, compared with 0.01 mm in the mid-infrared spectral region. The increased path length can be a benefit when analyzing larger samples. The maximum usable path length can be further increased by measuring three-quantum overtone and combination bands.

The relatively low absorptivities in the near-infrared spectral region allow light to be detected after many scattering events in inhomogeneous materials. The optical path in the sample is then mainly a function of the sample elastic scattering properties, which depend on physical properties such as particle size, hardness, and density. In most cases, elastic scattering degrades near-infrared chemical images. Not only is the size of the point spread function (transverse and axial) increased in a sometimes unpredicted way, but the distribution of optical path lengths also creates a nonlinear relationship between absorbance and analyte concentration.

Near-infrared chemical images can be created by measuring the spectrum of light transmitted through the sample, reflected from the sample, and emanating from a spatially offset position after diffuse reflection through the sample. All three techniques can produce images by mapping. The first two can also be carried out by global imaging. Spatially offset diffuse reflectance can be used for depth profiling as described earlier. One commercial application for near-infrared spatially offset diffuse reflectance is noninvasive, real-time determination of human brain oxygenation [31].

1.7 RAMAN SCATTERING

Raman scattering measures vibrational transition energies from about 17 to 4000 cm^{-1}, which is nearly equivalent to the entire spectral ranges covered by far- and mid-infrared spectroscopies together. Below about 17 cm^{-1}, the same physical effect is called Brillouin scattering. In contrast to previously described spectroscopies, Raman excitation frequencies can range from the ultraviolet to the near-infrared region, giving the technique an extra dimension of flexibility.

1.7.1 Spontaneous Raman Scattering

Raman scattering results from a process fundamentally different from that occurring in absorption or thermal emission spectroscopy. An inelastic collision of a photon with a molecule causes the photon to gain from or lose to the molecule one vibrational quantum of energy. A plot of the scattered intensity versus the energy difference between the incident and scattered photons ("Raman shift") yields the Raman spectrum. A Raman spectrum is similar to an infrared

absorption spectrum in that it reports the energy of some, but usually not all, of the molecular vibrations in the sample.

An energy level diagram illustrating Raman scattering for one vibration is shown in Figure 1.4. Light excites a molecule to a virtual state, represented by a horizontal dotted line. The virtual state is not a quantum mechanical stationary state. Rather, it is a distortion of the chemical bond by the electric field of the light. This distortion creates an induced dipole moment. The virtual state immediately relaxes back to a vibrational level in the ground state, emitting a photon. Photons resulting from the return to the original vibrational level have the same energy as before, and make up the elastically scattered (Rayleigh scattered) light. Photons resulting from a return to a vibrational level one vibrational quantum higher or one vibrational quantum lower than the original energy level make up the Raman scattered photons. Stokes Raman scattering produces Raman photons at energies lower than the excitation photon energy because one vibrational quantum of energy was left in the molecule. Anti-Stokes Raman scattering produces Raman photons at higher energy because a vibrational quantum of energy was taken away from the molecule. The frequency of Raman scattered light changes as the excitation frequency changes. It is the energy difference between the excitation frequency and the frequency of the Raman scattered light that does not change, for a given Raman band.

According to classical physics, Raman scattering can be viewed as resulting from the interaction of the oscillating electric field of light with the electrons in a chemical bond. As the optical electric field strength increases, it applies increasing force to move the bonding electrons away from

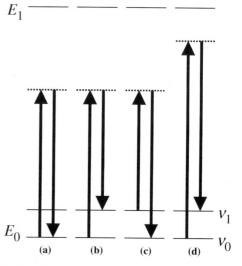

FIGURE 1.4 Energy level diagram illustrating spontaneous Rayleigh and Raman scattering. (a) Rayleigh scattering; (b) Stokes Raman scattering; (c) anti-Stokes Raman scattering; (d) Stokes Raman scattering using a different excitation wavelength.

their equilibrium position. The movement of the electrons creates a dipole moment and diminishes the intensity of the exciting light. As the optical electric field strength decreases, the bonding electrons move back toward their equilibrium position, reducing the induced dipole moment and increasing the intensity of the light. The light generated by the bonding electrons returning to their equilibrium position has the same frequency as the original light, and therefore the same energy, wavelength, and color. It does not have the same direction, however. This light is the elastically scattered light or Rayleigh scatter.

The intensity of Rayleigh scatter is proportional to the square of the chemical bond polarizability (a measure of how far electrical charge moves in response to an electric field). The polarizability of the chemical bond changes with nuclear displacement during a molecular vibration. As a result, the intensity of the elastically scattered light is modulated at the frequency of the molecular vibration or rotation. Such a modulation can be viewed as the sum of the original frequency plus two new "sideband" frequencies of light. The two sidebands are the Raman scattered light and possess frequencies equal to the elastically scattered frequency plus or minus the vibrational frequency of the chemical bond. Said another way, the intensity-modulated elastically scattered light is mathematically equivalent to the sum of three colors of light: the elastically scattered light and the two colors of Raman scattered light.

The classical description of Raman scattering in the previous two paragraphs is intuitive but incomplete. For example, it does not include a prediction of the Stokes to anti-Stokes Raman intensity ratio. Any ratio would be consistent with the model. A quantum mechanical analysis provides a basis to describe all aspects of Raman scattering, but is beyond the scope of this chapter. Detailed descriptions of both the classical and quantum mechanical models for Raman scattering can be found in Ref. 32. We will simply utilize results from the complete analysis.

The quantum mechanical model for Raman scattering was first described by Placzek [33]. His expression for Raman scattering intensity, I_R, is

$$I_R = \frac{2^4 \pi^3}{45 \cdot 3^2 \cdot c^4} \cdot \frac{h I_L N (v_0 - v)^4}{\mu v (1 - e^{-hv/kT})} \cdot \left[45 (\alpha_a')^2 + 7 (\gamma_a')^2 \right]$$

$$(1.7)$$

where h is Planck's constant, c is the speed of light, I_L is the excitation intensity, N is the number of scattering molecules, v is the molecular vibrational frequency in Hz, v_0 is the laser excitation frequency in Hz, μ is the reduced mass of the vibrating atoms, k is Boltzmann's constant, T is the temperature in Kelvin, α_a' is the mean value invariant of the polarizability tensor, and γ_a' is the anisotropy invariant of the polarizability tensor.

Equation 1.7 describes several aspects of Raman scattering that are of practical importance for chemical imaging. Raman intensity is proportional to the excitation intensity. The lasers that are used for Raman spectroscopy are often powerful enough to damage samples by excessive heating, so sample damage thresholds place an upper limit on the amount of sensitivity that can be gained by increasing the laser power, especially in point mapping applications. Raman intensity is proportional to the number of scattering molecules. This relationship is the basis for most quantitative and semiquantitative analysis using Raman scattering (e.g., concentration maps). Raman intensity is proportional to the fourth power of the Raman photon frequency $(v_0 - v)$. Raman sensitivity increases rapidly with increasing excitation frequency (decreasing wavelength) because the Raman photon frequency increases along with it.

The dipole moment induced by the optical electric field that elevates the molecule to the virtual state is in the same direction as the optical electric field, if the vibration has spherical symmetry. In this case, the polarizability is a scalar quantity. Chemical bonds in molecules lacking spherical symmetry may restrict the movement of electrons in some directions, causing the dipole induced by the optical electric field to point in a direction different from that of the optical electric field itself. This case requires the polarizability to be expressed as a tensor quantity with each Cartesian coordinate of the induced dipole moment depending on all three Cartesian coordinates of the optical electric field. The added information in a Raman spectrum due to this tensor nature of the polarizability has no analogue in absorption spectroscopy. The tensor nature of the polarizability is represented in Equation 1.7 by the splitting of the polarizability into a mean value invariant and an anisotropy invariant.

Experimentally, the tensor nature of the polarizability causes symmetric vibrations to yield Raman scattering with the same polarization as the excitation light, but the polarization of Raman scattering from nonsymmetric vibrations may be significantly different. Another consequence is that randomly oriented crystals excited with polarized light often have spectral bands at identical Raman shifts, but the relative band heights and areas from different crystals may vary considerably. Using unpolarized or circularly polarized excitation light can significantly reduce this effect by averaging two of the three tensor components, but does not fully eliminate the problem because one tensor component cannot be controlled.

Equation 1.7 can be rewritten in an analytically useful form similar to Beer's law:

$$I_R = (I_L \sigma_R X) P C \qquad (1.8)$$

where I_R is the measured Raman intensity, in photons per second, I_L is the laser intensity, in photons per second, σ_R is the absolute Raman cross section, in cm^2/molecule, X is the experimental constant, P is the sample path length, in cm, and C is the concentration, in molecules/cm^3.

Here, experimental factors, such as the efficiency of the optics and detector, are lumped together into a single constant, X. The molar absorptivity from Beer's law is replaced by the Raman cross section, which is a measure of the strength of a Raman band. Raman cross sections are usually tabulated in units of cm^2/molecule. Values on the order of 10^{-29}–10^{-30} cm^2/molecule are typical for a strong spontaneous Raman band scattered from a liquid or solid. Raman cross sections can also be expressed in the same units as molar absorptivity for a more direct comparison with infrared absorption. Table 1.5 shows Raman cross sections for several common functional groups. More extensive tables are available in Ref. 34. These cross sections are about 10 orders of magnitude smaller than mid-infrared molar absorptivities, illustrating how weak Raman scattering actually is. Fortunately, the sensitivity of Raman scattering is also proportional to the excitation intensity. Intense laser light sources are used in virtually all Raman chemical imaging applications to partially compensate for the low Raman cross sections.

Sensitivity, however, is not the only variable influencing detection and quantitative capabilities in Raman spectroscopy. Detection limits and limits to quantitative precision and accuracy are usually determined by noise rather than by sensitivity. Noise is any detected signal that is not wanted and that hinders the use of the signal that is wanted. The most serious noise source in Raman spectroscopy is fluorescence, usually from low levels of impurities. Many experimental and mathematical approaches have been explored and used to reduce the impact of fluorescence on Raman spectroscopy, but there is still room for substantial improvement. Another serious noise source is the Raman spectrum of the sample matrix containing the analyte of interest. A matrix, such as a solvent or excipient, is present at high concentration, so its Raman spectrum, or the uncertainties of its spectral intensities, may obscure the spectrum of an analyte present at much lower concentration.

Detection and quantitative capabilities are important in chemical imaging because they often help determine image contrast. For example, all Raman mapping or imaging must be at least semiquantitative since the relationship among pixels is a relative Raman intensity, which is proportional to concentration. Semiquantitative Raman imaging may map the approximate relative analyte concentration with no knowledge of the absolute concentration. But a high detection limit, and/or a low dynamic range, will still tend to obscure the presence of an analyte in an image. Another semiquantitative approach is to classify each pixel as consisting of only the major component detected in its Raman spectrum, and then to report analyte concentrations as the fraction of pixels classified as that analyte [36]. More accurate quantitative analysis is generally reserved for samples exhibiting minimal diffuse reflection so that Raman scattered light from neighboring pixels is negligible.

The basis for quantitative analysis using Raman scattering is Equation 1.8, which shows that the analyte concentration is proportional to its Raman intensity. Unfortunately, the Raman cross section, the instrument detection efficiency, and the path length are usually unknown. An image generated from uncorrected Raman intensity at an analyte band wavelength may provide a qualitative description of analyte distribution, but that distribution may be distorted by uncontrolled variables, such as focusing errors due to an uneven sample surface or path length variation due to variable degrees of Mie scattering from sample heterogeneity. Unknown Raman cross sections are normally addressed by building calibration curves from suitable standards. Uncertainty in instrument detection efficiency and path length can be addressed by ratioing the analyte Raman intensity to the Raman intensity of a material in the sample whose concentration is expected to be constant, or at least predictable. Mass balance may be used when no component of the sample is expected to have a uniform concentration as a function of location in the sample. Quantitative analysis using Raman spectroscopy is reviewed in Ref. 37.

TABLE 1.5 Raman Cross-Sections for Several Common Functional Groups

Band Descripton	Sample	Band Position (cm^{-1})	Cross Section (cm^2/(molecule sr) $\times 10^{30}$)	Cross Section (L/(mol cm) $\times 10^{10}$)
Ring stretch	Benzene liquid	992	28.0	169
CH stretch	Benzene liquid	3060	45.3	273
CC stretch	Cyclohexane liquid	802	5.2	31
CH stretch (all)	Cyclohexane liquid	2800–3000	43.0	259
CH stretch	CHCl$_3$ liquid	3032	4.4	26
CCl$_3$ asy stretch	CHCl$_3$ liquid	758	3.2	19
CCl$_3$ sym stretch	CHCl$_3$ liquid	667	6.6	40
NN stretch	Nitrogen gas	2331	0.4	3
OO stretch	Oxygen gas	1555	0.6	3
CCl$_4$ sym stretch	CCl$_4$	459	4.7	28

Compiled from Ref. 35.

Raman spectra are generally less temperature sensitive than far-, mid-, or near-infrared spectra. This is partly due to the reduced sensitivity of Raman scattering to hydrogen bonding, and partly due to sharper and less overlapping bands in Raman spectra, which are therefore less likely to be confounded by subtle temperature-induced changes. Nondestructive temperature effects are rarely a serious problem for Raman spectral imaging.

Table 1.6 lists Raman scattering characteristic frequencies of some common functional groups. Much more extensive tables are given in Refs 13 and 14. Tables such as these often provide additional information such as the strength of the Raman band or its polarization. They provide a good starting point for estimating the spectral location of Raman bands that may be analytically useful for chemical imaging. They are also helpful for assigning bands observed in spectra of known materials to chemical groups in the material.

Commercially available Raman libraries exist, but are not as extensive as those available for mid-infrared spectroscopy. In most Raman imaging applications, pure compounds that make up the sample are available, enabling the generation of small, custom libraries targeted to the individual imaging project. When pure components are not available, or when the pure components interact to create changed Raman spectra, chemometric methods can sometimes be used to extract spectra from image data that are similar to the pure component spectra.

There are several types of Raman spectroscopy in addition to spontaneous Raman scattering, each with its own capabilities, limitations, and unique experimental implementation. Some differ in the instrumental approach used to excite or collect Raman photons. Others differ in their use of optical fields, special substrates, or tuning to influence the physics of the scattering process. We conclude the section on Raman scattering by briefly reviewing the theory behind four forms of Raman scattering that are particularly useful for chemical imaging: resonance Raman scattering, coherent anti-Stokes Raman spectroscopy, surface-enhanced Raman spectroscopy (SERS), and Raman gain (loss) spectroscopy.

1.7.2 Resonance Raman Scattering

When the photon energy of the exciting radiation matches the energy of an electronic absorption band (rather than just the energy of transition to a virtual state), the intensity of some Raman bands is dramatically increased by as much as a factor of 10^6. This effect is called resonance Raman scattering [32, 39]. Totally symmetric vibrations that are similar to the changes in molecular geometry that occur during the transition from the ground electronic state to the first excited electronic state are strongly enhanced and dominate the Raman spectrum. In some cases, other vibrations may be enhanced, but the enhancements tend to be weaker. Resonance changes selection rules and depolarization ratios

TABLE 1.6   Raman Characteristic Frequencies for Several Common Functional Groups

Vibration	Shift (cm^{-1})	Group	Shift (cm^{-1})	Group
SS stretch	480–510	Dialkyl disulfides	432–467	Thiosulphate
CS stretch	620–715	Dialkyl disulfides	568–576	Tetraborate
Skeletal stretch	749–835	Isopropyl group	683–817	Iodate
CC stretch	837–905	n-Alkanes	710–745	Nitrate
Sym COC stretch	830–930	Aliphatic ethers	776–817	Bromate
CC stretches	950–1150	n-Alkanes	806–855	Persulfate
Sym SO_2 stretch	1188–1196	Alkyl sulfates	913–988	Phosphate
C_6H_5–C vibration	1205	Alkyl benzenes	914–943	Chlorate
CH_2 in-phase twist	1295–1305	n-Alkanes	933–952	Perchlorate
CH_2 twist and rock	1175–1310	n-Alkanes	956–1040	Sulfate
CH deformation	1330–1350	Isopropyl group	962–990	Sulfite
Ring stretch	1370–1390	Naphthalenes	1028–1045	Bicarbonate
Ring stretch	1385–1415	Anthracenes	1029–1069	Nitrate
CH_3 deformation	1465–1466	n-Alkanes	1051–1089	Carbonate
NH_2 scissors	1590–1650	Primary amines	1073–1090	Persulfate
C=O stretch	1700–1725	Aliphatic ketones	1320–1322	Nitrite
C≡C stretch	2100–2160	Alkyl acetylenes	2033–2162	Thiocyanate
C≡N stretch	2232–2251	Aliphatic nitriles	2044–2071	Ferrocyanide
SH stretch	2560–2590	Thiols	2070–2215	Cyanide
Sym CH_2 stretch	2849–2861	n-Alkanes	2102–2109	Ferrocyanide
Sym CH_3 stretch	2883–2884	n-Alkanes	3045–3120	NH_4^+

Compiled from Refs 13 and 38.

relative to those of nonresonance Raman scattering. For example, resonance Raman overtone and combination bands can be as strong as fundamental vibrations, at least for some small molecules, while in nonresonance spontaneous Raman scattering, such transitions would be forbidden.

Resonance Raman measurements can be challenging because exciting the analyte at an absorption band wavelength increases the chances for sample damage due to heating or photolysis. Spectral distortion due to self-absorption and high background intensity from fluorescence emission can also complicate resonance Raman measurements. Many resonance Raman spectral studies have been done using a flowing sample in order to minimize sample damage. Several reports have described ways to correct for spectral distortion due to self-absorption [40, 41].

The increased likelihood of fluorescence can be a significant problem for resonance excitation in the visible and near-infrared regions. However, Raman spectra, including resonance Raman, obtained with UV excitation can actually exhibit decreased fluorescence interference. This is because most of the UV excited fluorescence occurs in a spectral region separate from that of the UV excited Raman scattering, especially for excitation wavelengths below 250 nm. This benefit is being exploited in experimental groundwork to collect UV resonance Raman images of microorganisms on rocks and minerals for *in situ* planetary studies [42, 43].

Resonance enhancement decreases rapidly as the excitation frequency is tuned away from the electronic transition frequency. The reduction of the enhancement with detuning slows down far from resonance, though, so enhancements of 5–10 are not uncommon more than $1000\,cm^{-1}$ away from resonance. This lesser enhancement away from resonance is known as preresonance enhancement. This type of enhancement can be significant, yet has little effect on selection rules or depolarization ratios. Fluorescence, self-absorption, and sample damage due to absorption are less problematic as well.

Resonance Raman has been employed to improve the sensitivity and speed of acquisition in specific imaging applications, but has not, so far, become a widespread imaging technique. Imaging with resonance Raman works best for chromophores having a strong or unique (relative to the surrounding environment) absorption band coincident with an available excitation laser line. A number of reported imaging examples are based on resonance Raman scattering from porphyrin-containing molecules, such as hemoglobin, cytochromes, and hemozoin (malaria pigment). These have allowed label-free visualization of the distribution of heart tissue components [44], red blood cell infection [45], and heme-containing enzymes in immune response cells [46]. Similarly, resonance Raman imaging of carotenoid compounds in living human retinas has illustrated the complexity and variability of macular pigment distribution [47].

Resonance Raman studies can also be carried out at multiple excitation wavelengths. The intensity of a given Raman band as a function of excitation wavelength, called a Raman excitation profile, provides a spectrum similar to the electronic absorption spectrum of the transition responsible for the resonance enhancement. Other absorbing species do not contribute, so the Raman excitation profile can be used to resolve overlapping bands in absorption spectra. Full Raman spectra at several excitation wavelengths can be combined into a three-dimensional plot similar to a fluorescence excitation–emission plot. To the best of the authors' knowledge, there are not yet any published examples of Raman chemical imaging utilizing multiple excitation wavelengths. Such images would be difficult and time consuming to create with existing instrumentation, but could provide greatly enhanced chemical specificity.

1.7.3 Surface-Enhanced Raman Spectroscopy

The sensitivity of Raman scattering can be enhanced by several orders of magnitude in some cases by placing the analyte very close to a metal surface. The resulting surface-enhanced Raman process [48, 49] takes advantage of two types of interactions between the metal and the analyte, electromagnetic enhancement and a charge transfer mechanism. The total enhancement is the product of the two, with electromagnetic enhancement generally being much larger.

Electromagnetic enhancement is an antenna-like effect that increases the amplitudes of both the excitation and Raman scattered electric fields. It is caused by localized surface plasmons, collective oscillations of conduction electrons in the metal that are in resonance with the external optical electric fields. Gold and silver, and to a lesser extent copper, provide strong SERS enhancement because their localized surface plasmon resonance frequencies occur in the visible/NIR spectral region where Raman measurements are normally carried out, and because losses that damp the plasmon oscillations are small.

Surface plasmons cannot be excited by electromagnetic radiation when the metal surface is smooth and flat, but can be when the metal surface is roughened, or when the metal is a small particle. SERS has, therefore, most often been carried out on electrochemically roughened metal surfaces or on colloidal metal particles. More recently, precisely fabricated nanostructures have been used as SERS substrates, to reduce variation in the SERS enhancement factor and to study morphology effects on SERS enhancement.

Since the surface plasmon oscillation enhances the electric fields of both the excitation light and the Raman light, the SERS enhancement is proportional to the fourth power of the excitation electric field. An enhancement of the electric field by a small spherical particle may be on the order of 10, leading to a SERS enhancement on the order of 10^4. More complex nanostructures can produce much larger SERS enhancements. Optimum enhancement would occur if the optical frequency matched the peak in the surface plasmon resonance frequency. Since the excitation and Raman frequencies are not equal, however, optimum enhancement occurs when the peak of the surface plasmon frequency is between the Raman excitation frequency and the Raman scattered frequency.

Electromagnetic enhancement does not require the target molecule to touch the metal surface, but the enhancement falls off very rapidly with distance from the metal surface as shown by Equation 1.9:

$$I_{SERS} \propto \left(\frac{a+r}{a}\right)^{-10} \qquad (1.9)$$

where I_{SERS} is the SERS scattering intensity, a is the average size of field-enhancing features on the surface, and r is the distance from the adsorbate to the metal surface.

For example, one study [48] reported a factor of 10 decrease in SERS signal when the adsorbate was separated from a 12 nm enhancing particle by 2.8 nm. This level of depth resolution is far better than that of diffraction-limited microscopy at the same optical wavelength.

The charge transfer mechanism of SERS postulates the formation of an adsorbate–metal complex that allows the transfer of excitation and charge between the adsorbate and the metal. The charge transfer mechanism is difficult to verify because it would operate only on the first monolayer of adsorbates where electromagnetic enhancement is already very strong. Campion et al. [50] provided strong evidence for the charge transfer mechanism by observing a SERS enhancement factor of 30 on an atomically flat single crystal of copper, a situation where electromagnetic enhancement was expected to be small and well understood. A low-energy electronic absorption band appeared in the spectrum of the adsorbed molecule that was not present in the spectrum of the isolated molecule, further supporting a charge transfer mechanism.

SERS enhancements using spatially isolated silver nanoparticles are on the order of 10^6. But this enormous sensitivity enhancement pales in comparison to enhancements observed in silver colloidal aggregates. A small fraction of molecules adsorbed to silver colloidal aggregates have SERS enhancements on the order of 10^{14} [48]! This level of enhancement is sufficient for the measurement of single molecule Raman spectra. Studies are underway to better understand the nanostructure of these SERS "hot spots," which could lead to the ultimate in sensitivity for chemical imaging.

Several approaches to chemical imaging with SERS in addition to imaging an adsorbate on a metal surface have been reported. Gold nanoparticles have been injected directly into living cells to probe cellular molecules in their native environment. Reporter molecules attached to the gold nanoparticles have been used to make chemical images, such as pH images in single living cells [48]. Tip-enhanced Raman spectroscopy (TERS) moves a fine metal tip within a few nanometers of a surface, strongly enhancing Raman scattering [51]. TERS provides topographic images in addition to providing Raman maps with nanometer spatial resolution.

SERS sensitivity can be increased by using an excitation wavelength that is at or close to resonance with an adsorbate chromophore. Surface-enhanced resonance Raman spectroscopy (SERRS) provides an additional three or four orders of magnitude in sensitivity enhancement.

1.7.4 Coherent Anti-Stokes Raman Spectroscopy

CARS [52, 53] is another approach that dramatically enhances the sensitivity of Raman scattering, in some cases by as much as 10 orders of magnitude. Figure 1.5a shows an energy level diagram of the CARS process. Laser light, designated the pump field, excites a molecule from its ground vibrational state to a virtual state. Laser light at a second frequency, designated the Stokes field, stimulates the scattering of a Stokes photon from the virtual state, leaving the molecule in its first excited vibrational level. Laser light typically at the pump frequency, this time called the probe field, then excites the molecule from its first excited vibrational level to a new virtual state. Finally, the molecule returns to the ground vibrational state by scattering an anti-Stokes photon.

Unlike spontaneous Raman scattering, where each molecule in the sample scatters light independently of the

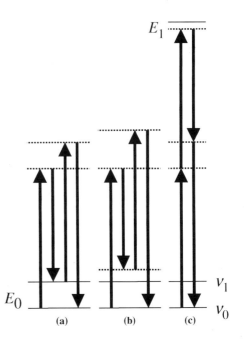

FIGURE 1.5 Energy level diagram illustrating CARS processes. (a) Resonant CARS; (b) CARS nonresonant background; (c) two-photon enhanced nonresonance CARS background.

other molecules, CARS uses optical fields to impose both a temporal and a spatial phase relationship between the molecules in the sample that scatter light. This nonrandom relationship imposed on the molecules is the reason for the "coherent" in the CARS acronym and distinguishes CARS from spontaneous anti-Stokes Raman scattering. CARS is an example of a process called four-wave mixing where four optical fields, the pump field, the Stokes field, the probe field, and the anti-Stokes field, exchange energy.

Since CARS is a coherent process, the energy exchange can build up with interaction distance as long as the four optical fields maintain a fixed phase relationship with each other. The speed of light through matter changes with wavelength due to the wavelength dependence of the refractive index, however, so effective phase matching occurs over a limited distance through the sample. This distance can be extended if the four optical fields propagate in slightly different directions such that the vector components of their velocities in a common direction are all equal. Then, as long as the four fields overlap spatially they will remain in phase. This noncollinear CARS geometry can be implemented experimentally by crossing the pump laser beam (which is also the probe laser beam) with the Stokes laser beam in the sample. A laser-like anti-Stokes beam then emerges from the sample at an angle different from the pump and Stokes beams. Phase matching can also be achieved in a collinear geometry if the beams are tightly focused. Tight focusing creates a distribution of propagation directions, which automatically satisfy the phase-matching condition.

The intensity of the CARS anti-Stokes signal, I_{AS}, is given by

$$I_{AS} \propto |\chi_3|^2 I_P^2 I_S \qquad (1.10)$$

where χ_3 is the third-order susceptibility, I_P is the intensity of the pump optical field, and I_S is the intensity of the Stokes optical field.

The third-order susceptibility describes the molecular bond polarization responsible for the CARS effect. Unlike spontaneous Raman scattering, where the scattered intensity is proportional to the number of vibrational oscillators, the CARS intensity is proportional to the square of the number of vibrational oscillators. The squared dependence has important implications for analytical applications of CARS described later in this book.

The third-order susceptibility consists of the sum of a resonant part that is enhanced by molecular vibrations and a nonresonant part due to the electronic response of the material that is not enhanced by molecular vibrations. Scattering from the resonant and nonresonant parts of the third-order susceptibility is illustrated in the energy level diagrams shown in Figure 1.5a and b, respectively. The squared third-order susceptibility in Equation 1.10, therefore, con-

tains three terms: a purely resonant term, a purely nonresonant term, and a mixed term. The resonant term is responsible for the vibrational spectrum that is normally desired. The nonresonant term provides a constant background that degrades CARS detection limits. The mixed term produces a derivative-like spectral peak shape that causes CARS bands to be redshifted with a dip on the high-energy side of the band. The amount of redshift and dip in a CARS band is proportional to the relative strength of the resonant and nonresonant parts of the third-order susceptibility.

The spatial dimensions of phase coherence between the CARS optical fields, and therefore the directionality of the anti-Stokes field, may be determined by the dimensions of the sample. Collinear CARS through extremely thin samples produces equal anti-Stokes intensity in the forward and backward directions. As the sample gets thicker, constructive interference increases the forward scattered intensity while destructive interference decreases the backward scattered anti-Stokes intensity, leading to much less backward scatter intensity for thick homogeneous samples. Anti-Stokes backscattering does occur in bulk samples, however, due to small objects ($\lambda_p/3$), sharp discontinuities in χ_3, and elastic scattering of the forward scattered anti-Stokes intensity.

Numerous schemes have been proposed to reduce the nonresonant background in order to improve the CARS signal. Two have proven to be especially enabling for CARS microscopy. Collection of the backscattered anti-Stokes intensity from thin samples (epi-CARS) greatly reduces the nonresonant intensity from the matrix. A second scheme is to use near-infrared pump wavelengths. The nonresonance anti-Stokes intensity can be enhanced if the pump wavelength is near a two-photon absorption band of the sample, as illustrated by the energy level diagram in Figure 1.5c. Fewer two-photon electronic absorption bands are accessible using near-infrared light, thereby reducing the nonresonant background for many samples.

Both qualitative and quantitative analyses of spectra are more difficult for CARS than for spontaneous Raman spectroscopy. The redshift and dip on the high-energy side of CARS bands distort CARS spectra by an amount that can vary depending on the experimental conditions. Further distortion can occur from interferences between resonances. For example, one band height can influence the height of a neighboring band. The quadratic dependence of CARS intensity on analyte concentration complicates the resolution of overlapping bands from different materials, as well as the use of standard multivariate analysis algorithms. These complications can be overcome, however, as evidenced by many semiquantitative and quantitative CARS publications.

Since the intensity of the anti-Stokes CARS signal is proportional to the intensity of the pump beam squared, pulsed pump lasers provide greater signal intensity than continuous-wave pump lasers having the same average power. For a fixed pump laser average power and pulse repetition rate, the peak optical power (and pulse bandwidth) increases with decreasing pulse width. The higher peak power causes a quadratic increase in nonresonant anti-Stokes CARS intensity as pulse duration shortens. The resonant anti-Stokes CARS intensity also increases with decreasing pulse width, but gradually saturates as the bandwidth of the laser pulse becomes larger than the Raman linewidth. Pulse widths of a few picoseconds tend to provide an optimal compromise for enhanced sensitivity of the resonance anti-Stokes intensity with minimal band broadening and nonresonance anti-Stokes CARS intensity.

The anti-Stokes CARS signal is proportional to the product of the pump laser power squared and the Stokes laser power, so the sensitivity to analyte concentration increases rapidly as the laser beams are focused. Most of the CARS signal, therefore, comes from the focal region, providing depth discrimination and the capability to do three-dimensional sectioning even in a collinear geometry.

One approach to CARS chemical imaging uses a single fixed wavelength for the Stokes laser. Anti-Stokes CARS intensity is collected as a function of spatial location to create a single-wavelength chemical image at video frame rates. The high image acquisition speed is truly enabling for many applications, but the use of a single wavelength is a significant limitation. While monochrome images have proven to be informative and valuable, they can be confounded by several different effects that can contribute to image contrast. Multiple images could perhaps be sequentially collected using different wavelengths, but registration of the different images may be problematic.

A second approach, broadband CARS [54], measures multiple anti-Stokes CARS wavelengths simultaneously by using a spectrally broad source for the Stokes beam. Acquisition times per pixel are increased for several reasons including distribution of Stokes beam intensity over many spectral resolution elements, less effective reduction of the nonresonance background, and increased detector readout time. Acquisition times are still much faster than those of spontaneous Raman spectroscopy, though, and are likely to improve in the near future.

1.7.5   Stimulated Raman Gain Spectroscopy

The first half of the CARS process, excitation to a virtual state from the ground vibrational level followed by stimulated emission from the virtual state to the first excited vibrational level, can be used as a real-time chemical imaging method [55]. This process transfers energy from the pump beam into the Stokes beam. The measurement of the increase in Stokes beam intensity is called stimulated Raman gain (SRG) spectroscopy, and the measurement of the decrease in pump beam intensity is called stimulated Raman loss (SRL) spectroscopy. Both are examples of stimulated Raman spectroscopy, and their names are used to distinguish them from the

older and less controlled form of stimulated Raman spectroscopy that used spontaneous Raman scattering from the sample to create the Stokes beam, rather than an external Stokes beam [56].

The change in pump beam intensity, ΔI_P, and the change in Stokes beam intensity, ΔI_S, are given by

$$\Delta I_P \propto -N\sigma_R I_P I_S, \quad \Delta I_S \propto N\sigma_R I_P I_S \quad (1.11)$$

where N is the number of molecules in the probe volume, σ_R is the Raman cross section, I_P is the intensity of the pump optical field, and I_S is the intensity of the Stokes optical field.

SRG and SRL spectroscopies are similar to CARS in that they are all coherent processes requiring phase matching between the pump and Stokes beams. In each case, the signal has a nonlinear dependence on total laser power, providing the opportunity for depth discrimination and three-dimensional sectioning. All three methods can be several orders of magnitude more sensitive than spontaneous Raman scattering. In fact, SRG spectra of molecular monolayers were reported as early as 1980 [57].

Unlike CARS, the signals from SRG and SRL are proportional to the number of molecules in the probe volume, making data interpretation and chemometric analysis much easier. There is no nonresonance background, so spectra look like familiar spontaneous Raman spectra. The lack of a nonresonance background also improves image contrast. Preliminary studies [55] showed an improved detection limit for retinol in methanol for SRL spectroscopy compared to CARS measurements made by the same group.

1.8   CHEMICAL IMAGE CREATION USING REMOTE SENSING

Remote sensing employs infrared emission and Raman scattering, as well as other forms of spectroscopy, to make chemical images of large targets. Remote sensing probably accounts for the vast majority of chemical images that have ever been collected. For example, the Moderate Resolution Imaging Spectroradiometer (MODIS) instruments on board both the Terra and Aqua satellites collect images with 36 visible and infrared bands (over the range of 0.4–14.4 μm) that map the entire surface of the Earth every 1–2 days, and have been doing so since 2002 [58].

Numerous imaging spectrometers have been flown on aircraft, placed in orbit around Earth and other planets, sent into deep space, and delivered to the surface of Mars. Table 1.7 summarizes a number of planet- and moon-observing instruments, including the spectral ranges of the data collected. Most of these spectral imaging data are available to the public from NASA.

Analysis of spectral images may involve traditional chemometrics, image processing, expert systems, extensive

TABLE 1.7   Examples of Remote Imaging Instruments Collecting Vibrational Data

Instrument	Target	Spectral Range (μm)				References
		Visible (0.4–0.78)	NIR (0.78–2.5)	Mid-IR (2.5–25)	Far-IR (25–1000)	
Airborne Visible/Infrared Imaging Spectrometer (AVERTS) Aboard aircraft since the mid-1990s and still in use	Earth	0.4 ————		———— 2.5		59
Moon Mineralogy Mapper (M^3) On India's Chandrayaan-1; 70 m resolution	Lunar surface	0.4 ————		———— 3		60
Visual and Infrared Mapping Spectrometer (VIMS) Cassini mission: Venus, Jupiter, Saturn, Titan, and so on	Planets and moons	0.35 ————		———— 5.1		61
Thermal Emission Spectrometer (TES) Mars Global Surveyor; 10^8 images of 3 km resolution	Mars			6 ————	———— 50	62, 63
Mliniature version of TES (Mini-TES) On Spirit and Opportunity rovers on surface of Mars	Mars			5 ———— 29		64, 65
Thermal Emission Imaging System (THEMIS) Orbiting on Mars Odyssey spacecraft	Mars	0.4 ———— 0.9		7---15		66
Compact Reconnaissance Imaging Spectrometer for Mars (CRISM) On Mars Reconnaissance Orbiter	Mars	0.36 ————		———— 4		67

Some instruments measure discrete bands, while others collect continuous spectra over the ranges specified.

spectral libraries, or a combination of approaches [68]. Common uses for the resulting chemical images of Earth and its atmosphere include applied geology and mineralogy, climate, agricultural, and pollution studies.

Abundant vibrational chemical images are being produced to investigate celestial targets, as well. Examples include instruments aboard NASA's Spitzer Space Telescope [69] and earlier COBE (Cosmic Background Explorer) [70], the collaborative Infrared Astronomical Satellite (IRAS) [71], the European Space Agency's Infrared Space Observatory (ISO) [72], and the Japanese AKARI [73]. Far-infrared wavelengths visualize clouds of cold dust and very cold molecular clouds, while shorter IR wavelengths (0.7–5 μm) can penetrate the dust to reveal astronomical objects hidden to visible and ultraviolet detectors. Ionic and molecular species in the universe, such as atoms and atomic ions, polyatomic ions and molecules, and aromatic molecules, are being mapped using emission bands throughout the vibrational spectrum from the near- to far-infrared region.

While the above applications all rely on passively emitted radiation, remote sensing can also be carried out using an excitation source, usually a laser. Laser radar or lidar (light detection and ranging) uses backscattered intensity from a pulsed laser as a function of time to measure absorption depth profiles in the atmosphere. Differential absorption lidar (DIAL) shifts the laser frequency onto and off an analyte absorption peak to derive the analyte absorption. An open-path FTIR array spectrometer has been reported that performs 3D profiling of atmospheric plumes in real time [74].

Raman remote sensing includes Raman lidar, which measures the intensity of Raman light detected from one or more analyte Raman bands. To quantify the analyte, the Raman band of nitrogen in the atmosphere is used as an internal standard. An advantage of Raman lidar is that its laser transmitter is simpler to automate than that for traditional lidar. Raman lidar has been extensively employed to make three-dimensional maps of water in the atmosphere extending dozens of kilometers. Other analytes such as NO, CO, H_2S, and hydrocarbons have also been mapped.

REFERENCES

1. Cotton, F. A. (1971) *Chemical Applications of Group Theory*, Wiley–Interscience, New York.

2. McHale, J. L. (1999) *Molecular Spectroscopy*, Prentice Hall, New Jersey.

3. Griffiths, P. R. (2002) Beer's law. In: Chalmers, J. M. and Griffiths P. R. (Eds.), *Handbook of Vibrational Spectroscopy*, Vol. 3, Wiley, West Sussex, UK, pp. 2225–2234.

4. Hecht, E. and Zajac, A. (1974) *Optics*, Addison-Wesley Publishing Company, Menlo Park, CA, pp. 72–84.

5. Mirabella, F. M. (1993) Principles, theory, and practice of internal reflection spectroscopy. In: Mirabella, F. M. (Ed.), *Internal Reflection Spectroscopy: Theory and Applications*, Marcel Dekker, New York, pp. 17–52.

6. Strangman, G., Boas, D. A., and Sutton, J. P. (2002) Non-invasive neuroimaging using near-infrared light. *Biol. Psychiatry* **52**, 679–693.

7. Okamoto, M. and Dan, I. (2007) Functional near-infrared spectroscopy for human brain mapping of taste-related cognitive functions. *J. Biosci. Bioeng.* **103**, 207–215.

8. Jobsis, F. F. (1977) Noninvasive, infrared monitoring of cerebral and myocardial oxygen sufficiency and circulatory parameters. *Science* **198**, 1264–1267.

9. Hoshi, Y. (2003) Functional near-infrared optical imaging: utility and limitations in human brain mapping. *Psychophysiology* **40**, 511–520.

10. McCreery, R. L., Fleischmann, M., and Hendra, P. (1983) Fiber optic probe for remote Raman spectrometry. *Anal. Chem.* **55**, 148–150.

11. Matousek, P. (2007) Deep non-invasive Raman spectroscopy of living tissue and powders. *Chem. Soc. Rev.* **36**, 1292–1304.

12. Mink, J. (2002) Infrared emission spectroscopy. In: Chalmers, J. M. and Griffiths, P. R. (Eds.), *Handbook of Vibrational Spectroscopy*, Vol. 3, Wiley, West Sussex, UK, pp. 1193–1214.

13. Lin-Vien, D., Colthup, N. B., Fateley, W. G., and Grasselli, J. G. (1991) *The Handbook of Infrared and Raman Characteristic Frequencies of Organic Molecules*, Academic Press, Boston, MA.

14. Socrates, G. (2001) *Infrared and Raman Characteristic Group Frequencies*, 3rd ed., Wiley, New York.

15. Shurvell, H. F. (2002) Spectra–structure correlations in the mid- and far-infrared. In: Chalmers, J. M. and Griffiths, P. R. (Eds.), *Handbook of Vibrational Spectroscopy*, Vol. 3, Wiley, West Sussex, UK, pp. 1783–1816.

16. Venyaminov, S. Y. and Prendergast, F. G. (1997) Water (H_2O and D_2O) molar absorptivity in the 1000–4000 cm^{-1} range and quantitative infrared spectroscopy of aqueous solutions. *Anal. Biochem.* **248**, 234–245.

17. Bertie, J. E., Jones, R. N., and Keefe, C. D. (1993) Infrared intensities of liquids. XII. Accurate optical constants and molar absorption coefficients between 6225 and 500 cm^{-1} of benzene at 25°C, from spectra recorded in several laboratories. *Appl. Spectrosc.* **47**, 891–911.

18. Bertie, J. E., Jones, R. N., Apelblat, Y., and Keefe, C. D. (1994) Infrared intensities of liquids. XIII. Accurate optical constants and molar absorption coefficients between 6500 and 435 cm^{-1} of toluene at 25°C, from spectra recorded in several laboratories. *Appl. Spectrosc.* **48**, 127–143.

19. Bertie, J. E., Lan, Z., Jones, R. N., and Apelblat, Y. (1995) Infrared intensities of liquids. XVIII. Accurate optical constants and molar absorption coefficients between 6500 and 800 cm^{-1} of dichloromethane at 25°C, from spectra recorded in several laboratories. *Appl. Spectrosc.* **49**, 840–851.

20. Goplen, T. G., Cameron, D. G., and Jones, R. N. (1980) Absolute absorption intensity and dispersion measurements on some organic liquids in the infrared. *Appl. Spectrosc.* **34**, 657–691.

21. Chan, K. L., Tay, F. H., Poulter, G., and Kazarian, S. G. (2008) Chemical imaging with variable angles of incidence using a diamond attenuated total reflection accessory. *Appl. Spectrosc.* **62**, 1102–1107.

22. Power, J. F. (1993) Photoacoustic and photothermal imaging. In: Morris, M. D. (Ed.), *Microscopic and Spectroscopic Imaging of the Chemical State*, Marcel Dekker, New York, pp. 255–302.

23. Bertie, J. E. and Lan, Z. (1996) Infrared intensities of liquids. XX. The intensity of the OH stretching band of liquid water revisited, and the best current values of the optical constants of $H_2O(l)$ at 25°C between 15,000 and 1 cm^{-1}. *Appl Spectrosc.* **50**, 1047–1057.

24. Pedersen, J. E. and Keiding, S. R. (1992) THz time-domain spectroscopy of nonpolar liquids. *IEEE J. Quantum Electron.* **28**, 2518–2522.

25. Wilk, R., Pupeza, I., Cernat, R., and Koch, M. (2008) Highly accurate THz time-domain spectroscopy of multilayer structures. *IEEE J. Sel. Top. Quantum Electron.* **14**, 392–398.

26. Bertie, J. E., Zhang, S. L., Eysel, H. H., Baluja, S., and Ahmed, M. K. (1993) Infrared intensities of liquids. XI. Infrared refractive indices from 8000 to 2 cm^{-1}, absolute integrated intensities, and dipole moment derivatives of methanol at 25°C. *Appl. Spectrosc.* **47**, 1100–1114.

27. Chan, W. L., Deibel, J., and Mittleman, D. M. (2007) Imaging with terahertz radiation. *Rep. Prog. Phys.* **70**, 1325–1379.

28. Shen, Y., Taday, P. F., Newnham, D. A., Kemp, M. C., and Pepper, M. (2005) 3D chemical mapping using terahertz pulsed imaging. *Proc. SPIE* **5727**, 24–31.

29. Weyer, L. G. and Lo, S.-C. (2002) Spectra–structure correlations in the near-infrared. In: Chalmers, J. M. and Griffiths, P. R. (Eds.), *Handbook of Vibrational Spectroscopy*, Vol. 3, Wiley, West Sussex, UK, pp. 1817–1837.

30. Buback M. and Vogele, H. (1993) *FT-NIR Atlas*, VCH, Berlin.

31. INVOS cerebral oximeter measures oxygen saturation of the blood in the brain noninvasively using near-infrared light, Somanetics Corporation, Troy, MI, USA.

32. Long, D. A. (2002) *The Raman Effect: A Unified Treatment of the Theory of Raman Scattering by Molecules*, Wiley, West Sussex, UK.

33. Placzek, G. (1934) Rayleigh-streuung und Raman-effekt. In: Marx, E. (Ed.), *Handbuch der Radiologie*, Vol. VI., No. 2, Acadeische-Verlag, Leipzig, pp. 205–374 (Translation: *The Rayleigh and Raman Scattering*, University of California Radiation Laboratory (UCRL) Trans. 526(L), 1962)

34. Schrotter, H. W. and Klockner, H. W. (1979) Raman scattering cross sections in gases and liquids. In: Weber, A. (Ed.), *Raman Spectroscopy of Gases and Liquids*, Springer, Berlin, pp. 123–166.

35. McCreery, R. L. (2000) *Raman Spectroscopy for Chemical Analysis*, Wiley–Interscience, New York.

36. Wang, A., Haskin, L. A., Lane, A. L., Wdowiak, T. J., Squyres, S. W., Wilson, R. J., Hovland, L. E., Manatt, K. S., Raouf, N., and Smith, C. D. (2003) Development of the Mars microbeam Raman spectrometer (MMRS). *J. Geophys. Res. E* **108**, 5005.

37. Pelletier, M. J. (2003) Quantitative analysis using Raman spectrometry. *Appl. Spectrosc.* **57**, 20A–42A.

38. Nyquist, R. A., Putzig, C. L., and Leugers, M. A. (1997) *Handbook of Infrared and Raman Spectra of Inorganic Compounds and Organic Salts: Raman Spectra*, Academic Press, San Diego, CA.

39. Smith, E. and Dent, G. (2005) *Modern Raman Spectroscopy: A Practical Approach*, Wiley, West Sussex, UK, Chapter 4.

40. Ludwig, M. and Asher, S. A. (1988) Self-absorption in resonance Raman and Rayleigh scattering: a numerical solution. *Appl. Spectrosc.* **42**, 1458–1466.

41. Womack, J. D., Mann, C. K., and Vickers, T. J. (1989) Correction for absorption in Raman measurements using the backscattering geometry. *Appl. Spectrosc.* **43**, 527–531.

42. Frosch, T., Tarcea, N., Schmitt, M., Thiele, H., Langenhorst, F., and Popp, J. (2007) UV Raman imaging—a promising tool for astrobiology: comparative Raman studies with different excitation wavelengths on SNC Martian meteorites. *Anal. Chem.* **79**, 1101–1108.

43. Tarcea, N., Harz, M., Roesch, P., Frosch, T., Schmitt, M., Thiele, H., Hochleitner, R., and Popp, J. (2007) UV Raman spectroscopy—a technique for biological and mineralogical *in situ* planetary studies. *Spectrochim. Acta A* **68**, 1029–1035.

44. Ogawa, M., Harada, Y., Yamaoka, Y., Fujita, K., Yaku, H., and Takamatsu, T. (2009) Label-free biochemical imaging of heart tissue with high-speed spontaneous Raman microscopy. *Biochem. Biophys. Res. Commun.* **382**, 370–374.

45. Bonifacio, A., Finaurini, S., Krafft, C., Parapini, S., Taramelli, D., and Sergo, V. (2008) Spatial distribution of heme species in erythrocytes infected with *Plasmodium falciparum* by use of resonance Raman imaging and multivariate analysis. *Anal. Bioanal. Chem.* **392**, 1277–1282.

46. Van Manen, H.-J., Kraan, Y. M., Roos, D., and Otto, C. (2004) Intracellular chemical imaging of heme-containing enzymes involved in innate immunity using resonance Raman microscopy. *J. Phys. Chem. B* **108**, 18762–18771.

47. Sharifzadeh, M., Zhao, D.-Y., Bernstein, P. S., and Gellermann, W. (2008) Resonance Raman imaging of macular pigment distributions in the human retina. *J. Opt. Soc. Am. A* **25**, 947–957.

48. Stiles, P. L., Dieringer, J. A., Shah, N. C., and Van Duyne, R. P. (2008) Surface-enhanced Raman spectroscopy. *Annu. Rev. Anal. Chem.* **1**, 601–626.

49. Kneipp, K. (2007) Surface-enhanced Raman scattering. *Phys. Today* (November), 40–46.

50. Campion, A., Ivanecky, J. E., Child, C. M., and Foster, M. (1995) On the mechanism of chemical enhancement in surface-enhanced Raman scattering. *J. Am. Chem. Soc.* **117**, 11807–11808.

51. Bailo, E. and Deckert, V. (2008) Tip-enhanced Raman scattering. *Chem. Soc. Rev.* **37**, 921–930.

52. Evans, C. L. and Xie, X. S. (2008) Coherent anti-Stokes Raman scattering microscopy: chemical imaging for biology and medicine. *Annu. Rev. Anal. Chem.* **1**, 883–909.

53. Cheng, J. (2007) Coherent anti-Stokes Raman scattering microscopy. *Appl. Spectrosc.* **61**, 197A–208A.

54. Okuno, M., Kano, H., Leproux, P., Couderc, V., and Hamaguchi, H. (2007) Ultrabroadband ($>$2000 cm^{-1}) multiplex coherent anti-Stokes Raman scattering spectroscopy using a sub-nanosecond supercontinuum light source. *Opt. Lett.* **32**, 3050–3052.

55. Freudiger, C. W., Min, W., Saar, B. G., Lu, S., Holtom, G. R., He, C., Tsai, J. C., Kang, J. X., and Xie, X. S. (2008) Label-free biomedical imaging with high sensitivity by stimulated Raman scattering microscopy. *Science* **322**, 1857–1861.

56. Ghaziaskar, H. S. and Lai, E. P. C. (1992) Stimulated Raman scattering in analytical spectroscopy. *Appl. Spectrosc. Rev.* **27**, 245–288.

57. Heritage, J. P. and Allara, D. L. (1980) Surface picosecond Raman gain spectra of a molecular monolayer. *Chem. Phys. Lett.* **74**, 507–510.

58. http://modis.gsfc.nasa.gov/about.

59. http://aviris.jpl.nasa.gov.

60. http://m3.jpl.nasa.gov.

61. http://wwwvims.lpl.arizona.edu.

62. Christensen, P. R., Anderson, D. L., Chase, S. C., Clancy, R. T., Clark, R. N., Conrath, B. J., Kieffer, H. H., Kusmin, R. O., Malin, M. C., Pearl, J. C., Roush, T. L., and Smith, M. D. (1998) Results from the Mars Global Surveyor Thermal Emission Spectrometer. *Science* **279**, 1692–1698.

63. http://tes.asu.edu/about/index.html#anchor_206million.

64. http://marsrover.nasa.gov/mission/spacecraft_instru_minites.html.

65. http://minites.asu.edu/aboutmer.html.

66. http://themis.asu.edu/faq.

67. http://crism.jhuapl.edu/CRISMfacts.php.

68. Clark, R. N., Swayze, G. A., Livo, K. E., Kokaly, R. F., Sutley, S. J., Dalton, J. B., McDougal, R. R., and Gent, C. A. (2003). Imaging spectroscopy: Earth and planetary remote sensing with the USGS Tetracorder and expert systems. *J. Geophys. Res. E* **108**, 5131–5175.

69. http://gallery.spitzer.caltech.edu/Imagegallery/chron.php?cat=Astronomical Images (Space telescope launched in 2003).

70. http://lambda.gsfc.nasa.gov/product/cobe/cobe_image_table.cfm (Satellite launched in 1989).

71. http://www.ipac.caltech.edu/Outreach/Gallery/IRAS/irasgallery.html (Satellite launched in 1983).

72. http://iso.esac.esa.int/science/ (Satellite launched in 1995).

73. http://sci.esa.int/science-e/www/object/index.cfm?fobjectid=39279 (Satellite launched in 2006).

74. Dupuis, J. R., Mansur, D. J., Engel, J. R., and Vaillancourt, R. (2008) Imaging open-path Fourier transform infrared spectrometer for 3D cloud profiling. *Proc. SPIE* **6954**, 69540N.

PART I

HARDWARE

2

RAMAN IMAGING INSTRUMENTATION

MATTHEW P. NELSON AND PATRICK J. TREADO

ChemImage Corporation, Pittsburgh, PA, USA

2.1 INTRODUCTION

Raman imaging combines the visual perception of digital imaging technology with the objective molecular information from Raman spectroscopy. In its most fundamental definition, a Raman image is an array of spatially accurate Raman spectra. Raman imaging is used synonymously with Raman chemical imaging [1], Raman hyperspectral imaging, and Raman molecular imaging. Closely related terminology includes multispectral imaging (<10 spectral bands), hyperspectral imaging (10–1000 spectral bands), and ultraspectral imaging (>1000 spectral bands).

A chief motivation for applying Raman imaging in materials characterization is that most materials are spatially heterogeneous in composition and structure. There is a fundamental need and desire to measure material properties in two and three spatial dimensions in order to fully understand material identity, compositional distribution, conformational distribution, and how these properties relate to performance. Raman imaging addresses this need by providing an efficient, intuitive means of visualizing the two- and three-dimensional architectures of materials in a nondestructive and noninvasive manner.

Raman imaging instruments are the tools used to determine where molecule-specific Raman scatter arises in highly heterogeneous materials. These instruments are used to generate hundreds, thousands, or even millions of independent, spatially resolved Raman spectra from the material of interest. Figure 2.1 shows the data structure known as a hyperspectral data cube consisting of X and Y (and even Z) spatial dimensions and a wavelength (i.e., Raman shift, cm^{-1}) dimension obtained during a typical Raman imaging experiment. The resulting hyperspectral data cube consists of a series of wavelength-specific images where each pixel contains a Raman spectrum associated with the material imaged at that location. The Raman spectra in the hypercube are generally processed to reveal image contrast that is chemical specific without the need for stains, dyes, or contrast agents. Consequently, there is often a reduced burden or even no need for sample preparation unlike many other material analysis tools.

The Raman spectra extracted from individual components compared to the mean spectral signature within the hyperspectral data cube in Figure 2.1d exemplify the analytical value that Raman imaging provides over traditional bulk spectroscopic measurements. The mean spectrum—representative of a bulk Raman spectral measurement—is very different from the localized spectral profile of individual particles. Spatial sampling allows more reliable identification in real-world mixtures by exploiting the natural spatial variation. It also allows the use of chemometric processing tools that can use the resulting spectral variability to identify the individual components.

Raman imaging instrumentation advancements over the years have paralleled progression in Raman spectroscopy instrumentation. Like Raman spectroscopy, Raman imaging has benefited from advancements in laser technology that provide a highly monochromatic source, multilayer dielectric and holographic dichroic filters that effectively remove Rayleigh scattered laser light, imaging spectrometers that isolate wavelengths of interest, highly sensitive charge-coupled device (CCD) imaging detectors, and powerful PCs having sufficient data storage capacities and processing speeds capable of handling the large data files obtained in a Raman imaging experiment. Other developments in Raman spectroscopy and Raman imaging instrumentation include

Raman, Infrared, and Near-Infrared Chemical Imaging   Edited by Slobodan Šašić and Yukihiro Ozaki

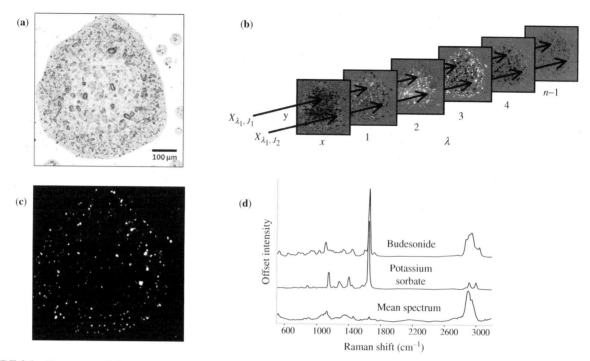

FIGURE 2.1 Hyperspectral data cube structure of a typical Raman imaging experiment. (a) Optical image of multiple components in a formulated nasal spray; (b) hypercube viewed as images as a function of wavenumber; (c) Raman image of budesonide (API) particles; (d) Raman spectra extracted from the hyperspectral data cube associated with budesonide, potassium sorbate, and the mean image spectrum from the entire hyperspectral data cube.

increased automation, improved ease of use, and enhanced hardware robustness and stability. Due to these instrumental improvements, there has been a dramatic increase in use and acceptance of Raman imaging for a whole host of industrial, forensic, and medical applications.

The focus of this chapter is on Raman imaging instrumentation. Here we detail technological approaches currently used in Raman imaging instrumentation, Raman instrumentation platform types, signal-enhanced Raman instrumentation that address one of the fundamental limitations of Raman, that is, low light levels, tandem Raman imaging instruments that benefit from orthogonal information obtained from multiple sensor types, and how to assess performance of a Raman imaging instrument.

2.2 RAMAN IMAGING INSTRUMENTATION TYPES

Over the past three decades, numerous instrumental designs for performing Raman imaging have been developed and commercialized. These instruments can be broadly classified as scanning (i.e., mapping) approaches or wide-field (i.e., global) source illumination approaches.

In this section, general comparisons will be made between scanning and wide-field approaches. A summary comparison of each approach is shown in Figure 2.2.

2.2.1 Scanning

The majority of Raman imaging since the development of the first Raman microprobes in the 1970s has been performed using scanning-based instrumentation [2]. In scanning methods, Raman "maps" of the sample surface are generated by focusing a laser beam source onto the surface of the sample and collecting one or more spectra at each spatial location using either a dispersive spectrograph or an interferometer [3]. Scanning methods can be further divided into point-by-point scanning and line scanning systems.

2.2.1.1 Point-by-Point Scanning   In point-by-point scanning (i.e., mapping) Raman instruments, a laser is focused to a small spot on the sample surface. A Raman spectrum is then acquired from each spatial position using a dispersive spectrograph equipped with an array detector or using an interferometer. Raman images are generated by raster scanning the sample through the laser beam using an X, Y (lateral scanning), Z (axial scanning) stage, collecting a spectrum at each position, and reconstructing the collected data in the form of an image (shown schematically in Figure 2.2). Often point-by-point scanning Raman imaging instruments are operated in a confocal configuration in which the source illumination and/or collected scattered light is focused through pinholes to localize the excitation and collection volumes from the sample. Using a confocal approach often

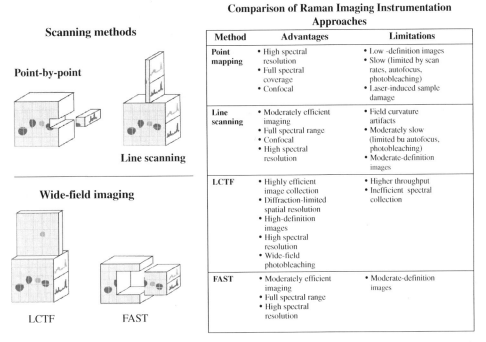

Comparison of Raman Imaging Instrumentation Approaches

Method	Advantages	Limitations
Point mapping	• High spectral resolution • Full spectral coverage • Confocal	• Low-definition images • Slow (limited by scan rates, autofocus, photobleaching) • Laser-induced sample damage
Line scanning	• Moderately efficient imaging • Full spectral range • Confocal • High spectral resolution	• Field curvature artifacts • Moderately slow (limited bu autofocus, photobleaching) • Moderate-definition images
LCTF	• Highly efficient image collection • Diffraction-limited spatial resolution • High-definition images • High spectral resolution • Wide-field photobleaching	• Higher throughput • Inefficient spectral collection
FAST	• Moderately efficient imaging • Full spectral range • High spectral resolution	• Moderate-definition images

Scanning methods — Point-by-point, Line scanning

Wide-field imaging — LCTF, FAST

FIGURE 2.2 Hyperspectral data cube construct and trade-offs associated with scanning-based and wide-field Raman imaging instruments.

reduces background fluorescence and helps reduce the effects of secondary scattering, but has the disadvantage of eliminating up to 98% of the captured Raman photons and further complication of poorly defined axial sampling.

A major limitation of point-by-point scanning approaches is lengthy experimentation times, T_p, since the duration is proportional to the number of pixels in the Raman image, n. Since 98% of the Raman signal is lost in confocal point scanning systems, even longer acquisition times are necessary. If sample fluorescence is a problem, additional time may be required to allow for photobleaching for each new spatial location. In the past, scan times of hours to days were not uncommon for point-by-point experiments. The lengthy acquisition times placed practical limits on the number of pixels collected to form an image, limiting its utility as a routine imaging tool for materials characterization. Today, however, advancements in scanning probe technology such as WiTec's continuous point-by-point scanning approach [4] have significantly reduced acquisition times associated with point-by-point Raman imaging. Individual spectra may be acquired in milliseconds making the acquisition of Raman images with thousands of spectra practical in a matter of minutes. The spatial resolution of point scanning systems is limited by the size of the laser spot ($\sim$1 μm) and the accuracy, reproducibility, and stability associated with the rastering stage. For light-absorbing materials having low damage thresholds, localized thermal expansion and photoinduced damage present further imaging challenges. This is especially problematic when attempting to find one laser power

density that is suitable for complex sample matrices having materials possessing a variety of damage thresholds.

Despite the inherent limitation of point-by-point scanning approaches to Raman imaging, the technique is mature and has been applied successfully in numerous applications [5–8]. Another advantage of point scan Raman imaging is the ability to efficiently collect an entire spectrum. This is particularly advantageous for work involving new material systems in which the underlying spectroscopy is poorly understood and a full spectrum may provide the required information. Having the full spectrum available also improves the chances that a chemometric process applied to the data will be able to pull out meaningful information.

2.2.1.2 Line Scanning An extension of the point-by-point scanning approach is line scanning [9, 10]. In line scanning, a laser beam is elongated in one dimension to form a line with the use of cylindrical optics or a scanning mechanism such as a moving mirror. The laser line that illuminates the sample is oriented parallel to the entrance slit of a dispersive spectrograph equipped with a two-dimensional CCD array detector. In this configuration, a spectrum may be captured for each row of the CCD detector, which corresponds to a spatial location along the length of the laser line. Like point scanning, mechanical scanning is required for image formation with line scanning approaches (shown schematically in Figure 2.2), but scanning is required only in the direction perpendicular to the laser line. As a result, the number of spatial channels is reduced by $\sqrt{n}$ decreasing the experiment time, T_1, such that

$$T_1 = \left(\frac{1}{\sqrt{n}}\right)T_p \qquad (2.1)$$

Equation 2.1 assumes that there is ample laser power available to maintain a comparable laser power density to the point scanning approach for a given experiment. The image spatial resolution for the line scanning approach differs for the x- and y-axes. The resolution in the direction parallel to the laser line is determined by the magnification of the collection optic and the pixel size associated with the detector. Along this direction, diffraction-limited spatial resolution is achievable. In the direction perpendicular to the laser line, however, spatial resolution is dictated by the laser beam width and the precision of the sampling stage rastering mechanics.

While not as widely used as its point scanning predecessor, line scanning Raman imaging systems are commercially available and have been utilized in numerous applications [9–12]. A relatively new development in line scanning technology is Renishaw's StreamLine™ Plus imaging that utilizes a time-interleaved sampling approach for reducing data acquisition times. Microscope optics are used to illuminate a line on the sample, which is moved relative to the sample by adjusting a motorized microscope stage securing the sample beneath an objective lens. Spectral data are read continuously from the detector as the grating motion is synchronized with the movement of the line across the sample [13].

2.2.2   Wide-Field Raman Imaging

A major motivation for performing Raman imaging is the need to rapidly and accurately characterize sample morphology and composition in two or three spatial dimensions and perhaps additional dimensions such as time or temperature. It may also be assumed that the higher the image fidelity (i.e., image definition), the higher the quality of the image and the more complete the accurate assessment of material constituents. Raman imaging methodologies have advanced in the past two decades to enable collection of high-resolution (spectral and spatial) data through what have become known as wide-field (i.e., global) imaging approaches.

In wide-field Raman imaging systems, the entire sample field of view is illuminated (i.e., globally illuminated) with laser light and analyzed in parallel. Numerous wide-field Raman imaging approaches have been demonstrated and will be described in greater detail (below). The majority of these approaches involve collecting images at discrete frequencies (shown schematically in Figure 2.2).

Since two spatial dimensions are collected simultaneously, the duration of a wide-field experiment, T_w, is proportional to the number of spectral channels, m, as opposed to the number of image pixels, n, as in point and line scanning approaches.

The experimentation time relationships between wide-field and point and line scanning approaches are shown in Equations 2.2 and 2.3. The wide-field imaging time saving is realized when m is less than n or $\sqrt{n}$ for point-by-point and line scanning, respectively. With these equations, it is assumed that the delivery and collection efficiencies of the optics are identical for each approach, the time delay between acquisitions (i.e., stage movement or filter tuning) is negligible, the laser power is adjusted so that the power density at the sample remains constant and below any damage threshold of the sample, and the image is square.

$$T_w = \left(\frac{m}{n}\right)T_p \qquad (2.2)$$

$$T_w = \left(\frac{m}{\sqrt{n}}\right)T_1 \qquad (2.3)$$

For most materials characterization application needs, a small number of spectral bands (typically <30) are adequate to provide sufficient chemical and spatial information for the analytes of interest. Reducing the number of spectral channels shortens the total experiment time and lowers the burden associated with data storage and computer processing requirements. In many materials characterization applications, a very small number of spectral bands (i.e., <5) may be required allowing rapid wavelength tuning to support dynamic or "real-time" Raman imaging, which is a capability not feasible with scanning methods. With scanning Raman methods, a reduction in the number of spectral channels has no real impact on experiment times since the entire spectrum is captured simultaneously. Time reduction with scanning methods can result mainly by reducing n, which negatively affects the spatial resolution or field of view of the resultant Raman image.

Spatial resolution for wide-field imaging is normally dictated by a convolution of diffraction theory, CCD pixel size, and magnification imparted by the imaging system at the CCD. The Rayleigh criterion-based diffraction-limited resolution (r) is given by

$$r = \frac{1.22\lambda}{2NA} \qquad (2.4)$$

where λ is the wavelength of light and NA is the numerical aperture of the light gathering optic. A spatial resolution of 250 nm has been demonstrated using an electrooptically tunable filter-based Raman imaging microscope system equipped with a 514 nm laser [14]. A spatial resolution of 250 nm, however, does not limit the ability to image even smaller features of brightly scattering materials. Figure 2.3, for example, shows a Raman imaging example from a SERS active gold surface acquired using a wide-field Raman

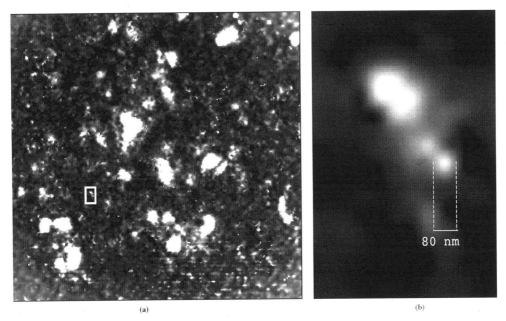

(a)   (b)

FIGURE 2.3   (a) Raman image of a SERS active gold surface acquired using a wide-field Raman imaging system. (b) Raman imaging reveals "hot spots" as small as 80 nm on the surface.

imaging system. Raman imaging reveals "hot spots" of analyte distribution on the surface. Features as small as 80 nm are resolvable.

Wide-field Raman imaging has been demonstrated for a wide variety of materials characterization applications with an emphasis on polymer [15–17], semiconductor [18], biomedical [19–21], pharmaceutical [23, 24], and homeland security applications [25–27].

2.3 WIDE-FIELD RAMAN IMAGING INSTRUMENTATION DESIGNS

A variety of instrumental developments have occurred over the past two decades that fall under the category of wide-field Raman imaging system designs. Such developments include fiber array assemblies [28–32], tunable laser and fixed [33] or rotating dielectric filters [34], acoustooptic tunable filters (AOTFs) [35–38], and liquid crystal tunable filters (LCTFs) [39–42]. In addition, wide-field Raman imaging techniques such as coherent anti-Stokes Raman scattering (CARS) and stimulated Raman will be discussed in Section 2.5.

2.3.1 Fiber Array Raman Imaging

A more recent approach to Raman imaging that has gained popularity in the last decade is based on fiber optic arrays. This technology has been termed fiber array spectral translators (FAST) [27], dimension reduction arrays [29–31], and fiber image compression [32]. Two spatial dimensions and one spectral dimension of data may be collected simulta-

neously with fiber arrays by focusing Raman light captured from a globally illuminated field of view onto the proximal end of a two-dimensional array of optical fibers. The distal end of the fiber array is then drawn into a linear array and inserted parallel into the entrance slit of a dispersive spectrometer equipped with an imaging format CCD detector. The fiber array enables two spatial dimensions of data to be reduced to a single dimension, which is then dispersed fiber by fiber along the vertical axis of the spectrometer entrance slit onto the CCD camera. In a single frame acquisition, all of the spatial and spectral information is obtained simultaneously. Software is then used to unravel the embedded spatial and spectral information by reconstructing the data into a hyperspectral data cube in which each pixel of the reconstructed image contains an associated, full Raman spectrum (shown schematically in Figure 2.4).

Fiber arrays have an inherent speed advantage over competing Raman imaging techniques for applications that require substantial spectral information but only limited spatial information. For reasonably bright Raman scatterers, the data acquisition time for a single field of view is comparable to the time it takes to acquire a single dispersive spectrum. Equations 2.5–2.7 show the relationship between the experiment time required using a fiber array Raman imaging system in comparison to point-by-point scanning, line scanning, and (wavelength scanning) wide-field Raman imaging instruments where ROI is the number of regions of interest acquired in the experiment. The assumptions that apply to Equations 2.2 and 2.3 also apply to Equations 2.5, 2.6. Equations 2.5–2.7 further assume that the laser power may be adjusted to provide comparable power density at each

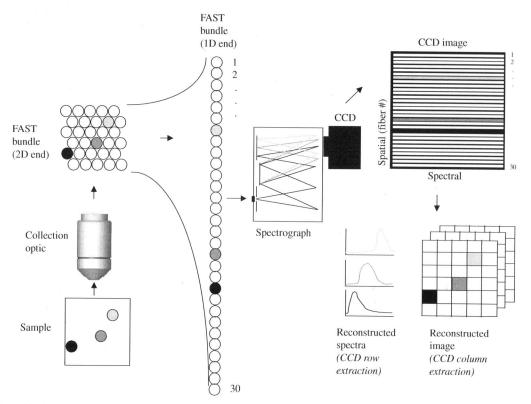

FIGURE 2.4 Wide-field Raman imaging instrument design incorporating a fiber array spectral translator. Raman light is focused onto a two-dimensional array of a fiber optic bundle that is reduced to a one-dimensional array on the distal end. The distal end is imaged through a spectrograph equipped with an imaging format CCD. Single CCD image frames may be reconstructed to provide wavelength-specific images and space-specific spectra.

spatial location such that the amount of Raman signal produced at each spatial location is equivalent for each technique.

$$T_{\text{FAST}} = \frac{T_{\text{p}}}{n} \times \text{ROI} \qquad (2.5)$$

$$T_{\text{FAST}} = \frac{T_{\text{l}}}{\sqrt{n}} \times \text{ROI} \qquad (2.6)$$

$$T_{\text{FAST}} = \frac{T_{\text{w}}}{m} \times \text{ROI} \qquad (2.7)$$

Other advantages of using fiber arrays for Raman imaging spectral resolution and spectral coverage are governed by the dispersion properties of the dispersive spectrometer used in the analysis. Fiber array image fidelity (number of pixels) is limited by the number of CCD detector rows. High-definition images may be obtained with this technique by collecting and reconstructing fiber array Raman images from multiple adjacent fields of view. Alternatively, multiple spectrometers each equipped with CCDs may be used to increase image definition. An alternate implementation of Raman imaging with fiber arrays is to use the data it produces to create a color map that is superimposed on a high spatial resolution gray-scale image of the sample acquired by a standard video camera. Another drawback of using fiber arrays for Raman imaging is susceptibility of pixel-to-pixel crosstalk as a result of imperfect image performance of modern spectrometers. Crosstalk may be minimized by intentional spacing of fibers or specialized fiber mapping arrangements in which adjacent fibers in the object field are also juxtaposed in the image field. Manufacturability of the fiber arrays is also a current challenge of the technique.

Fiber array Raman imaging has been demonstrated in several applications including analysis of microcomposites and biomaterials [32] and standoff detection of explosives [27].

2.3.2 Dielectric Interference Filter

Raman imaging with the use of dielectric interference filters was first demonstrated in the early 1990s. Batchelder et al. placed mechanically rotatable dielectric interference filters in the infinity-corrected path of a Raman microscope [34]. Raman images were generated by capturing images of a globally illuminated sample surface through the dielectric filters with an imaging format CCD detector. The center passband of the dielectric filters that had a reasonably narrow bandpass (20 cm^{-1}) could be tuned by mechanically rotating the filters. Numerous filters were required to tune across the entire Raman spectrum (Figure 2.5).

Use of dielectric interference filters for Raman imaging has several advantages including simplicity, relatively low-cost components, and spatial resolutions that approached diffraction limits. Drawbacks of this approach include image shift associated with the mechanical movement of the filters and bandpass position inhomogeneity across the field of view.

Fixed dielectric filters have also been used in conjunction with tunable lasers to perform Raman imaging [33]. This approach addresses key limitations of the angle-tuned dielectric interference filter approach to Raman imaging, but suffers from high laser source background since Rayleigh scatter rejection becomes a major challenge.

2.3.3 Acoustooptic Tunable Filter

AOTFs were first demonstrated as an imaging spectrometer for performing Raman imaging in the early 1990s [35, 36, 38, 43]. AOTFs are solid-state, no moving parts devices that operate based on the interaction of light with a traveling acoustic sound wave in an anisotropic crystal medium. A diffracted, narrow spectral bandpass of the incident light is created by applying an rf signal to the AOTF. Bandpass tuning and diffracted light intensity are achieved under computer control by changing the applied rf frequency and power, respectively (shown schematically in Figure 2.6).

Advantages of AOTFs include high optical throughput (40% for unpolarized light), variable spectral bandpass, broad spectral coverage (UV to mid-IR), and rapid tuning speeds ($\sim$100 μs). AOTFs possess characteristics that limit their applicability as a Raman imaging spectrometer, however, including broad spectral bandpasses (50 cm^{-1}—about 10 times worse than a typical Raman spectrometer) and spatial resolutions approximately 5 times worse than diffraction-limited conditions.

Despite the limitations of AOTFs for Raman imaging, these devices have been applied successfully in numerous applications including studies of polymer blend systems [36] and pathology analysis [37].

2.3.4 Liquid Crystal Imaging Spectrometers

For decades, Raman imaging instrumentation developers have aspired to build systems that provide users the ability to capture high spatial and spectral resolution images and to do so in a rapid, efficient, and cost-effective way. At present, it is the view of the authors that wide-field Raman imaging systems equipped with liquid crystal imaging spectrometers come closest to realizing those goals.

Most liquid crystal imaging spectrometers are capable of providing diffraction-limited spatial resolution with spectral resolutions rivaling that of a single-stage dispersive monochromator. These devices also provide high out-of-passband rejection efficiencies and broad free spectral ranges, operate in the visible and near-infrared portions of the spectrum (400–2500 nm), and have high overall etendues. Since these devices are electrooptical having no mechanical moving parts, they may be rapidly tuned with automatable random accessing under computer control. Limitations of liquid crystal imaging spectrometers include inefficient full spectral data collection and low to moderate throughput (in some designs).

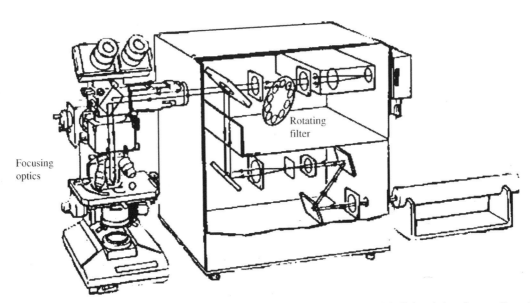

FIGURE 2.5 Wide-field Raman imaging instrument design incorporating mechanically rotatable dielectric interference filters in the infinity-corrected path of a Raman microscope.

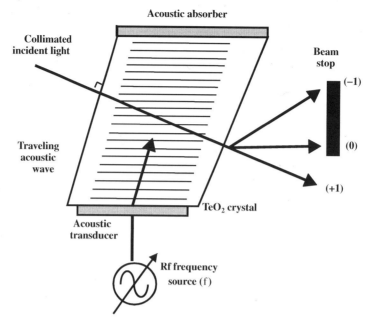

FIGURE 2.6 Schematics of an acoustooptical tunable filter. Collimated incident light is diffracted as it passes through the device creating a narrow spectral bandpass of light when an rf signal is applied to the AOTF.

In the following sections, we summarize five types of liquid crystal imaging spectrometers that have been implemented for Raman imaging.

2.3.4.1 Lyot LCTF

French astronomer B. Lyot introduced the first birefringent interference filter in 1944 [44]. Lyot filters are currently composed of a series of stages cascaded together, each consisting of fixed retardance birefringent elements bonded to a nematic liquid crystal (LC) wave plate (W) sandwiched and preferentially oriented between two parallel linear polarizers. The multiple Lyot stages cascaded together result in a single, narrow passband of light due to constructive and destructive interference. The center wavelength of the filter may be controlled electrooptically under computer control by applying a potential (V) to the LC wave plate. Filter retardance is determined by the thickness of each fixed birefringent element and doubles with each successive stage, resulting in a transmission spectrum having half the free spectral range and half the bandpass of the previous stage. The bandpass and free spectral range of the Lyot filter are determined by the thickest and the thinnest birefringent elements, which are fixed during manufacturing. Lyot filter designs are capable of spectral bandpasses ranging from 30 down to 0.05 nm.

The overall transmittance of a Lyot filter has a sinc function profile and is the product of the individual filter stage transmittances. A Lyot filter with a continuously tunable bandpass of 7.6 cm^{-1} and a free spectral range greater than 4600 cm^{-1} (500–650 nm) suitable for Raman imaging has been demonstrated with a peak transmittance of only 16%. The poor overall transmittance is a consequence of the relatively large number of highly absorbing polarizers and imperfect wave plate action [42].

Compared to interferometer technologies such as Fabry–Perot or Michelson interferometers, birefringence filters such as Lyot filters have significantly relaxed tolerances in precision requirements of the optical components and are much less sensitive to effects of vibration, temperature changes, and nonuniformity across the open optical aperture enabling large aperture devices to be manufactured supporting wide-field Raman imaging system designs. The largest drawback of the Lyot design is low optical throughput as a result of the large number of polarizers needed to achieve a given finesse (free spectral range divided by bandpass width). For Raman imaging applications, nearly 20 polarizers are required each having a transmission loss of approximately 10% [40].

2.3.4.2 Solc Filters

To address the low overall transmittance of Lyot filters, Solc proposed a fanout and folded filter design [45] that improves optical throughput by using only two polarizers in the design. Unfortunately in this design, the transmission spectrum has side lobes [46] that can cause low out-of-passband rejection efficiency making the Solc filter a poor choice for Raman imaging applications.

2.3.4.3 Evans Split Element Filters

Evans later proposed a split element filter design [47] that has balanced manufacturability, throughput, and side lobe characteristics. The Evans split element design reduced the number of polarizers to approximately one half the original Lyot designs resulting in a doubling of the optical throughput while maintaining a narrow spectral bandpass (9 cm^{-1}) and extending the free spectral range for Raman imaging applications [24, 47, 48].

Throughout the 1990s and early 2000s, Evans split element liquid crystal imaging spectrometers represented the state-of-the-art technology for wide-field Raman imaging applications. While the Evans filter properties exhibit acceptable performance for numerous materials characterization applications, application demands have driven the need for the development of imaging spectrometers having even higher optical throughput for extremely low light imaging applications, such as the detection of biothreat materials, and improved thermal stability for applications requiring extremely high spectral bandpass precision, such as polymorph discrimination.

2.3.4.4 Multiconjugate Filter   To address increasing demands for improved optical throughput and thermal stability, a new class of liquid crystal imaging spectrometer, multiconjugate filter (MCF), has been developed [49]. MCF combines the high transmittance associated with a Solc filter and the high out-of-band rejection efficiency of the Lyot filter. MCF differs from previous generation devices in that each stage of the filter has an increased finesse—1.5 times the finesse of Evans filters and 2 times the finesse compared to a typical Lyot design. Increasing the finesse of individual stages allows a filter to be manufactured having fewer stages to achieve an overall finesse comparable to earlier generation filters with greater optical throughput over a wider spectral range (shown schematically in Figure 2.7a).

MCF also differs from previous generation filters in how thermal stability is managed. Evans filters, for instance, use an active temperature correction strategy that is based on measuring capacitance in the LC variable retarder. MCF instead uses a passive temperature correction strategy to

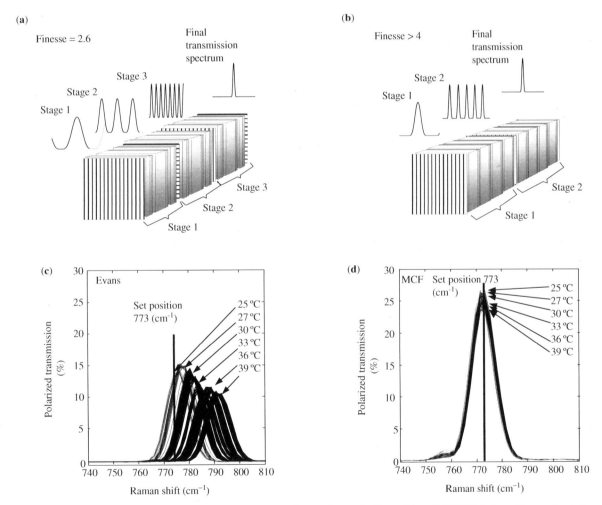

FIGURE 2.7   Schematic showing a comparison of the birefringent filter structure for Evans (a) and MCF (b) designs. Comparison of response to thermal shock for the Evans filter (c) and the MCF (d).

compensate for thermal effects on the device. This is achieved by precisely maintaining chamber thickness for each LC electrooptical component over a large range of temperatures. The result is a highly predictable and repeatable temperature behavior for each electrooptical component (Figure 2.7b).

Table 2.1 shows a side-by-side comparison of filter properties of MCF and the previous state-of-the-art Evans split element design.

2.3.4.5 Fabry–Perot Liquid Crystal Tunable Filter

For completeness, it is worth mentioning the use of Fabry–Perot liquid crystal imaging spectrometers for Raman imaging applications. A Fabry–Perot device having a $20\,cm^{-1}$ spectral bandpass has been demonstrated in a Raman imaging microscope [42]. While exhibiting excellent image quality, the Fabry–Perot device suffers from moderate spectral resolution, low out-of-band rejection, limited free spectral range, and small acceptance angle and requires extensive measures for maintaining thermal stability to avoid thermally induced bandpass drift.

2.4 RAMAN IMAGING INSTRUMENTATION PLATFORMS

Like normal Raman spectroscopy, Raman imaging had to wait for technological developments to become a practical tool that could later be commercialized. Most of the current Raman imaging instrument platforms benefit from the use of continuous-wave (CW) or pulsed laser sources to provide intense, highly monochromatic light, light directing optics such as fiber optics, lenses, and microscope objectives, multilayer dielectric or holographic optical filters for rejecting Rayleigh scattered light, wavelength selection devices including gratings from a dispersive spectrometer, interferometer, or electronically tunable narrow bandpass imaging

TABLE 2.1 Comparison of Previous State-of-the-Art Evans Split Element LCTF with Multiconjugate Filter

	Evans Split Element Filter	Multiconjugate Filter
Design	Green Raman	Green Blue Raman
Spectral range	500–720 nm	445–740 nm
Polarized peak transmission	14% @ 550 nm 24% @ 650 nm	26% @ 550 nm 32% @ 650 nm
Nominal FWHM @ 550 nm	0.30 nm (9.9 cm^{-1})	0.42 nm (13.8 cm^{-1})
Angular field of view	±3°	±3°
Tuning accuracy @ 550 nm	±0.1 nm (3.3 cm^{-1})	±0.05 nm (1.7 cm^{-1})

spectrometers, highly sensitive detectors such as CCDs, and PCs having the processing speeds and data storage capacity for collecting, processing, and displaying often very large data sets.

Raman imaging platforms have taken on numerous shapes, sizes, and capabilities over the years. Raman imaging instrumentation has proven effective in microscopic, macroscopic, flexible endoscopic, and even standoff telescopic analyses. This section will highlight examples of each.

2.4.1 Raman Imaging Microscopes

Currently, the most common platform for conducting Raman imaging has been the light microscope. Raman imaging performed in this manner provides high spatial resolution images allowing for localized analysis of material composition and structure from the millimeter to submicron scale.

Laser light is typically delivered to the microscope platform via optical fibers or through one or more direct beam designs. Confocal Raman designs often include the use of pinhole apertures in the laser delivery path to restrict the illumination volume near the sample surface. Line scanning systems typically use cylindrical optics while wide-field imaging systems use beam expanding optics to shape the beam prior to delivering the laser light to the sample. Optical filters may be required to condition the laser beam to isolate a particular laser wavelength of interest. Most designs deliver laser light to the sample through a microscope objective and utilize the same optic to collect the 180° backscattered Raman light from the sample. Additional optical filters such as dielectric interference filters or holographic notch filters are usually required to remove the Rayleigh scattered laser light from the detection channel. The filtered light is then presented to one or more light dispersion or frequency isolation devices including an interferometer, dispersive spectrograph, or tunable filter. Finally, the light is detected using one or more detector types depending on the wavelength range over which the Raman scattering occurs as well as the type of Raman imaging system being used. Figure 2.8 shows a point-by-point scanning Raman imaging result from an over-the-counter (OTC) analgesic tablet performed by Dr. Simon FitzGerald at HORIBA Scientific using an XploRA (HORIBA Scientific) confocal Raman microscope. The excitation laser wavelength was 638 nm and the grating was 1200 grooves/mm. The entire map is composed of 50,901 spectra, covering $1.8 \times 0.7\,cm^2$ at 200 ms per point in SWIFT™ mode. Data were analyzed using the modeling function in LabSpec 5 (HORIBA Scientific), which employs a direct classical least square (DCLS) algorithm. Three models that represent aspirin (Figure 2.8a), acetaminophen (Figure 2.8b), and caffeine (Figure 2.8c) are shown. Scores of individual spectra with respect to a model represent the spatial distribution in the sample of the chemical the model represents. Figure 2.8a shows the scores image of individual

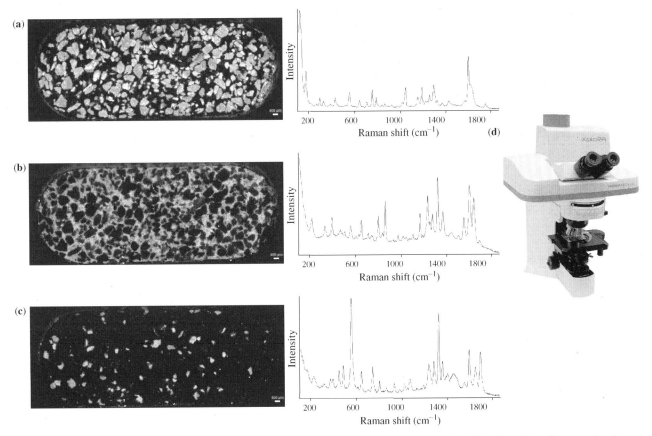

FIGURE 2.8   Commercial point-by-point scanning Raman imaging system (XploRA, HORIBA Scientific): (a) aspirin, (b) acetaminophen, and (c) caffeine scores images and respective Raman spectra associated with a pharmaceutical tablet, and (d) photograph of XploRA system.

spectra with respect to the aspirin model and represents the spatial distribution of aspirin in the tablet. The scores images representing the spatial distribution of acetaminophen and caffeine in the tablet are shown in Figure 2.8b and c, respectively.

Figure 2.9 shows results from a commercial Falcon II™ (ChemImage Corporation, Pittsburgh, PA) wide-field Raman imaging system used to characterize the budesonide active pharmaceutical ingredient (API) particle size distribution (PSD) in a sample of Rhinocort Aqua® nasal spray. Wide-field Raman imaging is currently undergoing validation as a method for characterizing API PSD in nasal aerosol and spray products as part of bioequivalence (BE) testing for new drug applications (NDAs) and abbreviated new drug applications (ANDAs) [23]. Figure 2.9a shows the maximum chord PSD of the Rhinocort Aqua® API, budesonide, associated with the Raman image shown in Figure 2.1c. Also shown is a photograph of the Falcon II™ wide-field Raman imaging system used to collect the data.

2.4.2   Raman Imaging Macroscopes

Until recently, Raman imaging has been constrained mainly to the microscopic world (i.e., 0.25–500 μm). As technolog-

ical improvements continue to be made in the area of high power output solid-state lasers, scaling Raman images up to the centimeter and even meter scales has suddenly become practical.

Figure 2.10 shows Raman image results collected using a hyperspectral contrast imager system (HCIS, ChemImage Corporation) for the macroscopic examination of forensic evidence under contract to the Technical Support Working

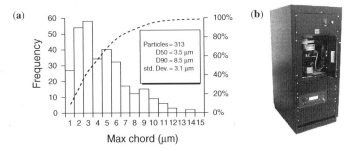

FIGURE 2.9   Commercial wide-field Raman imaging system (Falcon II™, ChemImage Corporation): (a) the maximum chord PSD of the Rhinocort Aqua® API, budesonide, associated with the budesonide particle Raman image shown in Figure 2.1 and (b) Falcon II™ system photograph.

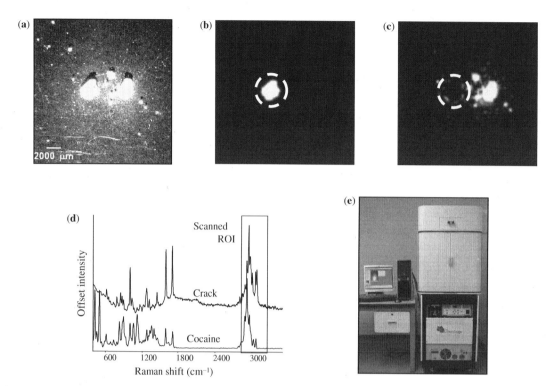

FIGURE 2.10 Raman imaging macroscope (HCIS, ChemImage Corporation): optical image (a) and macro-Raman image of "crack" (i.e., 2949 cm^{-1} Raman image (b) and powder cocaine (2800 cm^{-1} Raman image (c) forms of cocaine acquired using HCIS, library Raman spectra (d), and photograph of HCIS system (e).

Group (TSWG) for counterterrorism. This system is capable of operation from 350 to 1700 nm and provides absorption/reflectance, fluorescence/luminescence, and Raman hyperspectral image sets. The HCIS incorporates a 2 W, 532 nm laser and a liquid crystal imaging spectrometer for use in Raman imaging. Figure 2.10 shows macro-Raman imaging of "crack" and powder forms of cocaine acquired using HCIS. This result demonstrates the ability to identify and differentiate drugs of abuse on a macroscopic scale.

2.4.3 Raman Imaging Fiberscopes

Although most of today's Raman imaging instruments utilize research-grade light microscope technology as the image-gathering platform, the application of Raman imaging to *in situ* industrial process monitoring and *in vivo* clinical analysis is often made possible through the use of fiberscopes. Fiberscopes are an ideal platform for industrial and clinical applications since these settings often require equipment that is lightweight, compact, and rugged enough to operate in harsh, hard to reach environments.

The first Raman imaging fiberscope utilized an AOTF but suffered limitations inherent to AOTFs such as low spectral resolution [50]. A Raman imaging fiberscope design employing LCTF was later described [51]. In this design, a coherent imaging fiber bundle within the fiberscope was coupled to a video CCD for real-time video imaging of the analysis area

and an LC imaging spectrometer coupled with an imaging format CCD for Raman image acquisition. The fiberscope was engineered such that laser delivery and Raman image collection occurred in that same unit. The fiberscope tip contained the necessary optics to filter the scattered radiation generated by the interaction of the laser with the laser delivery fiber and the laser rejection optics essential for removing the Rayleigh scattered light. Design measures were also taken to enable operation of the device in high-temperature environments up to 315°C while maintaining high signal to background (S/B) Raman imaging performance.

Figure 2.11 shows an example Raman image of a pharmaceutical tablet acquired using a Raman imaging fiberscope. The figure shows a bright-field image (Figure 2.11a), grayscale Raman image (Figure 2.11b) revealing aspirin domains (bright regions), and a plot of the imaging spectrometer Raman spectra (Figure 2.11c) from region 1 (localized aspirin) and region 2 (excipient).

2.4.4 Raman Imaging Telescopes

In the past two decades, researchers have made great strides in demonstrating the utility of telescopes for performing standoff Raman measurements of solids, liquids, and gases at ranges up to 500 m [52]. Applications range from standoff Raman detection of minerals on planetary surfaces to remote detection of explosive residues to standoff detec-

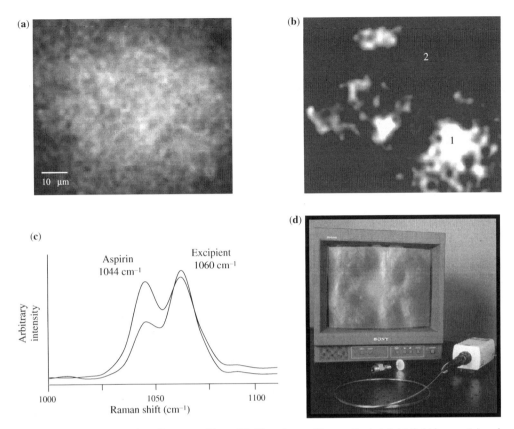

FIGURE 2.11 Commercial Raman imaging fiberscope (Raven™, ChemImage Corporation): bright-field image (a) and grayscale Raman chemical image (b) revealing aspirin (light) and excipient (dark) locations in an over-the-counter pharmaceutical tablet collected using a Raman fiberscope. LCTF spectra (c) from aspirin regions (1) and excipient regions (2). Photograph of Raven™ fiberscope (d).

tion of atmospheric gases [53]. More recently, researches have begun to capitalize on the added value that imaging provides.

The vast majority of standoff Raman systems utilize pulsed laser excitation with gated detection enabling detection of relatively weak Raman signatures in the presence of large ambient backgrounds and highly fluorescent materials. Typically, a nearly collimated laser beam pulse is directed toward a remote target. A portion of the Raman scattered radiation is captured and refocused by the optical tube assembly of the telescope. The focused light is normally conditioned by laser rejection filters, *f*-number matching lenses, and/or fiber optic couplers that present the light to the entrance slit of a dispersive spectrometer equipped with an intensified CCD detector. For Raman imaging, researchers introduce the use of tunable filters or FAST bundles [27].

Figure 2.12 shows a standoff FAST Raman chemical imaging example from a Styrofoam/high-density polyethylene (HDPE) target acquired using a standoff Raman chemical imaging sensor developed by ChemImage Corporation. The target consists of letters that spell the word "FAST" cut from Styrofoam plates and were adhered to a clear plastic (HDPE) storage bin. Data were acquired at a 30 m standoff

distance over a 17×6 montage where each region of interest in the montage captured 36 spectra simultaneously through the FAST bundle. The figure shows FAST reconstructions for Raman bands associated with the Styrofoam and HDPE, respectively. The data collection time was on the order of 20 min and covered an area of $225\,\mathrm{cm}^2$.

2.5 SIGNAL-ENHANCED RAMAN IMAGING INSTRUMENTATION TECHNOLOGIES

"Normal" Raman is inherently a low-sensitivity technique due to the small cross section at the point of analysis. In parallel with Raman instrumental developments have been technological advancements that help address this key limitation of Raman and Raman imaging. This section briefly introduces signal-enhancing techniques and associated instrumentation.

2.5.1 SERS Raman Imaging

Surface-enhanced Raman scattering (SERS) [54, 55] involves absorbing molecules onto a rough metal surface prior to analysis resulting in a much stronger Raman signal than

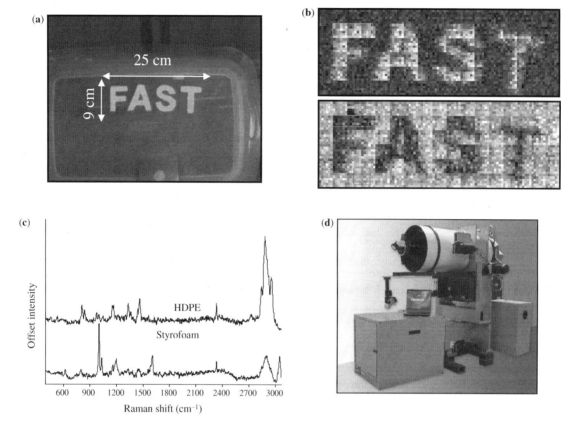

FIGURE 2.12 Raman imaging telescope (Raman-ST, ChemImage Corporation): digital image (a), 1002 cm^{-1} FAST Styrofoam Raman image (top, b), 2882 cm^{-1} FAST HDPE Raman image (bottom, b), and FAST Raman spectra (c) associated with each material. Photograph of Raman-ST sensor (d).

that obtained using normal Raman. To achieve the SERS enhancement, the excitation wavelength of the laser needs to be the same as, or close to, the plasma wavelength of the metal surface on which the molecules are absorbed, and the metal surface must be highly reflective and atomically rough. SERS Raman imaging may be performed using most commercially available Raman point, line, or wide-field imaging systems.

Recent work performed by Guicheteau et al. [56] demonstrated how SERS Raman imaging can greatly enhance conventional Raman imaging of biological materials. In this work, normal and SERS Raman imaging were performed on single biological spores in complex biological mixtures using a wide-field Raman imaging platform. Although detection was achieved using both approaches, a 3000–5000× enhancement was observed with SERS Raman imaging compared to normal Raman imaging. As a result of the signal enhancement provided by SERS, there was a significant improvement in data SNR, reduced laser power density demands thus lowering the chance of photoinduced sample damage, and significant reduction in acquisition times.

2.5.2 SERRS Raman Imaging

A technique that provides even greater signal enhancements over SERS alone is surface-enhanced resonance Raman scattering (SERRS) [57, 58]. SERRS is a single process that is a combination of SERS and resonance Raman scattering. Resonance Raman scattering takes place when the energy of the laser matches, or closely matches, the energy of an electronic transition in a molecule. This makes the Raman signal emitted from the molecule up to four times more intense than the enhancement obtained by SERS alone. Like SERS, SERRS Raman imaging may be achieved using commercially available Raman imaging systems.

In one study, SERS/pre-SERRS (780/633 nm laser excitation) point-by-point scanning and pre-SERRS line scanning and SERRS wide-field imaging using an LC imaging spectrometer were used to characterize Langmuir–Blodgett (LB) films of methacrylic homopolymer (HPDR13) deposited on silver island films [59]. In addition to the expected SERRS signal enhancement, the wide-field SERRS images revealed localized areas exhibiting high SERRS activity.

2.5.3 CARS Raman Imaging

Another Raman signal enhancement technique is CARS [60, 61]. CARS uses two synchronized, near-infrared pulsed lasers for sample excitation. These two pulsed lasers are at different wavelengths with one generating a pump field while the other generating a Stokes field. These two fields are focused on the sample at different angles and the subsequent interaction between the sample and the lasers creates an anti-Stokes signal that is much stronger than the signal that would be generated using conventional Raman spectroscopy. Current CARS Raman imaging instruments are typically bread-board optical bench setups.

Currently, CARS is being used in several application areas including imaging of biological cells and tissues. CARS Raman imaging has been used to study the amount of lipids and fats *in vivo* in real time [62]. With this advancement, there is promise that CARS could be used for real-time Raman imaging of proteins and DNA.

Although CARS is more sensitive than conventional Raman microscopy, nonresonant background created when using CARS limits its sensitivity and makes spectral feature assignments challenging.

2.5.4 SRS Raman Imaging

Researchers have recently developed a technique called stimulated Raman scattering (SRS) that helps address the incoherent background and sensitivity limitations of CARS [63]. Like CARS, SRS uses two lasers for sample excitation with one generating a pump field and the other generating a Stokes field. With SRS, Raman signal given off from a molecule is intensified when that molecule's vibrational frequency matches the frequency difference between the two excitation lasers. When there is no vibrational frequency to match with the excitation lasers' difference in frequency, SRS as well as nonresonant background does not occur. Like CARS, SRS Raman imaging instrumentation is typically found in academic settings in an optical bench configuration or an in-house configured microscope.

SRS has numerous applications in the biomedical imaging field including tissue and skin imaging as well as drug delivery monitoring [63, 64]. In one study, SRS was used to visualize multiple cell structures of varying size and had the ability to track drug delivery to exact cellular locations [63].

2.5.5 SORS Raman Imaging

Spatially offset Raman spectroscopy (SORS) [65–67] is another signal-enhancing technique recently developed that is capable of collecting weak, subsurface Raman signals. SORS collects Raman signals from laterally offset regions away from the point where the laser contacts the sample using optically breadboard components allowing for relative movement between the laser illumination and Raman collection areas. Once the spatially offset spectra have been obtained, multivariate data analysis is applied to resolve the depth-dependent spectral information. SORS suppresses fluorescence generated from the sample surface enabling the collection of very weak Raman signals that would otherwise be overshadowed by fluorescence.

SORS Raman imaging is finding utility in numerous fields including the medical and pharmaceutical industries. SORS has been used to analyze tablets through their packaging to determine if the drugs are counterfeit [66]. SORS Raman imaging is also being assessed in medical research as a potential highly specific, noninvasive diagnostic tool for breast cancer [66]. Methods are currently being developed that may allow analysis of breast tissue $\leq 10\,mm$ deep under the surface of the breast. SORS may be used to analyze the soft tissue and calcification deposits to determine if cancer is present based on the Raman signals emitted, dramatically reducing the number of needed biopsies [67].

2.6 TANDEM RAMAN IMAGING INSTRUMENTATION

2.6.1 Raman–SEM/EDS Imaging

A more complete understanding of a material's spatiochemical makeup often requires both elemental and molecular information. One useful approach to obtaining such information is through the combined use of Raman imaging with scanning electron microscopy coupled with energy dispersive (X-ray) spectroscopy (SEM/EDS) [68]. While Raman imaging provides an understanding of the molecular distribution of components, SEM/EDS [69–72] provides elemental information.

SEM is a type of electron microscope that employs a high-energy electron beam to raster a sample surface to produce high-resolution images (1–5 nm resolution) that detail surface topography and composition among other properties. The interaction of the electron beam with the atoms at or near the sample surface produces secondary electrons (SE), back-scattered electrons (BSE), X-rays, light (cathodoluminescence), specimen current, and transmitted elections, each requiring specialized detectors. Images produced by SEMs have a very large depth of field generating a three-dimensional appearance over a wide range of magnifications. BSE images are related to the atomic number of the specimen, therefore providing information about the elemental composition. X-rays characteristic of the sample are emitted when the high-energy electron beam dislodges an atomic inner shell electron resulting in the release of energy in the form of an X-ray as a higher energy electron fills the void. The analytical technique used to characterize these

X-rays to obtain both qualitative and quantitative elemental composition information about a sample is known as energy dispersive X-ray spectroscopy (EDS, EDX, or EDXRF). By generating an X-ray spectrum at each spatial location, EDS can be used to generate an elemental map of the sample surface.

Although the images produced by SEM/EDS contain better feature resolution, a wider range of magnifications, and a greater depth of field than those obtained using conventional light microscopes, SEM/EDS does not provide molecular information. Raman imaging provides a complementary means of characterizing the molecular spatial architecture of the material.

Figure 2.13 shows results from an integrated wide-field Raman imaging system (ChemImage Corporation) and SEM–EDS (Aspex) platform prototype developed under a DOE-funded Phase II STTR program. The figure shows a schematic of the integrated Raman/SEM–EDS system (e), a SEM secondary electron image (a) of an agglomerate on an ambient MOUDI sample, an EDS spectrum (b) of the agglomerate showing the presence of Ca, S, Al (background), and O, elemental chemical images (a) of the agglomerate created from the EDS spectrum, and a Raman image from the agglomerate collected at $1010 \, cm^{-1}$. Figure 2.13 also shows a comparison of the Raman dispersive spectrum collected from the agglomerate along with the library spectrum from $CaSO_4$ (d). The Raman spectrum along with the elemental dispersive spectroscopy verified the identity of the calcium sulfate particle.

2.6.2 Raman–MXRF Imaging

An alternative elemental mapping method to SEM/EDS is micro-X-ray fluorescence (MXRF). Unlike EDS, where X-ray fluorescence of atoms is produced upon the interaction with a focused electron beam, MXRF uses X-rays to generate element-specific fluorescence spectra of the sample. MXRF instruments use apertures to restrict the X-ray beam to a small spot on the sample that is then raster scanned relative to the sample surface by moving a stage that supports the sample. Spatial resolution is determined by a convolution of the apertured beam diameter and the precision of the stage movement. Elemental images are produced by recording the X-ray fluorescence signal using a detector such as a liquid nitrogen-cooled lithium-drifted silicon chip as the stage is translated in a raster scanning fashion.

In combination with Raman imaging, MXRF provides a means to produce a low-resolution elemental map of the sample surface that may be used to guide a more detailed, high-resolution molecular image of regions of interest. These

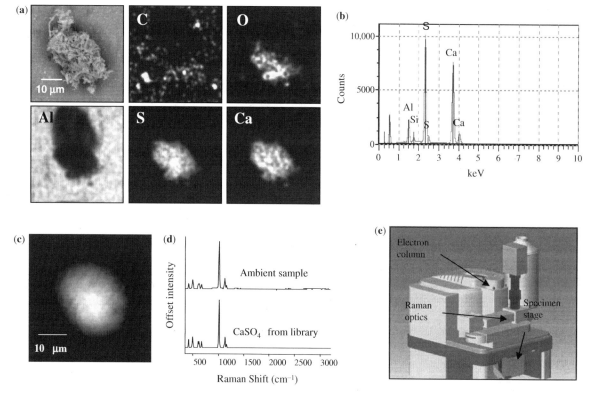

FIGURE 2.13 Integrated Raman imaging/SEM–EDS instrument: SEM secondary electron image and elemental chemical images for C, O, Al, S, and Ca (a) and $1010 \, cm^{-1}$ (sulfate) Raman image (c) of an agglomerate on an ambient MOUDI sample. EDS spectrum (b) of the agglomerate showing the presence of Ca, S, Al (background), and O. Raman dispersive spectrum (d) collected from the agglomerate showing a high correlation with library spectrum for $CaSO_4$. Three-dimensional rendering of integrated Raman imaging/SEM–EDS instrument (e).

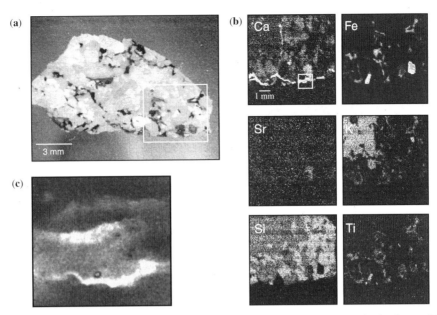

FIGURE 2.14 Combining Raman imaging and micro-X-ray fluorescence: bright-field image (a), single-element MXRF images (b), and a $1080\,cm^{-1}$ Raman image of a granite thin section (c).

integrated techniques are nondestructive and provide complementary chemical information [73].

An example of the combined use of MXRF and Raman imaging is illustrated in Figure 2.14. The figure shows a bright-field image of a granite thin section (a), single-element (Ca, Fe, K, Si, Sr, and Ti) MXRF images for the region highlighted in Figure 2.14a (b), and a $1080\,cm^{-1}$ Raman image associated with highlighted region of the MXRF calcium image in Figure 2.14b (c). The MXRF elemental images reveal a patchwork of localized elemental content consistent with the typical composition of granites including silica polymorphs (mainly quartz), feldspars, and amphiboles. The Raman image shown in the figure was generated from the $1080\,cm^{-1}$ band corresponding to calcium carbonate ($CaCO_3$). A good correlation is evident between the elemental image distribution for Ca obtained with MXRF and molecular image distribution for $CaCO_3$ obtained from the Raman image.

2.6.3 Raman–LIBS Imaging

Laser-induced breakdown spectroscopy (LIBS) provides elemental composition information about solid, liquid, and even gaseous samples [74–77]. LIBS instruments utilize a high-energy pulsed laser beam that is focused to a small spot at the surface of the specimen. The laser energy density is made high enough to ablate a small amount of material generating a high-temperature plasma consisting initially of mostly excited ionic species. A massive continuum emission is observed following the laser pulse due to Bremsstrahlung emission resulting from ion–electron recombination. After a few hundred nanoseconds, the plume begins to cool and the electron density decreases, resulting in predominately atom-

ic emission. Therefore, it is common practice for scientists to time gate their experiments using a dispersive spectrometer, delay generator, and intensified CCD detector to observe the emission of the laser-induced plume during a time window that minimizes the background continuum emission and maximizes the ionic or atomic emission.

When combined with Raman imaging, Raman–LIBS imaging enables complementary elemental, molecular, and structural information to be obtained from the sample. Since Raman and LIBS instrumentation share many of the same components, integrated instruments can be made with a small amount of additional cost. A limitation of combining LIBS with Raman is the fact that LIBS is a destructive technique. Therefore, Raman data are normally collected before the destructive LIBS data.

2.6.4 Raman–AFM Imaging

The atomic force microscope (AFM) [78–80] is one of the leading tools for imaging materials on the nanometer scale. With resolutions demonstrated at fractions of a nanometer, AFM is better than 1000 times the optical diffraction limit. When combined with Raman imaging, AFM imaging provides a means to probe topographical details of sample surface at the atomic level and chemical details of the sample on the micron scale.

An AFM gathers information by "feeling" the sample surface using a mechanical probe consisting of a microscale cantilever with a sharp tip (probe) at its end. Forces between the tip and the sample occur as the tip is brought in close proximity to the sample that causes the cantilever to be deflected. The deflection may be measured using laser

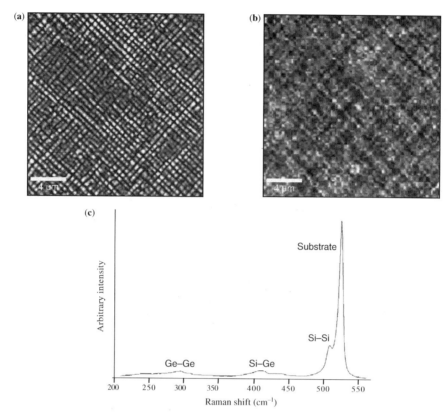

FIGURE 2.15   AFM (a) and Raman (b) imaging result from a 20 μm × 20 μm area of a SBR/PMMA polymer blend sample. The Raman image is associated with the Si–Si stretching mode of the SiGe layer at 502 cm^{-1}. The single Raman spectrum (c) reveals three peaks corresponding to the SiGe layer: the Ge–Ge (200–300 cm^{-1}), Si–Ge (380–450 cm^{-1}), and Si–Si (502 cm^{-1}) stretching modes along with the Si–Si (520 cm^{-1}) peak of the substrate.

deflection, optical interferometry, capacitive sensing, or piezoresistive AFM cantilevers. Most AFM instruments use a feedback mechanism to adjust the tip-to-sample distance so that a constant force may be maintained between the tip and the sample. Images may be achieved by mounting the sample on a piezoelectric tube that moves the sample in the z-direction for maintaining a constant force and in the x- and y-directions for scanning the sample or a "tripod" configuration of three piezo crystals, each responsible for scanning in the x-, y-, and z-directions.

Compared with SEM, AFM has several advantages including the ability to provide a true three-dimensional surface profile at the atomic level resolution without special sample treatments (such as metal/carbon coatings) and does not always require operation under vacuum. Disadvantages of AFM compared with SEM include susceptibility to image artifacts, limitations on scanning areas (approximately 150 μm × 150 μm), and limited abilities to measure steep walls or overhangs in the sample surface topography.

Figure 2.15 shows an AFM and Raman imaging result from a 20 μm × 20 μm area of a SBR/PMMA polymer blend sample obtained using a WiTec alpha300R confocal Raman microscope equipped with an AFM.

2.6.5   Raman–SNOM Imaging

Near-field scanning optical microscopy (NSOM/SNOM) [81–83] is a nanoscale imaging technique that operates beyond the optical diffraction limit by exploiting the properties of evanescent waves. Evanescent or nonpropagating fields exist only near the surface of the sample and carry the high-frequency spatial information about the object. By positioning the detector very close ($\ll \lambda$) to the sample surface, surface inspection may be accomplished with high spatial, spectral, and temporal resolving power. SNOM instruments typically consist of a laser light source coupled into an optical fiber, feedback mechanism, scanning tip, detector, and piezoelectric sample stage. Images are formed by rastering the detector across the sample using a piezoelectric stage operating at a constant height or with regulated height by using a feedback mechanism. Raman–SNOM is one of the most popular near-field spectroscopy techniques one may use to probe spectroscopically with subwavelength resolution for the identification of nanosized features with chemical contrast. Disadvantages of apertured Raman–SNOM include being limited to using very hot and blunt tips and by lengthy data acquisition times due to limited

Raman signal. Apertureless Raman–SNOM approaches and SERS Raman–SNOM approaches have been used to enhance Raman signals. Topological artifacts, however, make it hard to implement these techniques for rough surfaces. For optical images with spatial resolutions that surpass the diffraction limit, WiTec offers a commercially available SNOM add-on to their alpha300 R confocal Raman microscope.

2.6.6 Raman–IR Imaging

IR spectroscopy [84] and IR imaging [85, 86] are methods commonly involving the study of the absorption properties of materials in the IR portion of the electromagnetic spectrum. The IR region of the electromagnetic spectrum may be broken down to the near-infrared (NIR) spectral region (0.8–1.4 µm), mid-infrared (MIR) spectral region (1.4–30 µm), and the far-infrared (FIR) spectral region (30–1000 µm). This section will focus on IR instrumentation used to study fundamental vibrations and associated rotational–vibrational structure of molecules found in the MIR spectral region.

Molecules have specific frequencies at which they absorb MIR radiation, in turn causing the molecule to rotate or vibrate. Such energies are dictated by the mass of the atoms associated with the vibration or rotation, the associated vibronic coupling, and shape of the molecular potential energy surfaces associated with the molecule. For a particular vibration to be IR active, it must exhibit a change in the permanent dipole. This is distinct from Raman active modes in which there must be a change in the polarizability. The frequencies at which the absorption and associated molecular vibrations occur are characteristic of the bonding structure within the molecule, giving rise to information about the molecular structure of the sample.

Infrared spectra are obtained by examining the amount of transmitted or reflected light that occurs at various wavelengths when a sample is irradiated with infrared radiation. This is accomplished by splitting the IR source beam into two beam paths—one that interacts with the sample and one that is used as a reference (i.e., control) beam. A comparison of the transmitted or reflected light to the source beam gives rise to the infrared spectrum. This may be accomplished using a monochromatic beam, which changes in wavelength over time, or by using a Fourier transform interferometer-based instrument that measures all wavelengths at once. Raman and MIR are often thought of as competing techniques. By integrating Raman and MIR imaging [87], however, molecular images that detail sample morphology, composition, and structure may be generated in a way that takes advantage of the complementary chemical information that each method may provide based on selection rules. Both methods are nondestructive, noninvasive, and based on molecular vibrations that give rise to fingerprint spectra for inorganic and organic liquids, gases, and solids. MIR is particularly useful

as a low spatial resolution (>15 µm) imaging solution while Raman imaging is more suited for high spatial resolution imaging (>250 nm). MIR imaging is advantageous in that it is a more rapid imaging technique since it is not based on an inherently weak phenomenon like Raman.

2.6.7 Raman–NIR Imaging

NIR spectroscopy [88] and NIR imaging [89] are techniques that assess the overtone and combination vibrations of molecules that occur at energies consistent with the near-infrared portion of the spectrum (800–2500 nm). Compared to fundamental vibrations observed in the MIR spectral region, NIR molar absorptivities are typically small and spectral features are broad and less specific chemically. NIR spectroscopy and NIR imaging are particularly useful for applications that require quantitative information and for those applications that require chemical information into the material bulk as NIR can penetrate much further into the sample than MIR radiation. Chemometric (i.e., multivariate) calibration techniques such as principal component analysis (PCA) and partial least squares are often employed in quantitative NIR applications. NIR applications range from pharmaceutical tablet ingredient analysis to medical diagnostics such as blood sugar level assessment and food and agrochemical quality control.

NIR instrumentation is rather simplistic consisting of only a light source (i.e., incandescent lamps, quartz halogen lamps, or light-emitting diodes), a dispersive element (i.e., dispersive spectrometer, interferometer, or tunable imaging spectrometer), and a detector (i.e., CCD, InGaAs FPA, or PbS). When combined with Raman imaging, NIR imaging provides the ability to image large areas in short amounts of time since NIR is not a light-limited technique while Raman provides added chemical specificity [90–92].

Figure 2.16 shows a LCTF Raman and NIR imaging example of a model binary tablet composed of aspirin and lactose acquired on a Falcon Raman imaging microscope and a Condor NIR imaging macroscope system (ChemImage Corporation), respectively. Shown in the figure are aspirin (a) and lactose (b) NIR and Raman (inset images) of the model tablet along with representative NIR (c) and Raman (d) spectra associated with the tablet ingredients. In this binary example, the specificity associated with NIR is sufficient to discriminate the tablet ingredients without reliance on chemometric treatment of the data. The Raman and NIR results are qualitatively similar for the areas imaged that are in common for the two techniques. Subtle differences may be attributed to differences in the penetration depths between the Raman and NIR techniques.

As more ingredients are added into a tablet formulation, it becomes increasingly difficult to rely solely on NIR for component discrimination even with the use of chemometrics. Figure 2.17 shows an example of whole tablet FAST

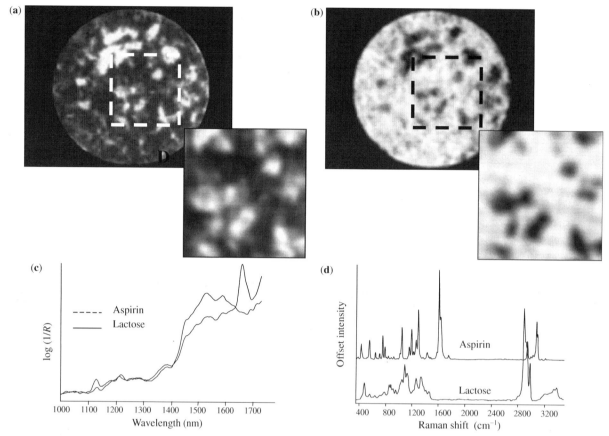

FIGURE 2.16 Aspirin (a) and lactose (b) NIR and Raman (inset images) of a model tablet along with representative NIR (c) and Raman (d) spectra associated with the tablet ingredients.

Raman and NIR imaging of Extra Strength Excedrin coated tablets (active ingredients per tablet: acetaminophen 250 mg (pain reliever), aspirin 250 mg (pain reliever), and caffeine 65 mg (pain reliever aid)) acquired on a Falcon Raman imaging microscope and a Condor NIR imaging macroscope system (ChemImage Corporation), respectively. Shown in the figure is a digital photograph of the tablet (a), Raman imaging of aspirin (b), acetaminophen (c), and caffeine (e), NIR PCA score images revealing acetaminophen (d) and caffeine (f) distributions, and representative FAST Raman spectra (g) and NIR absorption spectra (h) associated with individual domains of each active ingredient within the tablet. The FAST Raman image was acquired as a 28×28 montage over a $121\,mm^2$ area at an average rate of approximately 300 ms/pixel—nearly half of which was attributed to photobleaching and stage translation times. Lack of inherent specificity when compared to Raman can make near-IR imaging a challenge for a lot of tablet ingredients. Such was the case for aspirin in this Excedrin tablet example.

2.6.8 Raman–Fluorescence Imaging

Fluorescence imaging, like fluorescence spectroscopy [93, 94], is a method that analyzes fluorescence produced from a sample. Electrons in sample molecules are excited from lower energy ground electronic states to higher energy vibrational levels in excited electronic states using a light source such as a filtered xenon or mercury lamp. Electrons in the excited molecule will then relax to vibrational levels of the ground electronic state by one or more nonradiative and/ or radiative mechanistic paths including the release of a photon as is the case in fluorescence and phosphorescence. By exciting the sample at a constant wavelength and screening the different energies of the emitted photons, an emission spectrum (i.e., image) may be recorded in which the structure of the different vibrational levels of the sample molecules may be determined. Although not a very selective technique, fluorescence is an extremely sensitive technique. Single molecule detection has been demonstrated [95].

Fluorescence spectroscopy and imaging instruments consist of a light source (i.e., laser, photodiode, xenon arc, or mercury vapor lamp), a wavelength selection device (i.e., optical filter or dispersive element such as a grating monochromator or tunable imaging spectrometer), and a single-channel or multichannel detector such as a CCD. Most modern fluorimeters contain both an excitation monochromator and an emission monochromator allowing one to record both an excitation spectrum and a fluorescence spec-

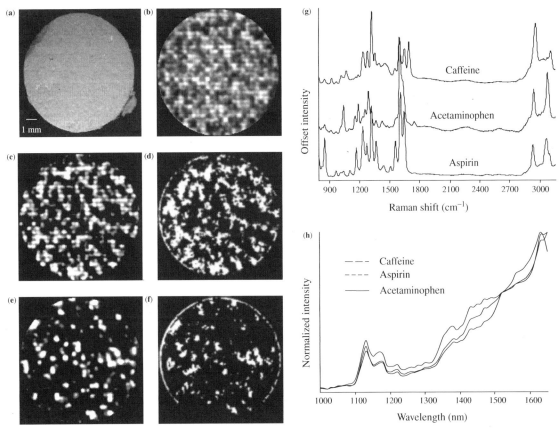

FIGURE 2.17   Whole tablet FAST Raman and NIR imaging of Extra Strength Excedrin coated tablets: digital photograph of the tablet (a); Raman imaging of aspirin (b), acetaminophen (c), and caffeine (e); NIR PCA score images revealing acetaminophen (d) and caffeine (f) distributions; representative FAST Raman spectra (g) and NIR absorption spectra (h) associated with individual domains of each active ingredient within the tablet.

trum. Fluorescence spectra are recorded by holding the excitation wavelength constant and measuring the emission wavelengths with the emission monochromator while excitation spectra are recorded by holding the emission wavelength constant and measuring the absorption spectrum with the excitation monochromator. Fluorescence imaging systems using tunable imaging spectrometers typically incorporate a filtered light source producing a constant excitation wavelength while images of the sample are recorded at various emission wavelengths using a CCD detector staring through the tunable filter. Fluorescence imaging provides a high-sensitivity means of rapid visualization and presumptive identification of fluorescence species in a sample. When combining fluorescence imaging with Raman imaging [26], Raman provides a more definitive determination of the material composition. Figure 2.18 demonstrates the combined utility of fluorescence and Raman imaging for targeting and detection of trace bioagents in a complex background. The sample in the figure consists of *Bacillus globigii* (*Bg*)-spiked outdoor ambient background particulate collected into an aqueous collection fluid buffer by a Joint Biological Point Detection System

(JBPDS) sensor and deposited onto an aluminum-coated microscope slide. Figure 2.18 shows a differential image contrast optical image (a) of a region of interest containing the elliptically shaped *Bg* endospores—a stimulant for *Bacillus anthracis* (Anthrax)—in the presence of the complex background matrix that was targeting using hyperspectral fluorescence imaging (b). The fluorescence imaging revealed the presence of the biological spores based on the characteristic autofluorescence and shape/size of the spores. Figure 2.18d shows representative hyperspectral fluorescence spectra of the *Bg* spores and background, respectively. Once targeted using fluorescence, Raman imaging is used for the purpose of presumptive identification of the biothreat stimulant. The Raman image in Figure 2.18c was based on a Euclidean distance classifier-based search against a predetermined spectral library. Despite the greater sensitivity inherent to fluorescence, Raman was able to detect a *Bg* endospore not detected by fluorescence (arrow in Figure 2.18c) due to the greater penetration depth associated with the Raman laser excitation wavelength (i.e., 532 nm) compared to the fluorescence excitation wavelength (365 nm).

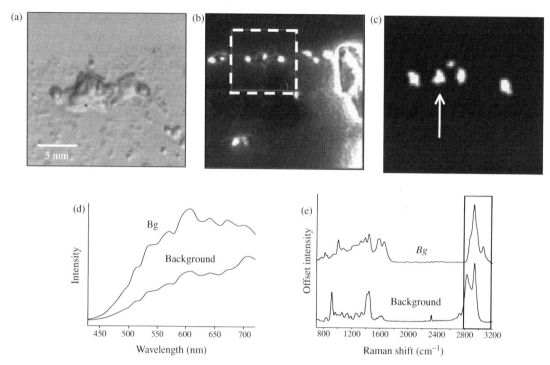

FIGURE 2.18 Combined utility of fluorescence and Raman imaging for targeting and detection of trace bioagents in a complex background: (a) differential interference contrast (DIC) optical image revealing elliptical *Bg* spores in the presence of background; (b) autofluorescence hyperspectral image used to target *Bg*; (c) C–H Raman image for presumptive identification of *Bg* spores; (d) fluorescence spectra associated with *Bg* spores and surrounding matrix; and (e) Raman library spectra of *Bg* and surrounding background surfactant matrix.

2.7 EXTRA-DIMENSIONAL RAMAN IMAGING INSTRUMENTATION

Industrial trends have begun to shift the utility of Raman imaging from the laboratory and academic settings to more at-line/online applications. The demands on the technology include adding extra dimensionality to the data (i.e., additional spatial and time dimensions) and operation in real-time configurations. This section highlights recent efforts in volumetric and dynamic Raman imaging.

2.7.1 Volumetric Raman Imaging (X, Y, Z, λ)

Depth-resolved Raman microspectroscopy is an active area of research in the Raman community [96, 97]. Many researchers employ optical confocal Raman microscopes to perform their depth-resolved analysis. While effective for localized evaluations of microchemistry, optical confocal methods are time consuming for creating high-definition images and are susceptible to complex optical effects that make spatially accurate depth information highly challenging to obtain. While susceptible to many of the same limiting optical effects suffered by confocal Raman, Raman imaging technology employing wide-field illumination provides a means for substantially reducing the time required to generate high-definition volumetric Raman images.

Figure 2.19 shows an example of volumetric Raman imaging acquired using a Falcon II™ Raman chemical imaging system (ChemImage Corporation). Raman images and dispersive Raman spectra were acquired as a function of depth through the sample on a Paramount polymer sample provided by Richard Mendelson et al. from Rutgers University, Newark, NJ (shown schematically in Figure 2.19d). The multilayer polymer system consists of layers of polyethylene terephthalate (PET), adhesive, ethylene vinyl alcohol (EVOH), and a mixture layer of low-density polyethylene (LDPE) and titanium dioxide (TiO_2). Reconstructed Raman images of the PET (a), EVOH (b), and LDPE/TiO_2 layers as well as Raman intensity profiles as a function of depth for the dispersive Raman data (e) and Raman imaging data (f) are shown in the figure.

2.7.2 Dynamic Raman Imaging (X, Y, T, λ)

Wide-field Raman imaging platforms in which all spatial elements of the image are collected in parallel enable the technology to transcend static imaging. Wide-field Raman imaging enables scientist to monitor dynamic processes, to analyze materials in motion, and to watch reactions in real time with visualization of changes in the material content. For the pharmaceutical industry, wide-field Raman imaging allows degradation studies and interacting studies of APIs

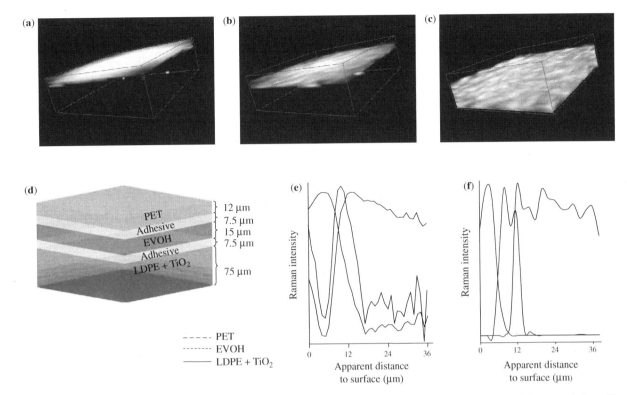

FIGURE 2.19 Volumetric Raman imaging of a Paramount polymer sample (d) provided by Richard Mendelson et al. from Rutgers University, Newark, NJ. Volumetric Raman image reconstructions of PET (a), EVOH (b), and LDPE/TiO$_2$ (c) layers. Component-specific Raman intensity profiles as a function of sample depth for the dispersive Raman channel (e) and wide-field Raman imaging channel (f) of a Falcon II™ Raman chemical imaging system (ChemImage Corporation).

and/or excipients in conjunction with monitoring dynamic processes to observe the change in material content in real time. The combined chemical and spatial information reveals subtle features of materials often unseen with traditional imaging techniques.

Figure 2.20 shows dynamic Raman imaging of acetaminophen dissolution and subsequent recrystallization. Time-dependent Raman images were acquired on a 532 nm Falcon II™ system (ChemImage Corporation) with an MCF tuned to 1617 and 1625 cm^{-1}. Raman image frames were acquired using 100 ms integration times. A bright-field reflectance image (a) shows an area of acetaminophen type I and type II polymorph mixture prior to dissolution with methanol. Figure 2.20b–i shows a time sequence of Raman images associated with the 1625 cm^{-1} band taken before experiment (b), at solvent introduction time (c), crystallization onset time (d), and 16 s (e), 25 s (f), 32 s (g), 54 s (h), and 116 s (i) after crystallization onset.

Crystallization rates were derived for type I and II acetaminophen polytypes from a linear fit of mass percentage recrystallized for acetaminophen type I polymorph crystal and acetaminophen polymorph type II crystal in the field of view imaged over the period of recrystallization time (Figure 2.20m). Crystallization rates were determined to be 2.5 and 0.93 ng/s for acetaminophen polymorph types I and

II, respectively. Polymorph type formation was confirmed using Raman imaging with 4 s integration time per Raman image frame immediately following the dynamic Raman imaging experiment. Figure 2.20 also shows a bright-field image of the region of interest following recrystallization (j), a 1625 cm^{-1} Raman image associated with the type II polymorph (k), and a 1617 cm^{-1} Raman image associated with the type I polymorph.

2.8 RAMAN IMAGING INSTRUMENTATION PERFORMANCE ASSESSMENT

As Raman imaging instrumentation has matured, it has become apparent that there was a need for an objective means of assessing instrument performance. In 2002, the ASTM E13.10 subcommittee "Molecular Spectroscopic Optical Imaging" was founded in order to address the need for standards and methods for measuring chemical imaging equipment performance. As part of ASTM E13.10, a Raman imaging task force was formed that focused on method and standard development for assessing Raman imaging performance. Such standards [98] and methods would be used to characterize day-to-day system performance, which could be used to alert the user of the need for maintenance and/or calibration and to determine the suitability of equipment for a specific analysis.

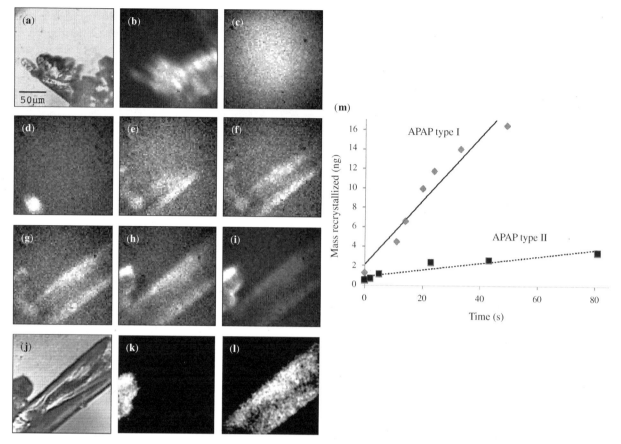

FIGURE 2.20 Dynamic Raman imaging using a wide-field Falcon II™ Raman imaging system (ChemImage Corporation): (a) bright-field reflectance image of acetaminophen (APAP) type I and type II polymorph mixture prior to dissolution with methanol; time sequence of Raman images associated with the 1625 cm^{-1} band taken before experiment (b), at solvent introduction time (c), crystallization onset time (d), and 16 s (e), 25 s (f), 32 s (g), 54 s (h), and 116 s (i) after crystallization onset; (j) postcrystallization bright-field reflectance image; (k) 1625 cm^{-1} Raman image associated with the type II polymorph; 1617 cm^{-1} Raman image associated with the type I polymorph; and plots of mass percentage recrystallized for APAP types I and II in the field of view imaged over the period of recrystallization time.

2.8.1   Standard Practice Development

The Raman imaging task force has drafted a standard practice document (currently up for ballot) entitled "Standard Practice for Evaluating the Performance of Raman Molecular Spectroscopic Optical Imaging Instruments" [99]. This practice provides a method to verify the overall performance of a Raman imaging instrument through use of an objective figure of merit (FOM) calculated from Raman imaging data collected from a proper test reference material.

2.8.2   Test Standards

The test standard shall consist, ideally, of a Raman active substrate with a pattern of a different Raman active material deposited on the surface of the substrate or impregnated within the substrate. The test standard should exhibit minimal diffuse scattering. In the event of a test standard having only one Raman active material, either the test pattern or the substrate may be Raman inactive. The test pattern shall be of

a type that provides several different line spacings, which bracket the expected spatial resolution of the Raman imaging instrumentation under evaluation. The spatial features on the test pattern must be traceable to NIST. Suggested patterns are the USAF 1951 or the NBS (NIST) 1963A Resolution Targets. The test standard shall consist of a substrate or target pattern comprised of Raman active material with known and well-characterized peak(s) in its spectrum. Figure 2.21b shows Raman spectra of monocrystalline silicon employed as the test standard substrate, as well as the response from the chromium test pattern deposited on the silicon surface.

2.8.3   FOM Calculations

In order to determine the FOM, the Raman image area (A), free spectral range ($v_{max} - v_{min}$), spatial resolution ($R_{Spatial}$), spectral resolution ($R_{Spectral}$), Raman spectral signal-to-noise ratio (SSNR), Raman image signal-to-noise ratio (ISNR), and data collection time (t_{Acq}) are measured using a proper

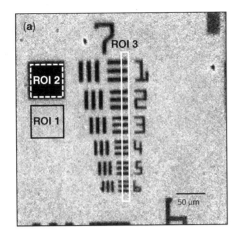
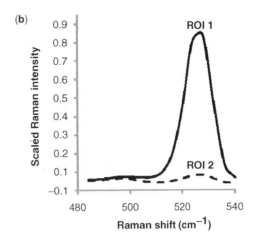

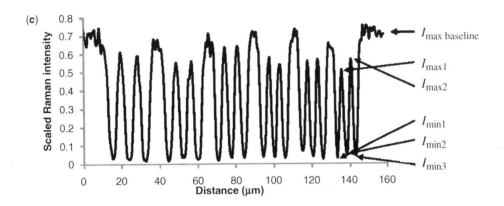

FIGURE 2.21 (a) Flat-field-corrected Raman image at 520 cm^{-1} from Cr on Si USAF 1951 Raman resolution target test standard collected using a wide-field Raman microscope (FALCON II™, ChemImage Corporation) employing an MCF imaging spectrometer. The Raman image was collected using a laser power of 475 mW at the sample, which corresponds to a laser power density of 1.1×10^4 W/cm^2 for the 10× (NA 0.3) microscope objective. The image was collected in 960 s over a spectral range of 60 cm^{-1} with a 4 cm^{-1} step size using an EM CCD detector (DU-897 iXon™, Andor, South Windsor, CT) operating at a gain of 200. (b) The mean Raman spectra of ROI1 and ROI2. (c) Intensity profile corresponding to ROI3 from Figure 2.21a.

test reference material.

$$\text{FOM} = \frac{A \cdot (v_{max} - v_{min}) \cdot \text{SSNR} \cdot \text{ISNR}}{R_{Spatial} \cdot R_{Spectral} \cdot t_{Acq}} \quad (\mu\text{m/s}) \quad (2.8)$$

2.8.3.1 Data Acquisition For instruments employing point and line mapping, Raman spectra of the target spectral region are collected over an image area in a systematic way in order to construct a Raman image hypercube. In order to achieve the highest possible spatial resolution, the laser spot area shall be minimized and the sample shall be translated relative to the laser illumination at fine spatial intervals.

For instruments employing wide-field illumination, a Raman hypercube spectral imaging data set is collected over a spectral range sufficient to measure the FWHM, wavenumber accuracy, and SSNR associated with the test standard Raman spectral band(s) of interest. Appropriate test standard

patterns are imaged in order to determine the spatial resolution and ISNR of the Raman imaging system.

In order to evaluate the anamorphic magnification of the Raman imaging system, both horizontal and vertical test patterns must be measured for both point scan and wide-field imaging systems.

2.8.3.2 Measurement of Image Area Image area (A) has units of μm^2 and is defined as

$$A = (P_X \cdot \text{DP}_X) \times (P_Y \cdot \text{DP}_Y) \quad (2.9)$$

where P_X and P_Y are the number of pixels and DP_X and DP_Y are the dimensions of the pixels in the X and Y spatial dimensions, respectively.

Figure 2.21a shows a flat-field-corrected Raman image at 520 cm^{-1} from Cr on Si USAF 1951 Raman resolution target test standard collected using a wide-field Raman microscope (FALCON II™, ChemImage Corporation) employing an MCF imaging spectrometer. The Raman image was collected using a laser power of 475 mW at the sample, which corresponds to a laser power density of 1.1×10^4 W/cm^2 for the $10 \times$ (NA 0.3) microscope objective. The image was collected in 960 s over a spectral range of 60 cm^{-1} with a 4 cm^{-1} step size using an electron-multiplying (EM) CCD detector operating at a gain of 200. The Raman image in Figure 2.21a consists of 350×350 square pixels of 0.585 μm/pixel, to yield the image area equal to 41,943 μm^2.

2.8.3.3 Determination of Free Spectral Range

The Raman free spectral range FSR is defined as

$$FSR = (\nu_{max} - \nu_{min}) = BP \cdot C \qquad (2.10)$$

where BP is the spectral bandpass of the Raman imaging spectrometer in the current measurement and C is the number of *spectral channels* corresponding to the number of independent spectral regions. The FSR in the Raman image in Figure 2.21 is equal to $540 - 480 = 60$ cm^{-1}.

2.8.3.4 Measurement of Spectral Signal-to-Noise Ratio

The Raman SSNR is defined as

$$SSNR = (I_{max} - I_{baseline})/\sigma_{baseline} \qquad (2.11)$$

where $(I_{max} - \bar{I}_{baseline})$ is the bias-corrected maximum Raman spectral band intensity value for the target region of interest and $\sigma_{baseline}$ is the standard deviation of the mean spectrum within a defined background region of the spectrum devoid of Raman spectral features associated with the target region of interest.

For the image shown in Figure 2.23a and corresponding Raman spectrum in Figure 2.23b, the SSNR is 310.9.

2.8.3.5 Measurement of Image Signal-to-Noise Ratio

The Raman ISNR is defined as

$$ISNR = \frac{\langle a \rangle_{Target\ ROI} - \langle b \rangle_{Background\ ROI}}{\sigma_{Background\ ROI}} \qquad (2.12)$$

where $\langle a \rangle_{Target\ ROI}$ and $\langle b \rangle_{Background\ ROI}$ are the mean Raman image intensity values for the target region of interest and the background region of interest, respectively, and $\sigma_{Background\ ROI}$ is the standard deviation of the signal in the background region of interest. ISNR of Raman image in Figure 2.23a is equal to 15.5 when using the mean pixel intensity within ROI1 (—) comprised of Raman active Si minus the mean pixel intensity within ROI2 (- - -) comprised of Raman inactive Cr divided by the standard deviation within ROI2.

2.8.3.6 Measurement of Spatial Resolution

Measurement of spatial resolution involves imaging a series of features of known size and determining the minimum line spacing that can be resolved by the system under evaluation. Contrast (modulation) is defined in Equation 2.13:

$$Contrast = \frac{i_{max} - i_{min}}{i_{max} + i_{min}} \qquad (2.13)$$

where I_{max} and I_{min} are the maximum and minimum intensity produced by the image, respectively.

Contrast is measured for the line pairs with higher numbers of lines per mm (spatial frequency) until either a set of line pairs is found where the image provides contrast between

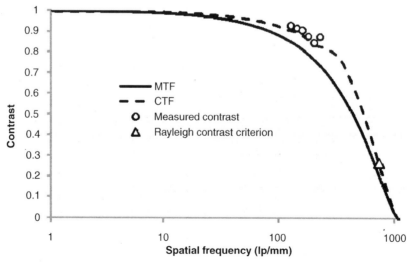

FIGURE 2.22 Determination of spatial resolution from the Raman image of the Raman test standard. A Raman CTF measured from the Raman test standard (♦) is compared to the theoretical MTF (—) and CTF (- - -). The diffraction-limited CTF shows the Rayleigh criterion contrast (26.5%) at 746 lp/mm spatial frequency (Δ) corresponding to a spatial resolution of 1.34 μm.

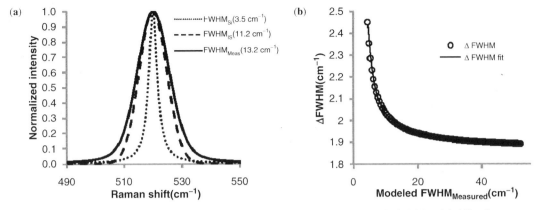

FIGURE 2.23 Determination of instrument bandpass from the Raman image of the test standard. (a) Modeled lineshapes are shown for the convolved peak (solid) with FWHM of 13.2 cm^{-1}, the natural Raman linewidth (dotted) of 3.5 cm^{-1}, and the bandpass of the imaging spectrometer (dashed) (11.2 cm^{-1}). (b) The difference (ΔFWHM) between the convolution-broadened FWHM and the imaging spectrometer FWHM as a function of varying measured FWHM follows a double exponential decay function.

adjacent lines in the set that is $\leq 26.5\%$ contrast (the Rayleigh criterion) or the smallest line pair features available on the test standard are imaged. If the smallest features are still resolved (contrast $\geq 26.5\%$), a contrast transfer function (CTF) plot is constructed and the spatial resolution is estimated from the extrapolation of the CTF plot.

Figure 2.22 shows a case based on the contrast measured for the Raman image shown in Figure 2.21a. Despite measurable degradation, the contrast does not fall below the Rayleigh criterion limit. The spatial resolution was estimated by modeling the theoretical CTF to the Rayleigh criterion. Alternatively, a linear extrapolation of the measured contrast may be employed to predict the Rayleigh criterion.

Modulation transfer function (MTF) for USAF 1951 target was constructed using Equation 2.14:

$$\text{MTF} = \frac{2}{\pi} \left[\arccos(\mu) - \mu \cdot \sqrt{1 - \mu^2} \right] \quad (2.14)$$

where μ is the normalized spatial frequency defined as

$$\mu = \frac{x}{x_0} = x \cdot \lambda \cdot f/\# = \frac{x \cdot \lambda \cdot \text{fl}}{D} = \frac{x \cdot \lambda}{2\text{NA}} \quad (2.15)$$

where x is the absolute spatial frequency, x_0 is the incoherent cutoff frequency, λ is the wavelength, $f/\#$ is the lens f-number, fl is the focal length, and D is the lean (entrance pupil) diameter. The MTF function may be converted to a CTF using a Coltman series expansion formula with 50 terms:

$$\text{CTF} = \frac{4}{\pi} \left[M(x) - \frac{M(3 \cdot x)}{3} + \frac{M(5 \cdot x)}{5} - \frac{M(7 \cdot x)}{7} \right.$$

$$\left. + \frac{M(9 \cdot x)}{9} - \frac{M(11 \cdot x)}{11} + \frac{M(13 \cdot x)}{13} - \cdots \right] \quad (2.16)$$

The Rayleigh criterion for the Raman image in Figure 2.23a was reached at a 746 lp/mm spatial frequency. This spatial frequency corresponds to the spatial resolution of 1.34 μm.

2.8.3.7 *Measurement of Spectral Resolution*

Spectral resolution is determined by comparing theoretical and measured peaks of a Raman active material. The average FWHM for a well-characterized spectral band is measured from multiple spatial locations on the test standard. For example, ROI1 in USAF 1951 resolution target (Figure 2.21a) is generated by a band corresponding to the first optical phonon mode of silicon at 520 cm^{-1} (Figure 2.21b). The measured FWHM in Figure 2.23b is 13.2 cm^{-1}, which is substantially broader than the 3.5 cm^{-1} natural Raman linewidth of silicon. The measured FWHM is also broader than the instrumental bandpass of the imaging spectrometer (11.2 cm^{-1}) that was measured independently using a collimated light source. Therefore, the measured FWHM is a convolution of the spectrometer bandpass and the linewidth of the Raman target.

The convolution of Raman spectral linewidth and the spectrometer bandpass leading to the broadening of the measured bands may be numerically modeled by generating a Lorentzian distribution with a FWHM equal to the natural linewidth of the test standard Raman band of interest. Multiple Gaussian distributions are then generated representing instrumental bandpass ($\text{FWHM}_{\text{ImagingSpectrometer}}$) and are then convolved with the Lorentzian distribution to represent measured Raman spectral resolution $\text{FWHM}_{\text{Measured}}$. An example of modeling results is shown in Figure 2.23a. The difference (ΔFWHM) between the convolution-broadened FWHM and instrumental bandpass plotted as a function of measured FWHM for the Cr on silicon example from Figure 2.21 is shown in Figure 2.23b. The plot follows a double exponential decay

TABLE 2.2   Figure of Merit Parameters for the Image in Figure 2.21a

	Value	Notes
Image area (μm^2)	41,943.0	204.8 μm × 204.8 μm
Wavelength range (cm^{-1})	60.0	480–540 cm^{-1} at 4 cm^{-1} step
SSNR	310.9	
ISNR	15.5	
$R_{Spectral}$ (cm^{-1})	11.2	
$R_{Spatial}$ (μm)	1.34	
Acquisition time (s)	960	16 image frames, 60 s per frame
FOM ($\mu m/s$)	841,722	

function (Equation 2.17).

$$\Delta FWHM = FWHM_{Measured} - FWHM_{Imaging\ Spectrometer}$$

$$= A \cdot e^{(-B \cdot FWHM_{Measured})} + C \cdot e^{(-D \cdot FWHM_{Measured})} + E \quad (2.17)$$

For the example in Figure 2.23, measured FWHM at 520 cm^{-1} is 13.2 cm^{-1}, which corresponds to the instrumen- tal bandpass of 11.2 cm^{-1}, which is in good agreement with an independent measurement of 11.2 cm^{-1}.

2.8.3.8   Determination of Acquisition Time   The Raman acquisition time t_{Acq} is defined as the total time required to collect the Raman image, inclusive of the time required to translate the sample in X, Y, or Z spatial dimensions, the time required to scan the imaging spectrometer, and the time to autofocus the sample, including settling time, the time to photobleach the sample, and the time to readout the detector. The Raman image in Figure 2.23a was collected for 60 s per frame over 16 frames, resulting in t_{Acq} of 960 s.

Table 2.2 summarizes the parameters used to calculate the FOM for the image shown in Figure 2.23a, to arrive at FOM of 841,722 $\mu m/s$.

2.8.4   A Practical Example

Figure 2.24 shows an example use of the FOM for assessing the quality of Raman images generated from a wide-field Raman imaging system (Falcon II™, ChemImage Corpora- tion) when comparing an Evans split element and MCF liquid crystal tunable filters [49]. In this scenario, an aliquot of an

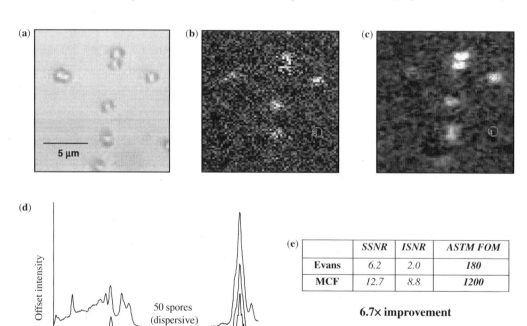

FIGURE 2.24   Comparison of liquid crystal tunable filter-based Raman image instrument performance using the FOM: (a) bright-field reflectance image of *Bg* spores deposited onto an aluminum slide; 1450 cm^{-1} Raman images collected using the Evans LCTF (b) and MCF (c); (d) single spore Raman spectra obtained from the hyperspectral data for the two filter types compared to a dispersive Raman spectra acquired from 50 spores; and image SSNR, image ISNR, and overall FOM calculation results comparing MCF and Evan split element tunable filter designs. These results suggest a 6.7× improvement of MCF over the Evans split element design.

aqueous solution containing *Bg* spores was deposited onto an aluminum slide, as shown in the bright-field image (Figure 2.24a). After 10 min of photobleaching of the sample to reduce the fluorescent background to a minimum, Raman images were collected with each LCTF type scanning from 800 to 3150 cm^{-1}. Also shown in the figure are raw 1450 cm^{-1} Raman images collected using the Evans LCTF (b) and MCF (c). Clearly, the noise on the image frame for Evans filter is higher than the noise on MCF image frame due to the lower throughput associated with this type of filter. Superior Raman spectral quality of MCF compared to the Evans filter design is again evidenced by the single spore Raman spectra obtained from the hyperspectral data for the two filter types compared to a dispersive Raman spectra acquired from 50 spores (d). A calculation of image SSNR, image ISNR, and overall FOM using the images shown in Figure 2.24b and c results in the MCF having an FOM value about 6.7 times higher than Evans filter measurement (d).

2.9 CONCLUSIONS AND FUTURE DIRECTION

Raman imaging instrumentation has advanced significantly since the first Raman images were obtained nearly three decades ago. Substantial improvements have been made in point scanning, line scanning, and wide-field Raman imaging platform designs. Raman imaging instrumentation has branched out from breadboard components on an optical table to microscope, macroscope, fiberscope, and even telescope systems that have moved the technology far beyond the research laboratory setting. Along with this development has come the expansion of the technology and associated applications ranging from submicron Raman imaging of biological spores to standoff macroscopic Raman imaging of homemade explosive devices. Integrating Raman imaging with other chemical imaging technologies has dramatically increased what is possible from a structural, elemental composition, and molecular composition materials characterization standpoint. With the formation of the ASTM E13.10 subcommittee "Molecular Spectroscopic Optical Imaging," focus has now been applied to Raman imaging instrument validation and performance assessment through standard reference material and standard practice and method development.

Future trends will inevitably continue to push instrumentation limits in areas such as improved spatial and spectral resolution. Application and market demands will thrust Raman instrumentation development to enable greater area coverage, increased data acquisition rates, improved data SNR, increased instrument stability and ease of use, and lowered instrument cost. Dynamic Raman imaging applications will require further development of wide-field Raman imaging technology in combination with tandem methods. With the realization of these continual improvements in Raman imaging instrumentation, new applications and uses of the technology will inevitably follow that will prove beneficial to mankind.

REFERENCES

1. Treado, P. J., Levin, I. W., and Lewis, E. N. (1992) Near-infrared acousto-optic filtered spectroscopic microscopy: a solid state approach to chemical imaging. *Appl. Spectrosc.* **46**, 553.

2. Delhaye, M. and Dhamelincourt, P. (1975) Raman microprobe and microscope with laser excitation. *J. Raman Spectrosc.* **3**, 33–34.

3. Boogh, L., Meier, R., and Kausch, H. (1992) A Raman microscopy study of stress transfer in high-performance epoxy composites reinforced with polyethylene fibers. *J. Polym. Sci. B* **30**, 325–333.

4. Continuous point-by-point scanning. Retrieved from http://www.witec.de/en/products/raman/alpha300r/.

5. Nadula, S., Brown, T., Pitz, R., and DeBarber, P. (1994) Single-pulse, simultaneous multipoint multispecies Raman measurements in turbulent nonpremixed jet flames. *Opt. Lett.* **19**, 414–416.

6. Yang, X., Ajito, K., Tryk, D., Hashimoto, K., and Fujishima, A. (1996) Two-dimensional surface-enhanced Raman imaging of a roughened silver electrode surface with adsorbed pyridine and comparison with AFM images. *J. Phys. Chem.* **100**, 7293–7297.

7. Stellman, C., Booksh, K., and Myrick, M. (1996) Multivariate Raman imaging of simulated and "real world" class-reinforced composites. *Appl. Spectrosc.* **50**, 552–557.

8. Brenan, C. and Hunter, I. (1994) Chemical imaging with a confocal scanning Fourier-transform-Raman microscope. *Appl. Opt.* **33**, 7520–7528.

9. Bowden, M., Gardiner, D., Rice, G., and Gerrand, D. (1990) Line-scanned micro Raman spectroscopy using a cooled CCD imaging detector. *J. Raman Spectrosc.* **21**, 37–41.

10. Jestel, N., Shaver, J., and Morris, M. (1998) Hyperspectral Raman line imaging of an aluminosilicate glass. *Appl. Spectrosc.* **52**, 64–69.

11. Markwort, L. and Kip, B. (1996) Micro-Raman imaging of heterogeneous polymer systems: general applications and limitations. *J. Appl. Polym. Sci.* **61**, 231–254.

12. Bowden, M. and Gardiner, D. (1997) Stress and structural images of microindented silicon by Raman microscopy. *Appl. Spectrosc.* **51**, 1405–1409.

13. StreamLine™ Plus Raman imaging. Retrieved from http://www.renishaw.com/en/9449.aspx.

14. Treado, P. and Nelson, M. (2001) Raman chemical imaging. *In: Handbook of Raman Spectroscopy from the Research Laboratory to the Process Line.* Marcel Dekker, New York, Chapter 5, pp. 191–249.

15. Schaeberle, M., Karakatsanis, C., Lau, C., and Treado, P. (1995) Raman chemical imaging: noninvasive visualization of polymer blend architecture. *Anal. Chem.* **67**, 4316–4321.

16. Garton, A., Batchelder, D., and Cheng, C. (1993) Raman microscopy of polymer blends. *Appl. Spectrosc.* **47**, 922–927.

17. Morris, H., Munroe, B., Ryntz, R., and Treado, P. (1998) Fluorescence and Raman chemical imaging of thermoplastic olefin (TPO) adhesion promotion. *Langmuir* **14**, 2426–2434.

18. Schaeberle, M., Tuschel, D., and Treado, P. (2001) Raman chemical imaging of microcrystallinity in silicon semiconductor devices. *Appl. Spectrosc.* **55**, 257–266.

19. McClelland, L., Stewart, S., Maier, J., Nelson, M., and Treado, P. (2005) Automated spectral acquisition: a smart biomedical sensor technology. In: *Proceedings of SPIE International Symposium on Smart Medical and Biomedical Sensor Technology III*, Vol. 6007.

20. Schaeberle, M., Kalasinsky, V., Luke, J., Lewis, E., Levin, I., and Treado P. (1996) Raman chemical imaging: histopathology of inclusions in human breast tissue. *Anal. Chem.* **68**, 1829–1833.

21. Sijtsema, N., Duindam, J., Puppels, G., Otto, C., and Greve, J. (1996) Imaging with extrinsic Raman labels. *Appl. Spectrosc.* **50**, 545–551.

22. Kline, N. and Treado, P. (1997) Raman chemical imaging of breast tissue. *J. Raman Spectrosc.* **28**, 119–124.

23. Doub, W., Adams, W., Spencer, J., Buhse, L., Nelson, M., and Treado, P. (2007) Raman chemical imaging for ingredient-specific particle size characterization of aqueous suspension nasal spray formulations: a progress report. *Pharm. Res.* **24**, 934–945.

24. Zugates, C. and Treado, P. (1999) Raman chemical imaging of pharmaceutical content uniformity. *Int. J. Vib. Spectrosc.* **2**, 4.

25. Tripathi, A., Jabbour, R., Treado, P., Neiss, J., Nelson, M., Jensen, J., and Snyder, A. (2008) Waterborne pathogen detection using Raman spectroscopy. *Appl. Spectrosc.* **62**, 1–9.

26. Kalasinsky, K., Hadfield, T., Shea, A., Kalasinsky, V., Nelson, M., Neiss, J., Drauch, A., Vanni, G., and Treado, P. (2007) Raman chemical imaging spectroscopy reagentless detection and identification of pathogens: signature development and evaluation. *Anal. Chem.* **79**, 2658–2673.

27. Wentworth, R., Neiss, J., Nelson, M., and Treado, P. (2007) Standoff Raman hyperspectral imaging detection of explosives. In: Antennas and Propagation Society International Symposium, IEEE, pp. 4925–4928.

28. Nelson, M., McLestar, M., Aust, J., and Myrick, M. (1996) Distributed sensing of fiber-optic arrays. In: *The Pittsburgh Conference and Exposition on Analytical Chemistry and Applied Spectroscopy*.

29. Nelson, M. and Myrick, M. (1999) Single-frame chemical imaging: dimension reduction fiber-optic array improvements and application to laser-induced breakdown spectroscopy. *Appl. Spectrosc.* **53**, 751–759.

30. Nelson, M., Bell, W., McLester, M., and Myrick, M. (1998) Single-shot multiwavelength imaging of laser plumes. *Appl. Spectrosc.* **52**, 179–186.

31. Nelson, M. and Myrick, M. (1999) Fabrication and evaluation of a dimension-reduction fiber-optic system for chemical imaging applications. *Rev. Sci. Instrum.* **70**, 2836–2844.

32. Ma, J. and Ben-Amotz, D. (1997) Rapid micro-Raman imaging using fiber-bundle image compression. *Appl. Spectrosc.* **51**, 1845–1848.

33. Puppels, G., Grond, M., and Greve, J. (1993) Direct imaging Raman microscope based on tunable wavelength excitation and narrow-band emission detection. *Appl. Spectrosc.* **47**, 1256–1267.

34. Batchelder, D., Cheng, C., Muller, W., and Smith, B. (1991) *Makromol. Chem. Macromol. Symp.* **46**, 171.

35. Treado, P., Levin, I., and Lewis, E. (1992) High-fidelity Raman imaging spectrometry: a rapid method using an acousto-optic tunable filter. *Appl. Spectrosc.* **46**, 1211–1216.

36. Schaeberle, M., Karakatsanis, C., Lau, C., and Treado, P. (1995) Raman chemical imaging: noninvasive visualization of polymer blend architecture. *Anal. Chem.* **67**, 4316–4321.

37. Schaeberle, M., Kalasinsky, V., Luke, J., Lewis, E., Levin, I., and Treado, P. (1996) Raman chemical imaging: histopathology of inclusions in human breast tissue. *Anal. Chem.* **68**, 1829–1833.

38. Goldstein, S., Kidder, L., Herne, T., Levin, I., and Lewis, E. (1996) The design and implementation of a high-fidelity Raman imaging microscope. *J. Microsc.* **184**, 35–45.

39. Morris, H., Hoyt, C., and Treado, P. (1994) Imaging spectrometers for fluorescence and Raman microscopy: acousto-optic and liquid crystal tunable filters. *Appl. Spectrosc.* **48**, 857–866.

40. Morris, H., Hoyt, C., Miller, P., and Treado, P. (1996) Liquid crystal tunable filter Raman chemical imaging. *Appl. Spectrosc.* **50**, 805–811.

41. Turner II, J. and Treado, P. (1997) LCTF Raman chemical imaging in the near infrared. In: *Proceedings of SPIE— Infrared Technology and Applications XXIII*, Vol. 3061, pp. 280–283.

42. Christensen, K., Bradley, N., Morris, M., and Morrison, R. (1995) Raman imaging using a tunable dual-stage liquid crystal Fabry–Perot interferometer. *Appl. Spectrosc.* **49**, 1120–1125.

43. Hoke, S., Wood, J., Cooks, R., Li, X., and Chang, C. (1992) Rapid screening for taxanes by tandem mass spectrometry. *Anal. Chem.* **64**, 971A– 981A.

44. Lyot, B. (1944) The birefringent filter and its application in solar physics. *Ann. Astrophys.* **7**, 3136.

45. Solc, I. (1965) Birefringent chain filters. *J. Opt. Soc. Am.* **55**, 621–625.

46. Saeed, S. and Bos, P. (2002). Multispectrum, spatially addressable polarization interference filter. *J. Opt. Soc. Am A* **19**, 2301–2312.

47. Evans, J. (1949) The birefringent filter. *J. Opt. Soc. Am.* **39**, 229–237.

48. Evans, J. (1958) Solc birefringent filter. *J. Opt. Soc. Am.* **48**, 142–143.

49. Wang, X., Voigt, T., Bos, P., Nelson, M., and Treado, P. (2006) Evaluation of a high-throughput liquid crystal tunable filter for Raman chemical imaging of threat materials. *Proc. SPIE* **6378**, 637808.

50. Skinner, H., Cooney, T., Sharma, S., and Angel, S. (1996) Remote Raman microimaging using an AOTF and a spatially coherent microfiber optical probe. *Appl. Spectrosc.* **50**, 1007–1014.

51. Smith, R., Nelson, M., and Treado, P. (2000) Raman chemical imaging using flexible fiberscope technology. In: *Spectral Imaging: Instrumentation, Applications, and Analysis. Proceedings of SPIE BIOS 2000 International Symposium on Biomedical Optics*, Vo. 3920, pp. 14–20.

52. Wu, M., Ray, M., Fung, K., Ruckman, M., Harder, D., and Sedlacek, A. (2000) Stand-off detection of chemicals by UV Raman spectroscopy. *Appl. Spectrosc.* **54**, 800–806.

53. Sharma, S., Lucey, P., Ghosh, M., Hubble, H., and Horton, K. (2003) Stand-off Raman spectroscopic detection of minerals on planetary surfaces. *Spectrochim. Acta A* **59**, 2391–2407.

54. Kneipp, K., Kneipp, H., Itzkan, I., Dasari, R., and Feld, M. (2002) Surface-enhanced Raman scattering and biophysics. *J. Phys.: Condens. Matter* **14**, R597–R624.

55. Campion, A. and Kambhampati, P. (1998) Surface-enhanced Raman scattering. *Chem. Soc. Rev.* **27**, 241–250.

56. Guicheteau, J., Christesen, S., Emge, D., Tripathi, A., and Jabbour, R. (2010) Bacterial mixture identification using Raman and surface-enhanced Raman chemical imaging. *J. Raman Spectrosc.*, DOI 10.1002/jrs.2601.

57. Cooper, S., Smith, W., Rodger, C., and White, P. (1997) SERRS—a sensitive spectroscopic technique. *Int. J. Vib. Spectrosc.* **1**(4), 68–84.

58. Surface-enhanced resonance Raman scattering (SERRS) from metal complexes. Retrieved from http://www.personal.dundee.ac.uk/~tjdines/Raman/research4.htm.

59. Constantino, C., Aroca, R., Mendonça, C., Mello, S., Balogh, D., and OliveiraJr., O. (2001) Surface enhanced fluorescence and Raman imaging of Langmuir–Blodgett azopolymer films. *Spectrochim. Acta A* **57**, 281–289.

60. Coherent anti-Stokes Raman scattering (CARS). Retrieved December 8, 2008, from the University of Exeter, School of Physics web site: http://newton.ex.ac.uk/research/biomedical/multiphoton/advantages/cars.html.

61. Evans, C. and Xie, X. (2008) Coherent anti-Stokes Raman scattering microscopy: chemical imaging for biology and medicine. *Annu. Rev. Anal. Chem.* **1**, 883–909.

62. Evans, C., Potma, E., Puoris'haag, M., Côté, D., Lin, C., and Xie, X. (2005) Chemical imaging of tissue *in vivo* with video-rate coherent anti-Stokes Raman scattering microscopy. *Proc. Natl. Acad. Sci. USA* **102**, 16807–16812.

63. Freudiger, C., Min, W., Saar, B., Lu, S., Holtom, G., He, C., Tsai, J., Kang, J., and Xie, X. (2008). Label-free biomedical imaging with high sensitivity by stimulated Raman scattering microscopy. *Science* **322**, 1857–1861.

64. Ozeki, Y., Dake, F., Kajiyama, S., Fukui, K., and Itoh, K. (2009) Analysis and experimental assessment of the sensitivity of stimulated Raman scattering microscopy. *Opt. Express* **17**, 3651–3658.

65. Eliasson, C. and Matousek, P. (2007) Raman Spectroscopy: spatial offset broadens applications for Raman spectroscopy. *Laser Focus World* 43.

66. Science and Technology Facilities Council, (2008) New laser technique promises better process control in pharmaceutical industry. *Science Daily.*

67. Stone, N., Baker, R., Rogers, K., Parker, A., and Matousek, P. (2007) Subsurface probing of calcifications with spatially offset Raman spectroscopy (SORS): future possibilities for the diagnosis of breast cancer. *Analyst* **132**, 899–905.

68. Nelson, M., Zugates, C., Treado, P., Casuccio, G., Exline, D., and Schlaegle, S. (2001). Combining Raman chemical imaging and scanning electron microscopy to characterize ambient fine particulate matter. *Aerosol Sci. Technol.* **34**, 108–117.

69. Johnson, D., McIntyre, B., Stevens, R., Fortmann, R., and Hanna, R. (1981) A chemical element comparison of individual particle analysis and bulk chemical methods. *Scanning Electron Microsc.* **1**, 469–476.

70. Casuccio, G., Janocko, P., Lee, R., Kelly, J., Dattner, S., and Mgebroff, J. (1983) The use of computer controlled scanning electron microscopy in environmental studies. *J. Air Pollut. Control Assoc.* **33**, 937–943.

71. Goldstein, G. I., Newbury, D. E., Echlin, P., Joy, D. C., Fiori, C., and Lifshin, E. (1981) *Scanning Electron Microscopy and X-Ray Microanalysis*, Plenum Press, New York.

72. Wittry, D. (1958). Resolution of electron probe microanalyzers. *J. Appl. Phys.* **29**, 1543–1548.

73. Schoonover, J., Weesner, F., Havrilla, G., Sparrow, M., and Treado, P. (1998) Integration of elemental and molecular imaging to characterize heterogeneous inorganic materials. *Appl. Spectrosc.* **52**, 1505–1514.

74. Lee, W., Wu, J., Lee, Y., and Sneddon, J. (2004) Recent applications of laser-induced breakdown spectrometry: a review of material approaches. *Appl. Spectrosc. Rev.* **39**, 27–97.

75. Cremers, D. A. and Radziemski, L. J. (2006) *Handbook of Laser-Induced Breakdown Spectroscopy*, Wiley, London.

76. Miziolek, A. W., Palleschi, V., and Schechter, I. (2006) *Laser Induced Breakdown Spectroscopy*, Cambridge University Press, New York.

77. Vadillo, J. and Laserna, J. (2004) Laser-induced plasma spectrometry: truly a surface analytical tool. *Spectrochim. Acta B* **59**, 147–161.

78. Humphris, A., Miles, M., and Hobbs, J. (2005) A mechanical microscope: high-speed atomic force microscopy. *Appl. Phys. Lett.* 86.

79. Sarid, D. (1991) Scanning Force Microscopy, *Oxford Series in Optical and Imaging Sciences*, Oxford University Press, New York.

80. Giessibl, F. (2003) Advances in atomic force microscopy. *Rev. Mod. Phys.* **75**, 949–983.

81. Synge, E. (1928) A suggested method for extending the microscopic resolution into the ultramicroscopic region. *Philos. Mag.* **6**, 356.

82. Hecht, B., Sick, B., Wild, U., Deckert, V., Zenobi, R., Martin, O., and Pohl, D. (2000) Scanning near-field optical microscopy with aperture probes: fundamentals and applications. *J. Chem. Phys.* **112**, 7761–7774.

83. Ash, E. and Nicholls, G. (1972) Super-resolution aperture scanning microscope. *Nature* **237**, 510.

84. McClain, B., Clark, S., Gabriel, R. L., and Ben-Amotz, D. (2000) Educational applications of IR and Raman spectroscopy: a comparison of experiment and theory. *J. Chem. Educ.* **77**, 654–660.

85. Katon, J., Pacey, G., and O'Keefe, J. (1986) Vibrational molecular microspectroscopy. *Anal. Chem.* **58**, 465A–481A.

86. Lewis, E., Treado, P., and Levin, I. (1994) Near-infrared and Raman spectroscopic imaging. *Am. Lab.* **26**, 16.

87. Sostek, R., et al. (1998) U.S. Patent 5,841,139.

88. Romanach, R. and Santos, M. (2003) Content uniformity testing with near infrared spectroscopy. *Am. Pharm. Rev.* **6**, 64–67.

89. Lewis, E., Lee, E., and Kidder, L. (2004) Combining imaging and spectroscopy: solving problems with near infrared chemical imaging. *Microsc. Today*, 8–12.

90. Rios, M. (2008) New dimensions in tablet imaging. *Pharm. Technol.* **32**, 52–62.

91. Clarke, F., Jamieson, M., Clark, D., Hammond, S., Jee, R., and Moffat, A. (2001) Chemical image fusion. The synergy of FT-NIR and Raman mapping microscopy to enable a more complete visualization of pharmaceutical formulations. *Anal. Chem.* **73**, 2213–2220.

92. Sasic, S. (2008) *Pharmaceutical Applications of Raman Spectroscopy*, Wiley, London.

93. Sharma, A. and Schulman, S. (1999) *Introduction to Fluorescence Spectroscopy*, Wiley–Interscience, New York, pp. 1–100.

94. Lakowicz, J. (1999) *Principles of Fluorescence Spectroscopy*, 2nd ed. Kluwer Academic/Plenum Publishers, New York.

95. Weston, K, Carson, P., Dearo, J., and Buratto, S. (1999) Single-molecule fluorescence detection of surface-bound species in vacuum. *Chem. Phys. Lett.* **308**, 58.

96. Everall, N. (2004) Depth profiling with confocal Raman microscopy. *Part I. Spectroscopy* **19**, 22–28.

97. Everall, N. (2004) Depth profiling with confocal Raman microscopy. Part II. *Spectroscopy* **19**, 16–27.

98. Kauffman, J., Gilliam, S., and Martin, R. (2008) Chemical imaging of pharmaceutical materials: fabrication of micropatterned resolution targets. *Anal. Chem.* **80**, 5706–5712.

99. ASTM, Standard WK12868, (2006) *New Standard Guide for Evaluating the Performance of Raman Molecular Spectroscopic Optical Imaging Instruments*, ASTM International.

3

FT-IR IMAGING HARDWARE

J. SELLORS AND R. A. HOULT
PerkinElmer, Buckinghamshire, UK

R. A. CROCOMBE
Thermo Fischer Scientific, Billerica, MA, USA

N. A. WRIGHT
Applied Instrument Technologies, Hamilton Sundstrand, Pomona, CA, USA

3.1 INTRODUCTION

The combination of IR spectroscopy with visible microscopy has been used in a wide range of analytical applications for more than 20 years. More recently, however, IR microspectroscopy has benefited from developments in IR detector arrays leading to a marked growth in FT-IR imaging technologies and applications. It is now a fairly simple task to obtain a high-quality IR spectrum from a sample region of around $20\,\mu$m in matter of seconds, and the ability to collect full IR images containing hundreds of thousands of pixels, where every image pixel contains a full range IR spectrum, is now available in many hundreds of laboratories worldwide. IR imaging hardware is not yet mature, but despite this, with today's state-of-the-art FT-IR imaging systems, the analysis time for many applications is limited not by the speed at which quality images are obtained, but by the data analysis or sample preparation techniques at the disposal of the operator.

Progress in commercial FT-IR imaging hardware development comes from various drivers, but two are particularly relevant: (a) the high popularity of single point IR microscopy systems that has fuelled the interest in technology and applications utilizing more rapid methods of data acquisition and (b) the development of multichannel array detectors that operate in the mid-infrared region for nonspectroscopic applications. These are fundamental to both the understanding of current FT-IR imaging technologies and probable developments in the near future.

3.1.1 Developments in IR Microscopy and Imaging Systems

Interest in obtaining IR spectra from small samples goes back to over 60 years; for example, a reported study of the structure of penicillin by Thomson [1] in the late 1940s used a prism-based dispersive spectrometer coupled with a beam condenser/microscope system. A commercial IR microscope system described in 1953 by Coates et al. [2] demonstrated quite respectable IR spectra from single fiber samples of less than $20\,\mu$m in diameter, recorded with 15 min scan times, and contained some design attributes that are still present in today's systems. However, it was not until early 1980s when the rapid uptake of commercial FT-IR systems and applications such as semiconductor microanalysis provided both the applications and technology interest to spur the growth in IR microspectroscopy. Today, all major manufacturers of laboratory FT-IR spectrometers include IR microscopes, some with automated point mapping systems, as well as imaging systems in their product portfolios.

Early (1980s) FT-IR microscope systems were mostly bolt-on accessories derived from optical microscope frames that were modified to support the collection of IR spectra from selected regions of interest (ROI). Later, systems were

Raman, Infrared, and Near-Infrared Chemical Imaging   Edited by Slobodan Šašić and Yukihiro Ozaki
Copyright © 2010 John Wiley & Sons, Inc.

designed from the outset to deliver improved IR spectroscopic performance, but were still constrained somewhat by the design characteristics of the FT-IR spectrometers to which they were coupled. As the technique became more popular during the 1990s, manufacturers adopted a more system-based approach to IR microscopy, resulting in performance, ease of use, and cost benefits. Various advances in the basic system hardware components have been realized in this timeframe, for example, the use of CCD (charge-coupled device) cameras in the visible imaging systems to assist the visual examination and selection of ROI in the sample, and improved illumination through use of light-emitting diodes (LEDs). Spectrometer design improvements have provided greatly improved FT-IR performance, but pivotal among these developments (as far as imaging is concerned) has been those in IR array detectors. Today's instruments are available as both add-ons to high performance FT-IR spectrometers and stand-alone units. An example of the latter approach are the iN™10 and iN™10MX IR microscope imaging systems, respectively, from Thermo Scientific Nicolet, introduced in 2008, which integrate spectrometer and microscope components into a single unit requiring no separate FT-IR bench.

This chapter outlines the common hardware systems used in laboratory FT-IR imaging. Because today's technology has evolved largely from single point microscopy systems, some materials covered are common to both microscopy and imaging (in fact many of today's systems include both point mapping and imaging capability in a single instrument). First, we consider the basic optomechanical components of the system. Next, array detector technology and approaches to its implementation will be discussed. For reasons of space and likely redundancy, the design of the FT-IR component of the system is not covered in detail. For this, there are numerous review articles in the literature, for example, for FT-IR systems and principles, the publication of Griffiths and de Haseth [3] or for a discussion on contemporary commercial interferometer designs, the article of Jackson [4]. It should be noted, however, that the basic spectrometer performance attributes *are* of paramount importance to overall system performance, and it is perhaps no surprise that the most successful FT-IR imaging systems are provided by manufacturers where both the spectrometer and imaging system are manufactured by single suppliers. Due to the differences in implementation among the various manufacturers of equipment, this chapter is not a comprehensive review of all the various design approaches, but rather an outline of some of the key design considerations relevant to the more popular systems available.

With *commercial* FT-IR imaging hardware, it is also worth mentioning the influence of commercial factors. If a systems designer were to develop an FT-IR imaging system from the ground up for highest performance only (however defined), it is likely that the resulting system would be unlike most current, commercially available systems. For example, current laboratory FT-IRs have fundamental design constraints imposed by requirements such as spectral resolution, compatibility with commercial macrosampling accessories, cost, and so on, which may be inconsistent with the design requirement or imaging. Designers need to strike the right balance of imaging performance, ease of use/manufacturing/maintenance/longevity, and a host of other factors that will not only carry the instrument through the crucial early market phase but also enable the instrument to successfully progress through subsequent stages in its life cycle to generate the required return on investment. In other words, for commercial success, imaging hardware design is strongly influenced by performance requirements but not necessarily dictated by them.

A further point to note is that although various system components will be outlined individually, this does not help to illustrate the effect of the strong interdependence of these components on total system performance. An overall systems engineering approach is crucial to good FT-IR imaging design and considering the performance attributes of components in isolation is but one part of the overall design.

3.2 SYSTEMS OVERVIEW

To understand the basic layout, an FT-IR imaging optomechanical system can be compared with that of a conventional optical microscope, noting the important differences. A typical microscope/imager schematic is shown in Figure 3.1; although different manufacturers offer some variants on this, it conveys the main elements under consideration. The "imager" is generally capable of viewing a sample at both visible and IR wavelengths. The use of suitable flip mirrors or dichroics (see later) enables both IR and visible radiation to illuminate the sample, and post-sample, to be directed to the appropriate detectors. In the system shown in visible operation, the illumination beam is produced by an LED and passes through a dichroic mirror that combines the IR and visible beams. A second LED beam is produced to illuminate the sample from above if required. The beam either passes through the sample or is reflected by the sample to be collected by the objective optic and then relayed to a visible camera or binocular viewing system. In IR image collection mode, the modulated IR beam from the spectrometer illuminates the sample from either above or below (i.e., reflection or transmission mode of operation). The energy is focused onto the sample in transmission mode by a focusing optic that can usually be adjusted. When a cassegrain system is employed (see later) in reflection mode, the beam is focused onto the sample through half of the objective cassegrain and collected using the other half of the system. In either mode, the IR energy, which is reflected or transmitted by the sample, is collected by the objective cassegrain, and

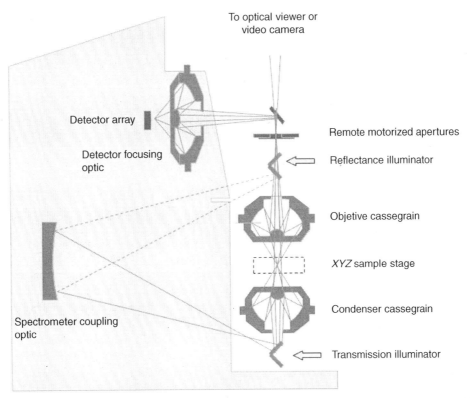

FIGURE 3.1 Typical IR microscope and imaging system layout. Solid line: input beam for transmission mode; dotted line: input beam for reflection mode. The spectrometer coupling optic accepts the output beam from the FT-IR spectrometer.

the image optionally magnified (by either passing through a second high magnification optics unit or by using an alternative cassegrain to change the magnification) if required. The system is usually arranged to refocus the beam after the sample (at the "intermediate focus plane"), where a set of adjustable aperture blades can be moved so as to restrict the sample field points that are ultimately projected onto the IR detector. Next, the beam is directed onto a fixed dichroic mirror, where the IR part is reflected onto the IR detector while the visible component passes through to the visible camera optics.

3.3 THE ROLE OF THE FT-IR SPECTROMETER

While discussions about IR imaging technology usually concentrate on the detector, the source and interferometer cannot be overlooked in their contribution to the overall system performance. "Garbage in–garbage out" applies here and the best detectors are unlikely to compensate for a poor IR source or interferometer. Fortunately, these components are well developed in laboratory FT-IR systems and though not optimal, current designs should not limit the performance of imaging systems. Detailed descriptions of how an FT-IR spectrum is generated are widely available in the literature (see, for example, Ref. iii). Two elements of the FT-IR that

are of particular relevance to imaging systems, however, are the IR source and the interferometer mode of data collection.

3.3.1 Source

Sources of continuous IR radiation (note that special components to modulate the source intensity are not required as the interferometer acts as a modulator) mostly consist of a resistively heated silicon carbide or ceramic element that operates at around 1100–1400K. In laboratory FT-IR, the combination of source size and interferometer optics is chosen to deliver a homogeneous image of about 6–10 mm diameter at the normal sample position. As signal-to-noise ratio is a key performance attribute of IR imaging systems, it might be expected that increasing source intensity (e.g., by raising its temperature) might be expected to increase the signal-to-noise level in an instrument, with all other things equal. However, consideration must be given to the interferogram (see later) shape, its dynamic range, and the associated analog-to-digital converter (ADC) sampling. In a high optical throughput laboratory FT-IR using a DTGS detector, the dynamic range of the interferogram is such that increasing the source intensity much further would have little improvement in signal-to-noise ratio because of insufficient dynamic range of the ADC to properly sample the noise in the system. Indeed, other factors come into play, such as

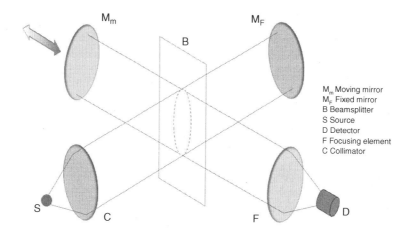

FIGURE 3.2 Principle of common FT-IR interferometers used in IR imaging systems.

stability and longevity, which generally discourage this practice. For microsampling, light losses due to the sample itself can be substantial, so the issue of inadequate ADC sampling no longer applies, and it might be considered beneficial to operate a brighter source element for imaging. However, with imaging using focal plane array (FPA) detectors, the contribution of the relatively limited dynamic range of the array detectors employed comes into play to counter this effect: that is, the detectors and associated electronics physically cannot handle the large signals encountered with laboratory FT-IR in any event. Current systems generally use standard FT-IR sources. At the time of writing although there are reports of some special sources with higher operating temperatures (at considerable cost), the current state of the art in source design is probably not a severe limitation of imaging systems.

There are examples of use of radiation from a synchrotron beamline to advantage due to the increased intensity and coherence of the source [5–7]. The impact of such a source on the performance of single point microscopy systems is beyond doubt, enabling higher signal-to-noise spectra from smaller sample areas. However, its potential advantage in large area imaging systems is less clear. Given the practicality issues and likelihood that the SNR is limited by components other than the source, these studies are of greater benefit from their applications perspective.

3.3.2 Interferometer

This technology is now relatively mature, having been available commercially for almost 30 years. Many manufacturers currently utilize fourth- or fifth-generation interferometer designs and their major design issues are well understood. Systems are now appreciably more stable, reliable, and perform with much higher signal-to-noise ratios than their 1980s counterparts. To understand the operation of FT-IR imaging systems, especially the signal collection requirements for the various detectors, it is necessary to

understand the modes of operation of the interferometers utilized.

All systems in common use are based on the Michelson interferometer, the fundamental design of which goes back to the 1890s [8, 9]. A schematic is shown in Figure 3.2. Light from the IR source S is collimated (usually by a relatively short-focus paraboloid mirror) and the collimated beam split into two nearly equal paths along the "arms" of the interferometer by a beamsplitter B. In the classical design, one beam is reflected back along its path by a fixed mirror M_F and the other beam reflected back from a second mirror M_m that moves parallel to the beam. Both beams recombine at the beamsplitter and the useful half is directed onto a detector D (the other half is returned to the source). Looking at the source from the detector (Figure 3.3), we see the two mirrors—one image moving relative to the other. When the beams have traveled different path lengths they arrive in different phases. Provided the two mirrors appear to be moving perfectly parallel to each other, the returning plane wavefronts will interfere constructively or destructively depending on whether the wavefronts are in or out of phase. The detector "sees" a signal as a function of the difference between the two path lengths and this signal at any point contains partial information about the entire source spectrum. This signal requires a Fourier transform to decode it. The plot of intensity versus optical path difference (OPD) of the two arms of the interferometer is known as the *interferogram*, and the point in the interferogram where OPD is zero is where all frequencies interfere constructively. This is the most intense part of the interferogram—the centerburst. The distance moved by the mirror, and hence OPD, is usually tracked by the same interferometry principle and the distance scale defined by the laser frequency acts as the reference for all wavelengths in the spectra. The key to good interferometer design is to keep the two beams perfectly aligned (as seen by the detector) as the OPD is varied. With a linear tracking mirror, this places a high demand on the engineering tolerances required in the bearings—near nanometer scale

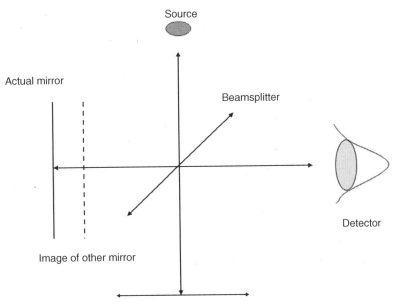

Source

Actual mirror

Beamsplitter

Detector

Image of other mirror

FIGURE 3.3 Detector view of fixed and moving mirror beampaths combining to form the interferogram in a scanning interferometer.

precision is required and the larger the distance over which this has to apply the bigger the problem. Today's interferometer variants are arrived at by various ingenious methods to enable these requirements to be met as far as possible within certain cost and reliability targets. While the discussion of these is interesting itself, it is beyond the scope of this chapter. The main point to note is that the OPD may be generated by one of two methods. In both cases, the interferogram is necessarily sampled at finite intervals of the OPD or time. In the so-called *continuous mode* of operation, the interferogram points are sampled during a continuous movement of the optical element, which generates the OPD*.[1] At the point of data collection, the OPD is changing. In an alternative mode of operation, the *step-scan* mode, the OPD is again generated as a function of distance but the signal is collected at a point when the OPD is not changing. Some interferometers operate in both step-scan and continuous-scan mode, while others are limited to continuous-scan operation only.

3.3.3 Continuous-Scan Mode

The characteristic frequency of the signal at a given wavenumber is a function of the velocity v at which the OPD is generated and the wavelength (or wavenumber σ)

$$F_v = v\sigma \ \text{Hz}$$

and known as the Fourier frequency corresponding to that wavenumber. The spectrum can be thought of as being

encoded by the interferometer at the various Fourier frequencies. The interferometer scans with an optical path difference velocity of typically between 0.1 and 5 cm/s so that the Fourier frequencies fall approximately within the audio frequency range and are easily digitized. Although it varies slightly between models, a common mirror speed is 0.16 cm/s, giving an OPD velocity of 0.32 cm/s, which is often referred to as a "5 kHz" scanning speed, based on the measured frequency of the reference He–Ne laser. In continuous-scan FT-IR, the common single detector signal sampling rates are around 5 kHz for room temperature detectors, 20 kHz for MCT detectors, and today as fast as 250 kHz for kinetics studies. Detectors generally work optimally only in certain frequency ranges, and the matching of interferometer speed to optimum detector operational frequency is an important performance consideration, as discussed in the section on array detectors below.

3.3.4 Step-Scan Mode

Many research-grade FT-IR spectrometers are also capable of operation in *step-scan mode*. Step-scan mode became more popular in the early 1990s, originally to facilitate time-resolved studies of repetitive events and modulation experiments [10], and proved to be an enabling capability for early imaging systems. In step-scan mode, the optical path difference is incremented and then held for a period of time; a 1 Hz step rate implies that the OPD is stepped every second leading to a constant OPD for just under 1 s, allowing for a short settling time. While held at that step, time- or frequency-independent measurements, including collecting a frame or co-added group of frames from an array detector, can be acquired. It has been found advantageous to scan both

---

[1] In practise, various components may be moved to generate the OPD, depending on the design, including linear movement of single plane or cube-corner optics, or rotary movement of mirror pairs.[iv]

arms of the interferometer in tandem. One arm mirror moves continuously over the entire OPD in exactly the same manner as in a continuous-scan interferometer, while the second arm is scanned over much shorter distances at the same speed (to keep the OPD constant) and then quickly returned to start the short sweep again, repeatedly, to generate an effective stepped motion of OPD as seen by the detector. The main point to note here is that at each point the IR modulation frequencies can now be effectively decoupled from the OPD velocity, that is, *no longer time/ speed dependent*. This provides (a) potentially much greater flexibility in the choice and operating conditions for the detectors and (b) potential for enabling much faster kinetics studies under favorable conditions. However, step-scan systems have their own issues due to the fact that moving systems have their own inherent inertia and mirror stabilization times can be significant in the overall measurement time, and the relative cost/complexity of such systems place them at a disadvantage compared with modern continuous-scan systems. In fact, the use of step-scan systems was a necessity for the early commercial FPA-based systems because of the data capture rates of these devices. Subsequently, the use of continuous-scan systems with small arrays [11] proved to be at least as effective and more efficient than the early FPA-based systems. It was not until the availability of subsequent generations of FPA detectors with higher data acquisition rates that the more popular continuous-scan interferometers could be used more effectively with these detectors.

3.4 OPTOMECHANICAL CONSIDERATIONS

As noted above, the fundamental requirement to operate in two different wavelength regions places certain design constraints on the system relative to a visible-only system. Optical microscopes operate over a relatively short-wavelength range ($\sim$0.5 μm) compared with IR systems ($>$12 μm) and are more readily corrected for the various aberrations present [12]. For example, in the IR system the relative importance of spherical aberrations is generally higher and aspheric or toroidal mirrors tend to be preferred. The IR component is usually all-reflecting, and the use of Schwarzschild Cassegrain optics (referred to as "cassegrains" in this chapter) is common. These are frequently found in the objective, condenser, and sometimes detector focusing elements of the system.

3.4.1 Cassegrain Optics

This double-spherical mirror system is widely employed and is broadly similar to that used in telescope systems (Figure 3.4). Three properties of cassegrains are particularly relevant, numerical aperture (NA), magnification, and working distance—and the correct balance of these properties is key to good system design. The numerical aperture of the objective is important for IR imaging. Its value is the sine of the angle v between the axis and most extreme ray passing through the image point and the system—that is, the greater the NA the greater the cone angle or light gathering power of the objective. For good images, each point in the sample needs to be focused to a point to avoid image blur (Figure 3.5). The blur is directly related to the point spread function (PSF) for the system, which is used to determine the projected spot size for a point source (or point in the sample). Even with perfect optics, a point source is spread out by virtue of the wave nature of light, which limits the spot size to around a wavelength due to diffraction. In fact, the projected spot size is inversely proportional to the cone angle, that is, the NA. The PSF for a circular optic is known as an Airy disc function (Figure 3.6),

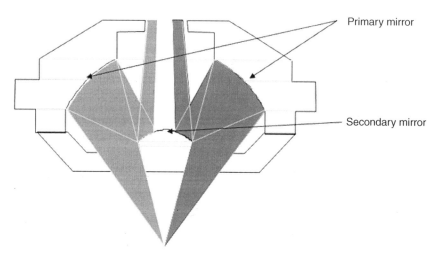

FIGURE 3.4 All reflecting "cassegrain" focusing system in common use in IR imaging systems. As an objective, in transmission mode, the full aperture is used. In reflection mode, illumination is through half of the optic and collection through the other half (shaded beam).

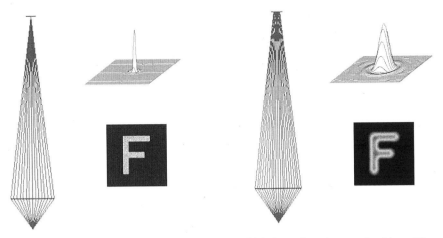

FIGURE 3.5 Image blur is directly related to the point spread function, which determines the spot size focused from a single sample point. A broader PSF implies greater image blur.

where the full width at half height (FWHH) is given by

$$\text{FWHH} = 0.61\lambda n \sin\theta = 0.61\lambda/\text{NA}$$

that is, about 1 wavelength for an NA of 0.6. Correct positioning of the primary and secondary optics in the cassegrain is also important to minimize aberrations and becomes more difficult at higher magnification. A common choice for objectives is to use an NA of around 0.6 with magnifications between 6× and 32×. Higher magnification, higher NA systems generally suffer the disadvantage of smaller working distance and image field. Another parameter to consider is the distance to the back focal plane of the objective. Some optical microscopes have no such back focal plane, that is, the output beam is effectively collimated, allowing various image-enhancing ancillary components such as filters or polarizers to be placed in the beam without having to be concerned about the location of the primary image plane. This design was introduced by Spectra Tech [13] in the late 1990s, and also used by Thermo in the Continuum™ systems, and Bio-Rad in the UMA 500/600 microscopes.

It is also worth noting that these on-axis cassegrain designs have a central obscuration zone due to the secondary (smaller) mirror that both reduces transmission and spreads the light further out to the outer rings of the PSF compared to an unobscured system [14].

The question of what is the optimum magnification of the cassegrain sometimes arises. Higher magnification is not necessarily better as a number of factors need to be considered, depending, for example, on the viewing system used (e.g., eyepieces or video camera) in the case of an objective, or detector pixel size in the case of the detector cassegrain. For example, if a binocular viewer is used, it helps to have a higher magnification objective, but field of view and working distances are compromised; however, a relatively low magnification provides a more convenient image size for the remote apertures that are used. Given the objective is a relatively costly and critical component of the system, one way to circumvent the need for multiple objectives to provide multiple magnifications is to employ a single fixed objective and magnify further along the optical path. This has been implemented, for example, by use of Z-fold optics [15],

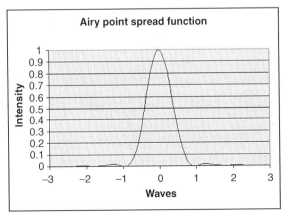

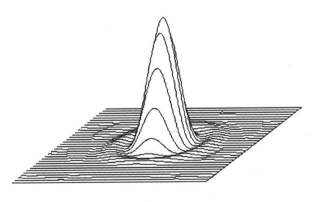

FIGURE 3.6 Airy point spread function for a circular optic.

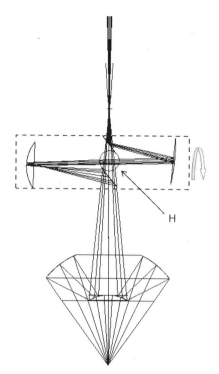

H

FIGURE 3.7 "Z-fold" magnification optics used to provide additional magnification for a given objective cassegrain configuration.

enabling rapid switching between magnifications without moving the objective cassegrain (Figure 3.7).

Magnification between sample and detector varies with detector used. For single point and small array detector focusing systems, high magnifications are generally not required. Given that the total noise contribution for a detector is proportional to the square root of the detector area, that is, to the linear size for a square pixel, it is desirable to completely fill the detector area with as much signal as possible to maximize signal to noise. There are also practical issues that limit the physical size of the detector element— currently of the order of about 20μm, although this continues to decrease. In fact, in some single point systems, the image is demagnified onto the detector. When the sample is imaged onto larger detector array, typical sample–detector magnifications project 6.25μm $\times$ 6.25μm at the sample onto actual detector pixel sizes of 40μm $\times$ 40μm, an overall magnification of only 6.4.

3.4.2 The Use of Apertures

An essential component of IR microscope systems, apertures play a slightly different role in imaging systems. As most commercial systems combine point microscope and imaging systems, their role will be described briefly here. In a single point system, apertures are placed before or after the sample to restrict light from unwanted regions of the sample reaching the detector. In some systems, apertures are placed *both* before and after the sample in a dual confocal arrangement to further reduce interference from light from outside the masked area [16], known as Redundant Aperturing™ by Spectra Tech. A further enhancement to this approach uses a single reflex aperture in a dual-pass configuration to restrict both sample input and output beams at the same time. This is a conceptually attractive approach to limiting the diffraction blur caused by apertures, but the advantages are partially offset by the practicalities of maintaining perfect alignment in a more complex optical system, and some applications (e.g., diffuse reflectance) tend to benefit from overfilling the sample with IR illumination, which is not possible with such a system. The point to note is that in IR microscopy, apertures play a major role in determining the *spatial resolution* of the system. Effects due to the aperture edges themselves and of the central obscuration zone of the cassegrain both contribute to pushing the energy distribution away from the center of the PSF, degrading achievable spatial resolution. The effects of various schemes have been described by Sommer and Katon. In the absence of sample effects (which unfortunately can be very significant!), apertures placed before and after the sample are more effective in reducing the problem than use of single apertures, but this dual aperture configuration requires considerably more care with alignment. In addition, in single aperture systems, in favorable situations, these diffraction effects can be partially overcome by over-aperturing at the sample, that is, setting the apertures slightly smaller than required to reduce the contributions due to diffraction.

In imaging systems using array detectors, the spatial discrimination is achieved by the individual pixels themselves, although apertures may be used for other reasons. The spatial resolution of these systems is sometimes incorrectly confused with the sample image pixel size. Provided the pixel size is small enough to ensure the sample image plane is sampled correctly, generally in accordance with the Nyquist criterion, the spatial resolution of the array system is determined mainly by the wavelength of light and NA of the system. For example, many systems today offer a sample image pixel size of 6.25μm $\times$ 6.25μm, but the spatial resolution achievable is somewhat worse than this—typically $>12.5 \mu$m, even in the absence of sample effects. A description of the factors determining spatial resolution in the various aperturing modes of single point systems and array systems is described in a technical note by Nishikida [17].

3.4.3 Visible Image System

To facilitate correct positioning of the sample prior to IR image collection, the sample is usually viewed and correctly positioned using a built-in optical microscope. Over recent years, the use of a digital imaging camera has become more popular than the use of binocular viewers, though some systems still provide both means of sample imaging. The fundamental background pertaining to design and construction of digital visible imaging systems is well described in the numerous resources available on the Web, particularly from the major suppliers of optical microscopes and camera systems. With respect to their application in FT-IR imaging systems, video cameras can provide adequate quality live image display with potential for much improved ease of use and productivity. Image capture software can allow graphics overlay and user interaction with the live image, enabling simple definition of IR image areas via "grow boxes" on the live sample image using the PC mouse. Additionally, the graphics image can be digitally enhanced and various software algorithms such as feature selection and particle size analysis can be readily invoked. Modern USB cameras further simplify the visible image system by eliminating the need for interfacing video capture cards at the PC. Given the typical sample areas that are imaged in the IR, the image fields required for live visible imaging are such that camera sizes of about 0.5–1.5 mega pixels are typically used. If larger visible image areas than that can be projected onto the detector array at the required spatial resolution are required, then the use of a computer-controlled mapping stage can be used to step the sample to build up a composite image by stitching together individual frames in a mosaic pattern. This type of operation is commonly referred to as "mosaicing" and while initially used to build up larger visible survey images of the sample, it is now used for both visible and infrared images where larger area sampling is required. Due consideration of the uniformity of the field of illumination across the detector is required in order to minimize artifacts at the boundaries between the individual "tiles" of the mosaic pattern. This applies to both IR and visible imaging, although various response correction schemes can be implemented in software to correct for this nonuniformity in illumination. This mode of data collection obviously places special requirements for precision and backlash in the sample stage mechanism in order to ensure the individual "tiles" are stitched together without overlap or gaps.

In most systems, some optical components (e.g., the objective) are common to both the IR and visible beam paths. At points where the visible and IR systems intersect, there is usually a suitable beam-switching mirror or increasingly with modern systems, a *dichroic* optic. Dichroic optics generally transmit the visible wavelengths and reflect the infrared wavelengths by suitable choice of an appropriate coating and substrate. Use of a dichroic optic offers the potential advantage of eliminating a movable mirror and hence improved alignment stability—always a good thing—at the expense of a reduction in visible transmission (which can usually be compensated for elsewhere). In addition, choice of dichroic coatings needs to be made with care if the system is to be used for spectroscopy in the NIR region where some coatings can have poor reflectivity in this region.

In addition to digital enhancement, various optical schemes can be employed to enhance the visible image, the most common being the use of simple polarizers. These techniques are well described in the visible microscopy literature; as mentioned previously, some designs incorporate a collimated beam from the back pupil of the objective to facilitate the addition of image-contrasting optics such as differential interference contrast (DIC) prisms. This "infinity correction" technique has been used quite extensively in advanced light microscope applications.

3.4.4 Sample Stage

With more systems using the technique of "mosaicing" to generate larger images, and an increasing demand for automation as the technique becomes more widely adopted, the use of a computer-controlled precision stage is becoming the norm in IR imaging systems. The sample stage is an important yet often overlooked attribute of a good imaging system, which is a little surprising given the very high demands placed on the part and its relative high cost (generally second only to the detector array).

3.5 DETECTORS FOR IR IMAGING

Over the past decade, the increased interest in IR microanalysis is largely due to the applications that have been made more accessible as a result of the development of systems employing array detectors. There are two main reasons for this: first, the parallel nature of data collection using detector arrays means that images can be generated much faster, limited largely by the detector arrays and associated electronics; second, with the removal of masking apertures and their contribution to diffraction effects, the possibility of higher spatial fidelity might be assumed compared with single detector systems. While the benefit of faster data collection certainly justifies the use of arrays, the improvements due to improved spatial resolution are not so clear in practice. To better understand the current status of the use of detector arrays in IR imaging, a brief history of their utilization will be outlined.

3.5.1 Early Developments

The potential of coupling infrared array detectors to Fourier transform spectrometers was realized more than 35 years ago [18], but this field was initially developed by the astronomy and remote sensing communities [19], and was largely overlooked by the analytical spectroscopic community. It took many years to achieve the first practical system [20]. The field progressed from linear array detectors, through small discrete arrays [21–23], and the first utilization of IR arrays in space for astronomy was on the IRAS mission in 1983 [24].

By 1985, astronomers had access to 32 × 32 element mercury cadmium telluride (MCT) and indium antimonide (InSb) hybrid arrays with direct readout multiplexing [25–28], while on the ground, a Fourier transform spectrometer of the classic French "cats-eye/step-and-integrate" design was interfaced to the Canada–France–Hawaii telescope by 1984, and equipped with a pair of 256 × 256 MCT arrays by 1993 [29, 30]. A laboratory step-scan interferometer was modified for airborne operation in 1993 [31], with more detailed descriptions published in 1995 [32, 33] and in the patent literature [34]. A 1997 SPIE monograph summarizes different approaches to imaging spectrometers at that time, and describes some of the technical details [35], while spaceborne imaging FT-IR spectrometers are discussed further by Beer [36].

As infrared array detectors became commercially available for civilian purposes, analytical studies became feasible, and the first paper that reported combining an infrared array detector and a Fourier transform spectrometer for chemical studies in a laboratory was published in 1995 [37]. This paper by Lewis et al. brought this technique to the attention of analytical chemists. By constructing hardware and software interfaces between commercially available components (step-scan FT-IR spectrometer, infrared microscope accessory, focal plane array, and data acquisition system), they developed a laboratory-based infrared spectroscopic imaging system, suitable for use in chemical and biological/biomedical applications. Commercial systems followed rapidly, using step-scan interferometers, first with InSb near-infrared (NIR) cameras, then with MCT mid-infrared cameras. With the development of faster readout arrays, and more powerful personal computers (PCs), the first commercial FT-IR imaging systems using conventional rapid-scan interferometers and FPAs were introduced in 2002. Since that time, the applications of this technique have grown very rapidly [38], although the pace of technological innovation has slowed.

Sampling of the detector signal is usually performed at even OPD displacement increments, defined by the He–Ne laser zero crossings, and considerable effort is made to maintain a constant mirror velocity [39]. The sampling rate and the spectral region over which unambiguous data are collected (*free-scanning spectral range*) are related. A sampling rate of once per cycle of the He–Ne reference signal, which is every other zero crossing, results in a free-scanning spectral range of 0–7900 cm^{-1}. Optical filtering is not usually required, because the loss of efficiency of the beamsplitter, combined with reduced source intensity, attenuates signals at higher infrared frequencies. However, if the sampling rate is halved, that is, data collected at every fourth zero crossing, corresponding to 2.5 kHz for the laser, the free-scanning spectral range is reduced to 0–3950 cm^{-1}. This is adequate for the mid-IR spectral region; however, optical filtering is required to prevent folding of information from the region 3950–7900 cm^{-1} back into the range 3950–0 cm^{-1}. The

ability to reduce the free spectral range results in two direct benefits for infrared imaging: slower data collection rates and smaller data files.

3.5.2 Infrared Array Detectors

The array detectors are manufactured by aerospace and defense companies, since historically the developments in IR radiation detection and temperature and emission measurements were targeted for end users such as astronomers, climatologists, and the military, all programs funded by the government. Military systems include recognition and surveillance, tank sight systems, and missile control. More recently, infrared array detectors have been considered as "dual technology" due to the growing number of alternate areas of application. The combination of passive operation and high sensitivity is leading to many commercial uses including environmental and chemical process monitoring and medical diagnostics. The potential for civilian and commercial applications has steadily grown in part due to a noticeable decrease in these high-cost technologies following initial development of the technology base.

Infrared array detectors have their own terminology for portions of the infrared spectrum, specified in wavelength (microns), and are based more on the response of detector materials, and atmospheric transmission windows, than on molecular vibrations (fingerprint, combinations, and overtones). In this scheme, *near-infrared* covers from 0.7 to 1.0 μm, from the edge of the visible to the silicon detector cutoff. *Short-wave infrared* (SWIR) is 1.0–3 μm (10,000–3300 cm^{-1}), covered by InGaAs detectors to about 1.7 μm (5900 cm^{-1}), and lead salt detectors (Pbs and PbSe) to 3 μm. There is a transmission window in the atmosphere from 3 to 5 μm (3300–2000 cm^{-1}) and this is termed *mid-wave infrared* (MWIR); the detector used widely here is InSb, with some PbSe and MCT applied as well. *Long-wave infrared* (LWIR) is used in several different ways: 5 to ~11 μm (2000 to ~950 cm^{-1}), where standard MCT arrays cut off, or the atmospheric window variously described as 7–14 μm (~1400 to ~700 cm^{-1}), or 8–12 μm (~1250 to ~850 cm^{-1}); here, MCTs and microbolometers are used. *Very long-wave infrared* (VLWIR) usually refers to wavelengths longer than ~11 μm to about 25 μm (400 cm^{-1}), and this region is covered by long-wavelength cutoff MCT [40, 41] and materials like arsenic-doped silicon impurity band conduction (Si:As IBC) [42].

MCT arrays operating at longer than 5 μm require cooling to liquid nitrogen temperatures and Si:As arrays to liquid helium temperatures. Imaging detectors used in military and remote sensing application usually employ Stirling cycle refrigerators, while in laboratory applications, they are mounted in pour-fill Dewars. Microbolometers are thermal detectors generally not requiring liquid nitrogen temperatures, and have a flat response across the spectrum [43].

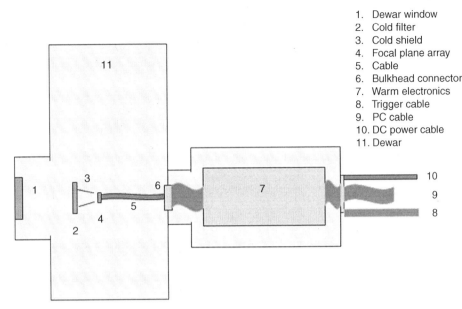

1. Dewar window
2. Cold filter
3. Cold shield
4. Focal plane array
5. Cable
6. Bulkhead connector
7. Warm electronics
8. Trigger cable
9. PC cable
10. DC power cable
11. Dewar

FIGURE 3.8   Block diagram of an infrared camera, incorporating a two-dimensional focal plane array. The focal plane array is attached to the cold finger of the Dewar and is therefore maintained at around 77K.

Infrared array detectors are sometimes classified as first, second, or third generation [44]. First generation generally refers to scanning linear arrays, often the MCT "common module" developed in the 1970s. Second generation arrays refer to two-dimensional staring detectors, integrated via bump bonds to a silicon-based readout integrated circuit (ROIC). The ROIC may contain one or more analog-to-digital (A/D) converters. Third generation arrays is a much less well-defined term [45, 46], but can include "intelligent" arrays with an integrated A/D for each pixel, two- or multicolor arrays with multiple detector elements per pixel, large numbers of pixels (2048 × 2048 and greater), and finally, the vision of eliminating the bump bonds and growing the detector material on a silicon substrate.

Infrared array detectors in general are monolithic or hybrid devices and operate either cooled or uncooled depending on the active material and performance characteristics of the array. Each array, based on its designed use, will have a different type of signal processing electronics. For the purposes of this chapter, we will concentrate on cooled detectors (MCT and InSb) that in their respective spectral ranges exhibit the greatest sensitivity and can operate at the fastest frame rates.

Mercury cadmium telluride (MCT or HgCdTe) detectors are widely used in infrared detection, and the history and technology of this material has been recently reviewed [47–51]. Today's state of the art in MCT arrays [52] include single FPAs with 4096 × 4096 pixels, and a mosaic of thirty-five 2048 × 2048 FPAs, for a device with a total of 147 million pixels [53]. Hoffman and Rogalski have described improvements analogous to Moore's law in semiconductors: the number of pixels has doubled every 19 months over the past

25 years. In semiconductors, larger wafers, smaller feature sizes, and larger die sizes became possible, while wafer fabs worked to reduce particle counts and other sources of defects, leading to increased yields.

3.5.3   Infrared Cameras

A *focal plane array*, which is the rectangular or square set of detectors, must be incorporated into a *Dewar* and built up into a *camera* (Figure 3.8), to be a useful device. In the case of InSb and MCT cameras, the FPA is located on a cold finger within a pour-fill liquid nitrogen Dewar. The FPA itself consists of the detector array, fabricated from InSb or MCT [54], and a *multiplexer* or *readout integrated circuit*, which is a silicon-based device.

Unlike visible region focal plane arrays using silicon as both a detector and the readout circuitry (e.g., charge-coupled devices), the detectors and readouts in infrared focal planes cannot be fabricated together, because the detector materials are incompatible with the temperatures used in integrated circuit manufacturing. Therefore, the detector array and the ROIC are fabricated separately and then integrated via *indium bump bonds*, which provide the electrical connection between the detectors and the ROIC.

To stop radiation, which does not come from the spectrometer from falling on the detector, a conical *cold shield* is placed in front of the FPA. *Optical filtering* can be used to limit the wavelengths of light falling onto the detector. In some cases, the optical filter can be the Dewar window itself; however, this means that the filter is at room temperature, and its emitted blackbody radiation will therefore fall on the focal plane. This can be a problem for detectors with long-wavelength response (MCT, and to a certain extent InSb). In these cases, it is useful to locate the bandpass filter within the Dewar and maintain it at a low temperature, to limit its long-wavelength emissions.

There is an electrical feedthrough, via a bulkhead connector, between the FPA and a set of electronics in the camera head. These electronics are at ambient temperature and are commonly called the *close proximity electronics* or *warm electronics*. These provide the biases, clocks, and control information necessary to run the FPA, and convert the FPA's analog output data to digital form for frame grabbers, giving an interface to the personal computer. Today, one or more A/D converters are in the warm electronics, and the connection to the PC may be via a video cable to a digital frame grabber card, or via FireWire.

A frame of data represents the infrared signals for each pixel at a certain optical retardation of the interferometer. In an FT-IR spectrometer, the classical approach is to read out the detector signal at a zero crossing of the He–Ne reference signal, ensuring that data points are recorded at even optical displacement increments [55]. To send a trigger to read out the frame at the appropriate time, a separate coax cable is normally run from the appropriate output on the spectrometer, directly to the camera. Ideally, the trigger signal to the camera should clear the array, initiate acquisition for the desired time, and then read out the whole array. This is true for a triggerable *snapshot* array, but a *rolling mode* array operates differently (see below). This trigger is often referred to as the *Jam sync* signal.

Cameras can operate in rolling mode where, for example, only two rows of the array are active at any time, followed by the next two, then the next two, and so on. A rolling mode camera is also free running: it is running continuously, and is not reset by a trigger, so that the start of data collection of a complete new frame is a short, but indeterminate, time after the trigger. The majority of cameras, however, operate in snapshot mode, where the whole array is integrated and read out at one time.

3.5.4 Indium Antimonide-Based Systems

The 128×128 InSb detector [56] used by Lewis et al., contained 16384 pixels, and neither the readout rate of the detector nor the speed of a 1995-vintage data system permitted rapid-scan operation. Therefore, in the 1990s, these imaging spectrometers had to be operated in step-scan [57] mode, with a simple electronic interface between the spectrometer and the camera. Following this work, Bio-Rad [58] commercialized an InSb-based imaging system in 1995. The InSb system was nonideal for scientific reasons, as the most specific and assignable spectroscopic information occurs in the mid-infrared region, but there were also technical issues; for instance, the signal was digitized using a 12-bit A/D converter, providing significantly less precision than con-

ventional FT-IR A/Ds and the digitization occurred remote to the detector Dewar, generating high noise levels.

3.5.5 The "Javelin" MCT Camera

In 1995, two-dimensional mid-infrared cameras were still a rarity in the civilian world. Notable individual systems included a 16×64 liquid helium-cooled, arsenic-doped silicon (Si:As) FPA [59], and a 256×256 MCT FPA [60], but the challenge was to locate an MCT array, and then a camera, and integrate it into a system that could be manufactured in volume, and therefore suitable for a commercial product. Bio-Rad commissioned Santa Barbara Focalplane (SBFP) to produce such a camera. A significant number of 64×64 MCT focal plane arrays were available at Santa Barbara Research Center [61] as a result of the Javelin antitank missile program [62]. SBFP obtained and integrated these detectors with a newer generation of electronics, including a 14-bit A/D converter, into their standard Dewar. The first mid-infrared imaging system using this camera was introduced in 1997 [63, 64].

The "Javelin" camera had its own peculiarities though. The focal plane was designed around 1992, and is actually contained in the missile itself, not the launcher. As such, it is not expected to be subject to multiple cooldown cycles, although the Javelin literature talks about achieving as many as 80 cooldown cycles. In a normal laboratory, this would correspond to only 4 months of operation, with one cooldown per working day! Because the thermal characteristics of MCT and silicon are different, multiple cooling cycles can induce *delamination*, where the indium bump bonds fracture and the MCT "peels away" from the ROIC.

The pixel pitch of the Javelin FPA is $61\,\mu m \times 61\,\mu m$, for a focal plane size of $4\,mm \times 4\,mm$. The pixel size allowed for a significant well capacity, quite suitable for the high-flux application of FT-IR-based spectral imaging. The spectral response starts above $5000\,cm^{-1}$, rolls off at about $1010\,cm^{-1}$, and is dead by about $950\,cm^{-1}$. Some Javelin cameras also exhibited "fixed pattern noise" in the images, most likely due to crosstalk between rows in the ROIC; this problem was addressed in later focal plane developments.

The readout frame rate of the Javelin array is specified at 180 Hz, operating in rolling mode. In this scheme, only two of the 64 rows are active at any time, one reading (collecting photons) and one writing (moving the charge into an active readout capacitor row). The integration time is not set independently, and therefore equals 1/32 of the frame time (the reciprocal of the frame rate). The rest of the FPA is essentially turned off for 31/32 of the time, giving a maximum effective duty cycle for the camera of only 3%. Although the Javelin camera as installed in the missile has a frame rate of 180 Hz, SBFP, using their own warm electronics, successfully ran it with a frame rate as high as 419 Hz, with 316 Hz being a typical operating rate. Thus, it was necessary for commercial instruments using the Javelin detector to use step-scan interferometers, although some rapid-scan experiments using high undersampling rates were performed [65, 66]. The development of this "Javelin camera" based imaging system initiated a second wave of imaging applications development [67–70], including the first imaging kinetics studies by Koenig [71–73].

3.5.6   Rapid-Scan FT-IR Imaging

To deal with the significant inefficiencies of the Javelin-based imaging FT-IR spectrometer, a systems design approach was used to match the speeds of all the components in the signal chain, including matching the flux levels, well capacity, and integration time of the array. The design goal of a rapid-scan system employing a two-dimensional array detector required new components all along the control and signal paths. In 1998, Bio-Rad contracted with SBFP to develop a new MCT camera. The key component was a high-speed ROIC, which interfaced to a new set of warm electronics containing four A/D converters. The ROIC was designed to accommodate arrays up to 128×128 pixels, with the more modern $40\,\mu m$ pixel pitch, and be triggerable, with snapshot readout and windowing (choice of the size and location of the pixels to be read out within the array). With a windowing ROIC, the readout speed of the FPA is inversely proportional to the number of active columns, so that if a 128×128 can be readout at "x" kHz, a 16×16 can be readout at "$8x$" kHz. The first product resulting from this program (128×128 camera, for a $5\,mm \times 5\,mm$ size FPA) was introduced in 2001 [74]. This camera had a frame rate of $\sim 1700\,Hz$ at the 128×128 frame size, and was operated in "fast step-scan" mode, with a step rate of $\sim 100\,Hz$. In 2002, the 64×64 pixel and smaller format cameras were introduced, with readout rates compatible with rapid-scan operation.

3.5.7   Use of Linear MCT Arrays

The cost, reliability, and performance issues of the early FPA-based systems prompted an alternative approach to FT-IR imaging, first by PerkinElmer, then other vendors, utilizing a small array that could be easily coupled with a rapid-scan interferometer. This provided the possibility of delivering spectroscopic performance on individual pixels, which was analogous with that of conventional FT-IR systems equipped with MCT detectors. The 16-element linear array of the PerkinElmer Spotlight™ was purpose built for FT-IR imaging and made from the highest quality, intrinsically pure MCT. Each detector element is connected with an individual gold wire connector as opposed to "bump bond," taking the signal out of the Dewar to a PCB for processing via a hermetically sealed and shielded cable. All 16 channels are processed in parallel at 32 kHz using individual integrated circuits with multiple ADCs, that is, all 16 channels

are reading photons effectively 100% of the time, (in contrast to the 3% duty cycle of the Javelin array at the time) resulting in much higher signal quality. On the optical side, imaging is simplified as a flat illumination field is now required over a much smaller area, making it easier to maintain uniform illumination and spatial resolution across the sample field. This combination of high-quality signal processing and ultrafast frame rate allowed the system to acquire high sensitivity IR images in impressively low times. This design approach was later adopted by Thermo Scientific and Jasco with some minor modifications. The original Spotlight™ detector measures a small $100\,\mu m \times 6.25\,\mu m$ (or $400\,\mu m \times 25\,\mu m$) area relatively quickly. This was later modified for larger frame areas. The signal processing quality is very high—such that in many cases a single scan, with an acquisition time of 0.2 s for all 16 elements, is all that is required to get good quality spectra. The sample stage is then stepped at up to 10 steps per second in $6.25\,\mu m$ or $25\,\mu m$ steps to quickly generate images of a user defined size at up to about 170 spectra per second in a so-called "push broom" mode of operation. At the time, this approach allowed both the use of a rapid scanning FT-IR spectrometer providing ease of use, better reliability, a lower purchase price, and lower ongoing maintenance costs. The image collection speed was limited by the interferometer OPD velocity rather than the detector, and the electronics system essentially limited by photon noise. To further improve efficiency of data collection, stage movement is synchronized with the interferometer, moving when the interferometer reaches each end of its travel and its velocity is momentarily zero. A further benefit in efficiency with small arrays pertains to background collection. Given that each detector pixel spectrum is usually processed as a ratio of the sample spectrum against a background spectrum, the time required to collect background spectra of a given quality for the detector array is considerably lower with the small array compared with the large FPA.

From a spectroscopic view, in addition to the impressive signal-to-noise performance of these systems, the use of photoconductive MCT in the linear arrays allows a wider wavelength range to be accessed. As we have seen, the conventional FPA's response does not extend much beyond about $1000\,cm^{-1}$ whereas the MCT in the Spotlight system comfortably reaches to about $720\,cm^{-1}$. In addition, the custom-designed detector array included a single element "mid-band" MCT detector alongside the array on the same substrate, allowing both imaging operations to $720\,cm^{-1}$ and single point microscopy to about $600\,cm^{-1}$ using a single Dewar.

3.6   SAMPLING MODES FOR IR IMAGING

The majority of laboratory mid-IR imaging systems are designed to provide transmission imaging by default, with options for various types of reflection measurement. Early applications were mostly transmission studies, and of these, biomedical applications were prevalent due to the pioneering hardware developments undertaken by Levin and coworkers at the National Institutes of Health in the United States in the 1990s. Since then, increased interest in other sampling techniques such as attenuated total internal reflection (ATR) and use of low-*E* slides for transflectance work has shifted the emphasis somewhat, although transmission remains the most popular technique. With current hardware, it is often the case that highest optical throughput and best signal-to-noise performance is usually obtained in transmission, although some systems can be better optimized for other modes of measurement if set up specifically for that purpose.

3.6.1   Transmission Sampling

Sample preparation methods are well described in the literature [75]; however, two points about transmission sampling are relevant to hardware. First, when samples are supported on an IR transmitting window or located in devices such as compression cells, they tend to shift the focus on the detector as a result of refraction effects. The shift depends on the sample/substrate/cell windows, but can be hundreds of microns. The relatively fast optics often used in the condenser and detector systems means this can have a considerable effect on the signal magnitude at the detector. To compensate for this, adjustment of the position of the focus is usually provided, either by adjustment of the position of the condenser or by change in its focal length. Second, in transmission imaging, the sample itself can be a major factor in determining the spatial resolution achievable. This is a consequence of the following: (a) in real systems, light enters the sample at a range of angles to the surface and (b) these light rays striking the boundaries between different components in an inhomogeneous sample (for example, two polymers of different refractive index in a laminate) can be redirected in various ways depending, for example, on whether they are transmitted (and subsequently refracted) or totally internally reflected at the boundary and subsequent boundaries. For these reasons, the spatial resolution of a system is often estimated not by using a real sample but by examining a response across a physical boundary such as knife edge.

3.6.2   Reflection Sampling

In many systems, both specular and diffuse reflection imaging can be readily performed with standard hardware used for transmission, but with the sample illuminated from the upper cassegrain by use of a flip mirror or beamsplitter positioned to intercept 50% of the incoming beam and direct into one half of the objective. The second half of the cassegrain collects the reflected beam that is subsequently relayed through the system onto the detectors. This is therefore at best 50% efficient compared with transmission mode. Furthermore, as only half of the collection geometry is used, the point spread function is broadened [76]. The use of reflection mode is popular for studying tissue sections using low-*E* slides to support the sample [77]. As the slide acts as an IR reflector, the beam passes through the sample twice, with the reflected signal undergoing a phase shift at the interface. This probably further confounds the quantitative interpretation of measured spectra.

Most simple specular reflection and transflection measurements are performed with the average incoming beam at an angle of about 20–30° (the minimal angle limited by the size and position of the primary cassegrain mirror), but reflection–absorption measurements at grazing angle to the sample may be performed by use of suitable objectives. Various modifications to the standard cassegrain layout have been described [78]. In some cases, the illumination is restricted to grazing incidence (typically 65–85° from the normal to the surface) by use of an appropriately placed aperture [79], or in another example used in Bruker systems [80], a system of additional mirrors inside the cassegrain deviate the beam to high incidence angle. This technique, used primarily in single point mapping systems, has demonstrated impressive measurements of layers of just a few nanometers thickness [81], but its use with FPAs is not widely reported.

3.6.3   Diffuse Reflection

The cassegrain objectives used for standard mid-IR imaging systems provide illumination and reflection cone angles that allow collection of both specular and diffusely reflected (in a geometrical sense) components of the reflected beam. Unlike some of the macroaccessory designs for diffuse reflection, there are generally no special hardware considerations to remove the front surface component (i.e., Fresnel component), and this can cause additional difficulties in interpretation of reflection spectra in the mid-IR. Combining this effect with the relatively high absorption in the mid-IR, diffuse reflection of neat samples is somewhat limited. However, this is not the case in the NIR region and much of this region (about 8000–4000 cm^{-1}) is readily accessible with some systems using MCT arrays. Systems using both small linear [82] and FPAs [83] can generate high-quality NIR diffuse reflection images in the longer wavelength NIR region. This is achieved using essentially the same imaging hardware but replacing the mid-IR source with a tungsten halogen source. Other hardware modifications for NIR diffuse reflection can include replacing the FT-IR beamsplitter material with one more efficient at NIR wavelengths, and modifying any optical filter, for instance, the long-pass filter in the FPA Dewar can be eliminated or replaced by a NIR region-specific short-pass filter.

The understanding of spatial resolution in diffuse reflection imaging is less clear. Not only does the incoming beam penetrate the sample to some considerable depth, it is likely to emerge at some displacement from the point of entry, due to the nature of diffuse reflection itself. This effect is more severe in the NIR region. Various qualitative estimates of these effects have been made [84, 85], and values of 50–100 μm penetration depth have been suggested—this value varying with a number of sample parameters such as particle size. This and typical NIR absorptivities suggest that the use of imaging optics with a magnification providing smaller than about 10 μm per pixel at the sample is unlikely to be beneficial for current applications of this method.

3.6.4 ATR Imaging

The ATR technique has become one of the most popular sampling techniques for macro-IR sampling [86] and there are now a number of devices variously utilized to generate ATR images. It is of course possible to generate ATR images using single point ATR microscope accessories by repeatedly removing the sample, repositioning, and rescanning repeatedly to build up an ATR "image," but this has obvious drawbacks apart from the huge time penalty, as possible cross-contamination and sample deformation issues come into play.

ATR imaging systems using array detector systems [87] present some considerable design challenges both from an optical and mechanical point of view. Current ATR imaging accessories are designed as add-on accessories for existing transmission/reflection systems and this in itself places some constraints on system design. Current methods include use of various size hemispheres, often of germanium, as the internal reflection element (IRE), with both FPAs and linear detector array systems operating in a "raster-scan" mode, and also the use of commercial macro-ATR accessories where the ATR element is imaged directly onto a FPA without the presence of a visible microscope system [88]. In addition to the well-documented sampling advantages (no requirement to prepare thin sections as with transmission), one particular advantage is a potential improvement in spatial resolution compared with transmission sampling for two reasons: (a) the sample is immersed in the IRE and the effective NA, and hence spatial resolution improves with the refractive index of the IRE and (b) the much lower sample penetration depths of typically less than 2 μm mean that the effect of the sample-induced resolution degrading mechanisms described above for transmission and diffuse reflectance modes is virtually eliminated.

The issue of sample contact is now of a different magnitude compared with most single point ATR designs, as now the contact has to be maintained over the whole area to be imaged, which can be very difficult for hard samples with less than very flat surfaces. For softer materials, variants on IRE shape have been employed, including a KRS-5 chevron

design used by Esaki et al. [89], but the design does not offer the magnification or increased NA advantage of a Ge hemisphere.

Using FPA systems, typically the sample is pressed against the IRE and the IRE/sample held stationary. The imaging area is determined by the field of view (FOV) of the camera (assuming all other things equal). With typical imaging optics, this FOV ranges from about 45 to 180 μm linear dimension with a sample image pixel size of about 1.5 μm using standard array sizes between 32×32 and 128×128 pixels. However, obtaining bigger ATR images is not simply a matter of increasing crystal size (and number of FPA pixels as required) with such systems, as efficient and uniform illumination becomes more difficult with increasing crystal size for the hemisphere arrangement in particular. That said, other crystal parameters such as radius of curvature can be adjusted to alter the illumination field. Increasing image area with a reduction of image pixel density relative to that of micro-ATR accessories has been achieved by use of commercial macro-ATR accessories [90] with areas up to about 4 mm^2 imaged. Working at the near-theoretical spatial resolution, however, generally limits the FOV to about 50 μm linear dimension using a 64×64 array.

An alternative approach is to move the IRE + sample relative to the beam Figure 3.9 in an off-axis configuration. Now, the FOV is determined not by the detector array size but to first order by the diameter (in the case of a hemisphere) or geometry of the IRE crystal. The PerkinElmer Spotlight system [91] utilizes an IRE tip diameter of about 500 μm, and Paterson and Havrilla [92] have demonstrated this design approach using a 25 mm diameter Ge hemisphere with a 16×1 array to provide an effective sample image area of about 2500 μm $\times$ 2500 μm with a 1.56 μm pixel size at the sample. Selecting optimal crystal parameters is nontrivial as parameters such as illumination field flatness, energy throughput, and radius of curvature of input and output IRE surfaces are somewhat interrelated. Furthermore, properties such as sample penetration depth and spatial resolution vary with offset from the center of the hemisphere, with optical aberrations increasing outward from the center. Careful matching of IRE crystal geometry to the other system components is fundamental to optimum performance. A further point is that without appropriate software correction, such rastering systems can provide artifacts due to nonidentical responses of the individual detector pixels. To illustrate an extreme case, consider a 16×1 array where the sample/ IRE is rastered in a direction perpendicular to the line defined by the detector pixels. If the response of detectors 1 and 16 is different, the resulting image will show striations at the points where the pixels 1 and 16 are adjacent to each other (i.e., at the stitching points). This is not limited to linear arrays and is worse with larger arrays. With FPAs, this effect can lead to both horizontal and vertical striations. This type of artifact is not eliminated by simply ratioing the image spectra with

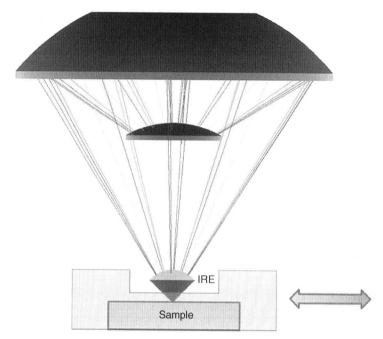

FIGURE 3.9 ATR imaging configuration. In this method of ATR image data collection, the sample is clapped against the IRE, and sample + IRE is moved relative to the beam to generate an image.

background spectra recorded from identical pixels and requires software correction and/or data collection schemes to minimize this effect.

3.7 IR IMAGING SPEED AND PERFORMANCE CONSIDERATIONS

As the detector technology and data acquisition methods have evolved, the factors affecting IR signal-to-noise and other spectroscopic performance of merit have become better understood for array-based systems. With the earlier step-scan-based systems, Koenig's and Levin's groups investigated various noise sources and possible approaches to data acquisition to improve S/N [93, 94]. Bhargava and Levin [95] and Srinivasen and Bhargava [96] discuss and compare FPA and linear array detector systems and issues around performance comparison between different imaging systems, for example, a figure of merit based on the time to collect a fixed number of IR image pixels at a given S/N for a given set of FT-IR data collection parameters (e.g., resolution, apodization, etc.) is proposed. As with the much simpler FT-IR spectrometers, however, meaningful and accurate direct comparisons between different manufacturers' instruments remain plagued with difficulties.

Today, the rapid-scan mode of data collection is generally preferred, with most manufacturers using detector arrays of 16×1 to 128×128. This permits kinetics studies of nonrepeatable events that occur on the timescale of seconds, for instance polymer dissolution. However, if an event is

reproducible and triggerable, it can be synchronized at each step of the mirror retardation, with faster time resolutions achievable using the step-scan mode. Time resolutions in the millisecond timescales have been demonstrated by Bhargava and Levin [97], who suggest that with the appropriate setup, time resolutions in the microsecond timescale are feasible.

REFERENCES

1. Barer, R., Cole, A. R. H., and Thomson, H. W. (1949) *Nature* **163**, 198.
2. Coates, V. J., Offner, A., and Seigler, E. H. (1953) *J. Opt. Soc. Am.* **43**, 984.
3. Griffiths, P. R. and de Haseth, J. A. (2007) *Fourier Transform Infrared Spectrometry*, 2nd ed., Wiley, Hoboken, NJ.
4. Jackson, R. S. (2006) *Handbook of Vibrational Spectroscopy*, Vol. 1, Wiley, pp. 264–282.
5. See ALS Infrared Beamlines Homepage, http://infrared.als.lbl.gov/content/home and references therein.
6. Carr, G. L., Hanfland, M., and Williams, G. P. (1995) *Rev. Sci. Instrum.* **66**, 1643.
7. Reffner, J. A., Martoglio, P. A., and Williams, G. P. (1995) *Rev. Sci. Instrum.* **66**, 1298.
8. Michelson, A. A. (1891) *Philos. Mag.* **31**, 256.
9. Michelson, A. A. (1927) *Light Waves and their Uses*, University of Chicago Press, Chicago (Phoenix edition, 1962).
10. Griffiths, P. R. and de Haseth, J. A. (2007) *Fourier Transform Infrared Spectrometry*, 2nd ed., Wiley, Hoboken, NJ, Chapters 19 and 21, pp. 53–54.

11. Spragg, R. A., Carter, R., Clark, D., and Hoult, R. (2001) *Performance and Applications of a Novel FT-IR Imaging System, FACSS Conference.*

12. Microscopy Resource Center, Olympus http://www.olympus-micro.com/primer/webresources.html and various references therein.

13. Schiering, D. W., Tague, T. J., Reffner, J. A., and Vogel, S. H. (2000) *Analusis* **28** (1), 46–52.

14. Messerschmidt, R. G. (1987) *The Design, Sample Handling, and Applications of Infrared Microscopes*, ASTM STP 949, Philadelphia, PA.

15. Hoult, R. A. and Carter, R. L. (2002) Dual magnification for imaging infrared microscope. U.S. Patent US2002034000.

16. Sommer, A. J. and Katon, J. E. (1991) *Appl. Spectrosc.* **45**, 1633.

17. Nishikida, K. Spatial resolution in infrared microscopy and imaging, Thermo Scientific Application Note No. 50717.

18. Potter, A. E., Jr. (1972) Multispectral imaging system. U.S. Patent 3,702,735.

19. SPIE, Bellingham, WA, www.spie.org.

20. Wells, W., Potter, A. E., and Morgan, T. H. (1980) Near-infrared spectral imaging Michelson interferometer for astronomical applications. In: *Infrared Imaging Systems Technology, Proc. SPIE* 226, 61–64.

21. Huppi, R. J., Shipley, R. B., and Huppi, E. R. (1979) Balloon-borne Fourier spectrometer using a focal-plane detector array. In: *Multiplex and/or High Throughput Spectroscopy, Proc. SPIE* 191, 26–32.

22. Smithson, T. (1994) Imaging emission spectroscopy. In: *9th International Conference on Fourier Transform Infrared Spectroscopy, Proc. SPIE* 2089, 530–531.

23. Villemarie, A., Fortin, S., Giroux, J., Smithson, T., and Oermann, R. (1995) An imaging Fourier transform spectrometer. In: Imaging Spectrometry, Proc. SPIE 2480, 387–397.

24. See http://irsa.ipac.caltech.edu/IRASdocs/iras.html and http://irsa.ipac.caltech.edu/IRASdocs/exp.sup/ch2/C4.html.

25. Gillett, F. C. (1995) Infrared arrays for astronomy. In: *Infrared Detectors for Instrumentation and Astronomy, Proc. SPIE* 2475, 2–7.

26. Norton, M., Kindsfather, R., and Dixon, R. (1995) Infrared (3-12 μm) narrowband and hyperspectral imaging review. In: *Imaging Spectrometry, Proc. SPIE* 2480, 295–313.

27. Rapp, R. J. and Register, H. I. (1995) Infrared imaging spectro-radiometer program overview. In: *Imaging Spectrometry, Proc. SPIE* 2480, 314–321.

28. Goetz, A. F. H. (1995) Imaging spectroscopy for remote sensing: vision to reality in 15 years. In: *Imaging Spectrometry, Proc. SPIE* 2480, 2–13.

29. Simons, D. A., Clark, C. C., Smith, S., Kerr, J., Massey, S., and Maillard, J.-P. (1994) CFHT's imaging Fourier transform spectrometer. In: *Instrumentation in Astronomy VIII, Proc. SPIE* 2198, 185–193.

30. Maillard, J.-P. (1997) Astronomical Fourier-transform spectroscopy of the 1990s. *Microchim. Acta Suppl.* **14**, 133–141.

31. Bennett, C. L., Carter, M., Fields, D. J., and Hernandez, J. (1993) Imaging Fourier transform spectrometer. In: *Imaging Spectrometry of the Terrestrial Environment, Proc. SPIE* 1937, 191–200.

32. Carter, M. R., Bennett, C. L., Fields, D. J., and Hernandez, J. (1995) Imaging Fourier transform spectrometer (LIFTIRS). In: *Imaging Spectrometry, Proc. SPIE* 2480, 380–386.

33. Bennett, C. L., Carter, M. R., and Fields, D. (1995) Hyper-spectral imaging in the infrared using LIFTIRS. In: *Infrared Technology XXI, Proc. SPIE* 2552, 274–283.

34. Bennett, C. L. (1996) Method for determining and displaying the spatial distribution of a spectral pattern of received light. U. S. Patent 5,539,518.

35. Wolfe, W. L. (1997) Introduction to imaging spectrometers. In: *SPIE Tutorial Texts in Optical Engineering*, Vol. TT25, SPIE Optical Engineering Press, Bellingham, WA.

36. Beer, R. (1992) *Remote Sensing by Fourier Transform Spectrometry*, Wiley, New York, Sections 2.6 and 5.

37. Lewis, E. N., Treado, P. J., Reeder, R. C., Story, G. M., Dowrey, A. E., Marcott, C., and Levin, I. W. (1995) FTIR spectroscopic imaging using an infrared focal-plane array detector. *Anal. Chem.* **67**, 3377.

38. Bhargava, R. and Levin, I. (Eds.) (2005) *Spectrochemical Analysis Using Infrared Multichannel Detectors*, Blackwell Publishing.

39. Recently introduced commercial FT-IR spectrometers mostly use a variant of the scheme described by Brault: Brault, J. W. (1996) *Appl. Opt.* **35**, 2981.

40. Ashcroft, A., Jones, C., Hipwood, L., Baker, I., Shorrocks, N., Knowles, P., and Weller, H. (2008) Recent developments in very long wave and shortwave infrared detectors for space applications. *Proc. SPIE*, **7106**, 71061L-1–71061L-11.

41. Chu, M., Gurgenian, R. H., Mesropian, S., Becker, L., Walsh, D., Kokoroski, S. A., Goodnough, M., and Rosner, B. (2004) Advanced planar LWIR and VLWIR HgCdTe focal-plane arrays. *Proc. SPIE*, **5167**, 159–165.

42. Love, P. J., Hoffman, A. W., Lum, N. A., Ando, K. J., Ritchie, W. D., Therrien, N. J., Toth, A. G., and Holcombe, R. S. (2004) 1K × 1K Si:As IBC detector arrays for JWST MIRI and other applications. *Proc. SPIE*, **5499**, 86–96.

43. Kruse, P. W. (2001) *Uncooled Thermal Imaging: Arrays, Systems and Applications*, SPIE Press, Bellingham, WA.

44. Kinch, M. A. (2007) *Fundamentals of Infrared Detector Materials*, SPIE Press, Bellingham, WA.

45. Rogalski, A. (2006) Competitive technologies for third generation infrared photon detectors. *Proc. SPIE* **6206**, 62060S-1–62060S-15.

46. Rogalski, A. (2006) Competitive technologies of third generation infrared photon detectors. *Opto-Electron. Rev.* **14**, 87–101.

47. Bajaj, J. (2000) State-of-the-art HgCdTe infrared devices. *Proc. SPIE*, **3948**, 42–54.

48. Norton, P. (2002) HgCdTe infrared detectors. *Opto-Electron. Rev.* **10**, 159–174.

49. Rogalski, A. (2003) HgCdTe infrared detectors—historical prospect. *Proc. SPIE* **4999**, 431–442.

50. Rogalski, A. (2002) Infrared detectors: an overview. *Infrared Phys. Technol.* **43**, 187–210.

51. Norton, P. R. (1999) Infrared detectors in the next millennium. *Proc. SPIE*, **3698**, 652–665.

52. Hoffman, A. W., Love, P. J., and Rosbeck, J. P. (2004) Mega-pixel detector arrays: visible to 28 μm. *Proc. SPIE*, **5167**, 194–203.

53. Sprafke, T. and Beletic, J. W. (2008) High performance infrared focal-plane arrays for space applications. *Opt. Photon. News* **19** (6), 22–27.

54. For a background on infrared detector materials, see Kinch, M. A. (2007) *Fundamentals of Infrared Detector Materials*, SPIE Press, Bellingham, WA.

55. Griffiths P. R. and de Haseth, J. A. (2007) *Fourier Transform Infrared Spectrometry*, 2nd ed., Wiley, Hoboken, NJ, Chapter 3.

56. Lockheed Martin Santa Barbara Focalplane (SBFP), Goleta, CA, http://www.sbfp.com/.

57. Griffiths P. R. and de Haseth, J. A. (2007) *Fourier Transform Infrared Spectrometry*, 2nd ed., Wiley, Hoboken, NJ, pp. 312–320.

58. Bio-Rad Spectroscopy Division (Cambridge, MA). In 2002, Bio-Rad sold this business to Digilab, LLC (Randolph, MA), and in 2006 the spectroscopy assets of Digilab were in turn acquired by Varian (Walnut Creek, CA).

59. Lewis, E. N., Kidder, L. H., Arens, J. F., Peck, M. C., and Levin, I. W. (1997) Si:As focal-plane array detection for Fourier transform spectroscopic imaging in the infrared fingerprint region. *Appl. Spectrosc.* **51**, 563–567.

60. Kidder, L. H., Levin, I. W, Lewis, E. N., Kleiman, V. D., and Heilweil, E. J. (1997) MCT focal-plane array detection for mid-infrared FT spectroscopic imaging. *Opt. Lett.* **22**, 742.

61. At that time, Santa Barbara Research Center. Now Raytheon Vision Systems, Goleta, CA, http://www.raytheon.com/capabilities/products/ScanningIR/.

62. The Javelin close combat/anti-armor weapon system program is a venture between Raytheon and Lockheed Martin. See http://www.raytheon.com/capabilities/products/javelin/. By 2006, Raytheon had produced 30,000 Javelin missiles.

63. Crocombe, R. A., Wright, N., Drapcho, D. L., McCarthy, W. J., Bhandare, P., and Jiang, E. Y. (1997) FT-IR spectroscopic imaging in the infrared 'fingerprint' region using an MCT array detector. In: *Microscopy and Microanalysis,* Vol. 3, Supplement 2, *Proceedings,* Springer-Verlag, New York, pp. 863–864.

64. Wright, N. A., Crocombe, R. A., Drapcho, D. L., and McCarthy, W. J. (1998) The design and performance of a mid-infrared FT-IR spectroscopic imaging system. *Am. Inst. Phys. Proc.* **430**, 371–372.

65. Snively, C. M., Katzenberger, S., Oskarsdottir, G., and Lauterbach, J. (1999) Fourier-transform infrared imaging using a rapid-scan spectrometer. *Opt. Lett.* **24**, 1841–1843.

66. Huffman, S. W., Bhargava, R., and Levin, I. W. (2002) Generalized implementation of rapid-scan Fourier transform infrared spectroscopic imaging. *Appl. Spectrosc.* **56**, 965–969.

67. Chalmers, J. M., Everall, N. J., Hewitson, K., Chesters, M. A., Pearson, M., Grady, A., and Ruzicka, B. (1998) Fourier transform infrared microscopy: some advances in techniques for characterization and structure-property elucidations of industrial material. *Analyst* **123**, 579–586.

68. Marcott, C. and Reeder, R. C. (1998) Industrial applications of FT-IR microspectroscopic imaging using a mercury-cadmium-telluride focal-plane array detector. In: *Infrared Technology and Applications XXIV, Proc. SPIE* 3436, 285–289.

69. Marcott, C., Reeder, R. C., Paschalis, E. P., Tatakis, D. N., Boskey, A. L., and Mendelsohn, R., (1998) Infrared microspectroscopic imaging of biomineralized tissues using a mercury-cadmium-telluride focal-plane array detector. *Cell. Mol. Biol.* **44**, 109–115.

70. Colarusso, P., Kidder, L. H., Levin, I. W., Fraser, J. C., Arens, J. F., and Lewis, E. N. (1998) Infrared spectroscopic imaging: from planetary to cellular systems. *Appl. Spectrosc.* **52**, 106A–120A.

71. Koenig, J. L. and Snively, C. M. (1998) Fast FT-IR imaging: theory and applications. *Spectroscopy*, **13** (11), 22–28.

72. Snively, C. M. and Koenig, J. L. (1998) Application of real time mid-infrared FTIR imaging to polymeric systems. 1. Diffusion of liquid crystals into polymers. *Macromolecules* **31**, 3753–3755.

73. Oh, S. J. and Koenig, J. L. (1998) Phase and curing behavior of polybutadiene/diallyl phthalate blends monitored by FT-IR imaging using focal-plane array detection. *Anal. Chem.* **70**, 1768–1772.

74. The camera was named "Lancer" by Bio-Rad.

75. Sommer, A. J. (2006) Mid-infrared transmission microspectroscopy. In: *Handbook of Vibrational Spectroscopy*, Vol. 2, Wiley, pp. 1370–1385.

76. Lewis, L. and Sommer, A. J. (1999) *Appl. Spectrosc.* **54**, 324.

77. Story, G. M., Marcott, C., and Dukor, R. K. (1999) A method for analysis of clinical tissue samples using FT-IR microspectroscopic imaging. In: *Microscopy and Microanalysis*, Vol. 5, Springer-Verlag, New York, p. 69.

78. See, for example, Sting, D. W. (1989) Grazing angle microscope. U.S. Patent 4,810,077;Simon, A. (1999) Grazing angle microscope. U.S. Patent 6,008,936.

79. Reffner, J. A., Alexay, C. C., and Hornlein, R. W. (1991) *8th International Conference on FT-IR Spectroscopy* SPIE Vol. 1575.

80. *'Grazing Angle Objective'* Bruker Product Note, Bruker Optics.

81. Katon, J. E. and Sommer, A. J. (1992) *Anal. Chem.* **64**, 931A.

82. Spragg, R. A., Hoult, R., and Sellors, J. (2002) Comparing near-IR and mid-IR microscopic reflectance FT-IR imaging. *FACCS Conference*.

83. Miseo, E., Weston, F., and Leonardi, J. (2008) Extending the range of MCT focal-plane arrays–based imaging systems to

near IR imaging applications. In: *Molecular Spectroscopy Application Notebook*, Varian Inc, Randolph, USA.

84. Hudak, S., Haber, K., Sando, G., Kidder, L. H., and Lewis, E. N. (2007) *NIR News* **18**, 6.

85. Spragg, R., Locke, T., Hoult, R., and Sellors, J. (2003) In: Davies, A. and Garido-Varo, A. (Eds.), *Proceedings of the 11th International Conference on NIR Spectroscopy* NIR Publications, Chichester, UK.

86. Harrick, N. J. (1967) *Internal Reflection Spectroscopy*, Harrick Scientific Corporation, Pleasantville, New York.

87. Burka, E. M. and Curbelo, R. (2000) Imaging ATR spectrometer. U.S. Patent 6,141,100.

88. Chan, K. L. A. and Kazarian, S.G. (2003) *Appl. Spectrosc.* **57** (4), 381.

89. Esaki, Y., Nakai, K., and Araga, T. (1995) *R&D Rev. Toyota CRDL* **30**, 57.

90. Crocombe, R. A., Wright, N. A., and Bhandare, P. (1999) Applications and sampling techniques for mid-IR spectroscopic imaging. In: *12th International Conference on FT Spectroscopy*, Tokyo, Japan.

91. Canas, A., Carter, R., Hoult, R., Sellors, J., and Williams, S. (2006) Mid-IR ATR imaging using a linear detector array system. *FACSS Conference*.

92. Patterson, B. M. and Havrilla, G. J. (2006) *Appl. Spectrosc.* **60**, 11.

93. Bhargava, R., Scharberle, M. D., Feranndez, D. C., and Levin, I. W. (2001) *Appl. Spectrosc.* **55**, 1079.

94. Snively, C. M. and Koenig, J. L. (1999) *Appl. Spectrosc.* **53**, 170.

95. Bhargava, R. and Levin, I. W. (Eds.) (2005) *Spectrochemical Analysis Using Infrared Multichannel Detectors*, Blackwell Publishing, Sheffield, UK, Chapter 1.

96. Srinivasen, G. and Bhargava, R. (2007) *Spectroscopy* **22** (7), 30–31.

97. Bhargava, R. and Levin, I. W. (2003) *Appl. Spectrosc.* **57** (4), *357;* see also Bhargava, R. and Levin, I. W. (2004) *Appl. Spectrosc.* **58** (8) 995; Bhargava, R. and Levin, I. W. (2004) *J. Phys. Chem. A* **108** (18), 3896.

4

TECHNOLOGIES AND PRACTICAL CONSIDERATIONS FOR IMPLEMENTING NEAR-INFRARED CHEMICAL IMAGING

E. Neil Lewis and Linda H. Kidder
Malvern Instruments, Columbia, MD, USA

4.1 INTRODUCTION

Near-infrared chemical imaging (NIRCI) has evolved from a novel analytical technique to one in routine use in a variety of settings. Please refer to other chapters for explorations of NIRCI applications in biomedical, pharmaceutical, food, and polymer fields of study. Ignoring the spatially resolved nature of the resulting data, NIRCI can be viewed as highly parallel near-infrared spectroscopy (NIRS), where tens to tens of thousands of NIR spectra are acquired simultaneously. Therefore, NIRCI can be expected to provide qualitative and quantitative analytical information that is, at the least, equivalent to NIRS. In practice, however, *significantly more* insight into the functionality of heterogeneous materials may be obtained using NIRCI rather than NIRS. The reason for this is twofold. As NIRCI data are collected in a spatially resolved manner, that is, the user knows from where on a sample a NIR spectrum is derived, novel information about the spatial distribution of sample components and the correlation of this with performance can be investigated. Second, the large number of spectra acquired, and the ability to characterize the resulting distribution of values across a sample, provides statistical insights into chemical heterogeneity, a characteristic that is also known to impact sample performance.

Chemical imaging instrumentation has become more economically accessible and robust because of iterative advances in hardware, computational capabilities, electronics, and data processing strategies. Its increased accessibility and ability to solve real-world problems has acted as a catalyst for the discovery of further uses. It is likely that as imaging solves more problems, the techniques will be more routinely deployed in laboratory, QA/QC, and process monitoring environments.

Although there has been a parallel path in the development of NIR imaging spectrometers for space and remote sensing applications, this chapter will focus on terrestrial laboratory and process systems. An excellent source of general background information that encompasses early developments in both fields can be found in Refs 1 and 2.

4.2 NEAR-INFRARED SPECTROSCOPY

The analytical near-infrared (NIR) region spans the range from approximately 700 to 2500 nm, where absorptions arising principally from overtones and combination bands of O–H, N–H, and C–H stretching and bending fundamentals are found. In comparison to the fundamental modes found in the mid-infrared spectral region (MIR: 2500–25,000 nm, 4000–400 cm^{-1}), sample absorptivities can be one to many orders of magnitude less. This minimizes the need for extensive sample preparation whose principal purpose is to avoid saturation of absorbed radiation. Samples can be examined through packaging materials, and the technique can be used to examine hydrated samples, within limits. Intact samples are usually characterized using diffuse reflectance, and the technique is particularly useful for performing rapid, reproducible, and nondestructive analyses of known materials [3]. Although less preparation in comparison to the MIR may be needed in presenting samples to the spectrometer, data analysis can often be relatively more extensive. The absorption bands for the overtone and combination modes tend to be broader and more overlapped than for the fundamental modes, and so multivariate methods are typically used to separate spectral signatures of sample components.

4.3 NEAR-INFRARED CHEMICAL IMAGING

Unlike its MIR and Raman imaging counterparts, NIR microscopy was relatively rarely applied in the laboratory. In *Infrared Microspectroscopy: Theory and Applications* published in 1988 [4], *near-infrared* appears only twice in the index, and in the 1995 volume, *Practical Guide to Infrared Microspectroscopy*, the term does not appear in the index *at all* and is not explicitly covered by any of the individual chapters [5]. In the *Handbook of Near-Infrared Analysis* published in 1992 [3], there is no mention of microscopy, mapping, or imaging in the index. The reason for the relatively late development and application of NIRCI, compared to the MIR and Raman counterparts, perhaps has to do with the typical application areas of traditional NIR spectroscopy. NIRS was largely targeted toward agricultural and industrial applications for the determination of bulk, averaged properties [3]. Industry mindset was that traditional NIRS applications would not benefit from the ability to explore the spatial distribution of sample components. Scattered early work describes data collection of *spatially resolved* NIR spectra [6–19], but the real proliferation of NIRCI started approximately in 2001 with the commercialization of global imaging and line mapping implementations: (Spectral Dimensions—now Malvern Instruments—Columbia, MD, USA), (PerkinElmer, Shelton, CT, USA), (Specim, Oulu, Finland). These first commercial systems mostly targeted pharmaceutical, agricultural, and waste sorting applications [20–23]. By the mid-2000s, once imaging platforms became widely available, and the utility of the approach in the pharmaceutical and agricultural areas became apparent, NIRCI has become more widely used [24–27].

In a simplistic way, NIRCI can be viewed as just an "expanded" version of standard NIRS, where multiple NIR spectra are acquired in parallel or in series. The near-infrared spectral information in a NIRCI data set is comprised of the same spectral absorbance bands, molecular scattering, diffraction effects, instrument line shapes, and so on, as for single point NIR spectra. Therefore, spectral preprocessing (dividing through by a background data set, baseline subtraction, multiplicative scatter correction, derivatives, etc.) and multivariate data analysis approaches are as critically important for the correct interpretation of NIRCI data as they are for single point NIR spectra.

There is, of course, a difference in practice. Conventional NIR spectrometers average the spectral signatures across the sample area into one spectrum, whereas NIRCI spectrometers preserve spatial distinctions, collecting multiple unique NIR spectra across the same area. Depending on the technique employed, tens to tens of thousands of NIR spectra comprise a data set, and the resulting spatially resolved data can be applied to understanding the heterogeneity of a single sample or for high-throughput analysis of multiple samples. The statistical information provided by having so many data points enables the application of unique quantitative analysis methods. The combination of spatial and statistical information enables NIRCI to also be a *primary analytical method*—for instance, in calculating the size of chemical domains or the thickness of a coating cross section. Figure 4.1 presents an example of this capability. Figure 4.1a shows a single wavelength image (at 2080 nm) that distinguishes several layers within a pharmaceutical granule. The thickness of three layers has been calculated by individually thresholding and binarizing each, and then determining the distribution of thickness values across the layer. The results of this calculation are displayed in Figure 4.1b as a 1-pixel thick line located along the outer edge of its corresponding layer. The coating thickness is indicated by the *shade or color* of this 1-pixel outline—in the color representation red shows areas of thicker coating, whereas blue indicates thinner coatings. The actual coating statistics can be represented by displaying a histogram of coating values. The distribution statistics for these three layers are shown in Figure 4.1c. The mean coating thicknesses for these three layers are 387.3 μm, 64.3 mm, and 61.9 mm, respectively. These values are determined *a priori* from the chemical image. This is in contrast to standard NIR spectroscopy that is exclusively employed as a secondary analytical method, in which calibration models must first be derived to establish correspondence between NIR spectroscopy and a correlative gold standard. These points are discussed in more detail below.

4.3.1 Chemical Imaging Data Cube: Hypercube

Chemical image data sets can be represented as a three-dimensional cube spanning one wavelength and two spatial dimensions. This data construct is often called a hypercube, and each element within the cube contains the spectral intensity response measured at a particular spatial location and wavelength [1]. The hypercube can be explored as a series of spectra where the spatial location is known and preserved, or alternatively, as a series of images at specific wavelengths. However, much of the novel information provided by a hypercube combines both spectral and spatial information, for example, correlating sample function with the spatial interrelationship or size of components. A variety of uni- and multivariate (chemometric) data processing

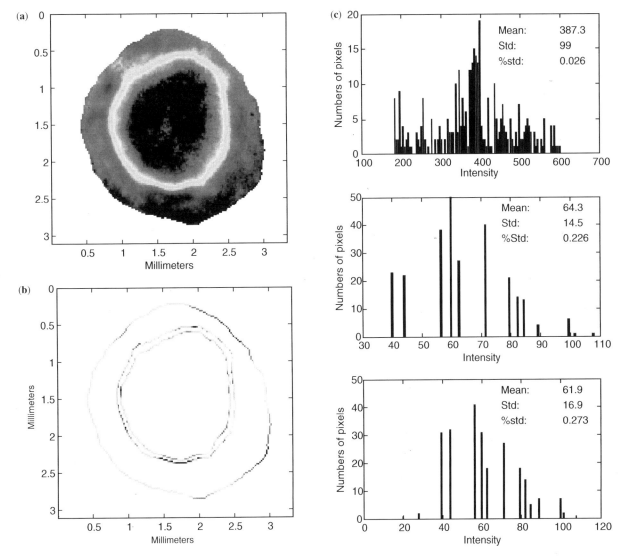

FIGURE 4.1   (a) Single wavelength image (2080 nm) highlighting multiple chemically distinct layers within a cross-sectioned pharmaceutical granule. (b) Coating thickness distribution traces summarizing the thickness of three visible layers of the granule. The trace outlines the outer edge of the layer and the shade or color gradation gives the thickness value—in the color representation red represents a relatively thicker layer and blue thinner. For the black and white representation, variations in shade are indicative of coating thickness differences. (c) Histogram representations of the coating thickness distributions of the three layers. Values given are in micrometer. The mean coating thicknesses for the first three layers are 387.3, 64.3, and 61.9 μm, respectively.

techniques are available to explore this type of information; Ref. 26, and its included references, provides a thorough survey of this area.

4.3.2   Multispectral and Hyperspectral

The terms multispectral and hyperspectral originated from the remote sensing community and are also used by the analytical imaging community. The terms indicate the number of wavelengths in a particular data set, with hyperspectral data containing more than multispectral data. Although there is no absolute, generally accepted limit, commonly "hyperspectral" implies tens to several thousands of contiguous but narrow wavelength channels, whereas "multispectral" implies several to tens of wavelengths, with no requirement for the selected wavelengths being contiguous or narrow. NIRCI implemented with an interferometer or an imaging monochromator (grating, prism, or a combination of dispersive elements) will collect hyperspectral data. Imaging with a series of bandpass filters provides a multispectral approach. A tunable filter (liquid crystal tunable filter (LCTF) or acousto-optic tunable filter (AOTF)) or tunable source (optical parametric oscillator (OPO)) may collect data in either multi- or hyperspectral mode, based on how many channels are measured, and this can be tailored to solve particular analytical problems.

4.3.3   Statistical Analyses and Spatial Heterogeneity

NIRCI data contain the same spectral information that is recorded with a conventional NIR spectrometer taken over the same sample area, but breaks the information apart as a function of spatial location. Therefore, a NIR spectral parameter in a NIRCI data set is not defined by a single value, but spans a range based on spatial location. As a simple representative example, Figure 4.2 presents an image and corresponding histogram derived from a NIRCI data set of a pharmaceutical tablet containing ~22,400 NIR spectra. A multivariate approach (partial least squares, PLS II) was employed, with the library built using representative "active pharmaceutical ingredient (API)" and "excipient" data points from within the NIRCI data set. Each of the ~22,400 NIR spectra is "scored" using the PLS model derived from the library, and the higher the score, the more of that component present in that pixel location. By arranging the score results by spatial (x, y) coordinates as shown in Figure 4.2a, the variation in relative concentration (abundance) of the API across the tablet is visualized. The histogram representation of the image shown in Figure 4.2b provides a statistical analysis of the PLS score results. The mean value of this distribution is 0.4495, indicating that the mean API abundance determined using this method is ~44.95%. Note that the model and the result are generated using only a single NIRCI data set. A standard NIR analysis of the same sample would give a single number—an API concentration—based on an average across the entire sam-

TABLE 4.1   Relative API Contribution Across 22,400 Locations on a Single Tablet   [See Figure 4.2b]

Normalized PLS Score	Range of Number of Pixels
0.6–1.0	514
0.5–0.6	5941
0.4–0.5	9010
0.3–0.4	5860
0.0–0.3	1077

ple. In addition to providing a mean value, Table 4.1 presents the relative amount of API present in a tablet across ~22,400 different spatial locations. For NIRCI, the question we ask is not simply "how much" but rather "how much at a particular location." NIRCI presents a distribution of values, not just a single one.

There are a variety of analytical and statistical tools that can be used to efficiently analyze these data, and in many instances, subjective and qualitative interpretation can be replaced with automated, robust, quantitative, and reproducible statistical analyses [28]. In addition to determining "calibrationless" component abundance information, one of the most obvious benefits of obtaining thousands of spatially resolved NIR spectra in a single data set is that the spatial heterogeneity of a sample can be characterized. This characterization can provide significant insight into the functionality of complex materials, both naturally occurring and man-made. Table 4.2 presents the mean, standard deviation, skew, and kurtosis of the histogram representation of the image presented in Figure 4.2b. The skew and kurtosis values can be used to characterize component heterogeneity within the sample, providing a reproducible method of

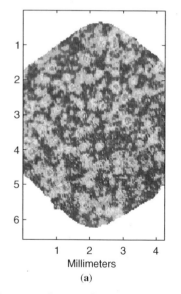

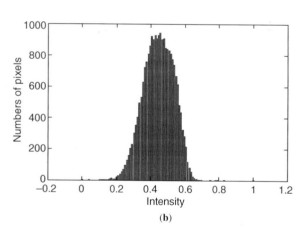

FIGURE 4.2   Presents an image and corresponding histogram derived from a NIRCI data set of a pharmaceutical tablet comprised of ~22,400 NIR spectra. A multivariate approach (partial least squares, PLS II) was employed, and the resulting image highlights the API. By arranging the score results by spatial (x, y) coordinates as shown in (a), the variation in relative concentration (abundance) of the API across the tablet is visualized. The histogram representation of the image shown in (b) provides a statistical analysis of the PLS score results. (See the color version of this figure in Color Plate section.)

TABLE 4.2 Statistical Analysis of 22,400 PLS Score Values Derived from a Single Tablet [See Figure 4.2b]

# Pixels	22402
Mean pixel value	0.4495
STD of distribution	0.0871
Skew	−0.1745
Kurtosis	−0.2248

analyzing this sample characteristic from NIRCI data. Reference 28 provides details on how to assess heterogeneity of samples using these statistical parameters.

In addition to providing a quantitative assessment tool to assess sample mixing, the statistical component of NIRCI analysis can also alleviate calibration and correlative issues that arise for single point NIR spectroscopy. Often, variation among spectra contained within a NIRCI data set, or obtained from pure components, can be used to create a multivariate model. As mentioned previously, the PLS analysis for the sample presented in Figure 4.2 is entirely self-contained. The library was constructed from pixels within the data set (spectra that most closely represented pure component API and excipient spectra), and a "classification" approach to generating a PLS model was utilized. A "classification" model uses a library composed of pure component spectra, rather than a series of spectra spanning a known range of concentrations. These data enjoy a "statistical robustness" due to the large number of individual pixels representing the sample. The resulting model is applied to the sample data to produce score images relating the abundance of the library components. The intensity of each pixel reflects how much of that component is predicted to be present at that spatial

location, and the brighter the pixel, the more content predicted at that location, as seen in Figure 4.2a. As a result, NIRCI can solve many problems spatially without first having to create a separate series of calibration samples that are typical for single point measurements [29].

4.3.4 High Throughput

An advantage of NIRCI is the ability to perform high-throughput measurements on samples, in effect ignoring the spatial dependence of the information [30–32]. Rather than focusing on characterizing the heterogeneity of a single sample, in high-throughput mode the average NIR spectrum of multiple (hundreds to thousands) individual samples can be compared. This can be tremendously useful in a QA/QC environment where large numbers of samples need to be screened or evaluated. Figure 4.3 presents an example of examining a sample in "high-throughput" mode. Unlike the previous examples presented in Figures 4.1 and 4.2, the image in Figure 4.3 highlights the chemical composition of individual pharmaceutical granules, not information relating to the distribution of components within a single granule. Spatial distribution has no relevance; the granules could be lined up in rows and columns or distributed randomly—the same answer would be achieved. The image in Figure 4.3a is created by overlaying images at four distinct wavelengths, once for each component: citric acid 2220 nm, sucrose 2080 nm, flavoring 1940 nm, and acetaminophen 1670 nm. A binary image at each of these marker bands is created, in which the chemical species of interest is thresholded to the value of "1" and anything else is set to "0". These separate binary images are overlaid to generate a four-color composite, where each color represents a separate chemical entity.

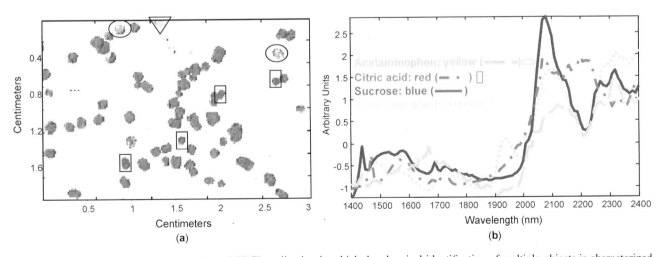

FIGURE 4.3 Example of a high-throughput NIRCI application in which the chemical identification of multiple objects is characterized. (a) A composite binary image made from single channel images at wavelengths characteristic of the four identified sample components: citric acid 2220 nm, sucrose 2080 nm, flavoring 1940 nm, and acetaminophen 1670 nm. (b) Single pixel representative spectra of the four components. Using size and number of the particles, approximate component abundances may be estimated from this image. This results in the following approximation of component abundance: sucrose 88%, citric acid 7%, acetaminophen 3%, and a flavoring component 2%.

TABLE 4.3   Granule Abundance by Component

Component	Component Abundance (%)
Sucrose	88
Citric acid	7
Acetaminophen	3
Flavoring	2

Using size and number, an approximate "abundance" calculation can be performed, the results of which are presented in Table 4.3.

Because the NIR spectra of these four components are well separated, it is possible to generate this image by collecting as few as six wavelengths—one for each component and two wavelengths to allow for baseline subtraction. Using an NIRCI system that employs a random access wavelength filter, it is possible to collect these images in almost real time. The field of view can be optimized depending on the desired sample size, and multiple fields of view can be stitched together to acquire enough particles to provide robust sampling statistics.

4.3.5   Calibration for Chemical Imaging Systems

Accepted strategies for calibration, instrument qualification, and performance verification are important for the implementation of NIRCI as a routine analytical method. Because NIR mapping/imaging systems employ multiple detector elements (pixels) to gather data, a truly quantitative calibration would include an individual calibration of each pixel as a unique detector. The need to characterize the signal-to-noise of each pixel has been presented [33]. Although many of the same procedures used to calibrate traditional NIR spectrometers can be applied to NIRCI, the requirement for standard reference materials (SRMs) that are homogeneous (relative to the size scale resolved by the imaging system) can make the implementation of these procedures difficult. If a standard reference material is not spatially homogeneous at the spatial resolution of the NIRCI system, then each detector will measure something *slightly different*. In characterizing a large number of pixels, one must be able to make the assumption that each pixel is "seeing" an equivalent scene; therefore, the development of truly homogeneous standard reference materials is critical to individually calibrating each detector. It is important that existing NIR spectroscopic standards be evaluated for use in NIRCI applications, primarily by characterizing the extent of spatial heterogeneity.

As an example, Figures 4.4–4.6 show NIRCI data from three reference materials commonly used in NIRS: Spectralon™ (Labsphere, North Sutton, NH, USA) impregnated with carbon black used to establish detector linearity and two different SRMs used for wavelength calibration (NIST 1920a and NIST 2036). Figure 4.4 shows a near-

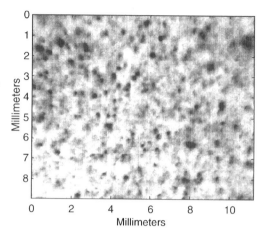

FIGURE 4.4   Near-infrared chemical image of Spectralon™ standard (80% reflectivity), taken with magnification of ~40 μm/pixel. Dark spots are carbon black particles that are clearly resolved at this magnification.

infrared chemical image of the 80% reflectivity Spectralon reflectance standard, which clearly shows the individual particles of carbon black that are used to reduce the overall reflectivity from 99%. This type of standard cannot be used to characterize the linearity of each pixel in a multielement detector, as the individual pixels are not exposed to an equivalent scene.

Figure 4.5 shows NIRCI data of a NIST 1920a SRM used for wavelength calibration [34]. The spectra seen in Figure 4.5a from different spatial locations within the sample highlight real chemical differences across what should be (or may be assumed to be) a homogeneous standard. The corresponding image (at 1530 nm) shows the extent of the spatial/chemical heterogeneities readily apparent with a 40 μm/pixel magnification. This SRM is also not sufficiently homogeneous to permit the characterization of individual pixels on a multielement detector. Figure 4.6 shows results taken with the same system magnification of the NIST 2036 SRM [35]. This standard, a rare earth oxide glass, is highly homogeneous at 40 μm/pixel magnification.

This difference is not only unique to NIRCI, but also impacts Raman- and MIR-based chemical imaging and mapping systems. Once a "uniform" scene has been measured (imaged), the response of tens to tens of thousands of detectors needs to be analyzed. This can be accomplished using a statistical approach, evaluating a mean response and the standard deviation around the mean, and then implementing a method to accommodate pixels (if any) that fall outside of acceptable limits. Approaches to this include excluding pixels entirely or replacing them with nearest neighbor averages. The definition of "acceptable limits" may change depending on the analytical requirements of specific applications and should be determined using scientific principles.

Very often, because spatially homogeneous standard reference materials are not available, or simply because

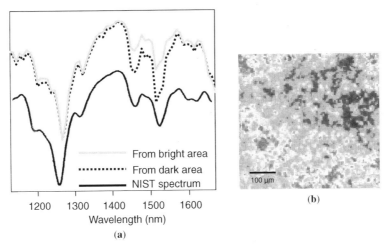

FIGURE 4.5 (a) Spectra from bright and dark areas in the data set compared with the NIST published spectrum of the 1920a SRM. Differences in the spectra are associated with true chemical differences, that is, different concentrations of the rare earth oxide powders at different locations in the SRM. (b) Normalized image at 1530 nm of NIST 1920a NIR standard reference material. Differences in intensity across the image reflect the spatial localization of the different rare earth oxide powder constituents of the sample. This SRM is not spatially homogeneous at this magnification (~40 mm/pixel).

a calibration of each pixel is not required to obtain a qualitative result, values obtained for the calibration of all pixels/detectors are averaged to a single value. However, it is important to consider that there is a *distribution of responses*

(e.g., signal-to-noise ratio) and *not just a single value* for a multidetector spectrometer. Averaging these values throws away important information. Efforts are underway to develop strategies, standard materials, and methods applicable

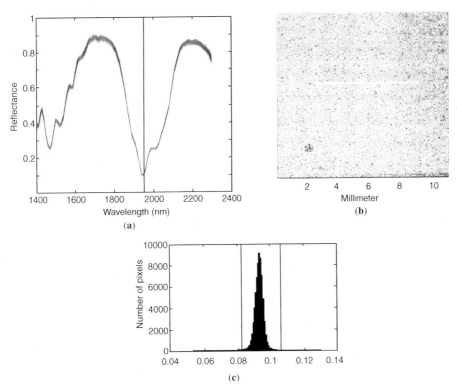

FIGURE 4.6 (a) ~200 overlaid single pixel spectra extracted across the line in the image in panel (b). The sample is NIST 2036 SRM. (b) NIR chemical image (at 1945 nm) of the same sample showing the high level of chemical and spatial homogeneity at 35 mm/pixel magnification. (c) A histogram distribution of the reflectance values at 1945 nm. The mean reflectance is 0.0942 with a standard deviation of 0.0026. These values provide a quantitative metric of chemical homogeneity.

to imaging. In addition to independent efforts, an ASTM subcommittee on Molecular Spectroscopic Optical Imaging (E13.10) has been working toward the development of consensus standards for NIR, Raman, and MIR imaging systems.

Basic instrument calibration of traditional NIR spectrometers consists of *x*- and *y*-axis (wavelength and intensity, respectively) calibration. Other parameters (such as spectrometer noise, limits of detection, linearity, etc.) are helpful in determining that an instrument is performing as specified [36]. Imaging-only considerations for method validation include spatial resolution, field of view and magnification, uniformity of illumination, and the extent of optical aberrations. An instrument qualification should be structured to make sure that the system is working to reasonable specifications; that is, it is fit for purpose. This type of check is performed upon installation, and perhaps after a repair or upgrade. There are several publications that deal with the unique issues associated with the calibration of NIRCI systems [37, 38].

A similar but different consideration is performance verification, in which the general performance of an instrument is verified. This level of system verification may occur on a daily basis, or as a quick check before data are collected. Again, it should be emphasized that data obtained from a NIRCI system represent a distribution of values across the number of pixels/detectors being used, and that reporting these values as a mean and accompanying standard deviation provide valuable information about the system performance.

The high-throughput capabilities of NIRCI discussed previously (see Section 4.3.4) enable calibration samples or pure components to be in the same field of view as the sample. Several groups have taken advantage of this capability for calibrating NIRCI systems [30–32, 38, 39]. Creating single data sets that contain both sample and reference materials simplifies many aspects of the experiment. The data are "internally" consistent because reference and sample data are measured simultaneously by the same instrument— temporal variations in instrument response are automatically accounted for. This minimizes the need to apply transfer of calibration procedures to analyze samples on other instruments [30–32]. As an example, Figure 4.7 shows a NIRCI data set in which three pure components are positioned in the same field of view as the sample. The pure components are three rectangular compacts of white powder, and the contrast in the image is derived from multivariate analysis that assigns shades based on a multivariate response. Data were collected on the sample and the pure components at the same time.

4.4 INSTRUMENTATION

4.4.1 Collection Modes

NIRCI can be collected one spectrum at a time (point mapping), one line at a time (line mapping), or one image at a time

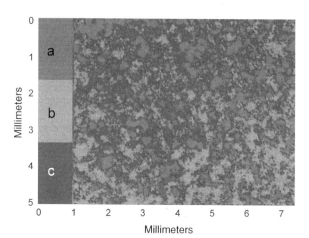

FIGURE 4.7 Chemical image of a three-component mixture: (a), (b), and (c) are the areas holding the pure components and (d) is the area comprising the mixture. In this case, the pure (white powder) sample components were pressed into rectangular compacts and placed in the same field of view as the mixture sample. Applying a PLS model based on the pure materials results in the three-channel composite image shown. This highlights the inherent chemical heterogeneity of the mixture and the chemical differences between each of the compacts. (See the color version of this figure in Color Plate section.)

(global imaging). In the remote sensing world, these modes are known as whisk broom, push broom, and staring, respectively. Table 4.4 attempts to enumerate common implementations of these three data collection methods, giving references to some of the corresponding seminal work.

As with most analytical techniques, there are a number of practical considerations that influence the suitability of specific chemical imaging instrumentation and methods for particular applications. To proceed rationally in developing a method, the analytical problem must be defined, and the data requirements to derive an analytical solution must be well characterized. Factors such as the desired spatial resolution, number of individual measurements, magnification, sample size, the amount of time available for data acquisition, the type of sample, whether the sample is stationary or moving, will all influence selection of the optimal technique and sample presentation. Although a 2D imaging approach will work equally well for a high-throughput, low magnification QA/QC screening application as it will for a microscopy application (such as the detection and identification of impurities within a single sample), in reality different experimental implementations (different modes) might be employed for those two applications. The following sections describe common implementations of NIRCI and specific components that comprise these instruments.

4.4.1.1 Point Mapping   As mentioned earlier, in contrast to its MIR and Raman counterparts, NIR microspectroscopy

TABLE 4.4   Varieties of NIRCI Implementations and Corresponding Early Publications

Device for Wavelength Separation	Detector	Source	Scan Sample or Wavelength	Source/Image Filtered	Early Reference
Point mapping					
Grating spectrograph	Point	Broadband	Both	Source	[6] (1990): Fiber probe, micropositioner, grating spectrometer, PbS detector, 1100–2500 nm
Grating spectrograph	Array	Broadband	Sample	Source	[13] (1997): Fiber probe, grating spectrograph, CCD detector, 508–1026 nm
FT-NIR Michelson interferometer	Point	Interferometer	Both	Source	[15] (1998): IR microscope, Michelson interferometer, 7500-4200 cm^{-1}
Tunable OPO	Point	Tunable OPO	Wavelength	Source	[16] (1998): OPO tunable source, PbS detector, fiber probe, 1400–4100 nm
Line mapping					
FT-NIR Michelson interferometer	Linear array	Interferometer	Both	Source	[23] (2003): IR microscope, Michelson interferometer, 7500-4200 cm^{-1}
Prism/grating/prism spectrograph	Array detector	Broadband	Sample	Image	[10] (1992): prism/grating/prism spectrograph, CCD, 650–1100 nm
Grating spectrograph	Array detector	Broadband	Sample	Image	[9] (1992): grating spectrograph, NIR tube camera, 900–1900 nm
Global Imaging					
FT-NIR Michelson interferometer	Array detector	Interferometer	Wavelength	Source	[12] (1996): Michelson interferometer, InSb FPA, 1000–5500 nm
Tunable filter (AOTF)	Array detector	Broadband	Wavelength	Image	[8] (1992): AOTF/CCD, 400–1200 nm [40] (1994): AOTF/InSb, 1000–2400 nm
Tunable filter (AOTF)	Array detector	Broadband	Wavelength	Source	[17] (1998): AOTF/InGaAs FPA/ source filtering, 1000–1700 nm
Tunable filter (LCTF)	Array detector	Broadband	Wavelength	Image	[20] (2001): LCTF, InGaAs FPA, 1100–1700 nm.
Hadamard pattern mask	Point detector	Broadband	Both	Source	[11] (1995): Hadamard transform/ FT-IR spectrometry Michelson interferometer, 7500–4200 cm^{-1}
Digital micromirror array	Point detector	Broadband	Both	Source	[41] (2000): digital mirror array/ grating spectrometer/point detector
Tunable laser (OPO)	Array detector	Tunable OPO	Wavelength	Source	[14] (1997): OPO, InSb FPA
Linear variable filter	Array detector	Broadband	Sample	Image	[42] (2004): linear variable filter, NIR FPA, 1500–2300 nm
Interference filter	Array detector	Broadband	Wavelength	Image	[7] (1990): filters/CCD, 400–1100 nm

studies, with no mapping component, are relatively unusual [43–46]. Similarly, there are only a few early references to NIR point mapping, and these were implemented using a grating spectrometer coupled to a fiber microprobe [6, 13], by illuminating the sample with a tunable OPO [16], and coupling a FT-NIR interferometer to a NIR microscope [15, 47]. Figure 4.8 shows a schematic implementation of point mapping with a grating spectrometer (a) and an interferometer (b).

NIR point mapping was implemented by Lodder et al. [6] by coupling the output of a grating spectrometer into a fiber optic using a compound parabolic concentrator. The resulting spot of collimated light ($\sim$0.74 mm^2) was directed onto the sample surface, and the spectrometer scanned from 1100 to 2500 nm. A single point PbS detector mounted off-axis at the proximal (input) end of the fiber was used to collect diffusely scattered light resulting from the interaction with the sample. The fiber probe could be moved in 10 µm increments along

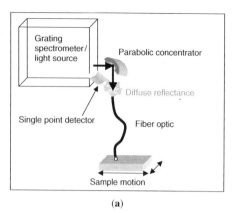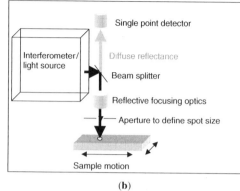

(a)   (b)

FIGURE 4.8 Schematic of point mapping NIRCI system in which the output of a grating spectrograph is directed onto a sample through a fiber optic. The output of an interferometer is condensed to a small spot on a sample for point mapping NIRCI.

the sample using a micropositioning stage. Another point mapping approach implemented by Lodder et al. [16] was to illuminate a sample through a fiber-optic bundle with the narrowband light output of a Nd:YAG-pumped OPO. This system generated tunable NIR light from 1.4 to 4.1 μm, with an effective power of 3.3 MW. In this implementation, the fiber bundle was split into illumination and collection fibers, and light scattered from the sample was returned to a PbS detector. In both these implementations, the size of the sample spot is defined by the output of the illuminating fiber.

NIR point mapping can also be implemented by coupling the output of a FT-NIR interferometer to an optical microscope that utilizes all reflective optics to eliminate chromatic aberration. The microscope is used to view and position the sample and, most importantly, to condense the incoming radiation into a small spot using optics and masking apertures. After interaction with the sample, the light is collected by the microscope and focused onto a standard single point detector. The spatial extent from which data are collected is defined by the size of the illumination spot and the projection of the detector through the collection optics onto the sample. The size of the illumination spot is determined by adjusting masking apertures before and, in some cases, after interaction with the sample [4].

In the point mapping approaches discussed above, by translating the sample in small increments along both x and y spatial axes, an area on the sample can be imaged. Image fidelity is determined by illumination spot size (via a fiber or microscope), step increment in the x and y directions, and the total number of points that are collected.

NIR point mapping, relatively uncommon to begin with, has been superseded by line mapping and global imaging. This is due to the much shorter data collection times, higher image fidelity, and relative economy available in these other implementations.

4.4.1.2 Line Mapping

Line mapping approaches fall between point mapping and global imaging in terms of the

number of spatial locations that can be collected simultaneously. These techniques sweep a line across the sample, and hence the "push broom" moniker. The most common implementations utilize either an FT-NIR interferometer or an imaging spectrograph based on a dispersive optical element. Figure 4.9 shows a schematic implementation of line mapping with a spectrograph.

FT-NIR-based line mapping is implemented in the same way as for the point mapping approach discussed above, but the sample is illuminated over an area, and the resulting light is focused through a linear aperture onto a linear array. The area of the sample from which data are collected is determined by the number of detector elements, typically 16 or 32 pixels, their size and spacing, and overall system magnification. By translating the sample along both x and y spatial axes, a larger area on the sample can be imaged.

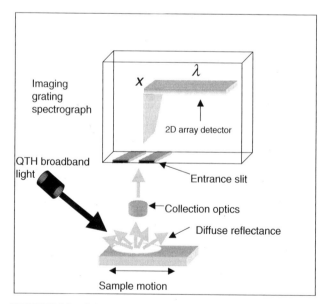

FIGURE 4.9 Schematic of line mapping NIRCI system where the wavelength dispersing element is an imaging grating spectrograph.

Data can be collected with unequal x and y spacing, or the distance that the translation stage moves can be matched to the detector element spacing imaged upon the sample to create a data set with the spacing for the two axes being equal. Data are collected more rapidly than for point mapping approaches. It is still a mapping technique though and does necessitate moving the sample along the x and y axes to build up area coverage across a sample.

A typical line mapping spectrograph approach to NIRCI is implemented as follows: the sample is illuminated with broadband light and after interacting with the sample, the resulting diffusely reflected NIR signal is collected with optics (lenses or fiber optics) and focused onto the input slit of the spectrograph. The dimensions of the input slit and the magnification of the input optics define the spatial extent of the sample field of view (which is linear). This line of light is accepted into the spectrograph and dispersed across a 2D NIR focal plane array (FPA). Ultimately, one axis of the FPA encodes the spatial information from the sample, while the orthogonal axis records the spectral information from the line image. Additional spatial information is built up by scanning the sample under the input optics, or less commonly by scanning the input optics across the sample [48], or moving the spectrograph and camera combination across a stationary sample [42].

4.4.1.3 Global Imaging

4.4.1.3 Global Imaging   The most common global imaging approach for NIRCI is to employ an image quality LCTF. A broadband source illuminates the sample, and light that has interacted with the sample is collected with a series of imaging optics, and passed through the tunable filter, before being imaged onto a NIR FPA. The sampling area is highly configurable, typically ranging from 3×3 mm ($\sim 10 \times 10\,\mu m$ per spectrum) on a side to $10 \times 10\,cm$ ($\sim 330 \times 330\,\mu m$ per spectrum) for standard commercial configurations. Figure 4.10 shows a schematic of this implementation.

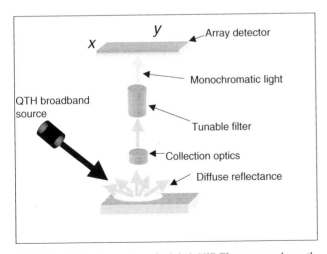

FIGURE 4.10   Schematic of global NIRCI system where the wavelength dispersing element is a tunable filter.

The AOTF, although it is more commonly employed for remote sensing applications, has also been used as an imaging tunable filter for global NIRCI in the laboratory [8, 17, 40]. Employed as an image filter [8], it is implemented much like the LCTF described above, where the diffusely reflected light from a sample that has been illuminated with broadband light is wavelength filtered upon passing through the AOTF. The resulting narrow wavelength image is focused onto a NIR FPA. The AOTF has also been used as a source filter [17], in which broadband light is wavelength filtered by the AOTF before impinging on the sample.

FT-based global imaging is not commonly implemented, particularly in comparison to its FT-MIR global imaging counterpart. The approach is identical to FT-NIR point and line mapping, except that the output of the interferometer is used to illuminate the entire sample area. For commercially available systems, the standard image area is typically $400 \times 400\,\mu m$ or $4 \times 4\,mm$. Although initially implemented using a step-scan interferometer [49], most commercial instruments now employ a rapid scan approach [50].

Another source filtered approach is global NIRCI with OPO laser illumination [16, 51]. The narrow wavelength output of an OPO is directed onto a sample. Optics collect the resulting diffusely scattered light after interaction with the sample and focus this onto a NIR FPA. The output of the OPO is scanned across a desired wavelength range, building up the data cube one wavelength at a time. The OPO light can be delivered onto the sample through refractive optics or fiber optics over a broad range of sample areas determined by the optics for sample illumination and the collection/focusing optics that transfer the resulting light onto the FPA.

A novel approach using a Hadamard encoding mask was implemented as a relatively inexpensive way to perform chemical imaging [11]. A later improvement on the original concept was the use of a digital micromirror array (DMA) to create the Hadamard pattern [41]. These approaches were never commercialized.

4.4.2 Illumination

For remote sensing applications, broadband illumination is most often provided by the Sun. In a laboratory setting, an alternate source of NIR light must be employed. This creates a bit more flexibility in how the illumination is provided to the sample, broadband (like the Sun) or prefiltered in narrow wavelength bands. These implementations are referred to as "image" and "source" filtering, respectively. This is comparable to the distinction of post- and predispersive systems for traditional NIRS [3]. For "image filtered systems," broadband NIR light impinges on and interacts with the sample, and the resulting light is wavelength filtered before being focused onto a detector. Most grating/prism-based systems, as well as tunable filter approaches, are image filtered. In contrast, FT-NIR-based NIRCI systems are source filtered, as broadband light is passed through an interferometer before interacting with the sample. Collecting NIRCI data by illuminating the sample with the narrowband output from an OPO is also a source filtered implementation.

No matter the chosen implementation, there are several basic criteria that must be met to optimize illumination for mapping and imaging approaches. Sample illumination should be intense enough to provide adequate signal across the sampling area, but not so intense as to damage the sample. It should also be relatively uniform across the field of view, as uneven illumination can cause the spectral quality of data collected at different points on the sample to vary significantly. Variations in signal-to-noise characteristics of data across maps and images can be particularly problematic because many of the processing methods applied to these data make inherent assumptions about the equivalence of all the data. When "viewing" chemical maps or images, the quality of the underlying spectral data may not be apparent. These considerations are mostly relevant for global imaging or line mapping systems since both criteria—appropriate power and uniformity—are usually met for point mapping systems. However, temporal drift in illumination power across separate data points in point maps has the same effect as uneven spatial illumination for line mapping and global imaging. For global imaging with OPO illumination, the variation in pulse-to-pulse power across the scanned wavelengths has a similar effect.

For interferometer-based imaging and mapping systems, a broadband quartz tungsten halogen (QTH) source is most common, covering the range 14,000–2,800 cm^{-1} (0.7–3.5 μm). The source is coupled to a FT-NIR spectrometer and the resulting modulated light is used to irradiate the sample. Originally designed for single point spectroscopy, conventional FT-NIR sources are not ideal for imaging, particularly for larger samples where total power with uniform distribution can become one of the dominant performance-limiting factors. Using multiple filtered sources for FT-based spectroscopic imaging presents practical and economic limitations since multiple spectrometers are required.

For tunable filter global imaging (LCTF or AOTF), and imaging spectrograph-based (grating/PGP) line imaging, the most common implementation is also a QTH source, but rather than being filtered through an interferometer first, it impinges directly (without modulation) and illuminates the entire sample area. The illumination delivered to the sample is easily optimized by the addition of more sources, if required. NIR global imaging is particularly amenable to studying large sample areas because of this flexibility.

Global imaging employing narrow wavelength illumination of a sample with an OPO tunable laser has only recently been commercially available. (OPOTEK, Inc., Carlsbad, CA, USA) OPOs are tunable across the entire NIR spectral range, with a tuning bandwidth of ∼3 nm from 1400–2500 nm. Very high powers (2.5 mJ peak energy) can be delivered in a single pulse, so expanding the beam using appropriate optics to illuminate large sample areas is possible. The sampling area illuminated is flexible, although care must be taken when probing *small* sample areas to appropriately attenuate the output power to prevent sample damage. Because of the nonuniformity in pulse-to-pulse power, it is necessary to have calibrated intensity standards in the field of view to appropriately scale the resulting diffuse reflectance. However, there is also spatial variation (speckle) that varies from pulse to pulse, and it can be difficult to adequately flat field the resulting image. Strategies employed to optimize the illumination uniformity include employing a high-frequency wobble on the fiber to blur the speckle and inserting an integrating sphere in the path between the OPO and the sample [52].

4.4.3 Optics

Collection and image formation optics are critical in the performance of an imaging instrument. Chromatic aberration, working distance, magnification, numerical aperture, depth of penetration, spatial resolution, and field of view are important attributes that should be considered.

One of the chief differences in the optical implementation of NIRCI systems is whether reflective or refractive optics are used: reflective optics eliminate chromatic aberration but are relatively bulky, whereas refractive optics are extremely flexible and compact, but are achromatic only over a relatively narrow wavelength range. Reflective optics predominant for FT-based NIRCI systems, and refractive optics are favored for most other implementations. FT-NIRCI systems are based on FT-MIR imaging systems, where the wavelength range that must be spanned is an order of magnitude greater than in the NIR. Only reflective optics have the achromatic performance to operate in both the NIR and the MIR, so the most practical solution for FT-based NIRCI is to employ reflective lenses.

In contrast, NIR global imaging systems that use tunable filters and PGP spectrographs employed for line mapping are tailored to specific and narrower wavelength ranges and may therefore employ appropriately designed refractive optics. Refractive optics offer greater flexibility in field of view, as there tend to be more "off-the-shelf" options. However, when compared to the visible region where a microscope is designed to provide good imaging performance over only ∼300 nm (see Table 4.5), the NIR that extends about ∼1800 nm creates a more significant optical design challenge.

4.4.4 Wavelength Filters

The selection of "spectrometer" (wavelength filter) is the most significant factor impacting the overall performance and behavior of a NIRCI system as it determines optical throughput, spectral resolution, and in numerous cases the

TABLE 4.5 Typical Image Formation Optic-Type Used, Based on Instrumental Wavelength Range

Spectral Region	Wavelength Limits	Wavelength Range	Optic Type
Visible	400–700 nm	300 nm	Refractive
NIR	700–2500 nm	1800 nm	Reflective (FT)/ refractive (PGP, tunable filter)
MIR	2500–25,000 nm	22,500 nm	Reflective

useable spectral range. Thus, a discussion of the available options will be a primary focus of this chapter.

4.4.4.1 Liquid Crystal Tunable Filter

An LCTF is a Lyot filter in which the wavelength is electronically tunable [53]. It is a solid-state device with no moving parts and is used for global NIR imaging. A typical wavelength range accessible to these filters is 950–2450 nm, although two filters in tandem are required to cover this range. One filter can be used in the range 950–1900 nm, and the other covers the range 1000–2450 nm. The spectral resolution of these filters is constant with wavenumber but variable with wavelength and is typically 7 nm at 1650 nm and 9 nm at 1900 nm.

The tuning time between discrete wavelengths is ∼100 ms, independent of the size of the wavelength step. The advantages of these filters are ruggedness (no moving parts and solid state), random wavelength access, rapid tuning, and excellent image quality with an on-axis (direct-view) optical configuration. Tunable filters operate in either hyperspectral or multispectral mode and can be tuned rapidly. This opens up the possibility of real-time imaging, in which a few analytical relevant wavelengths are collected in a few seconds.

The LCTF requires collimated input that acts to limit the numerical aperture of the detection system. Also, the filters are polarization sensitive, and as the device functions as a blocking filter, the transmission varies from <5% to ∼30% across the spectral range. Because of the polarization sensitivity of the LCTF, specular reflectance (i.e., glare from shiny surfaces) can be eliminated by polarizing light used to illuminate the sample.

4.4.4.2 Interferometers

Michelson Fourier transform interferometers are used for point mapping, line mapping, and global imaging. A series of data points are acquired as the moving mirror translates through the optical retardation defined by the wavelength resolution stipulated: the higher the resolution, the greater the overall mirror displacement. An interferogram is acquired for each spatial location, and the spectra are recovered via a fast Fourier transform (FFT). The shorter the wavelengths of light used, the more closely spaced the points must be in order to avoid aliasing, that is, a mixing of these shorter wavelengths into the longer wavelength data [54]. Compared to the MIR where undersampling

is possible, data sets in the NIR can have twice to three times the number of data points for the same spectral resolution (in cm⁻¹). NIR imaging data sets can therefore be quite large, especially if acquired over thousands of spatial locations.

If high spectral resolution data are required, an interferometric technique is the method of choice, as it will provide the highest available spectral resolution. However, as discussed earlier, for the NIR range, these higher spectral resolutions are often not necessary. Also, if the desired information is contained in only a few wavelengths, the interferometer may be at a disadvantage, since there is no capacity to operate in multispectral mode. Also, as the operation of commercial NIRCI interferometers is based on a moving mirror approach, they are best suited for environmentally controlled (laboratory) settings and are not as mechanically robust as the other filtering approaches addressed here.

4.4.4.3 Grating/Prism/Grism

There are several variations in the implementation of this approach: single diffracting elements such as planar gratings and prisms have been successfully employed, and compound dispersing elements such as a grating–prism (grism) or prism–grating–prism (PGP), as well as convex gratings [55], have been designed to overcome spatial and spectral distortions (smile and keystone, respectively), and to reduce the overall optical complexity and path length. Both grism and PGP implementations permit a "direct-view" configuration, with a linear optical path through the spectrograph, permitting the construction of a compact, linear imaging spectrograph [2, 56].

One commercially available PGP NIR imaging spectrograph spans the wavelength range 1000–2500 nm, and with a 30 μm slit provides a spectral resolution of 8 nm. This implementation is very robust, with no moving parts, and has been used in process environments. As with an interferometer, a PGP spectrograph has no capacity to operate in multispectral mode, but as typical data collection times can be quite rapid, this may not be an important consideration.

4.4.4.4 Acousto-Optic Tunable Filter

The AOTF is a solid-state, electronically tuned wavelength filter. The filter is comprised of an optically clear crystal, typically TeO_2, for operation in the NIR (400–1900 nm), to which a piezo-acoustic transducer has been bonded. By applying an rf frequency to the transducer, an acoustic wave is propagated through the crystal that sets up the equivalent of a transmission grating within it. The wavelength is tuned by varying the applied rf frequency. The time to tune between wavelengths is theoretically on the order of 5 μs, but 20 μs is more commonly achieved in practice [57]. The bandpass of the filter varies with wavelength, spanning 5–20 nm from 900–1700 nm. Like the LCTF, the device is polarization sensitive and requires collimated input, and because the center wavelength can be randomly accessed, an AOTF can operate in multispectral or hyperspectral mode. The input aperture is typically quite small, which can further reduce throughput. However, the filter is a robust, solid-state device and suitable for process environments. Although the optical path of an AOTF is not "direct view," relatively compact systems have been designed [58, 59].

4.4.4.5 Bandpass Filters

Filter wheels holding a selection of bandpass filters can be used for spectral imaging. While it is possible to tune the bandpass of a dielectric filter by tilting it, such a solution is relatively inflexible and may introduce image shift as well. Without this option however, one is left with only a series of discrete wavelengths. Tuning speed is limited by the mechanical operation of translating the filters in and out of the optical path. If only a few wavelengths are required to solve an analytical problem, this approach can be very efficient, but it provides limited flexibility for method development.

4.4.5 Detectors

More information about the development and characteristics of NIR detectors may be found in Refs 1 and 59. Detectors used for single point mapping are the same as those used for single point spectrometers. There are a variety of detectors, and typical systems might contain MCT, InSb, or indium gallium arsenide (InGaAs). Linear arrays used either in dispersive spectrographs or as line detectors in FT-NIR mapping systems are more typically MCT, InGaAs, and PbS. NIR global imaging systems and PGP line mapping systems require the use of two-dimensional detectors called focal plane arrays. NIR global imaging instruments commonly employ InSb arrays in a 320×256 (or 240) ($\sim$1100–5000 nm) format and InGaAs in a 320×256 (or 240) format ($\sim$900–1700 nm) range. Although most commonly employed for the MIR, MCT arrays (320×256 format) have also been developed to operate in the NIR ($\sim$1–2.5 μm, 1000–4000 cm^{-1}). Depending on wavelength range, and intended application, these detectors operate in uncooled, temperature stabilized, TE cooled, Stirling cooled, or liquid nitrogen cooled modes.

4.5 OPTIMIZING EXPERIMENTAL SUCCESS: PRACTICAL CONSIDERATIONS

4.5.1 Spatial Resolution and Magnification

Despite the similarities between conventional and imaging spectroscopies, there are particular considerations unique to imaging. For example, it is confusing when one refers to the "resolution" of an imaging spectrometer. Unlike standard spectrometers where resolution refers exclusively to *spectral* resolution, imaging spectrometers are characterized by both *spectral* and *spatial* resolution.

The concepts of spatial resolution and magnification are often confused, and the terms used interchangeably. Spatial resolution describes the smallest size objects that can be distinguished and is influenced by immutable constraints such as the diffraction limit of light, design considerations such as aberrations introduced by the optics, and even experimental considerations such as depth of penetration [61], sample type, and wavelength of probe radiation. Magnification, on the other hand, describes the size of an image relative to the original object. Simply switching to a higher magnification optic will not permit particles that are smaller than the spatial resolution of the system to be optically resolved. Magnification beyond the spatial resolution of the system is simply oversampling or "empty magnification" and only serves to reduce the area of a sample that can be imaged at the limit of resolving power of the optical system. In summary, the smallest size particle that can be clearly distinguished is ultimately determined by the system spatial resolution and not by the magnification employed.

4.5.2 Detection Limit

The detection limit for imaging spectroscopy is quite different from that for bulk spectroscopy and is highly sample dependent because in imaging, dilution occurs on a pixel-by-pixel level, rather than across the entire. In other words, if the spectrum of a trace particle is detectable in a single pixel, that component will be above the detection limit for a NIRCI system. This does not hold for a standard NIR measurement where the spectral responses of the contaminant and other sample components are averaged. If, however, the same amount of that trace component is dispersed equally throughout a sample, rather than being localized at a single pixel, NIRCI no longer has an advantage in detecting the contaminant. Therefore, detection limits of chemical imaging techniques are strongly influenced by particle size and the chemical and spatial heterogeneity of the sample.

4.5.3 Sampling and Samples

The utility of imaging lies in the ability to resolve spatial heterogeneities in solid-state samples. Imaging liquids or even suspensions have limited use, as constant sample motion or mixing at the molecular level serves to average spatial information. One possible exception is the use of high-throughput capabilities of imaging to screen multiple liquid or suspension samples, an application that relies on parallel acquisition of thousands of spectra to compare differences between samples, rather than exploring spatial heterogeneity within a single sample. Also, there is no benefit in imaging a truly homogeneous solid sample, as a standard spectrometer will generate the same information. Of course, the definition of the "extent" of homogeneity is ultimately dictated by the sample type, the spatial resolution of the imaging system, and also the magnification that is employed.

4.5.4 Data Analysis and Chemometrics

Because image contrast is based on intrinsic sample characteristics (the near-infrared spectral signature of the components), images are derived without tagging or staining of the sample. Contrast can be generated from many different spectroscopic parameters: peak height, peak area, wavelength shift, baseline variations (scattering differences), and other measures derived from multivariate analyses.

Data analysis methods for chemical imaging data sets typically begin with the same steps as for single point spectroscopy; preprocessing is utilized to separate chemical and physical effects, unless of course an analysis is *based on* physical effects (such as scattering differences seen as baseline variations). The next step involves the segmentation of components within an image, that is, being able to adequately separate sample components of interest. If the sample is comprised of components whose spectral features are well separated, it is possible to proceed with a univariate (single wavelength/marker band) approach. However, given the nature of diffuse reflectance NIR spectroscopy, spectral overlap is typically significant. Also, a univariate approach utilizes only a small fraction of available data, and in many cases a multivariate approach can improve results by including a greater proportion of the data. There are differences in how this approach is implemented for traditional NIRS compared to NIRCI. Building a multivariate model for traditional NIRS most typically commences with gathering spectra from a series of reference samples whose component concentrations vary to mimic the range expected from the unknowns. These spectra are used to create a library and subsequently to determine an appropriate quantitative multivariate model directly relating spectral profiles to component concentrations. An alternate approach is most often employed for NIRCI, in which "qualitative" classification models are developed from a library composed of *pure component* spectra. The inherent variation in "pure component" spectra across a large number of individual detectors provides inherent statistical robustness. By employing the "qualitative" classification approach, it is possible to estimate quite reasonably relative component abundance, using only pure component spectra as a starting point.

Once segmented through uni- or multivariate methods, for samples that are significantly heterogeneous, standard image processing tools such as morphological filtering and particle statistics may be applied directly. It may even be possible to derive pure component reference spectra for "qualitative" classification directly from the "unknown" sample data set, in which case there is no need to acquire additional "calibration" data. A model is developed and subsequently applied to the same data set.

As individual pixel spectra become increasingly well mixed, analytical strategies tend to converge with more traditional methods, and also to more heavily rely on multi- rather than univariate methods. However, constructing reliable mixture calibration data sets (rather than building a library of pure component spectra and applying a qualitative approach) for chemical imaging can be challenging, since as discussed in the Section , each pixel must see exactly the same composition for this to be a valid approach. Achieving homogeneity at the distance scale of NIR imaging (typically ~30–40 μm) is challenging. Because of this, the application of "qualitative" classification methods is still the favored approach although the spectra used to create the library are gathered from separate pure component data—from a standard spectrometer, separate data sets, or an isolated area in the field of view.

4.6 CONCLUSIONS

In contrast to its MIR and Raman counterparts, NIRCI did not evolve from point mapping approaches. Given the emphasis by the single point NIR spectroscopic community on bulk and averaged properties, the potential capabilities of acquiring mapping and imaging NIR data sets were not widely recognized. As different technological innovations emerged throughout the 1990s, a variety of groups began to explore the potential of line mapping and global implementations of NIRCI. The field and its applications began to blossom in the early 2000s with the commercial availability of global imaging (LCTF) and line mapping (PGP and FT-NIR) NIRCI instruments.

Since that time NIRCI has been applied to a variety of laboratory- and process-based areas, and recognition of the unique capabilities of NIRCI systems has grown. NIRCI has the capability to qualitatively and quantitatively characterize the chemical distribution within a sample, providing unique insights into bulk sample properties not easily characterized by techniques that average physical and chemical information. The increasing sample throughput of the instruments, coupled with the enhanced sophistication of data analysis and calibration strategies, has ensured that the resulting conclusions based on imaging data are more robust.

The available experimental implementations provide a variety of options, and user choice can be guided by aligning system attributes with the experimental goals. NIRCI systems can be optimized for high-spectral resolution laboratory work, or repetitive process monitoring of samples moving on a web, and any number of scenarios in between.

REFERENCES

1. Colarusso, P., Kidder, L. H., Levin, I. W., Fraser, J. C., Arens, J. F., and Lewis, E. N. (1998). Infrared spectroscopic imaging:

from planetary to cellular systems. *Appl. Spectrosc.* **52**, 106A–120A.

2. Aikio, M. (2001) Hyperspectral prism-grating-prism imaging spectrograph, PhD Thesis, University of Oulu, Publ. 435, VTT Publications, Espoo.

3. Burns, D. A. and Ciurczak, E. W. (Eds.) (2007) *Handbook of Near-Infrared Analysis*, CRC Press, Boca Raton, FL.

4. Messerschmidt, R. G. and Harthcock, M. A. (Eds.) (1988) *Infrared Microspectroscopy: Theory and Applications*, Marcel Dekker, New York.

5. Humeck, H. J. (Ed.) (1995) *Practical Guide to Infrared Microspectroscopy*, Marcel Dekker, New York.

6. Lodder, R. A., Cassis, L. A., and Ciurczak, E. W. (1990). Arterial analysis with a novel near-IR fiber-optic probe. *Spectroscopy* **5**, 12–16.

7. Taylor, S. K. and McClure, W. F. (1990) NIR imaging spectroscopy: measuring the distribution of chemical components. In: Iwamoto, M. and Kawano S. (Eds.) *Proceedings of the 2nd International NIRS Conference*, Korun, Tokyo, 1989, pp. 393–404.

8. Treado, P. J., Levin, I. W., and Lewis, E. N. (1992). Near-infrared acousto-optic filtered spectroscopic microscopy: a solid state approach. *Appl. Spectrosc.* **46**, 553–559.

9. Robert, P., Bertrand, D., Devaux, M. F., and Sire, A. (1992) Identification of chemical constituents by multivariate near-infrared spectral imaging. *Anal. Chem.* **64**, 664–667.

10. Aikio, M. (1992) An optical component. Finnish Patent 90,289.

11. Bellamy, M. K., Mortensen, A. N., Hammaker, R. M., and Fateley, W. G. (1995). NIR imaging by FT-NIR-HT spectroscopy. *NIR News* **6**, 10–12.

12. Lewis, E. N., Gorbach, A. M., Marcott, C., and Levin, I. W. (1996). High-fidelity Fourier transform infrared spectroscopic imaging of primate brain tissue. *Appl. Spectrosc.* **50**, 263–269.

13. Munro, H., Novins, K., Benwell, G., and Mowat, A. (1997). Interactive exploration of spatially distributed near infrared reflectance data. In: Proceedings of GeoComputation '97 and SIRC '97, Otago, New Zealand, August 1997, pp. 345–353.

14. Dempsey, R. J., Cassis, L. A., Davis, D. G., and Lodder, R. A. (1997) Near-infrared imaging and spectroscopy in stroke research: lipoprotein distribution and disease. In *Imaging Brain Structure and Function: Emerging Technologies in the Neurosciences, Ann. N. Y. Acad. Sci.* **820**, 149–169.

15. Hammond, S. V. (1998) NIR microspectroscopy and the control of quality in pharmaceutical production. *Eur. Pharm. Review* **3**, 4–51.

16. Cassis, L. A., Yates, J., Symons, W. C., and Lodder, R. A. (1998). Cardiovascular near-infrared imaging. *J. Near Infrared Spectrosc.* **6**, 21A–25A.

17. Tran, C., Cui. Y., and Smirnov, S. (1998). Simultaneous multispectral imaging in the visible and near-infrared: applications in document authentication and determination of chemical inhomogeneity of copolymers. *Anal. Chem.* **70**, 4701–4708.

18. Martinsen, P. and Schaare, P. (1998). Measuring soluble solids distribution in kiwifruit using near-infrared imaging spectroscopy. *Postharvest Biol. Technol.* **14**, 271–281.

19. Lu, R. and Chen, Y. R. (1998). Hyperspectral imaging for safety inspection of food and agricultural products. *Proc.* SPIE **3544**, 121–133.

20. Lewis, E. N., Carroll, J. E., and Clarke, F. M. (2001). NIR imaging: a near-infrared view of pharmaceutical formulation analysis. *NIR News* **12**, 16–18.

21. Lawrence, K. C., Windham, W. R., Park, B., and Buhr, R. J. (2001). Hyperspectral imaging for poultry contaminant detection. *NIR News* **12**, 3–6.

22. Kulcke, A., Gurschler, C., Spöck, G., Leitner, R., and Kraft, A. (2003) On-line classification of synthetic polymers using near-infrared spectral imaging. *J. Near Infrared Spectrosc.* **11**, 71–81.

23. Clarke, F. and Hammond, S. V. (2003) NIR microscopy of pharmaceutical dosage forms. *Eur. Pharm. Rev.* **1**, 41–50.

24. Reich, G. (2005) Near-infrared spectroscopy and imaging: basic principles and pharmaceutical applications. *Adv. Drug Deliv. Rev.* **57**, 1109–1143.

25. Taghizadeh, M., Gowen, A., O'Donnell, C. P., and Cullen, P. J. (2008) NIR chemical imaging for the food industry. In: Heldman, D. R. (Ed.), *Encyclopedia of Agricultural, Food, and Biological Engineering*, Taylor and Francis, New York.

26. Gendrin, C., Roggo, Y., and Collet, C. (2008) Pharmaceutical applications of vibrational chemical imaging and chemometrics: a review. *J. Pharm. Biomed. Anal.* **48**, 533–553.

27. Gowen, A. A., O'Donnell, C. P., Cullen, P. J., and Bell, S. E. J. (2008). Recent applications of chemical imaging to pharmaceutical process monitoring and quality control. *Eur. J. Pharm. Biopharm.* **69**, 10–22.

28. Lyon, R. C., Lester, D. S., Lewis, E. N., Lee, E., Yu, L. X., Jefferson, E. H., and Hussain, A. S. (2002). Near-infrared spectral imaging for quality assurance of pharmaceutical products: analysis of tablets to assess powder blend homogeneity. *AAPS PharmSciTech* **3**, article 17, 1–15.

29. Lewis, E. N., Schoppelrei, J. W., Lee, E., and Kidder, L. H. (2005) Near-infrared chemical imaging as a process analytical tool. In: Bakeev, K. A. (Ed.), *Process Analytical Technology*, Blackwell Publishing, New York, p. 201.

30. Lewis, E. N. (2002) High-throughput infrared spectroscopy. US Patent 6,483,112.

31. Lee, E., Huang, W. X., Chen, P. Lewis, E. N. and Vivilecchia, R. V. (2006). High-throughput analysis of pharmaceutical tablet content uniformity by near-infrared chemical Imaging. *Spectroscopy* **21**(11), 24–33.

32. Dubois, J. D., Lewis, E. N., Fry, F. S., and Calvey, E. M. (2005) Bacterial identification by near-infrared chemical imaging of food-specific cards. *Food Microbiol.* **22**, 577–583.

33. Kidder, L., Lewis, E. N., Lee, E., and Haber, K. S. (2003) Approaches to standards for qualifying vibrational spectroscopic imaging instrumentation. Presented at ICAVS-2, University of Nottingham, UK.

34. Standard Reference Material 1920a, Certificate of Analysis, National Institutes of Standards and Technology, 2004. http://ts.nist.gov/MeasurementServices/ReferenceMaterials/archived_certificates/1920a.Feb%2020,%202004.pdf.

35. Choquette, S. J., Duewer, D. L., Hanssen, L. M., and Early, E. A. (2005), Standard reference material 2036 near-infrared reflection wavelength standard. *Appl. Spectrosc.* **59**, 496–504.

36. Near infrared spectrophotometry, in *United States Pharmacopeia Official Compendia of Standards*, General Chapter <1119>, 2003, 2388–2391.

37. Polder, G., van der Heijden, G. W. A. M., Keizer, L. C. P., and Young, I. T. (2003). Calibration and characterization of imaging spectrographs. *J. Near Infrared Spectrosc.* **11**, 193–210.

38. Geladi, P. L. M. (2007) Calibration standards and image calibration. In: Grahn, H. F. and Geladi, P. (Eds.), *Techniques and Applications of Hyperspectral Image Analysis*, Wiley, New York, pp. 203–220.

39. Dempsey, R. J., Cassis, L. A., Davis, D. G., and Lodder, R. A. (1997) Near-infrared imaging and spectroscopy in stroke research: lipoprotein distribution and disease. In *Imaging Brain Structure and Function: Emerging Technologies in the Neurosciences*, Ann. N. Y. Acad. Sci. **820**, 149–169.

40. Treado, P. J., Levin, I. W., and Lewis, E. N. (1994) Near-infrared spectroscopic imaging microscopy of biological materials using an infrared focal-plane array and an acousto-optic tunable filter (AOTF). *Appl. Spectrosc.* **48**, 607–615.

41. DeVerse, R. A., Hammaker, R. M., and Fateley, W. G. (2000) Realization of the Hadamard multiplex advantage using a programmable optical mask in a dispersive flat-field near-infrared spectrometer. *Appl. Spectrosc.* **54**, 1751–1758.

42. Lewis, E. N., Strachan, D. J., and Kidder, L. H. (2004) High-volume on-line spectroscopic composition testing of manufactured pharmaceutical dosage units. US Patent, 6,690,464.

43. Blob, A., Rullkötter, J., and Welte, D. H. (1987) Direct determination of the aliphatic carbon content of individual macerals in petroleum source rocks by near-infrared microspectroscopy. In: Mattavelli, L. and Novelli, L. (Eds.), *Advances in Organic Geochemistry, Proceedings of the International Meeting on Organic Chemistry*, Pergamon Press, Oxford. *Org. Geochem.* **13**, 1073–1077.

44. Laughlin, R. G., Lynch, M. L., Marcott, C., Munyon, R. L., Marrer, A. M., Kochvar, K. A. (2000) Phase studies by diffusive interfacial transport using near-infrared analysis for water (DIT-NIR). *J. Phys. Chem. B* **104**, 7354–7362.

45. Benedetti, E., Galleschi, F., D'Alessio, A., Ruggeri, G., Aglietto, M., Pracell, M., and Ciardelli, F. (1989). Microscopic FT-IR analysis of blends from functionalized polyolefins and polyvinyl chloride or polystyrene. *Chem. Macromol. Symp.* **23**, 265–267.

46. Hill, S. L. and Krishnan, K. (1988) Some applications of the polarized FT-IR microsampling technique. In: Messerschmidt, R. G. and Harthcock, M. A. (Eds.), *Practical Spectroscopy Series*, Dekker, New York, p. 116.

47. Clarke, F. C., Jamieson, M. J., Clark, D. A., Hammond, S. V., Jee, R. D., and Moffat, A. C. (2001). Chemical image fusion: the synergy of FT-NIR and Raman mapping microscopy to enable a more complete visualization of pharmaceutical formulations. *Anal. Chem.* **73**, 2213–2220.

48. Mao, C. (2000) Focal plane scanner with reciprocating spatial window. US Patent 6,166,373.

49. Lewis, E. N., Treado, P. J., Reeder, R. C., Story, G. M., Dowrey, A. E., Marcott, C., and Levin, I. W. (1995) Fourier transform spectroscopic imaging using an infrared focal-plane array detector. *Anal. Chem.* **67**, 3377–3381.

50. Snively, C. M., Katzenberger, S., Oskarsdottir, G., and Lauterbach, J. (1999) Fourier-transform infrared imaging using a rapid-scan spectrometer. *Opt. Lett.* **24**, 1841–1843.

51. Marcott, C., Story, G. M., Dowrey, A. E., Grothaus, J. T., Oertel, D. C., Noda, I., Margalith, E., and Nguyen, L. (2009). Mining the information content buried in infrared and nearinfrared band shapes by temporal, spatial, and other perturbations. *Appl. Spectrosc.* **63**, 346A–354A.

52. Rice, J. (2004) Testing spectral responsivity of IR cameras. Presented at Workshop on Thermal Imaging Research Needs for First Responders.

53. Morris, H. R., Hoyt, C. C., and Treado, P. J. (1994) Imaging spectrometers for fluorescence and Raman microscopy: acousto-optic and liquid crystal tunable filters. *Appl. Spectrosc.* **48**, 857–866.

54. Griffiths, P. R. and de Haseth, J. A. (1986) *Fourier Transform Infrared Spectrometry*, Academic Press, New York.

55. Mouroulis, P. (1998) Low-distortion imaging spectrometer designs utilizing convex gratings. *Proc. SPIE* **3482**, 594–601.

56. Newport Richardson Technical Note 5, Grisms (grating prisms). http://gratings.newport.com/information/technotes/technote5.asp.

57. Goldstein, S. R., Kidder, L. H., Herne, T. M., Levin, I. W, and Lewis, E. N. (1996) The design and implementation of a high-fidelity Raman imaging microscope. *J. Microsc.* **184**, 35–45.

58. Gupta, N. (2006) Fiber-coupled AOTF spectrometers. *Proc. SPIE* **6083**, 174–185.

59. Gat, N. (2000) Imaging spectroscopy using tunable filters: a review. *Proc. SPIE* **4056**, 50–64.

60. Lewis, E. N., Kidder, L. H., Lee, E., and Haber, K. (2006). Near-infrared spectral imaging with focal plane array detectors. In: Levin, I. W. and Bhargava, R. (Eds.), *Spectrochemical Analysis Using Infrared Multichannel Detectors*, Blackwell, New York, p. 28.

61. Hudak, S. J., Haber, K. H., Sando, G., Kidder, L. H., and Lewis, E. N. (2007) Practical limits of spatial resolution in diffuse reflectance NIR chemical imaging. *NIR News* **18**, 6–8.

5

DATA ANALYSIS AND CHEMOMETRICS FOR HYPERSPECTRAL IMAGING

PAUL GELADI

Unit of Biomass Technology and Chemistry, Swedish University of Agricultural Sciences, Umeå, Sweden

HANS GRAHN

Department of Neuroscience, Division of Behavioral Neuroscience, Karolinska Institute, Stockholm, Sweden

MARENA MANLEY

Department of Food Science, Stellenbosch University, Stellenbosch, South Africa

5.1 INTRODUCTION

Data analysis and chemometrics for hyperspectral imaging is a vast topic and can easily become confusing. This is due to the many chemometrics methods available and the diversity of vibrational spectroscopic techniques where chemometrics is and can be applied. Still, in this chapter, sufficient material is presented for the reader to get an overview and pick up useful morsels. A simple NIR (near-infrared) imaging example is used throughout the chapter to illustrate the concepts in a didactical manner.

This section starts with some of the definitions of hyperspectral images and their mosaics, including data file formats. The topic of different types of image resolution is introduced. Calibration standards and standardization are an important part of this section. Different modes of hyperspectral imaging exist and each of them may require a different way of tackling the data. Section 5.2 describes some classical univariate techniques, without too much emphasis on all details, which may help in image cleaning or transformation of images. The true heart of the matter in chemometrics is in Section 5.3, where local models, image cleaning, spectral transformation, principal component analysis (PCA), multivariate curve resolution (MCR), image regression, discriminant regression, artificial neural nets (ANNs), and clustering/classification are described.

5.1.1 Digitized Images, Multivariate Images, and Hyperspectral Images

Data analysis is about analyzing results of measurements. These are by definition several measurements (a single number cannot be analyzed further) and these can be arranged in arrays of data, for example, vectors, matrices, and three-way arrays.

Digitized images are arrays of numbers. A grayscale or intensity image, called a B/W image, is a matrix consisting of lines (or rows) and columns (Figure 5.1). Each element in the matrix is a grayscale value for a certain position (a pixel position) and together these create the image impression if observed on a computer monitor, a piece of paper, or a sheet of film. In the early days, the sizes of the images were limited to powers of 2; thus, sizes of, for example, 64×64, 128×128, 256×256, and 512×512 were used, that is not the case anymore. For analog television, the PAL standard size is 625 lines of which only 575 are used. For digital television, sizes of 576×720, 720×1280, and 1280×1920 are standard [1–3]. Digital cameras for still color photography may have a range of sizes and very often huge color images can be made. The data matrices mentioned here are only virtual and the images are stored as data files. For color images shown on screens, three intensity values are available for each pixel position in the matrix to give the blue, green,

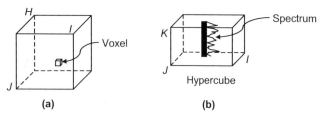

(a)   **(b)**

FIGURE 5.1 (a) A three-dimensional image has indices $h = 1, \ldots, H$ for depth and $i = 1, \ldots, I$ and $j = 1, \ldots, J$ for the planar dimensions. Each position (h, i, j) has an associated number or intensity called a voxel. For two-dimensional images the depth dimension collapses into a plane and the voxel becomes a pixel with indices (i, j). (b) The most common hyperspectral images have two planar position dimensions $i = 1, \ldots, I$ and $j = 1, \ldots, J$. Each pixel position has a K-dimensional vector or spectrum associated to it. The position (i, j, k) contains a number or intensity. There are both spatial and spectral neighbors. The data structure is also called a hypercube.

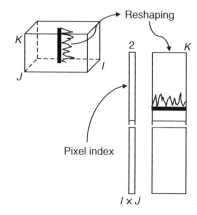

FIGURE 5.2 For most calculations, the hypercube can be reshaped into an image data matrix. This is acceptable as long as the pixel indices are kept.

and red values. These are often called the color channels because they are spectral bands and not single wavelengths. In the printing world, all colors are synthesized by using cyan, yellow, and magenta dots. As an array, color images are three-way arrays consisting of three layers.

Any image that has more than three channels could be called a multivariate image and early on, multivariate images with 4–20 channels were often used. With the introduction of the AVIRIS system for airborne imaging, the term hyperspectral was introduced and the AVIRIS had 224 channels. Nowadays, hyperspectral images usually have more than 100 channels (Figure 5.1). Hyperspectral images are three-way arrays made up of rows (lines), columns, and channels. Hyperspectral and multivariate images are also stored as data files.

A hyperspectral image has spectral properties in the variable dimension. This means there is a lowest wavelength (or wavenumber), a highest wavelength, and that the wavelengths in between are spread reasonably evenly. In other words, the variables are contiguous. This is illustrated in Figure 5.1. A wavelength or wavenumber is, however, an idealized concept. In reality, there is a bandpass of the monochromator or filter. Therefore, the terms *bands* or *channels* are often used, where the 1000 nm channel means the band centered around 1000 nm with the typical bandwidth for the monochromator/filter used.

In early days of image analysis, pixels were integers, values between 0 and 255 (2^8 possibilities to express an intensity). In radiology, integers between 0 and 4095 (2^{12} possibilities) are often the standard values. Newer systems have an even higher intensity resolution. It is also possible to average images in order to reduce noise. This creates decimals, usually expressed as real or double precision numbers.

An upcoming, but not yet very common, technique is three-dimensional imaging. A three-dimensional image has

three pixel coordinates and the pixel is called a voxel (Figure 5.1). The need for advanced computers and viewing equipment limits the wider spread of 3D techniques [4, 5] (see Chapter 6).

As a general definition of a hyperspectral image, one could use the following: an array with two (or three) pixel (voxel) coordinates, where each pixel (voxel) is represented by a vector of at least 100 elements of measured values and each element in the vector is single or double precision (Figure 5.1). A further requirement is that the elements in the vector can be ordered and are spread rather evenly. In many instances, the hypercube may be reshaped into an image data matrix (Figure 5.2). If the pixel coordinates are kept available this is an excellent way of handling images. A typical hyperspectral image may thus have tens of thousands of pixels.

Sometimes more than one hypercube need to undergo the same data analysis procedure. In such a case, it is handy to make a mosaic of many hypercubes. This could be a temporal sequence of hypercubes. Mosaicing is also the solution to some sampling problems when not enough objects (e.g., cereal grains) fit into one image (field of view of camera). Figure 5.3 gives a schematic view of the mosaic technique. Mosaics of a few hundred thousands of pixels are possible.

Mosaic

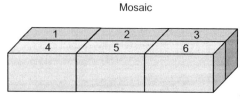

FIGURE 5.3 For analyzing more than one hypercube with the same model, it is possible to make a mosaic. For example, the numbers in the figure may belong to a time sequence.

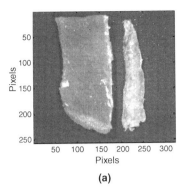 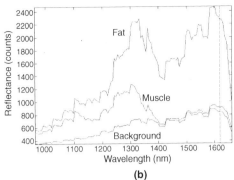

(a) (b)

FIGURE 5.4 A grayscale image at 1302 nm of the dried meat and fat sample (a) and three typical spectra (fat, muscle, and background) in A/D converter counts (b).

An example of a NIR hyperspectral image of size 256×320 with 118 wavelength bands from 960 to 1662 nm in 6 nm steps is shown in Figure 5.4. This is a grayscale image at 1302 nm of a piece of dried meat (a product known as biltong), consisting of muscle and fat, as well as a piece of only fat to be used as a reference. The pieces of meat and fat were positioned on a dark background (silicon carbide sandpaper). Some typical spectra (as detector current) are also shown in Figure 5.4. This image will be referred to throughout the chapter to illustrate some chemometrics techniques and principles.

5.1.2 Image Data Files

The choice of format for storing image data should basically be decided on a few principles: the choice of compression (if needed), accessibility to the source code, that is, is the open source principle valid, and how the data can be accessed by various applications for multivariate analysis.

Many hyperspectral file formats are derived from remote sensing, where the ENVI (environment for visualizing images) standard has prevailed. Vibrational spectroscopy instrument manufacturers commonly use similar formats customized to their hardware and software. In hyperspectral imaging, files are large, therefore the storage and file formats have to be designed accordingly. The ENVI application uses three different formats, that is, the band sequential (BSQ), band interleaved by line (BIL), and band interleaved by pixel (BIP) [6]. BSQ and BIL are designed specifically for tape storage. As with many other formats, the header is separated from the data that allows the description of the channels, bands, and other information about the images. BIP stores each line in an image sequentially. Each spectral vector λ_1 to λ_K is kept as a vector where the pixels are lined up after each other. Figure 5.5 illustrates the principle. Pixels are ordered according to their indices. For simpler images, for example, results from a chemometrics analysis, the formats TIFF (tagged image file format) [7], PNG (portable network graphics), and BMP (bitmap) are common. Another file

format in imaging is the JPEG (joint photographic expert group) format, only recommended for publication, but not for analysis.

5.1.3 Types of Image and Data Resolution

Resolution can mean different things in an image. There is spatial resolution: the bigger the image (megapixels!) the better the spatial resolution. Then there is spectral resolution (resolution in the wavelength range) and numerical resolution (each pixel in a wavelength band can be a short integer, a long integer, a floating point, or double precision). An image sequence measured over time may also have temporal resolution. In some hyperspectral images the resolution in each pixel approaches that of a high-quality spectrometer, but in many cases a compromise is made to get a better spatial resolution.

The human eye has very good spatial resolution, but limited intensity (maximum 32 gray levels) and almost no spectral resolution, while most calculations require a very high precision. This duality should be taken into account in all image analyses. All calculations should be carried out with the data in the most precise form, but for visual inspection rounding errors can be made.

5.1.4 Standardization and Standards

The human eye is very flexible and can recognize many things in an image regardless of small errors in, for example, intensity, contrast, gamma, and white balance. This approximate

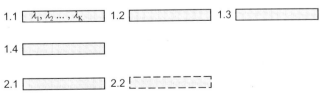

FIGURE 5.5 An illustration of the band interleaved by pixel format.

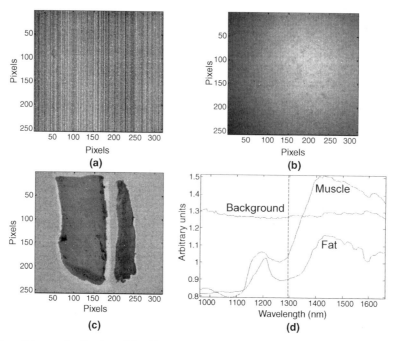

FIGURE 5.6 An illustration of the standardization of the dried meat image indicated in Figure 5.4. (a) The dark current image; (b) the 25% reflectance standard image; (c) the image converted to absorbance units at 1302 nm; (d) typical spectra for fat, muscle, and background after conversion.

visual recognition is not very reproducible. Correct calculations require precise, accurate, and reproducible values. This requires reproducible measurement equipment, though not all imaging setups are stable enough. Therefore, calibration objects with known properties have to be imaged and used in correction algorithms. An example used in NIR and visible reflectance imaging cameras is dark and white references that can be used to calculate pseudoabsorbance values [8, 9]. It is also very important to use geometrical references such as rulers or tiled patterns with known tile size to confirm correct pixel geometry [10]. Some instrument manufacturers do the standardization during the image scanning while others rely on the researcher to do the appropriate standardization.

The standardization to absorbance is illustrated in Figure 5.6 using the dried meat NIR hyperspectral image as an example. When comparing the spectra in Figure 5.6 with those in Figure 5.4, one notices that the spectra have now acquired the more familiar shape that is usual for NIR absorbance spectra; the images though look unfamiliar now (compared to the digital image).

5.1.5 Imaging Techniques Used for Hyperspectral Imaging

In remote sensing, the classical AVIRIS instrument makes hyperspectral images in the visual and near-infrared wavelength ranges: 224 bands between 400 and 2500 nm. Visual and near-infrared images can, however, also be made in the

laboratory. Similarly, it is possible to make hyperspectral Raman and FT-IR images [11]. Mass spectrometry techniques such as secondary ion mass spectrometry (SIMS) and time of flight (ToF)-SIMS also provide hyperspectral images. Hubbard [12] gives an overview of methods using electrons, ions, X-rays, gamma rays, infrared, and atomic force for characterizing surfaces. Many of these methods can give multivariate data and images.

It is important to know that the types of image cleaning and preprocessing applied are dependent on the imaging technique used. The demands on the multivariate chemometrics models are sometimes also dependent on imaging modus used. The presentation of analysis results may be mode dependent too. A special challenge is combining different imaging modes. Lau et al. [13] show an example of the same painting analyzed by micro-Raman and scanning electron microscope-based X-ray fluorescence.

5.2 OPERATIONS ON GRAYSCALE IMAGES

A grayscale image is a matrix of numbers and this allows many types of calculations. The traditional image analysis literature covers these calculations or operations [14–20]. The importance of such image operations is that they can be used to clean hyperspectral images before an analysis is attempted or to clean or analyze the results after a hyperspectral analysis. Some of the classes of operations are radiometric operation, local neighborhood operations, global operations, many-image arithmetic, and warping. These can be used in sequences or in the adaptive mode.

1. Radiometric operations are pixel-per-pixel operations and can be intensity, contrast, and gamma settings, as well as inversion. They are often used for preprocessing individual variable slices in a hyperspectral image.

2. Local neighborhood operations take into account the fact that each pixel has neighbors. A typical example is median filtering, where each pixel intensity is replaced by the median of itself and a group of neighbors. The median filtering operation is sometimes used for guessing the values of missing, dead, or incomplete pixels.

3. Global operation images can be transformed as a whole, such as in Fourier, cosine, slant, Haar, or Hadamard transform. This allows filtering in the Fourier domain.

4. Arithmetic operations are pixel-per-pixel arithmetic on two or more images, that is, addition, subtraction, multiplication, and division. Subtraction is often used for removing shading. Division can be used for creating ratios. This is popular in geology.

5. Warping is a general expression for geometrically correcting many images to a common geometrical base and includes rotation, mirroring, linear and sometimes nonlinear compression, or expansion. This is also called registration.

Interesting software is the freeware ImageJ [21]. It contains example images and can carry out all the techniques from 1 to 5. It is very instructive to test these. Also most software for handling photographic images contain elements of the above-mentioned techniques.

5.3 CHEMOMETRICS FOR HYPERSPECTRAL IMAGES

Basic information on the analysis of multivariate and hyperspectral images can be found in Refs 11 and 22–24. Remote sensing and satellite imaging use hyperspectral images and relevant information can be found in the literature [25–31]. The field of chemometrics is huge and only selected techniques will be explained in detail. For the techniques not discussed in detail, recent literature references will be given. Many of the older literature for chemometrics combined with multivariate imaging and hyperspectral imaging can be found in Refs 22 and 23. A good review of the chemometrics literature is given by Gendrin et al. [32] as well as Nicolaï et al. [33]. A tutorial on chemometrics for spectral data can be found in Ref. 34.

A general overview of the sequence between using a hypercube or mosaic as an input and producing multiple outputs is given in Figure 5.7, where the input is a hypercube or a mosaic. On this input array, a number of data analysis techniques can be applied, often in some logical sequence starting with cleaning, background removal, and shading correction, followed by exploratory analysis and ending with making classification or regression models. This sequence can be gone through once or in many loops. The results are presented as images, plots, and tables. Because of the huge amount of data, visualization is of utmost importance. Tables are used only for presenting general model parameters.

Factor analysis is a general term for models that contain a bilinear decomposition in factors. The most important factor analysis models are principal component analysis (Section 5.3.4) and multivariate curve resolution (Section 5.3.5). They are also maximally different in goals and algorithms. There are a huge number of modified factor models possible. These are only mentioned and not explained. Regression models can be used to build a linear

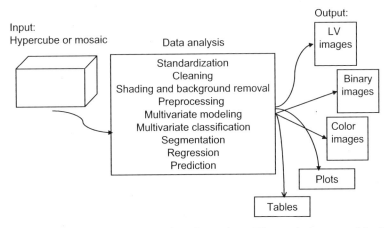

FIGURE 5.7　Using a hypercube or mosaic as input, a large number of operations of data analysis are possible, depending on the problem to be solved. The output may be color or false color images, latent variable images, binary/segmented images, and plots or tables.

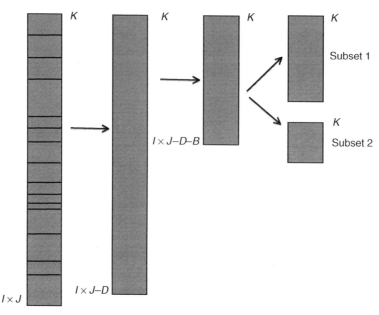

FIGURE 5.8   Many images contain harmful or unwanted pixels that have to be removed before data analysis. This reduces the image data matrix by D number of rows. Sometimes background pixels need to be removed. This further reduces the image data matrix by B number of rows. The final model is made on the image data matrix with $I \times J - D - B$ rows. Sometimes a preliminary analysis indicates that a further split in subsets is needed.

relationship between a spectrum and a concentration. These are described in Section 5.3.6 while Section 5.3.7 refers specifically to discriminant regression. A good alternative for nonlinear regression modeling is the artificial neural net as described in Section 5.3.8. Some special techniques for clustering are mentioned in Section 5.3.9.

5.3.1   Local Models

Because of the huge amount of pixels, not all have to be used. There is rarely any application where all available pixels are used or needed. It is always good to get rid of bad, erroneous, and background pixels and to build the data analysis models on a clean subset. This also means that the data array is not a hypercube anymore (Figure 5.8). Extensive use of local models forms the main difference between standard (non-image) chemometrics and image chemometrics. In image chemometrics, even a small subset of an image or mosaic may contain thousands of objects and allow a useful chemometrics model.

5.3.2   Exploratory Chemometrics for Hypercube Cleaning

It is not meaningful to look at all channels in a hyperspectral image. The tens of correlated channels make this exercise too time consuming and ineffective. Even making pseudocolor images of triplets of channels is not easy enough. For this reason, even exploratory studies need multivariate analysis. PCA may be used to construct 5–15 principal component

(PC) score images that give the essence of what is in the hypercube (Figure 5.9). The PC score images are uncorrelated, which is an advantage. An even more powerful technique is to make PC scatter plots of the scores. In these score plots, clusters, gradients, and single pixels (often outlier pixels) can be seen very clearly. Behrend et al. [35] describe a simple cleaning method for hyperspectral Raman images.

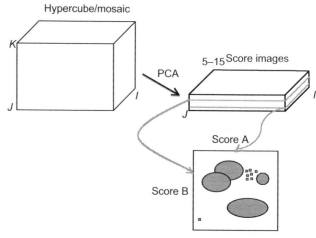

FIGURE 5.9   A hypercube or mosaic contains too many correlated images to make studying them meaningful. A simple PCA calculation can produce a smaller data array of uncorrelated score images, which is easier to overview. Even better is making score plots as these can show clean clusters, overlapping clusters, gradients, or outlier pixels.

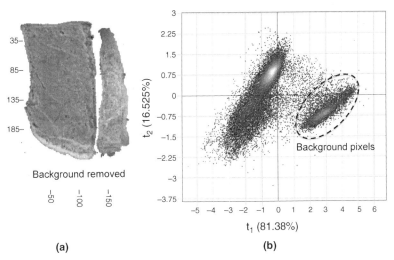

35–
85–
135–
185–

Background removed

–50
–100
–150

(a)

t_2 (16.525%)

Background pixels

t_1 (81.38%)

(b)

FIGURE 5.10 (a) PC3 score image of the dried meat with background removed. (b) A PCA score plot with the background pixels to be removed indicated by the ellipse.

By interactive operation between score images and score plots, disturbing or harmful pixels can be removed to produce a cleaned image. The same operation can also be used to make subsets. Examples of how this is done can be seen in Chapter 13. At this stage, there is no need to look at PCA loadings for spectral interpretation. This should be done when final models on image subsets are made. Figure 5.10 shows a score plot of the first two principal components calculated on mean-centered data for the meat image. In this score plot, a region of interest (ROI) (the background) is indicated with an ellipse. Another cluster obviously contains the muscle and fat, but overlapping. By removing the background cluster from the calculations simpler chemometrics models are made possible. Figure 5.10 also shows the third principal component image of the meat example after the dark background and some regions at the edges were removed. This image subset is the one that will be analyzed further.

5.3.3 Filtering and Preprocessing

Because the pixels in a hyperspectral image are spectra, many spectral preprocessing, error correction, or image improvement techniques can be used. The transformations are not different from what is used in bulk vibrational spectroscopy. Siesler et al. [36] give a good overview of transformation techniques for NIR spectra. Reference 37 can also be consulted for more information about preprocessing of NIR hyperspectral images. Raman preprocessing is described later on in Chapter 9. Transformation, correction, or enhancement techniques can be made in different ways and for different reasons. Corrections can be made on the image planes using classical univariate image analysis techniques. They can be made on spectra using spectral correction techniques or they can be made on an image data matrix based on statistical considerations (Figure 5.11).

As mentioned before, all classical univariate image analysis techniques (see Section 5.2) for error correction and image improvement can be used on each of the image planes in a hyperspectral image. Examples are noise reduction by median filtering or Fourier or wavelet transformation. Some transformations are based on statistical considerations. Columnwise removal of the mean (Figure 5.3) is used as a standard and division by the error or other standard deviation is sometimes used. If nonlinearities are expected, some nonlinear transformations such as taking logarithms can be attempted. These operations are all columnwise in the image matrix.

In NIR or VIS-NIR spectroscopy, measurements are often carried out in the modes of diffuse transmission or diffuse reflection. This creates effects on particle size, particle roughness, and others. The general effect of this is to add a baseline to the spectra. The main form of this baseline is an offset combined with a slope. Many spectral transformation techniques aim at removing these to better

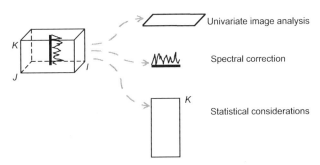

K
I
J
Univariate image analysis
Spectral correction
K
Statistical considerations

FIGURE 5.11 Corrections and transformations on hyperspectral images can be made on image planes (classical univariate image analysis), spectral vectors (spectral corrections), or on whole matrices (based on statistical considerations).

describe the chemical information. The transformations are used row-wise in the image data matrix (Figure 5.3). Spectral smoothing can be used to improve noisy spectra. In order to remove background, first derivatives can be calculated to remove offsets. Second derivatives can be used to remove both offsets and slopes. Using the Savitzky–Golay [38] transformation, the taking of derivatives and smoothing in a window are carried out at the same time. Other transformations used to remove offset and slope and to increase chemical information are multiplicative scatter correction (MSC) [39] and standard normal variate (SNV) [40]. A very simple but sometimes successful transformation is removing noisy wavelengths.

It is important to mention that the transformations should be made on the properly cleaned image subset. Some of the transformations proposed increase errors and cause confusion if bad, erroneous, or irrelevant pixels are left in the data set.

In a study where the application of hyperspectral imaging for damage detection on the surface of mushrooms was investigated, the spectral variability caused by sample curvature and nonuniform light scattering was emphasized [41]. Curvature, cusps, and edge effects, contributing to inhomogeneous lighting or nonuniform light scattering are often factors to be addressed and corrected for when analyzing food samples. Gowen et al. [41] investigated four different spectral preprocessing techniques in an attempt to effectively remove spectral variation due to external factors while retaining the spectral features that would allow sample characterization. The application of MSC and mean normalization (divide each spectrum by its maximum intensity value) were shown to be the most efficient preprocessing methods to decrease variability due to curvature present in the samples.

5.3.4 Principal Component Analysis

Once a problem-related image subset has been made by exploratory study as explained in Section 5.3.2 and transformed as in Section 5.3.3, a PCA model can be applied. The model made is

$$\mathbf{X} = \mathbf{TP}' + \mathbf{E} \qquad (5.1)$$

where $\mathbf{X}$ ($L \times K$) is the cleaned and preprocessed image subset, $\mathbf{T}$ ($L \times R$) a matrix with R score vectors, $\mathbf{P}$ ($K \times R$) a matrix with R loading vectors, and $\mathbf{E}$ ($L \times K$) the residual. The prime is used to indicate the transpose of a matrix. Equation 5.1 is also shown in Figure 5.9, but with $\mathbf{T} = \mathbf{C}$ and $\mathbf{P} = \mathbf{S}$. The properties are that the sum of squares (SS) of $\mathbf{E}$ is minimized and that the vectors in $\mathbf{T}$ and $\mathbf{P}$ are orthogonal. Because the pixel indices are saved, it is possible to reshape the score vectors in $\mathbf{T}$ into score images. It is also possible to use the information in $\mathbf{E}$ to make residual images. The loading vectors in $\mathbf{P}$ can be used to make spectral interpreta-

tions. The selection of R, the number of components to be used, is not very easy, but usually some practical cutoff limit can be found. This is easier for images as for other data, because of the possibility to visually study the score images and score plots (Figure 5.7). One may select R too high and then it is easy to find a cutoff where the score images or score plots become too noisy and to select a new R. Sometimes residual images based on $\mathbf{E}$ can be used for this purpose.

Because of the orthogonality of the loading vectors in $\mathbf{P}$, negative loadings have to occur, making the spectral interpretation difficult. Real spectra never have negative parts. More on PCA for images can be found in two books [22, 23]. See also Chapters 9 and 13.

An important diagnostic is the sum of squares. By rewriting Equation 5.1, the following is obtained:

$$\mathbf{X} = \mathbf{t}_1 \mathbf{p}_1' + \mathbf{t}_2 \mathbf{p}_2' + \mathbf{t}_3 \mathbf{p}_3' + \cdots + \mathbf{t}_R \mathbf{p}_R' + \mathbf{E} \qquad (5.2)$$

where $\mathbf{t}_r$ and $\mathbf{p}_r$ with $r = 1, \ldots, R$ are the numbered score and loading vectors from the matrices $\mathbf{T}$ and $\mathbf{P}$, respectively. By setting the SS of $\mathbf{X}$ to 100%, the SS for each term in the sum can be calculated and they also sum up to 100%:

$$SS_X = SS_1 + SS_2 + SS_3 + \cdots + SS_R + SS_E \qquad (5.3)$$

with $SS_1 = \mathbf{t}_1' \mathbf{t}_1, \ldots$. Equation 5.3 has the property $SS_1 \geq SS_2 \geq SS_3 \geq \cdots \geq SS_R$. The first component has the largest SS followed by the remaining components in decreasing order. The SS of unused components is the residual sum of squares SS_E. It should be noted that SS_E is dependent on the choice of R, but it is assumed to become small. Plotting the SS values against the component number is sometimes used to find a cutoff value for R, but in image data analysis with very small SS values it can become less meaningful because of the huge number of pixels and local phenomena. In general, one may say that the usual statistical diagnostics are less used and that interpretation of the score images, residual images, and score plots can be done visually with good results.

Figure 5.12 shows the PCA (after mean centering) results after background removal (as indicated in Figure 5.10) of the dried meat image. The score plot of PC1 and PC3 shows two clear clusters now. These correspond to muscle tissue and fat tissue. Figure 5.12 also shows an attempt at segmenting fat and muscle as two segmented images. Segmentation was done by interactively drawing polygons in the score plots and mapping the selected pixels to image space. This segmentation would have been impossible to carry out in the image space. In Figure 5.13, selected spectra from the centers of both clusters (fat and muscle tissue) in Figure 5.12 are shown. Also the loading of the third principal component is shown. The loading is dominated by the typical (positive) fat peak around 1212 nm. Other peaks are also related to fat or to the different types of water binding in fat and muscle tissue. The study of spectra and loadings is only meaningful after the

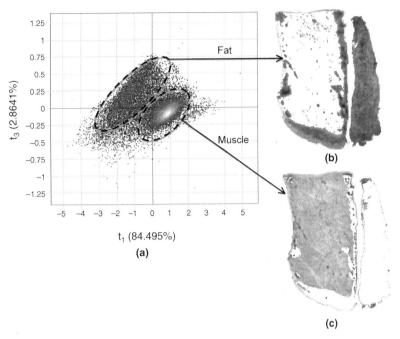

FIGURE 5.12   (a) The PC3 versus PC1 score plot with clusters for muscle and fat pixels delineated. The segmented classes for fat (b) and muscle (c). Some misclassification at the edges can be observed as a result of shading. The fat cluster is more extended than the muscle tissue cluster indicating more statistical variation in the fat material.

removal of background and other disturbing regions and segmentation and identification of the clusters.

5.3.5   Multivariate Curve Resolution

Curve resolution or multivariate curve resolution, unmixing, or endmember (a few other names also exist) analysis are all factor analysis models [32, 42–44]. The MCR models are calculated on a cleaned image subset but without any statistical preprocessing or derivatives that can produce negative results. An important assumption made is that concentrations and spectra are never negative. The model made is

$$\mathbf{X} = \mathbf{CS}' + \mathbf{E} \qquad (5.4)$$

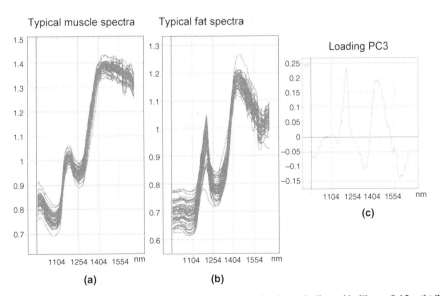

FIGURE 5.13   (a) Typical spectra of muscle tissue from the center of the muscle cluster indicated in Figure 5.12a. (b) Typical fat spectra from the center of the fat cluster indicated in Figure 5.12a. (c) For comparison, PC3 loading is added.

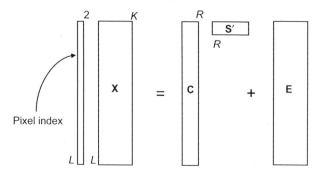

FIGURE 5.14 A general decomposition of a hyperspectral image or mosaic subset with L ($L \gg K$) spectra. The matrices $\mathbf{X}$, $\mathbf{C}$ (scores), and $\mathbf{E}$ (residual) all use the pixel index. The matrix $\mathbf{S}$ contains the loadings. The columns of the matrices $\mathbf{C}$ and $\mathbf{E}$ can be reshaped into images because the pixel index is available. R is the number of components calculated.

where $\mathbf{X}$ ($L \times K$) is the cleaned and preprocessed image subset, $\mathbf{C}$ ($L \times R$) is a matrix with R concentration vectors, $\mathbf{S}$ ($K \times R$) is a matrix with R pure spectra vectors, and $\mathbf{E}$ ($L \times K$) is the residual. This equation is also illustrated in Figure 5.14. Once a value for R is selected, alternating calculations of $\mathbf{C}$ and $\mathbf{S}$ are tried to minimize $\mathbf{E}$, while all values in $\mathbf{C}$ and $\mathbf{S}$ are kept nonnegative. For the correct value of R, this calculation usually converges. One may also study the final values of $\mathbf{S}$ because they are pure spectra and can be identified. Because pure spectra are nonnegative, the curve resolution model is good for spectral studies and identification. Curve resolution is only possible if the linear mixing model holds. This is true in some cases for IR and Raman and especially for TOF-SIMS data, but rarely for NIR data. A numerical difference between PCA and MCR is that PCA results can be calculated with great precision in a reproducible manner. MCR results depend on when the MCR iterations are stopped and can be variable. Some authors have tried to minimize this effect [42, 43]. With PCA, one may increase R without changing the previously calculated components. With MCR, all components have to be recalculated if R is changed. There is no specific order of the components in MCR. Once good values for $\mathbf{C}$ and $\mathbf{S}$ are found, the pure spectra can be studied for spectral interpretation and the concentrations can be used for making concentration map images. In some cases, the residual image can be used for diagnostic purposes.

MCR seems to work ideally for ToF-SIMS data and reasonably well for IR and Raman data and for NIR data from pharmaceutical samples. It does not work at all for NIR on food and agricultural products, probably because of issues related to wavelength-dependent penetration depth and scattering.

Besides PCA and MCR, also other factor models can be made. Independent component analysis (ICA) is some-

times used [45]; also factor rotation could be an alternative.

5.3.6 Multivariate Image Regression

A regression model can be built between a spectral vector and an external variable, for example, a concentration:

$$\mathbf{y} = \mathbf{Xb} + \mathbf{e} \qquad (5.5)$$

where $\mathbf{y}$ ($L \times 1$) is a mean-centered vector of concentrations, $\mathbf{X}$ ($L \times K$) a mean-centered image matrix or image subset, $\mathbf{b}$ ($K \times 1$) a vector of regression coefficients, and $\mathbf{e}$ ($L \times 1$) a vector of regression residuals (Figure 5.15). This is known as multivariate image regression (MIR). More can be found in Refs 23 and 46–48. The model is often built as a partial least squares (PLS) model, but alternatives exist. Without going into too much detail, PLS uses R components (just like a factor model) to calculate $\mathbf{b}$ and avoids a lot of pitfalls that other methods for calculating $\mathbf{b}$ have.

Equation 5.5 can also be expressed in terms of SS:

$$SS_y = SS_{Xb} + SS_e \qquad (5.6)$$

or as a coefficient of determination:

$$R^2 = 1 - SS_e/SS_y \qquad (5.7)$$

Here, SS_y is set to 100%, SS_e is the residual SS in percent, and SS_{Xb} is the model SS. Any acceptable model would have a R^2 or SS_{Xb} of at least 65% and often $SS_{Xb} > 90\%$ (another font is used for R^2 to distinguish it from R, the number of components).

One of the problems of Equation 5.5 is that the y-values have to be known accurately and with high precision. In remote sensing, this is called the ground truth. It is very expensive to find the ground truth for every pixel in an image, so a well-selected subset is used. In the laboratory, the ground

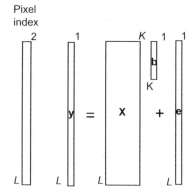

FIGURE 5.15 The regression equation for an image data matrix with L pixels. The pixel index vector is valid for $\mathbf{X}$, $\mathbf{y}$, and $\mathbf{e}$.

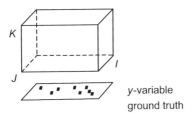

FIGURE 5.16 For some pixels or regions in the hypercube, there may be ground truth (a remote sensing term) or wet chemical information available. This allows the building of regression models.

truth is replaced by wet chemical analysis (Figure 5.16). As a result, **b** is often calculated with a well-selected smaller number of points.

Once **b** is known, it is valid for all the spectra in a hyperspectral image **Z** similar to **X** in Equation 5.5:

$$y_{hat} = Zb \qquad (5.8)$$

where y_{hat} are the predicted concentrations, **Z** the mean-centered image data matrix, and **b** as in Equation 5.5. The values in y_{hat} then form a concentration map because they have pixel coordinates.

The concentration map is basically an intensity image with dark for low concentration and bright for high concentration. It may also be color-coded using a range of colors, for example, blue (cold) for low concentrations and red (hot) for high concentrations. With up to three constituents, complex color maps can be made using the red, green, and blue channels [37, 46]. In these, each color nuance can be interpreted as a certain mix of the three constituents.

An important aspect of regression modeling is quality testing and this should be done using a test set with known y-values that is not used for model building. Cross-validation is sometimes used, but for images there are so many pixels available that a proper test set can always be made. Some authors use a smaller test set for model building and a larger validation set for determining the statistical properties of the prediction.

X_t is the test set image data matrix and y_t is the vector containing the ground truth or wet chemical values:

$$y_{hat} = X_t b \qquad (5.9)$$

This is the same as Equation 5.8, but now for the test set. The residual can be calculated as follows:

$$f = y_t - y_{hat} \qquad (5.10)$$

where **f** is the test set residual. It can be used to calculate a root mean square error of prediction (RMSEP) value:

$$RMSEP = [f'f/J]^{0.5} \qquad (5.11)$$

J is the number of test set pixels. RMSEP has the form of a standard deviation. It is the average prediction error that can be achieved and should be kept low.

5.3.7 Discriminant Regression

Sometimes it is necessary to make a model that gives maximum discrimination between two classes of pixels. This can be done as in Equation 5.5, but by filling the vector **y** with dummy variables: -1 and $+1$ (sometimes 0 and 1, but because of mean centering this does not matter). When used with PLS regression, this technique is called PLS discriminant analysis (PLS-DA). When used for prediction as in Equation 5.9, the y_{hat} values should ideally be -1 or 1. This is not usually the case, but they become a histogram of values around -1 and $+1$. One could then choose 0 as a cutoff for the class membership. The values in y_{hat} can also be used as class membership images because the pixel coordinates are available.

5.3.8 Artificial Neural Networks

A shortcoming, of all regression models shown earlier, is that they are *linear* regression models and not very adapted to dealing with nonlinearities. Artificial neural networks are better adapted for doing this. The reason for this is that they use nonlinear transformations and inner layers. A weak point of ANNs is the slow calculation with many input variables. Some authors have solved this by using a limited number of latent variables as inputs [49–51]. ANNs are very flexible and many varieties exist.

Figure 5.17 shows a typical neural network with one hidden layer. The layers consist of nodes. The nodes in the input layer represent wavelength bands or latent variables and an offset value. Inside each node, a nonlinear transformation (e.g., sigmoidal) takes place. The arrows represent weights connecting the nodes, thus removing an arrow is the same as putting its weight to zero. The output layer could be a concentration or class membership. By backpropagation, the weights are continuously adjusted until the relationship between inputs and outputs is as wanted, for example, prediction of a concentration from a spectral vector input. ANNs are slow to train, but very quick for prediction of concentrations or class memberships once the training is done.

5.3.9 Clustering and Classification

Clustering is the activity of separating the pixels of a hypercube in clusters or classes. This can be done in two ways: supervised and unsupervised. The unsupervised methods of clustering make no assumptions in advance. An algorithm is used to find a certain number of clusters that are maximally separated according to some criteria. In

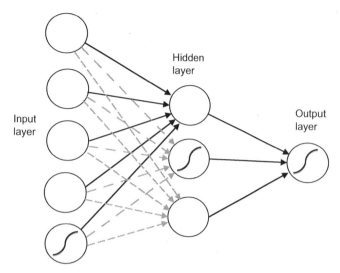

FIGURE 5.17 A typical (simplified) artificial neural network structure with one hidden layer. The circles are called nodes and may contain a nonlinear transformation (only shown for a few nodes). All nodes are connected between input and hidden layers and between hidden and output layers.

supervised clustering, some advance knowledge about the pixels is available and this knowledge is used in a clustering algorithm. The discriminant regression in Section 5.3.6 is a supervised clustering because it is known in advance which pixels are −1 and which are +1. The PCA- and MCR-based analyses in Sections 5.3.3 and 5.3.4 are unsupervised because the clusters are formed by the data themselves.

Clustering uses distances, similarities, and dissimilarities in multivariate Euclidean and non-Euclidean space. There are many techniques and algorithms for clustering and the nomenclature is confusing. Gan et al. [52] describe hierarchical, fuzzy, center-based, search-based, graph-based, grid-based, density-based, model-based, and subspace clustering methods. Xu and Wunsch [53] take up hierarchical, partitional, neural net-based, kernel-based, sequential, density-based, and grid-based clustering. Another interesting book is by Mirkin [54]. An overview of clustering methods is given by Omran et al. [55]. Clustering methods can handle many pixels, but not so many variables. In most cases, the data need to be reduced to a smaller number of latent variables by PCA, MCR, or something similar in advance. Once an acceptable clustering model is found, it can be used for classification. It is almost always possible to get 100% correct classification for a training set. Therefore, it is absolutely necessary to use a test set to evaluate the real classification accuracy of a model. An interesting tutorial is by Tran et al. [56]. An interesting new development is the use of support vector machines (SVM) and radial basis functions (RBF) [57].

5.4 CONCLUSION

Because of the huge number of pixels available, hyperspectral images or mosaics are ideal for the application of chemometrics in all its guises. Even more important is that chemometrics is absolutely necessary for handling the huge amount of data produced by imaging equipment. Some important advice could, however, be given. First, work only on properly cleaned images. Camera and optical errors should be removed and not modeled. The same goes for background unrelated to the problem studied. Second, consider spectral preprocessing as it can result in huge improvements. Finally, use image subsets. Sometimes a simple model built on a properly selected subset shows more interesting details than a sophisticated model built on a large data set.

ABBREVIATIONS

ANNs	Artificial neural networks
BIL	Band interleaved by line
BIP	Band interleaved by pixel
BMP	Bitmap
BSQ	Band sequential
ENVI	Environment for visualizing images
FT-IR	Fourier transform infrared
ICA	Independent component analysis
IR	Infrared
JPEG	Joint photographic expert group
MAF	Maximum autocorrelation factor
MCR	Multivariate curve resolution
MIR	Multivariate image regression
MLR	Multilinear regression
MRI	Magnetic resonance imaging
MSC	Multiplicative scatter correction
NN	Neural network
NIR	Near infrared
PLS-DA	Partial least squares discriminant analysis
PC	Principal component
PCA	Principal component analysis
PLS	Partial least squares
PNG	Portable network graphics
RBF	Radial basis function
ROI	Region of interest
RMSEP	Root mean square error of prediction
SECV	Standard error of cross-validation
SIMS	Secondary ion mass spectrometry
SNV	Standard normal variate
SS	Sum of squares
SVM	Support vector machines
TIFF	Tagged image file format
ToF-SIMS	Time of flight secondary ion mass spectrometry
TSS	Total soluble solids

DEFINITIONS

artificial neural network: a general name for a network of nodes and internode weights that can be trained for a specific task.

backpropagation: a popular method of training an ANN.

bad pixel: a pixel that is nonresponsive (dead) or gives a highly nonlinear response. These are camera-based bad pixels.

bands: same as channels.

channels: the K wavelength or wavenumber intervals in a spectrum, usually identified as the center wavelength/wavenumber.

classification: using the results of a (cluster) model to put pixels/pixel groups into classes.

cleaning: removing bad/erroneous/background pixels before further data analysis.

clustering: using similarity or dissimilarity in multivariate space to make groups of pixels.

coefficient of determination: a diagnostic for regression models, close to 1 for perfect models and below 0.65 for bad models.

concentration map: a score vector from curve resolution that is reorganized into an image.

cross-validation: a naive validation method that often gives misleading results.

dependent variable: y-variable or response variable in a regression model.

dummy variable: a variable set to -1 or 1 (or 0 or 1) in a discriminant regression model.

endmember analysis: same as multivariate curve resolution, mainly used in geology.

erroneous pixel: a pixel that has spectroscopic errors that cannot be easily corrected, such as specular reflection, extreme shading, and edge effects. These errors are not camera based.

factor model: a statistical model that decomposes an image data matrix into a number (R) of factors and a residual matrix. The R factors each consist of a score vector (column) and a loading vector (row). By constraining the residual and the score/loading pairs, different types of models can be allowed.

global operations: operations that transform whole images into new images or data matrices/vectors.

grayscale image: an $I \times J$ array of intensities/numbers.

hidden layer: a layer between the input and output layers in an ANN.

hypercube: another name for the hyperspectral image data collected.

hyperspectral image: an array of $I \times J$ pixels (sometimes $H \times I \times J$ voxels) where each pixel (voxel) is a K-dimensional vector representing a spectrum ($K > 100$).

image arithmetic: arithmetic operations on whole image planes, usually involving one or two images.

image data matrix: a hypercube reorganized into a pixel $\times$ channel matrix with pixel positions preserved. If no confusion is possible, "data matrix" can be used.

image plane: a grayscale image extracted from the hypercube for one channel.

image subset: a specific well-chosen subset of an image data matrix. If no confusion is possible "subset" can be used.

independent variable: X-variable in a regression model.

input layer: a collection of nodes, one for each input variable in an ANN.

intensity image: a grayscale image.

latent variable: another name for a score from PCA or FA.

loading: a vector in a factor model that describes spectral information.

local neighborhood operations: operations that take into account neighbor values for each pixel operated on.

mode: a physical technique for collecting a multivariate/hyperspectral image. NIR, Raman, FT-IR, MRI, CAT scan, and ToF-SIMS are, for example, all imaging modes.

mosaic: a combined hypercube made of hypercubes taken on different occasions or of different samples.

multilayer perceptron: an older name for ANN.

multivariate curve resolution: a factor model that minimizes the residual sum of squares by making R nonnegative scores and loading pairs.

multivariate image: same as a hyperspectral image, but sometimes with K being smaller ($K < 20$).

node: a data collection and nonlinear transformation point in an ANN.

nonnegativity: the principle that spectra and concentrations cannot become negative, used as the basis for MCR modeling.

output layer: a collection of nodes, one for each output variable in an ANN.

principal component analysis: a factor model that minimizes the residual sum of squares after each component and creates R orthogonal scores and loading pairs.

radiometric operations: image operations on pixels that ignore neighbor pixel values.

registration: also called warping.

regression coefficient vector: a vector of regression coefficients, one coefficient for each channel (wavelength or wavenumber band).

residual image: an image of pixelwise standard deviations that visualizes which parts of the image are well or badly modeled by a factor model.

residual (matrix): the part of an image data matrix that does not fit the model.

response variable: y-variable or dependent variable in a regression model.

root mean square error of prediction: average standard deviation of prediction, based on test set residual.

score: a vector in a factor model that describes pixel/concentration information.

score image: a score vector reorganized into an image with the help of pixel indices.

standardization: expressing the elements of the hypercube in spectroscopic units instead of currents measured by the camera hardware. This operation requires standards, but ensures that all images are reproducible independent of camera and illumination source instabilities.

supervised: used for clustering when the wanted results are known as external information.

test set: a data set used for testing the regression equation and not used for regression model building.

training set: a data set used to train a regression/ANN model.

validation set: a test set used for testing prediction results.

unmixing: same as MCR.

unsupervised: used for clustering when the data generate the clusters without using external information.

warping: also called registration, geometrically fitting two images to each other.

***y*-variable**: dependent or response variable in a regression model

REFERENCES

1. Robin, M. and Poulin, M. (2000) *Digital Television Fundamentals*, 2nd ed., McGraw Hill, New York.

2. Poynton, R. (2003) *Digital Video and HDTV Algorithms and Interfaces*, Morgan Kaufmann, San Francisco, CA.

3. Arnold, J., Frater, M., and Pickering, M. (2007) *Digital Television: Technology and Standards*, Wiley, Hoboken, NJ.

4. Toriwaki, J. and Joshida, H. (2009) *Fundamentals of Three-Dimensional Image Processing*, Springer, Dordrecht.

5. Nikolaidis, N. and Pitas, I. (2001) *3-D Image Processing Algorithms*, Wiley, New York.

6. Schowengerdt, R. A. (2007) *Remote Sensing: Models and Methods for Image Processing*, 3rd ed., Academic Press, Burlington, MA.

7. TIFF, http://partnersrrdrr.adoberrdrr.com/public/developer/en/tiff/TIFF6.pdf.

8. Geladi, P., Burger, J., and Lestander, T. (2004) Hyperspectral image calibration: calibration problems and solutions. *Chemometr. Intell. Lab. Syst.* **72**, 209–217.

9. Burger, J. and Geladi, P. (2005) Hyperspectral NIR image regression. Part 1: calibration and correction. *J. Chemometr.* **19**, 355–363.

10. Geladi, P. (2007) Calibration standards and image calibration. In: Grahn, H. and Geladi, P. (Eds.), *Techniques and Applications of Hyperspectral Image Analysis*, Wiley, Chichester, pp. 203–220.

11. Salzer, R. and Siesler, H. (Eds.) (2009) *Infrared and Raman Spectroscopic Imaging*, Wiley-VCH, Weinheim.

12. Hubbard, A. (Ed.) (1995) *The Handbook of Surface Imaging and Visualization*, CRC Press, Boca Raton, FL.

13. Lau, D., Villis, C., Furman. S., and Livett, M. (2008) Multispectral and hyperspectral image analysis of elemental and micro-Raman maps of cross-sections from a 16th century painting. *Anal. Chim. Acta* **610**, 15–24.

14. Gonzalez, R. and Wintz, P. (1977) *Digital Image Processing*, Addison-Wesley, Reading, MA.

15. Rosenfeld, A. and Kak, A. (1982) *Digital Picture Processing*, 2nd ed., Vol. 2, Academic Press, New York.

16. Serra, J. (1982) *Image Analysis and Mathematical Morphology*, Academic Press, London.

17. Pratt. W. (1978) *Digital Image Processing*, Wiley, New York.

18. Kriete, A. (Ed.) (1992) *Visualization in Biomedical Microscopies: 3-D Imaging and Computer Applications*, Wiley-VCH, Weinheim.

19. Gonzalez, R. and Woods, R. (1992) *Digital Image Processing*, Addison-Wesley, Reading, MA.

20. Pitas, I. (2000) *Digital Image Processing Algorithms and Applications*, Wiley, New York.

21. Freeware ImageJ, http://rsbwebrrdrr.nihrrdrr.gov/ij/.

22. Geladi, P. and Grahn, H. (1996) *Multivariate Image Analysis*, Wiley, Chichester.

23. Grahn, H. and Geladi, P. (Eds.) (2007) *Techniques and Applications of Hyperspectral Image Analysis*, Wiley, Chichester.

24. Chang, C. (Ed.) (2007) *Hyperspectral Data Exploitation: Theory and Applications*, Wiley, Hoboken, NJ.

25. Lillesand, T., Kiefer, R., and Chipman, J. (2008) *Remote Sensing and Image Interpretation*, 6th ed., Wiley, Hoboken, NJ.

26. Sabins, F. (1978) *Remote Sensing Principles and Interpretation*, Freeman, San Francisco, CA.

27. Campbell, J. (2007) *Introduction to Remote Sensing*, Guilford Press, New York.

28. Rencz, A. (Ed.) (1999) *Remote Sensing for the Earth Sciences*, Wiley, New York.

29. Asrar, G. (Ed.) (1989) *Theory and Applications of Optical Remote Sensing*, Wiley, New York.

30. Howard, J. (1991) *Remote Sensing of Forest Resources: Theory and Application*, Chapman & Hall, London.

31. Kalacska, M. and Sanchez Azofeifa, G. (Eds.) (2008) *Hyperspectral Remote Sensing of Tropical and Subtropical Forests*, CRC Press, Boca Raton, FL.

32. Gendrin, C., Roggo, Y., and Collet, C. (2008) Pharmaceutical applications of vibrational chemical imaging and chemometrics: a review. *J. Pharm. Biomed. Anal.* **48**, 533–553.

33. Nicolaï, B., Beullens, K., Bobeleyn, E., Peirs, A., Saeys, W., Theron, K., and Lammertyn, J. (2007) Nondestructive measurement of fruit and vegetable quality by means of NIR spectroscopy: a review. *Postharvest Biol. Technol.* **46**, 99–118.

34. Geladi, P. (2003) Chemometrics in spectroscopy. Part 1. Classical chemometrics. *Spectrochim. Acta B* **58**, 767–782.

35. Behrend, C., Tarnowski, C., and Morris, M. (2002) Identification of outliers in hyperspectral Raman image data by nearest neighbor comparison. *Appl. Spectrosc.* **56**, 1458–1461.

36. Siesler, H., Ozaki, Y., Kawata, S., and Heise, H. (2002) *Near Infrared Spectroscopy: Principles, Instruments, Applications,* Wiley-VCH, Weinheim.

37. Burger, J. and Geladi, P. (2006) Hyperspectral NIR image regression. Part II. Dataset preprocessing diagnostics. *J. Chemometr.* **20**, 106–119.

38. Savitzky, A. and Golay, M. (1964) Smoothing and differentiation of data by simplified least squares procedures. *Anal. Chem.* **36**, 1627–1639.

39. Geladi, P., Macdougall, D., and Martens, H. (1985) Linearization and scatter-correction for near-infrared reflectance spectra of meat. *Appl. Spectrosc.* **39**, 491–500.

40. Barnes, R., Dhanoa, M., and Lister, S. (1989) Standard normal variate transformation and de-trending of near-infrared diffuse reflectance spectra. *Appl. Spectrosc.* **43**, 772–777.

41. Gowen, A., O'Donnell, C., Taghizadeh, M., Cullen, P., Frias, J., and Downey, G. (2008) Hyperspectral imaging combined with principal component analysis for bruise damage detection on white mushrooms (*Agaricus bisporus*). *J. Chemometr.* **22**, 259–267.

42. Gallagher, N., Shaver, J., Martin, E., Morris, J., Wise, B., and Windig, W. (2004) Curve resolution for multivariate images with applications to ToF-SIMS and Raman. *Chemometr. Intell. Lab. Syst.* **73**, 105–117.

43. Berman, M., Phatak, A., Lagerstrom, R., and Wood, B. R. (2009) ICE: a new method for the multivariate curve resolution of hyperspectral images. *J. Chemometr.* **23**, 100–116.

44. Jones, H., Haaland, D., Sinclair, M., Melgaard, D., Van Benthem, M., and Pedroso, C. (2008) Weighting hyperspectral image data for improved multivariate curve resolution results. *J. Chemometr.* **22**, 482–490.

45. Miyakoshi, M., Tomiyasu, M., Bagarinao, E., Murakami, S., and Nakai, T. (2009) A phantom study on component segregation for MR image using ICA. *Acad. Radiol.* **16**, 1025–1028.

46. Burger, J. and Geladi, P. (2006) Hyperspectral NIR imaging for calibration and prediction: a comparison between image and spectrometer data for studying organic and biological samples. *Analyst* **10**, 1152–1160.

47. ElMasry, G. and Wold, J. (2008) High-speed assessment of fat and water content distribution in fish fillets using online imaging spectroscopy. *J. Agric. Food Chem.* **56**, 7672–7677.

48. Nicolaï, B., Lotze, E., Peirs, A., Scheerlinck, N., and Theron, K. (2006) Non-destructive measurement of bitter pit in apple fruit using NIR hyperspectral imaging. *Postharvest Biol. Technol.* **40**, 1–6.

49. Lasch, P., Clem, M., Hänsch, W., and Naumann, D. (2006) Artificial neural networks as supervised techniques for FT-IR microspectroscopic imaging. *J. Chemometr.* **20**, 209–220.

50. Fartifeh, J., Van der Meer, F., Atzberger, C., and Caranza, E. (2007) Quantitative analysis of salt-affected soil reflectance spectra: a comparison of two adaptive methods (PLSR and ANN). *Remote Sens. Environ.* **110**, 59–78.

51. Pu, R., Gong, P., Tian, Y., Miao, X., Carruthers, R., and Anderson, G. (2008) Invasive species change detection using artificial neural networks and CASI hyperspectral imagery. *Environ. Monit. Assess.* **140**, 15–32.

52. Gan, G., Ma, C., and Wu, J. (2007) *Data Clustering: Theory, Algorithms, and Applications*, SIAM, Philadelphia, PA.

53. Xu, R. and Wunsch, D. (2009) *Clustering*, Wiley, Hoboken, NJ.

54. Mirkin, B. (2005) *Clustering for Data Mining: A Data Recovery Approach*, Chapman & Hall/CRC Press, Boca Raton, FL.

55. Omran, M., Engelbrecht, A., and Salman, A. (2007) An overview of clustering methods. *Intell. Data Anal.* **11**, 583–605.

56. Tran, T., Wehrens, R., and Buydens, L. (2005) Clustering multispectral images: a tutorial. *Chemometr. Intell. Lab. Syst.* **77**, 3–17.

57. Fernandez-Pierna, J., Baeten, V., and Dardenne, P. (2006) Screening of compound feeds using NIR hyperspectral data. *Chemometr. Intell. Lab. Syst.* **84**, 114–118.

PART II

BIOMEDICAL APPLICATIONS

6

BIOMEDICAL APPLICATIONS OF RAMAN IMAGING

MICHAEL D. MORRIS AND GURJIT S. MANDAIR

Department of Chemistry, University of Michigan, Ann Arbor, MI, USA

6.1 INTRODUCTION

Human and animal tissues provide new and exciting prospects for the application of Raman imaging. Small structural features and compositional differences can be imaged with resolution that is comparable to that of optical microscopy and provide Raman spectral markers for a variety of disease states, such as brain cancer, gastrointestinal disorders, macular degeneration, and dental caries. Multivariate analyses are widely applied to Raman images of tissue to enhance image contrast and potentially visualize pathology before morphological changes become apparent. Raman images have also been used to develop morphological models to diagnose certain cancers, such a breast cancer *in situ*.

With fast computer processing and efficient CCD cameras and filters, high-fidelity Raman images of fresh tissues and single cells can now be obtained, as can images from specimens that have been fixed, stained, labeled, or embedded for conventional light microscopy. There is also considerable progress made toward the development of fiber-optic Raman *in vivo* techniques for defining brain or skin tumor margins for surgical treatment. Particularly exciting are the prospects of combining Raman imaging with other biomedical imaging modalities, such as infrared, fluorescence, ultrasound, acoustic impedance, and coherence optical tomography. Besides obtaining spatially localized compositional, structural, and functional data, multimodal approaches offer the first real opportunity to evaluate the specificity of Raman as a biomedical diagnostic. Our research group has recently demonstrated that bone Raman scatter from overlying soft tissue layers could be recovered noninvasively and reconstructed to produce Raman tomographic images using methods adapted from fluorescence diffuse tomography.

In this chapter, we will survey Raman imaging applications to a wide range of tissues, cells, and biofluids. This chapter will review the biomedical insights gained from earlier Raman mapping and spectral studies to provide the reader with some historical perspective and highlight the current status of Raman imaging in the field. More important, we hope that this broad overview of the field will serve as a guide to those who wish to enter the biomedical field or expand their studies into new areas.

6.2 BRAIN

6.2.1 Malignant Glioma and Tissue Necrosis

Brain tissue changes profoundly in structure and function during development and aging; however, these patterns can be dramatically altered by the growth, progression, and invasion of primary brain tumors [1, 2]. Malignant glioma is one of the most aggressive types of primary brain tumors, accounting for 45–50% of all brain tumors reported in patients. The prognosis for patients with malignant glioma is poor with median survival ranges of 9–12 months [3, 4]. The treatment for gliomas is surgical resection to the largest extent possible, followed by adjuvant radiotherapy and chemotherapy [5]. Optical imaging techniques that clearly define tumor margins during surgical resection would be of immense value to both basic and clinical neuroscience. Raman microspectroscopy holds considerable promise as an *in vivo* technique for diagnosing, grading, and defining tumor margins [6–8]. Early Raman mapping studies identified glycogen as one of the major polysaccharide components of vital glioblastoma tissues collected from

Raman, Infrared, and Near-Infrared Chemical Imaging   Edited by Slobodan Šašić and Yukihiro Ozaki
Copyright © 2010 John Wiley & Sons, Inc.

20 glioma patients [6]. By performing cluster analysis, spatial distribution of glycogen could be visualized. Microscopic crystal-like cholesterol inclusions and calcified deposits were also identified and localized to the necrotic region of glioma tissue. More important, biochemical information contained in the Raman spectrum enabled necrotic tissue to be discriminated from vital glioblastoma tissue. This finding would be extremely useful for grading tumors *in vivo*. To avoid underestimating tumor grades in which necrotic tissues was inadequately sampled, high wavenumber (HWVN) Raman maps of human glioblastoma tissues were produced [9]. Compared to vital glioblastoma, surrounding necrotic tissues give lower DNA and stronger cholesterol ester signal contributions.

Recent multivariate Raman imaging studies have shown that necrotic tissue could be distinguished from the proliferative and invasive activities of rat glioma cells, as well as from different brain anatomical structures, such as the corpus callosum and cortex [7]. In nontumor tissues, lipid content was highest in the corpus callosum but decreased gradually toward the cortex. A similar trend was observed in tumor tissues but the total lipid content was reduced as a result of demyelination. The lowering of myelin lipid content may have been caused by glioma cells requiring more energy to grow compared to the surrounding tissues. Clusters associated with proliferative and invasive activities of glioma cells were identified and correlated with the tumor-promoting activities of the Ki-67 and MT1-MMP proteins, respectively. Validation of proliferative and invasive Raman spectral markers by immunohistochemical methods is extremely invaluable, as is the identification of necrotic and perinecrotic zone clusters that have been linked with a poorer clinical outcome. Clusters associated with the accumulation of plasma proteins in response to edema were also identified and localized to tumor tissues and the tissues adjacent to the tumors.

6.2.2 Brain Metastases and Meningioma

Combined FTIR imaging and fiber-optic Raman mapping techniques have also been used to detect suspected brain metastases of malignant melanomas in mice injected with lung or skin cancer cell lines [10, 11]. Fiber-optic Raman mapping could detect the presence of suspected tumors in 2 mm thick tissue sections with sizes ranging from 240 μm × 240 μm (4 pixels) to 1.1 mm × 1.2 mm (90 pixels). The locations of the tumors were confirmed by consecutive histological staining. In addition, the absence of tumors from Raman maps collected from the reverse side of the tissue sections provided some information on tumor depth penetration. Raman spectra taken from the suspected tumor sites were dominated by melanin bands as they were resonance enhanced in the interval 400–1800 cm^{-1}. In contrast, FTIR images collected from the same tissue section in the interval

2750–3050 cm^{-1} showed differences only in brain tissue morphologies. Although fiber-optic Raman mapping were performed *ex vivo*, the flexibility of the probe may allow *in vivo* mapping during surgical resection. In contrast, FTIR imaging would be more suited for small specimens or specimens with known tumor locations. FTIR and/or Raman techniques have been used to study other brain tumor types and to visualize tissue hemorrhage, calcified deposits, carotene inclusions, and increased nucleic acid contributions [8, 12, 13]. For example, Figure 6.1 shows chemical images collected from human meningioma tumor tissues *via* a wide-view Raman imaging system [8]. In this figure, carotene inclusions in a 43 × 37 μm^2 tumor area is imaged in 50 min using a lateral resolution of 0.54 μm (Figure 6.1b). By using single band Raman imaging at 1581 cm^{-1}, tissue hemorrhage could also be visualized (Figure 6.1d).

6.3 BREAST

6.3.1 Breast Cancer

The potential benefits of using Raman spectroscopy to diagnose breast cancer have been studied by several research groups [14–20]. Visualization of breast tissue microstructural features is a key first step toward the development of a noninvasive Raman imaging technique for the identification and classification of breast cancer in a clinical environment. In one early study, high-definition chemical images of

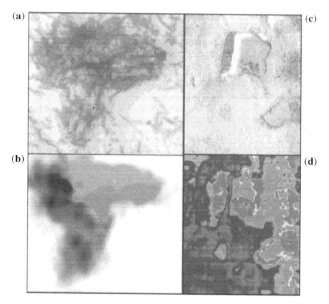

FIGURE 6.1 Raman imaging of a meningioma tumor section. (a) Photograph (size 43 × 37 μm^2) and Raman chemical image of carotene inclusion. (b) Photomicrograph (c, size 1.8 × 2.4 mm^2) and a single band Raman image at 1581 cm^{-1} to visualize hemorrhage (d). Reprinted from Ref. 8 with permission from Wiley. (See the color version of this figure in Color Plate section.)

thin chicken breast tissues were obtained using a Raman microscope equipped with an liquid crystal tunable filter (LCTF) [17]. Component discrimination based on univariate (ratiometric) technique was used to visualize the distribution of lipid and protein components. However, image contrast could be improved by using multivariate techniques, such as classical least squares analysis. Other multivariate techniques based on principle component analysis (PCA) have also been used to identify lipid and carotenoid components in lymph node biopsies collected from patients diagnosed with breast cancer [21].

In another study, Raman microimages collected from 60 human breast tissue biopsies were used to create a morphological model that could eventually be used to develop an algorithm to diagnose female breast cancer *in situ* [22]. Although morphological modeling requires more advanced knowledge of the specimen compared to PCA and other multivariate methods, it affords greater insight and accurately reflects the physiochemical changes associated with disease diagnosis [23]. Figure 6.2 shows the morphologically derived Raman images for collagen, cell cytoplasm, and cell nuclei from a normal breast duct [22]. A comparison with the serial stained section shows that the Raman images correlate well with the tissue architecture. The model used a linear combination of these Raman basis spectra together with those derived from fat (mostly triolein), cholesterol-like deposits, β-carotene, calcium hydroxyapatite, calcium oxalate dihydrate, and water. Although the value of the model is dependent on the signal to noise ratio of the spectra being fitted, the model overcomes the problems associated with

overfitting as the majority of the basis spectra are morphologically derived, especially collagen that is known to be present in human tissue in many different forms. However, basis spectra from synthesized or commercially available chemicals, including those derived from deparaffinized tissue sections, can be used when necessary. The nine morphological and/or chemical basis spectra used in the morphological model are sufficient to explain the major spectral features associated with normal, diseased, and cancerous breast tissue biopsies. In contrast, only three basis spectra are needed to model normal and transformed human breast epithelial cells [24]. In this approach, DNA, RNA, and proteins were extracted from the cell nuclei and used to construct spectra-fitting models to visualize compositional changes associated with tumorigenesis at the subcellular level.

6.3.2   Breast Tumor Progression Models

Breast pathologists frequently use tumor progression models to describe a series of abnormal changes that occur in breast ductal tissue over time. The model often includes hyperplasia of usual type (HUT), atypical hyperplasia (ADH), ductal carcinoma *in situ* (DCIS), invasive ductal carcinoma (IDS), and finally metastasis [25, 26]. Although there are clear morphological differences between the pathological groups, the models may not always be linear. Raman maps collected from 50 histochemically graded breast ductal biopsies were used to find a possible progressive biochemical link within a four-group proliferation model (HUT, ADH, DCIS, and IDC) and a four-DCIS pathological model (low-grade, intermediate-grade, and high-grade noncomedo, as well as high-grade comedo) [27]. Linear discriminant analysis (LDA) that maximizes variance between groups and minimizes variance within groups [21] was applied to the proliferation and DCIS models. Spectral analysis performed on the proliferation model showed that the relative lipid levels in breast tissues with HUT were similar to normal breast tissue [27]. As pathology progressed, fat levels appeared to vary as it was substituted by higher collagen contributions. There was poor discrimination of DCIS between the other pathological groups within the proliferation model. The DCIS group appeared to be associated more with HUT and IDC than its supposed precursor ADH. Although other biochemical changes were observed, the proliferation model did not appear to follow a general pattern as pathology progressed. In contrast, spectral differences were observed between the mean spectra of the first three grades of DCIS in the DCIS pathological model, whereas the mean spectra of high-grade comedo DCIS group were quite distinct, owing to some interductal necrosis. The DCIS pathological model appeared to follow a pattern within its own group. This study highlights some of the major challenges still faced by breast pathologists and spectroscopists in classifying or fitting

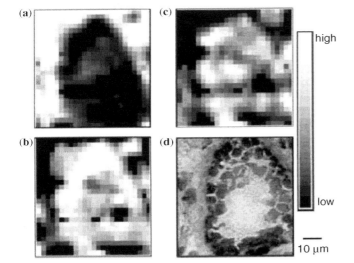

FIGURE 6.2   Raman images of normal breast duct (a–c) with corresponding serial stained section (d). Each image represents the contribution of a specific morphological element to the region being studied: (a) collagen, (b) cell cytoplasm, and (c) cell nucleus. Reprinted from Ref. 22 with permission from Wiley.

biochemical pathological progression to models that may not be linear or exhibit extreme nonlinearity.

6.3.3 Breast Implant Materials and Pathology

Raman spectroscopy is increasingly being used to answer important medical questions surrounding the composition and pathology of breast implant materials [28–30]. One of the earliest applications of Raman imaging to breast implant tissues employed an acoustic-optic tunable filter (AOTF) [31]. AOTF Raman imaging was used to detect Dacron polyester inclusions in a histopathologically graded breast implant capsular tissue biopsy. Dacron polyester patches are often used to attach silicone implants to the chest wall during reconstructive and cosmetic surgery. Although breast tissue lipid and protein spectral features are masked by high tissue fluorescence, polymer inclusions of $\sim$15 μm in diameter could be imaged by taking the ratio between the Dacron polyester Raman image at 1615 cm^{-1} and the background image at 1670 cm^{-1}. As shown in Figure 6.3, a high-fidelity Dacron chemical image within a 230 × 230 pixel area was acquired using a 10 min integration time. Confocal Raman mapping techniques have also been used to distinguish between protein and Dacron fibers from a surgical suture site, as well as spatially resolve silicone and polyurethane particles associated with silicone explants [32].

6.4 GASTROINTESTINAL TRACT

6.4.1 Barrett's Esophagus and Esophageal Adenocarcinoma

Barrett's esophagus, a condition of long-term gastroesophageal reflux disease, is characterized by the incomplete replacement of normal squamous mucosa in the lower esophagus by columnar-lined mucosa of intestinal origin [33, 34]. Patients with intestinal metaplasia are at increased risk of developing dysphasia and, in some cases, esophageal adenocarcinoma [35, 36]. The prognosis for patients with esophageal cancer is poor, as the tumor is highly invasive and can spread to other parts of the body by metastasis [37]. Patients with low-grade dysphasia (LGD) are advised to undergo routine endoscopic biopsy surveillance, whereas patients with high-grade dysphasia (HGD) may require more intensive surveillance with surgical resection or endoscopic ablation [38]. The benefits of using such surveillance approaches, however, still remain unproven, owing to the problem with interobserver variability and inadequately defined markers of dysplasia and early carcinogenesis [39–42]. Several emerging optical diagnostic techniques are currently been investigated and in some cases validated against gold standard histopathological approaches to address some of these issues [43–46]. For example, in

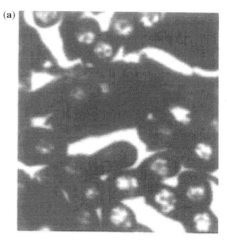

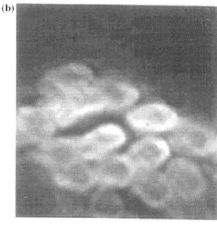

15 μm

FIGURE 6.3 High-definition images of Dacron polyester in human breast implant capsular tissue: (a) bright-field reflectance image and (b) background ratioed Raman image (1615 cm^{-1}/ 1670 cm^{-1}); 10 min integration, 20× (NA = 0.46) objective. Reprinted from Ref. 31 with permission from the American Chemical Society.

a consensus pathology and Raman spectroscopic study involving esophageal biopsies collected from 44 patients undergoing surveillance for Barrett's esophagus, sensitivities and specificities of 73–100% and 90–100% were obtained, respectively [47]. Multivariate analysis techniques were used to create spectral classification models to allow the objective prediction of pathology. In a following study, pseudocolor Raman (score) maps were generated and used to visualize the biochemical changes in histopathologically graded HGD and adenocarcinoma biopsies [48]. Mean spectra taken from selected regions of the maps showed that HGD and adenocarcinoma sites were associated with a high DNA, oleic acid, and actin level, while the relative glycogen levels were greatly reduced compared to normal squamous mucosa. More recently, the use of CCD cameras with zero readout times has enabled Raman images of frozen esophageal tissue sections to be acquired 3–7 times faster than point mapping approaches, thereby making the implementation of Raman imaging for histological screening in clinical environment a future possibility [49, 50].

6.4.2 Colon Wall Structure and Composition

The colon is the final area of the GI tract charged with recapturing electrolytes and water from food products [51]. Structurally, the colon wall consists of four basic layers, namely, the mucosa, submucosa, muscularis externa, and serosa [52, 53]. The mucosa, the innermost layer of the colon, is lined with adsorptive and secretory epithelial cells. These cells are supported by connective tissues and nerve cells from the underlying submucosa. Beneath the submucosa is the muscularis externa that contains circular and longitudinal tissues of smooth muscle. These muscle tissues assist in the movement of food products along the GI tract by means of wave contractions called peristalsis. Covering the muscularis externa is the outermost serosa layer that provides the necessary lubricating fluids to minimize friction associated with peristalsis.

The mucosa have been the focus of numerous *ex vivo* and *in vivo* Raman spectroscopic studies involving colon cancers [18, 54–56]. The first reported application of Raman imaging to colon cancer involved the human colonic carcinoma cell line HT29 [23]. In this study, cell membrane, nucleus, and cytoplasm were identified when spectra of HT29 cells were fitted with chemical spectra of phosphatidylcholine, DNA, cholesterol linoleate, triolein, and actin, the later being morphologically derived from "cell cytoplasm" of breast ductal tissues. Raman spectroscopic and imaging studies, however, have seldom been used to characterize tissues and nerve cells underlying the mucosa, such as ganglia nerve cells found in between the circular and longitudinal muscle tissues. The absence of ganglia cells is related to Hirschsprung's disease, a relatively common pediatric illness with symptoms that include feeding intolerance, abdominal distension, and chronic constipation [57, 58].

Combined Raman and FTIR imaging techniques have recently been used to visualize the biochemical composition of colon wall tissue section obtained by colostomy from a neonate with an anorectal malformation [53]. Unsupervised multivariate (cluster) analyses were applied to the segmented spectral data sets to yield 14 color-coded Raman and FTIR images and spectra that were assigned and compared. The epithelium and mucus secretion on top of the mucosal layer were the first cluster memberships identified from the Raman image acquired using a step width of 10 μm. The corresponding site-matched FTIR data convincingly showed that the mucus was produced by epithelial glands and comprised of polyglycosylated peptides. Moreover, the epithelium gave rise to two identical DNA clusters, albeit with slightly different signal contributions. These differences were attributed to spatial variation in epithelial cell turnover activities. Besides the elevated DNA and mucus levels, epithelial spectra gave lower collagen contributions compared to that of connective tissue and muscle tissue. An additional muscle tissue constituent, termed muscularis mucosa, was identified between the mucosa and submucosa tissues. Raman and FTIR images of longitudinal muscle tissues could be distinguished from circular muscle tissues by the lower collagen contribution in the later tissue constituent. By using a step width of 2.5 μm, Raman images of subcellular features of ganglia could be spatially resolved. This is clearly illustrated in Figure 6.4 in which the Raman image yielded more cluster memberships compared to those derived from the corresponding diffraction-limited FTIR image. The Raman image and single point spectra analysis showed that the ganglia substructure was comprised of lipid, DNA, and RNA components that could be readily distinguished from circular muscle tissues and the collagen-rich fibrous septa surrounding the ganglion.

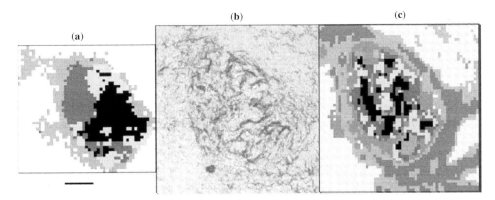

FIGURE 6.4 FTIR microscopic image (a), photomicrograph (b), and Raman microscopic image (c) of ganglia. The Raman image consist of 59 × 59 spectra (c) recorded with a step size of 2.5 μm. Colors represent cluster memberships: fibrous septa (red, orange), circular muscle layer (yellow), subcellular features (black, magenta, blue, and cyan), and transition between ganglion and fibrous septa (green). Bar = 20 μm. Reprinted from Ref. 53 with permission from Wiley-VCH Verlag GmbH & Co. (See the color version of this figure in Color Plate section.)

6.5 URINARY

6.5.1 Bladder Outlet Obstruction

Raman mapping has mainly focused on identifying the structural composition or biochemical changes associated with normal or diseased urinary tissues and cells. Historically, the first example of Raman mapping to urinary tissue involved the characterization of a layer composition of a healthy guinea pig bladder tissue in the 400–2000 cm^{-1} spectral range [59]. By subdividing the spectral data set into groups of similar spectra using cluster analysis, Raman pseudocolor maps of the urothelium, lamina propria, and muscle layer could be obtained. The lamina propria gave higher collagen signal contributions, which was confirmed by immunohistochemical analysis. In contrast, the muscle layer was dominated by high actin and myosin signal contributions, whereas the urothelium contained strong fatty acid signal contributions. Similarly, bladder clusters have been identified in Raman maps collected from patients in the HWVN spectral range (~2400–3800 cm^{-1}) [9]. This is an important finding as the HWVN region is far removed from signal interference from fused silica fibers and thus would simplify the collection and data analysis of *in vivo* fiber-optic Raman measurements [60]. HWVN Raman maps collected from submucosal regions showed evidence of colocalization between smooth muscle and fibrocollagenous tissue [9]. A similar colocalization of smooth muscle and collagen fibers was identified in the fingerprint Raman maps collected from damaged guinea pig bladder tissue in response to partial outlet obstruction [61]. This implied that the infiltration of collagen fibers in the bladder wall tissue was the result of outlet obstruction and was confirmed by histology. Clusters of normal muscle and damaged muscle were also identified. Difference spectra taken between these clusters gave an almost pure spectrum of glycogen that indicated the accumulation of glycogen in damaged muscle tissue areas.

6.5.2 Bladder Cancer

Raman maps of nontumor and tumor bladder tissues collected from 15 patients subjected to cluster analysis yielded from three to seven clusters per map [62]. For each cluster, a cluster-averaged spectrum (CAS) was calculated and classified by LDA as nontumor or tumor. Nontumor CAS contained higher collagen content, while CAS from tumor areas were characterized by high lipid, nucleic acid, protein, and glycogen content. However, two of the nontumor CAS were classified as tumor by the LDA model. Raman maps showed that CAS from highly inflamed tissues also contained high nucleic acid, and thus it would not be possible in all cases to discriminate between CAS of inflammation tissue areas and CAS from tumor areas. Nevertheless, 84 of the 90 CAS (93%) were correctly classified with 94% sensitivity and 92% specificity.

6.5.3 Testicular Microlithiasis

Testicular microlithiasis is a rare clinical condition characterized by multiple calcifications (microliths) scattered randomly throughout the testicular parenchyma [63]. Although this condition has questionable significance as a marker for testicular cancer, it is readily detected by radiological and histological methods [63, 64]. Microliths arise from an accumulation of cellular debris forming a central calcified core, surrounded by concentric layers of connective or stratified collagen fibers. Recent Raman mapping studies provided invaluable insights into the molecular composition of gonadal microlithiasis and its surrounding tissues in both malignant and benign specimens [65]. The structures (clusters) of the testicular parenchyma with microliths adjacent to a germ cell tumor (seminoma) obtained by Raman mapping closely followed the morphology observed in stained tissue sections. Raman spectra of microliths showed evidence for hydroxyapatite and protein-like components. This suggests that a protein-like material was captured in the microlith. In contrast, the tissue surrounding the microliths was rich in lycopene and glycogen, while the basal membrane of the seminiferous tubule contained higher collagen component. The presence of glycogen within all the malignant specimens and its noticeable absence in the only one benign lesion pointed toward the pathogenic role of the precursor cells of germ cell neoplasm in the gonads.

6.5.4 Kidney Glomerulus

In the kidney, the glomerulus is responsible for the clearance of waste products and therefore is a key site for many renal dysfunctions [66]. The application of Raman imaging techniques for detecting aberrations in renal architecture and function would be of considerable interest to the clinician. For example, Raman molecular imaging (RMI) can generate spatially accurate, reagentless images of the human glomerulus tissue [67]. The Raman images were acquired using a wide-field imaging microscope integrated with a multiconjugate filter (MCF). The MCF tunable filter enabled images to be created at specific wavelengths, thereby increasing throughput and rejecting unwanted wavelengths and stray light. As shown in Figure 6.5, the optical contrast between the kidney glomerulus bright-field image and the Raman images acquired at 1450, 1650, and 2930 cm^{-1} was clearly evident. In addition, by using a fluorescence MCF tunable filter, autoflourescence images (or spectra) collected from the stained tissue sections could be linked to specific regions in the Raman spectral image (or spectra).

6.5.5 Prostatic Cancer Cells

The combined approach of RMI and fluorescence imaging techniques has been used to visualize single prostate cancer

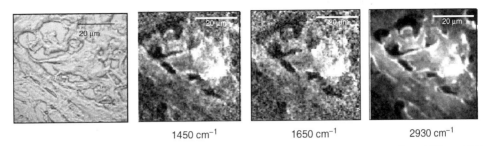

1450 cm⁻¹   1650 cm⁻¹   2930 cm⁻¹

FIGURE 6.5 Bright-field image of kidney glomerulus with select Raman image frames collected at 1450, 1650, and 2930 cm⁻¹. Reprinted with permission from Ref. 67.

cells labeled with fluorescent dye or nanocrystals [67, 68]. Confocal Raman techniques have also been used to analyze and visualize the intrinsic biochemical composition of two different human prostatic cell line populations, namely, PNT1A (normal prostate) and LNCaP (prostate adenocarcinoma) [69]. Raman score maps generated by PCA could be identified with the overall average cellular spectrum, as well as to visualize the distribution of DNA/protein, cytosol, nucleic acids, and a variety of different biochemicals linked to subtle differences between the cell types. For example, the malignant cells contained a higher β-sheet conformation, whereas the benign cells exhibited a higher, more stable, α-helical conformation. Furthermore, the malignant cells displayed a higher DNA content compared to benign cells.

6.6 SKIN

6.6.1 Basal Cell Carcinoma

Basal cell carcinoma (BCC) is one of the most studied malignancies of the skin, owing to its prevalence amongst Caucasians with a history of excessive sun exposure [70]. Epidemiological data estimate an incidence rate of 800,000 cases of BCCs in the United States every year with an overall incidence rate that is increasing worldwide by 3–10% per annum [70, 71]. Although slow growing and rarely metastatic, BCC can cause significant local destruction to sun-exposed areas of the body, such as the head or neck. For BCCs occurring near anatomically sensitive sites, such as the nose and eyes, examination of frozen pathology during Mohs surgery is recommended. Mohs surgery is an effective tissue-sparing procedure for BCC with a cure rate of 98–99% [72]. However, for large and complex cases, Mohs surgery is time-consuming for both the surgeon and the pathologist. Rapid detection of BCC tumor margins using real-time intraoperative optical methods may minimize the need for frozen histology and expedite Mohs surgery [71, 72]. Recently, confocal Raman system was used to map 15 frozen BCC sections collected from 15 patients [73]. Multivariate statistical analyses performed on the Raman maps were used to determine the tumor margins in frozen, unfixed tissue sections. Morphologically, BCCs resemble

hair follicle structures [70] and their locations coincided with the nodular features observed in the Raman maps and histology. In the Raman maps, BCC tumor margins could be delineated from the surrounding nontumorous dermal tissue. The BCC spectra contained higher lipid and nucleic acid content. The high BCC nucleic acid content reflected the high cell density of the tumor compared to the surrounding nontumorous tissue. In four Raman maps, dermal tissue adjacent to the tumor appeared to be poor in collagen, which was consistent with the theory that matrix metalloproteinases played a pathogenic role in BCC, degrading the collagen-rich dermal tissue. In addition, 3 of the 15 frozen sections appeared to contain a dense chronic inflammatory infiltrate. Spectral difference studies showed that the infiltrate contained higher fatty acid and aromatic amino acid contributions, whereas collagen levels were reduced compared to normal dermal tissue spectra. Although subtle differences were noticed between BCC and epidermal tissue spectra, three of the epidermal spectra were misclassified and predicted as BCC. Based on logistic regression modeling, a sensitivity of 100% and specificity of 93% were achieved for BCC. Micro-Raman spectroscopy has been used to visualize the distribution of protein in normal human skin melanocyte cells [74], whereas stimulated Raman scattering (SRS) microscopy used the CH_2 band at 2845 cm⁻¹ to visualize the lipid-rich domains in mouse skin tissue sections [75].

6.6.2 Wound Healing

The epidermis presents the first physical barrier to wound injury. When breached, the epidermis initiates the wound healing process to prevent bacterial infection and restore homeostasis [76]. Wound healing is a highly dynamic process, involving the timed and balanced activity of inflammation, proliferation, reepithelialization, and remodeling [77, 78]. Such processes can be followed by spectroscopic techniques in a time-dependent manner. For instance, combined IR and confocal Raman techniques followed changes that occurred during the reepithelialization of human excisional wounds maintained in cell culture for a period of 6 days [77]. Raman images of unwounded and

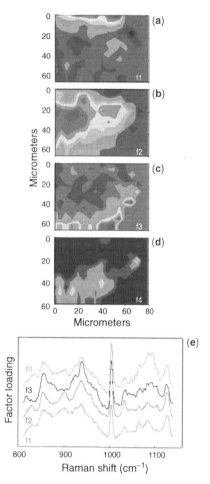

FIGURE 6.6 Factor analysis of a confocal Raman data set delineates skin regions near a wound edge 0.5 days after wounding. Factor analysis was conducted over the 800–1140 cm^{-1} region yielding four loadings that map to anatomically distinct regions in skin. (a) The spatial distribution of scores for f1 highlights the stratum corneum region of the skin, rich in keratin-filled corneocytes and lipids. (b) Factor loading 2 shows high scores in the underlying epidermal region, while high scores for f3 (c) reside near the dermal–epidermal boundary region. (d) The size, location, and spatial distribution of several smaller regions with high scores for f4 are identified as cell nuclei. (e) Factor loadings reveal several spectral features specific to the microanatomy of the epidermis in human skin. Reprinted from Ref. 77 with permission from Wiley. (See the color version of this figure in color plate section.)

wounded regions of the skin were collected in the interval of 800–1140 cm^{-1}. At 12 h post-wound injury, Raman (score) images in Figure 6.6 showed the spatial distribution of the stratum corneum, the underlying epidermal, and the dermal–epidermal boundary region, as well as the distribution of several smaller regions identified as cell nuclei. Keratin was identified as the major protein component of the stratum corneum and epidermal layers, while the dermal–epidermal layer was dominated by collagen. Lipids and DNA components were confined to the stratum corneum and cell nuclei,

respectively. In the time-dependent Raman images, variations in elastin spatial distribution was observed near the wounded area of the skin. Based on gene microarray studies, the sudden decrease in elastin levels observed during the second day of the wound healing process was attributed to increased expression of cathepsin S, matrix metalloproteinase-7, and the suppression of lysyl oxidase activity.

6.7   OCULAR

6.7.1   Age-Related Macular Degeneration

Macular pigment (MP) in the human retina is composed of three carotenoids, lutein, zeaxanthin, and *meso*-zeaxanthin [79]. These carotenoids are concentrated within the macula lutea region of the retina, as well as the retinal depression called the fovea. The fovea contains the highest density of cone photoreceptors that are essential for high-acuity color vision [79, 80]. MPs are potent antioxidants and are thought to protect the retina against oxidative stress in response to age-related macular degeneration (AMD), a leading cause of irreversible blindness in the elderly ($\geq$65 years old). A variety of methods have been used to assess MP in the human retina, of which resonance Raman imaging (RRI) is a recently developed *in vivo* method. MP carotenoids are stereoisomers, each containing long conjugated polyene chains, thereby giving rise to a prominent C=C stretching Stokes Raman band around 1524 cm^{-1}. This band is resonance enhanced in the blue-green spectral range, with a peak centering around ~527 nm [81]. The 1524 cm^{-1} band can be used to measure MP concentrations in human retina and has been validated against chromatographic methods using model systems, such as excised human donor eyecups [79]. RRI performed on 17 healthy volunteers showed significant intersubject variations in MP concentrations, symmetries, and spatial extent. The spatial distribution of MP in one RRI image was similar to that observed in the associated fluorescence-based image in which the MP level peaked at the center and fell rapidly toward the outer region of the macula. The RRI images of healthy volunteers were categorized into four main groups that could be distinguished from each other, as well as from RRI images collected from three elderly patients with pathological changes in the retina and/or vitreoretinal interfaces. For example, Figure 6.7 shows RRI images of a 57-year-old healthy male and a 70-year-old female diagnosed with the mild form of dry AMD. In the healthy male patient, the MP distribution consists of a narrow central peak but with a noticeable disruption to the outer MP ring structure. In contrast, the MP ring structure in the AMD case was broken up with a relatively high central peak and crosslike spokes. In addition, Figure 6.7 shows MP distributions in the left and right eyes of a 62-year-old female after detachment of the vitreous in the right eye. Detachment

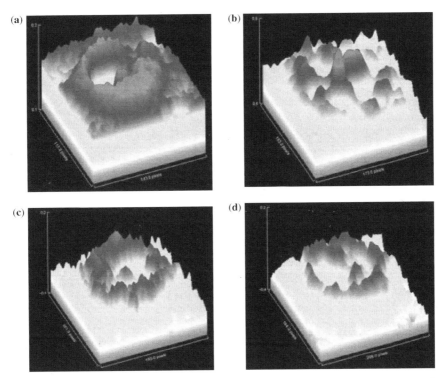

FIGURE 6.7   RRI images of three subjects with ringlike MP distributions. (a) A 57-year-old healthy male with MP distribution consisting of a narrow central peak and a surrounding strong, nearly rotationally symmetric distribution. MP levels in the ring are slightly higher than at the center and feature a noticeable disruption/offset at the "2 o'clock" position. (b) A 70-year-old female diagnosed with a mild form of dry AMD, showing a weak, broken ring structure with central high MP density and crosslike spokes. MP distributions in left (c) and right eye (d) of a 62-year-old female measured after detachment of the vitreous in the right eye. Six months prior to detachment, RRI images revealed the same ringlike MP pattern with a central spike in both eyes. Detachment of the vitreous apparently caused the formation of a double-peak MP structure inside the MP ring in this subject. Reprinted from Ref. 79 with permission from the Optical Society of America.

of the vitreous apparently caused the formation double-peak MP structures inside the MP ring.

6.7.2   Cholesterol and Cataracts

Cholesterol is a major lipid component of ocular tissue, and alterations in lens lipid composition and content have been implicated in the formation of cataracts [82]. Raman imaging approaches have been employed to indirectly image the distribution of cholesterol in healthy rat eye slices incubated with filipin, a cholesterol-binding fluorescent antibiotic [83]. Raman spectrum of filipin is characterized by an intense signal at $1586\,cm^{-1}$ and negligible interference from signals arising from lens proteins and lipids makes it ideal for imaging. The filipin Raman image obtained by subtracting the Raman scattered light at $1510\,cm^{-1}$ (background) from the Raman scattering at $1586\,cm^{-1}$ (filipin) showed a honeycomb structure identical to the one in the bright-field image. The appearance of the honeycomb structure in the filipin Raman image was attributed to the high cholesterol concentration ($\sim$7 mM) in the lens fiber membranes compared to that inside the fibers of the eye lens. The honeycomb structure was not observed in Raman images of rat lens

incubated in the absence of filipin. Moreover, the Raman image obtained using the $1450\,cm^{-1}$ CH_2 and CH_3 signal ($1450\,cm^{-1}$ minus $1510\,cm^{-1}$) showed that lens protein distribution was homogeneous. In another study, the relative filipin intensity was highest in cataractous region of the human eye lens, which equated to a high cholesterol concentration (with unesterified 3β-OH groups) compared to the healthy portion of the eye lens [84]. In contrast, the protein distribution as measured by the relative intensity of the intrinsic phenylalanine band at $1004\,cm^{-1}$ was reduced in the cataractous region, but the total amount of protein was not different. In single human cells, nonresonant Raman imaging experiments showed that the distribution of nuclear protein within eye lens epithelial cells was also homogeneous [85]. The later Raman image was obtained using the intense protein CH_2 and CH reporter bands around $3000\,cm^{-1}$.

6.7.3   Human Tear Fluid

Numerous proteomic reviews have indicated that changes in human tear fluid protein composition can be used to diagnose the disease status of ocular surfaces and to better understand the complex chemistry behind tear fluid deficiencies [86, 87].

Recently, a novel drop coating deposition Raman spectroscopy (DCDRS) technique was used to examine the composition of tear fluid samples collected from three healthy human volunteers [88]. In DCDRS, tear fluids were deposited onto hydrophobic surfaces and upon drying yielded a thick amorphous ring with fern-like patterns at the center. Raman maps obtained from the dried tear deposits revealed an inhomogeneous distribution of protein, urea, bicarbonate, and lipid components. The positioning of the protein, urea, and bicarbonate components was related to their relative solubilities and concentrations, while the lipid component was found not in solution but as insoluble debris in the drying droplet. The DCDRS method could potentially provide a fingerprint of protein (and lipid) composition, while the ferning pattern could give an empirical measure of the "quality" of the tears.

6.7.4 Eye Structure and Morphology

Raman mapping has been used to elucidate the microstructural organization in mouse eye slices after the rapid freezing *via* an *in vivo* cryotechnique [89]. Freeze-dried specimens were also embedded in resin, sectioned, and stained with toluidine blue to allow the tissue morphology to be compared against those obtained by Raman microscopy. Together with a light microscopic image and in-depth spectral study, the Raman map contained four typical spectral patterns that were color coded to reflect the ocular skeletal muscles, scleral connective tissues, choroids/pigment epithelium, and rod/cone photoreceptor layer. It was found that the choroid and pigment layers contained melanin, whereas the sclera and photoreceptor layers predominantly comprised of hemoglobin and rhodopsin proteins, respectively. This methodology was also used to visualize oxygen saturation in blood vessels of living animals [89].

6.8 CARDIOVASCULAR

6.8.1 Atherosclerotic Plaques

In cardiovascular research, numerous invasive and noninvasive imaging strategies have been sought to allow the timely identification of atherosclerotic plaques before disruption or rupture [90–92]. Although a vast majority of atherosclerotic plaques are asymptomatic, the rupture of these so-called "vulnerable plaques" can result in a cardiac event. Vulnerable plaques appear as a thin fibrous cap overlying a large neurotic lipid core (40% of the entire plaque). Raman imaging has often been used to visualize atherosclerotic plaques in *ex vivo* tissue specimens. In one study, Raman cluster and autofluorescence imaging techniques were used to spatially separate fluorescent ceroid deposits in atherosclerotic plaques from the surrounding nonfluorescent atheroma [93]. Ceroids are

final products of lipid oxidation and spectroscopically give strong hemoglobin and cholesterol ester signals. This supported the hypothesis that iron and heme formed complexes with intravascular lipoproteins, thereby stimulating the oxidation and initial formation of these ceroid deposits.

Other Raman studies have spatially resolved protein-rich and fat-rich regions in mice models of atherosclerosis [94]. By taking the ratio between the collagen and elastin bands at $1000\,cm^{-1}$ and $1015\,cm^{-1}$, smooth muscle of the media (middle layer) and collagen of the adventitia (outermost layer) could be separated [95]. Large arteries also contain a morphologically distinct collagen–proteoglycan-rich layer, the intima, with a sheet of elastic fibers of the internal elastic lamina (IEL) on the peripheral side. Based on the morphological modeling of mildly atherosclerotic human tissue, Raman images of cholesterol, foam cells, and necrotic core could be localized to the intima (Figure 6.8), while smooth muscle cells was found predominantly in the media [23]. By comparing the Raman image of the internal elastic lamina with the associated phase contrast image, fenestration patterns could be observed.

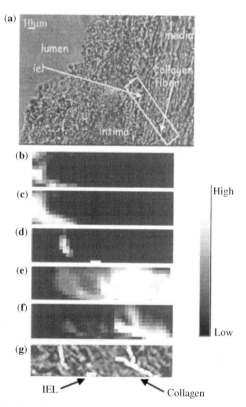

FIGURE 6.8 Phase contrast images (a and g) of a mildly atherosclerotic artery, with the IEL and collagen fibers highlighted in (g). Also shown are the Raman images of cholesterol (b), foam cells and necrotic core (c), IEL (d), smooth muscle cells (e), and collagen (f). Key morphological features, such as the fenestration of the IEL, can be observed. Reprinted from Ref. 23 with permission from Wiley.

Fenestration of the elastic lamina is known to be associated with atherosclerosis. Another hallmark of atherosclerotic disease progression, which was supported by Raman imaging, was the infiltration of smooth muscle into the intima.

Although Raman imaging can identify atherosclerotic plaques in *ex vivo* specimens, collecting *in vivo* single Raman spectra is currently complicated by poor tissue penetration, long acquisition times, and background fluorescence from blood products [90–92]. With the development of more efficient probe designs, compact diode lasers, and CCD detectors, *in vivo* detection might involve the use of a catheter-based Raman probe combined with intravascular ultrasound (IVUS) imaging [96, 97].

6.9 LUNG

6.9.1 Bronchial Wall Structure and Composition

The lung is a complex branching organ evolved to conduct air across multiple spatial scales, starting at the large scale from the bronchi to the lobular bronchioles and the terminal bronchioles. These terminal bronchioles eventually bifurcate into smaller alveolar units where gas exchange with blood ultimately occurs [98]. The first reported pseudocolor Raman images of a bronchial wall segment outlined its chemical composition and microscopic structure [99]. Structurally, the bronchial wall is comprised of a ciliated columnar epithelium with underlying lamina propria and submucosa. These structures (or clusters) were captured by the pseudocolor Raman images, together with the nuclear epithelial and cytoplasmic epithelial clusters associated with the bronchial epithelium. The nuclear epithelial cluster appeared to contain lower lipid and protein content and higher DNA content compared to the cytoplasmic epithelial cluster. A cluster associated with the liquid coating of the bronchial epithelium, termed bronchial mucus, was also identified. The bronchial mucus, which plays a critical role in defending the respiratory tract from inhaled airborne particles, contained a major lipid component, triolein. Triolein was only visible in frozen tissue sections as it was washed away during the staining process. Similar lipid signal contributions were detected in the submucosal region, which indicated that bronchial mucus was primarily produced by the submucosal glands and transported to the epithelial surface *via* gland ducts. The submucosal region was comprised of variable amounts of smooth muscle, cartilage, and fibrocollagenous stroma with variable quantities of glands and gland ducts. Subsequent Raman spectral studies showed that fibrocollagenous stroma and smooth muscle contained higher collagen and actin/myosin signal contributions, respectively, whereas cartilage tissue was dominated by variable amount of sulfated glycosaminoglycan and collagen signal contributions.

6.9.2 Congenital Lung Disorders

Lung development begins in embryonic life and continues several years postnatally [100]. Lungs exposed to environmental toxins during this developmental period, as well as during adolescence, may result in altered lung function and/or increased risk of respiratory disease in later life. The feasibility of using Raman and FTIR imaging methods to detect biochemical changes associated with congenital cystic adenomatoid malformation (CCAM), a rare but curable prenatal lung disorder, was investigated. CCAMs are benign nonaerating and nonfunctioning lung tissue masses arising from the overgrowth of the terminal bronchioles and reduction in alveolar growth [101]. Prenatal CCAM etiology is best classified by sonography, but histology is often needed to make a more definitive diagnosis [102]. From histology, CCAM contains fewer holes, whereas normal lung tissue contains sponge-like aerating morphologies. In the Raman study, a high spatial resolution of ~10 μm was necessary to distinguish CCAMs from normal tissue and to ensure that the Raman clusters obtained were continuous [101]. One of the four clusters associated with red blood cell content was lower in CCAMs compared to normal tissue, while clusters associated with lipids and smooth muscle were either unchanged or not important. However, the lipid cluster obtained by FTIR imaging was an important diagnostic criterion, the content being higher in CCAMs. Although Raman spectroscopy positively identified phosphatidylcholine as one of the major lipid components of CCAMs, FTIR showed that lung mucus contained glycogen. The Raman findings reported in this study appeared to be considerably different when compared to that reported on normal human bronchial tissues [99] and thus highlight the complexity of analyzing lung tissue structure and composition across multiple spatial scales.

6.10 BONE

6.10.1 Bone Microstructure and Composition

Bone is a highly specialized connective tissue that performs essential metabolic and weight-bearing functions, as well as adapting to changes in mechanical stress to which the organ is exposed in everyday life [103]. Bone may be viewed as a heterogeneous composite material comprised of cross-linked collagen fibrils interspersed with hydroxyapatite-like mineral crystallites of various sizes, shapes, orientations, and compositions [104, 105]. Raman microspectroscopy is well placed for investigating these metric parameters and has provided invaluable insights into bone microstructure and compositional changes associated with aging, disease, and trauma [106–110].

With modern quantum efficient CCD detectors and advanced signal recovery techniques, Raman images of different bone microarchitectures can be obtained with minimal interference from bone tissue fluorescence and polymer embedding reagents [111–113]. For example, Raman imaging combined with factor analysis was used to visualize phosphate (ν_1 PO_4) and monohydrogen phosphate (ν_1 HPO_4) gradients in trabecular (spongy) bone and cortical (compact) bone at 3 μm spatial resolutions [106]. Canine trabecular bone tissue was embedded in polymethylmethacrylate (PMMA), while a transverse section of human cortical bone was imaged fresh without embedding. In trabecular bone, independent PO_4 and HPO_4 factors at corresponding band positions of 958 and 1000 cm^{-1} were generated. In mature bone, PO_4 species were localized to the highly mineralized trabecular struts. However, mineralization at the edge of the struts was incomplete, as evident by the higher HPO_4 signal contributions from the same location. This showed that even in mature bone there were bone remodeling sites in which new bone could form. Bone remodeling was clearly observed in Raman images acquired from immature trabecular bone struts in which the HPO_4 region appeared to extend 20 μm into the mineralized PO_4 regions. Small amounts of PMMA resin were found penetrating the newly remodeled bone, which provided further evidence for incomplete mineralization of the organic matrix. In contrast, Raman score images obtained from bone forming around blood vessels (osteons) in cortical bone yielded only a single mineral factor containing both PO_4 and HPO_4 species. The uniformity of the ratio image of the two phosphate species showed that they tracked each other and that cortical bone was less frequently modeled compared to trabecular bone.

Because of the heterogeneity of bone, polarized light is often needed to provide a contrast between different bone tissue components, such as the alternating lamellae structures formed around the osteons. Similarly, Raman contrast images obtained from human osteon lamellae tissues showed that amide I and ν_1 PO_4 band intensities were sensitive to the orientation and polarization direction of incident light [114]. This finding provided invaluable insights into the structural organization of cortical bone, as well as highlighting some of the erroneous conclusions that may be obtained if ν_1 PO_4, amide I, or their ratios were used to calculate bone compositional properties. In contrast, the amide III, ν_2 PO_4, ν_4 PO_4, or their ratios are less orientation dependent and therefore would provide a more accurate description of bone compositional properties. The organization of osteon lamellae at different polarization directions is best illustrated by the 3D Raman contrast images shown in Figure 6.9 [115]. When polarization direction is perpendicular, lamellar bone structure is invisible (Figure 6.9a); however, when polarization direction turns parallel, the lamellar becomes visible (Figure 6.9b). Furthermore, by spatially fusing osteon lamellae Raman images with acoustic impedance images *via* nanoindentation landmarks, site-matched compositional, structural, and elastic information can be obtained [116]. These types of studies are important in elucidating whether

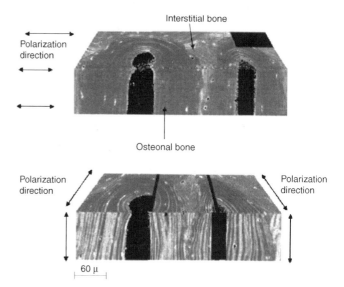

FIGURE 6.9 (a) 3D view of ν_1 PO_4/amide I ratio for different polarizations of the incident laser beam as indicated in the figure by double arrows. (b) Same lamellae show different contrast depending on polarization direction of the beam in panel (a). Reprinted from Ref. 115 with permission from Elsevier.

the alternating lamellar structures are designed toward protecting osteons from catastrophic failure [117].

6.10.2 Craniosynostosis

Raman hyperspectral imaging has also been widely used to study the pathology behind craniosynostosis in which the fibrous tissues between the cranial sutures fuse prematurely [118–120]. Craniosynostosis is a severe craniofacial birth defect and is believed to be caused by the expression of fibroblast growth factor receptor (FGFR) transcripts at the osteogenic front in developing calvaria bone. The pathology of the disease can be mimicked under tissue cell culture conditions by treating normal fetal day 18.5 mouse calvarial sutures with high FGF2 concentrations. Mineral and matrix Raman score images of FGF2-treated sutures and control sutures were obtained [119]. By dividing the mineral score image by the matrix score image, relative mineral-to-matrix ratios (MTMR) were calculated. No significant differences between the MTMR for FGF2-treated sutures and that for control sutures were found, owing to the short duration of the experiment (~48 h). However, the MTMR for FGF2-treated suture bone regions were between 1.5 and 2.0 times higher compared to the control bone suture regions. This provided evidence that the principal site of FGF2-induced craniosynostosis was the osteogenic front, or the tips of the bones. In a more detailed study, the suture region of normal fetal day 18.5 mice was found to contain a mineral factor that spectroscopically resembled octacalcium phosphate [120]. Compared to the characteristic apatitic ν_1 PO_4 band around 957–962 cm^{-1}, the octacalcium phosphate band is broader, less carbonated, and has been proposed as one of the mineral precursors in the formation of mature bone.

6.10.3 Bone Fragility

Bone fragility can be broadly defined as the susceptibility to fracture as a result of failed material and structural properties, rather than a simplistic reduction in the amount of bone [121]. For instance, bone fragility in osteogenesis imperfecta (OI) results from abnormal collagen type I synthesis in the organic matrix portion of bone that in turn disrupts mineral deposition and mineral crystalline size. Diseases of high bone remodeling, such as osteoporosis, produce disturbances in bone material composition and microarchitecture, leading to reduced bone strength, increased bone fragility, and susceptibility to fracture [121, 122]. On the other hand, the accumulation of bone microdamage in healthy young adults as a result of repetitive mechanical loading on the weight-bearing portions of the bone can also contribute to bone fragility [104]. Some spectroscopic measures of bone fragility have been identified and spatially resolved by multivariate Raman imaging techniques. For example, Raman images collected from human cortical bone specimens identified a mineral factor around $952\,cm^{-1}$ ($v_1\,PO_4$) that was localized to the interstitial tissue, away from the osteonal tissue [123]. The $952\,cm^{-1}$ band is usually associated with poorly crystalline, disordered (amorphous) calcium phosphate and its presence may be attributed to damage to the bone at some point in its history. A similar band was also detected in mouse models of osteogenesis imperfecta at $952\,cm^{-1}$ and in the diffuse region of microdamaged bovine bone, albeit at a higher band position of $956\,cm^{-1}$ [110, 124]. Mineral factors with bands between 963 and $964\,cm^{-1}$ have been identified in mouse models of osteogenesis imperfecta and partially damaged/microdamaged bovine bone [110, 113, 124]. The $963–964\,cm^{-1}$ band in mouse models of OI and damaged bone were interpreted as being more stoichiometric and less carbonated. Whereas in microdamaged bovine bone, it was attributed to a phase transformation and/or amorphization.

Abnormal changes to the organic component of bone have also been identified by Raman imaging. To simulate failed bone material and structural properties, a cylindrical indenter was used to mechanically deform bovine cortical bone specimens with loads of up to $1.2\,GPa$ [108]. In high-loaded intents, single organic matrix factors were obtained from the control (nonindented) and indented areas, whereas the edges of the indented region yielded two matrix factors. Raman images taken from the edges of the indent showed increase in the low-frequency component of the amide III band and high-frequency component of the amide I band. These changes indicated the rupture of collagen cross-links as a result of shear forces exerted by the indenter passing through cortical bone. Remarkably, no evidence of ruptured collagen cross-links was observed at the center of the indent, which indicated that only compression of the organic matrix occurred at this location.

Trabecular bone, which is found primarily in the spine, hip, and wrist, is greatly impacted by osteoporosis because it is remodeled up to a magnitude faster compared to cortical bone [125]. Raman studies on trabecular bone have shown that women who sustained osteoporotic hip fractures gave higher carbonate/amide I ratios compared to women without fractures [109]. These studies also showed that carbonate/phosphate ratios in cortical bone biopsies were elevated in osteoporotic women. The later finding may provide indirect evidence for recent bone remodeling activity and/or attempts made by the tissue to repair the damage. The ability to measure cortical bone carbonate/phosphate ratios, noninvasively, would be a significant first step toward the detection and monitoring of bone fragility and fracture healing in living human patients. Despite the technical challenges of collecting weak bone Raman scatter through layers of skin, muscle, fat, and other connective tissues, bone Raman spectra have been recovered to depths of ∼5 mm using ring/disk fiber-optic probe collection geometries [126]. By using more sophisticated fiber-optic illumination/collection geometries and reconstruction methods adapted from fluorescence diffuse tomography, Raman tomographic images could be obtained to depths of 24–45 mm below the skin [127]. Figure 6.10 shows the Raman tomographic images of an *ex vivo* canine bone tissue superimposed on a three-dimensional reconstruction of a canine limb section.

6.11 TEETH

6.11.1 Dental Caries

The pattern of caries development is changing, with increasing prevalence of smaller carious lesions with slower progression rates [128, 129]. This shift has rendered many conventional visual and visuotactile detection methods, such as dental radiographs and dental explorers, less useful as more sensitive methods with higher specificity would be needed to identify them [129, 130]. This situation is further complicated by lesions that develop just below the interproximal contact site (e.g., between adjacent teeth), an area that is difficult to examine by the clinician. Several optical detection methods are currently been developed to identify these carious lesions and to quantify the degree of mineral loss to ensure that the correct dental intervention is implemented.

In one such study, a multimodal approach of Raman microspectroscopy and optical coherence tomography (OCT) was used to detect dental caries extracted from orthodontic patients [130]. Raman microspectroscopy furnished biochemical and structural information on tooth

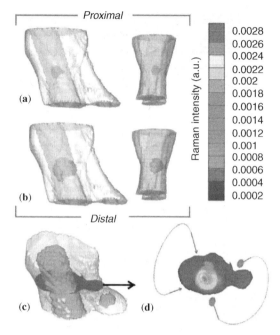

FIGURE 6.10 Raman tomographic images of canine bone tissue. (a) Medial and anterior views of soft tissue mesh (white) and bone surface mesh of tibia and calcaneus (turquoise) overlaid with 50% contrast isosurface of the reconstructed three-dimensional (3D) Raman image of bone (green). (b) Same view as (a) overlaid with 10% contrast isosurface of the reconstructed Raman image of bone (blue). (c) 3D mesh of limb section (white), including bone (turquoise), illustrating location of the cross section (blue) containing the highest Raman scatter intensity. (d) Raman intensity at cross section in (c) in pseudocolor overlaid on the micro-CT image of the bone, showing range of illumination (red arrows) and collection (green dots) positions. Reprinted with permission from Ref. 127. (See the color version of this figure in Color Plate section.)

enamel, while OCT imaging provided morphological and depth information on carious enamel. For example, a compounded OCT depth image of a tooth surface containing two clinically confirmed incipient lesions exhibited increased light backscattering compared to sound enamel. Light backscattering was detected to a depth of ~290 μm at carious locations and was attributed to the increased porosity of the underlying enamel surface. Examination by Raman microspectroscopy identified changes to hydroxyapatite, the major mineral component of tooth enamel. It was found that the intensity of the 960 cm^{-1} peak, which corresponded to the v_1 PO$_4$ stretching mode in hydroxyapatite, was different for carious and sound enamel. Similar changes to the PO$_4$ vibrations in the 350–700 cm^{-1} (v_2, v_4) and 800–1200 cm^{-1} (v_3) spectral ranges were also observed. From Raman imaging studies, the intensity ratio of the 1043 cm^{-1} (v_2) and 959 cm^{-1} (v_1) bands appeared to reach a maximum at the carious region of the tooth. The increased ratio intensity was attributed to changes in enamel crystallite morphology and/or orientation of the enamel rods as a result of demineralization during caries development. This hypothesis was

supported by subsequent polarized Raman analysis in which carious enamel exhibited a higher degree of depolarization and reduced anisotropy compared to sound enamel [131]. Raman polarization and anisotropy images acquired from a sample area measuring 880 μm × 715 μm showed that lesion location, size, and possible severity could be visualized. Furthermore, numerical Raman measures of depolarization and anisotropy, including those values derived from OCT, could be obtained [131, 132]. Numerical Raman anisotropy measures of sound and carious enamel have also been obtained using a fiber-optic Raman probe, and ex vivo carries were detected with 100% sensitivity and 98% specificity [133]. Such measures would allow dental clinicians to evaluate the degree of caries severity and to ensure that the correct conservative treatment was implemented.

With increased advances in fiber-optic Raman and CCD camera technologies, large areas of tooth enamel surfaces and sections can now be imaged, for example, by using a hollow optical fiber probe with a glass ball lens to furnish a high-contrast Raman image of a lesion within a 4.5 mm × 4.5 mm sample area [134]. The lesion was imaged using the intensity ratios of the v_4 PO$_4$ bands at 443 and 446 cm^{-1}. The authors envisioned that by using bundled hollow optical fibers, wider enamel tooth surfaces could be imaged more rapidly. On the other hand, the advent of CCD cameras with zero readout times has enabled whole carious human tooth sections to be imaged in less than an hour [50]. The imaged tooth section measuring 9 mm × 16 mm contained over 84,000 spectra. Carious enamel could be readily discriminated from the surrounding sound enamel and dentin owing to its high fluorescence signal contribution. A horizontal fluorescent feature that coincided with the cementoenamel junction (CEJ) of the tooth was also identified. Figure 6.11 shows Raman images of a smaller 1.5 × 3.4 mm^2 cross section that was obtained in 27 min and contained over 42,000 spectra. In this figure, carious enamel exhibited weaker polarization dependence compared to sound enamel. Raman images based on the v_1 PO$_4$ bandwidths and band positions could also be created. Remarkably, a 4–8 cm^{-1} bandwidth difference between enamel and dentine regions was observed; however, the significance of this finding has yet to be fully investigated.

6.11.2 Dental Restoration

In dental restoration, acid etching of the dentine layer is often necessary to expose the underlying intertubular dentin fibrils, in preparation for bonding with a resin-based composite filling. This procedure involves the diffusion of resin monomers into the "wet" demineralized dentin zone, followed by entanglement with the collagen fibrils to yield adhesive/dentin (a/d) interfaces with high tensile bond strengths [135, 136]. However, the bond strength of the interface is not maintained over time and the reasons behind this

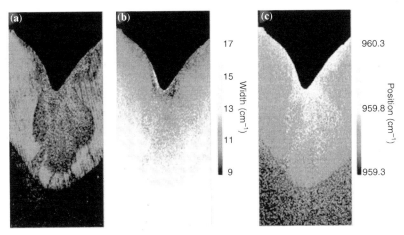

FIGURE 6.11   (a) A composite croos-polarized image of a tooth with a carious defect. The defect appears darker compared to sound enamel due to a weaker polarization dependence. The bandwidth and the position of the 960 cm^{-1} peak in the dental caries region are shown in (b) and (c), respectively. Reprinted from Ref. 50 with permission from Laurin Publishing Co., Inc.

decrease remain unclear [136, 137]. Recently, Raman microscopic imaging was used to examine the distribution of resin monomer and mineral in the intertubular region of the adhesive/dentin interface [136]. The adhesive system used in this study consists of hydrophilic HEMA and hydrophobic BisGMA resins. Raman imaging of the adhesive/dentin interface revealed a partially demineralized dentin layer that could be distinguished from the adhesive resins. This was achieved by using the characteristic dentin mineral ν_1 PO$_4$ band at 961 cm^{-1} and the adhesive resin CH$_2$ and C–O–C bands at 1453 cm^{-1} and 1113 cm^{-1}, respectively (Figure 6.12). The adhesive resins readily penetrated into

the demineralized dentin tubules and spread into the intertubular regions through open tubular channels. In comparison to HEMA, BisGMA resisted diffusion into the "wet" demineralized dentin, owing to its hydrophobic functionality. The incomplete infiltration of BisGMA into the demineralized dentin matrix may account for the decrease in bond strength of the adhesive/dentin interface observed in aqueous environments over the long term. Similar results were reported in a following multivariate-based Raman imaging study, together with the detection of chemical and structural changes to the adhesive/dentin interface not previously identified by univariate-based statistical methods [137].

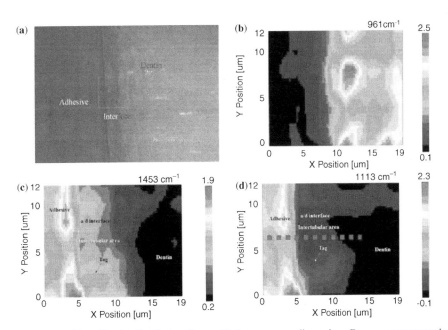

FIGURE 6.12   (a) Visible image of the adhesive/dentin interface with the corresponding micro-Raman spectroscopic images: (b) 961 cm^{-1} (phosphate), (c) 1453 cm^{-1} (CH$_2$), and (d) 1113 cm^{-1} (C–O–C). The spectral mapping was recorded from sites corresponding to the demarcations noted on the visible image. Reprinted with permission from Ref. 136. (See the color version of this figure in Color Plate section.)

Raman microspectroscopy has also been used to visualize the adhesive/dentin interface formed *via* the Silorane adhesive system [138].

6.11.3 Dentinoenamel Interfaces

Raman microspectroscopy has also proved useful in visualizing mineral/protein distributions and stress fields across dentinoenamel interfaces [139–142]. For example, confocal Raman system was used to image the microscopic stress fields across a dentinoenamel junction (DEJ) of a sectioned cattle tooth under increasing compressive loads [141]. By utilizing the piezospectroscopic property of hydroxyapatite under high loads, the ν_1 PO_4 vibration undergoes a spectral shift that can be measured. Spectral shifts of $\sim 2.45 \pm 0.12$ cm^{-1}/GPa have been reported for synthetic hydroxyapatite loaded under uniaxial stress [141]. Given that the enamel layer was highly mineralized and that it was inclined with respect to the normal axis of the tooth, a complicated stress field pattern was observed mainly at this location. Further, the DEJ appeared to be stress free and did not separate or delaminate when higher compressive loads were applied. This observation clarified the micromechanical role of the DEJ, which was to release microscopic stress across the interfacial zone when loaded in compression. Raman microspectroscopy has also been used to visualize stress fields in bone tissues and artificial joints [141], as well as nanoindentation-induced residual stress fields in human tooth enamel [143].

6.12 CONCLUSIONS

This chapter has summarized the progress made in the application of Raman imaging to a diverse variety of specialized tissue types in their normal healthy and diseased states. The collection of Raman images from tissue areas spanning several centimeters, however, still poses a serious challenge to the early diagnosis of disease in a clinical environment. This challenge cannot be underestimated as the discrimination of disease tissues across multiple spatial scales is required in order to meet the requirements of disease diagnosis. Although this challenge can be met by current instrumentation, the results of randomized clinical trials have yet to appear in the literature.

In parallel with the significant advances made toward the development of fiber-optic Raman probes for *in vivo* diagnostic applications, multimodal approaches that combine Raman with other optical imaging modalities have already made a significant impact on biomedical research. Multimodal approaches may in the future allow surgeons to use spatially localized images to guide them during surgery and to evaluate the efficacy of therapeutic interventions.

ACKNOWLEDGMENT

The authors are most grateful to Ms. Mekhala Raghavan for useful comments and suggestions to this chapter.

REFERENCES

1. Del Sole, A., Falini, A., Ravasi, L., Ottobrini, L., De Marchis, D., Bombardieri, E., and Lucignani, G. (2001) Anatomical and biochemical investigation of primary brain tumours. *Eur. J. Nucl. Med.* **28**, 1851.

2. Toga, A. W. and Thompson, P. M. (2003) Temporal dynamics of brain anatomy. *Annu. Rev. Biomed. Eng.* **5**, 119.

3. Chen, W. and Silverman, D. H. S. (2008) 'Advances in evaluation of primary brain tumors. *Semin. Nucl. Med.* **38**, 240.

4. Maher, E. A., Furnari, F. B., Bachoo, R. M., Rowitch, D. H., Louis, D. M., Cavenee, W. K., and DePinho, R. A. (2001) Malignant glioma: genetics and biology of a grave matter. *Genes Dev.* **15**, 1311.

5. Chen, W. (2007) Clinical applications of PET in brain tumors. *J. Nucl. Med.* **48**, 1468.

6. Koljenovic, S., Choo-Smith, L. P., Schut, T. C. B., Kros, J. M., van den Berge, H. J., and Puppels, G. J. (2002) Discriminating vital tumor from necrotic tissue in human glioblastoma tissue samples by Raman spectroscopy. *Lab. Invest.* **82**, 1265.

7. Amharref, N., Bejebbar, A., Dukie, S., Venteo, L., Schneider, L., Pluot, M., and Manfait, M. (2007) 'Discriminating healthy from tumor and necrosis tissue in rat brain tissue samples by Raman spectral imaging. *Biochim. Biophys. Acta* **1768**, 2605.

8. Krafft, C., Sobottka, S. B., Schackert, G., and Salzer, R. (2006) Raman and infrared spectroscopic mapping of human primary intracranial tumors: a comparative study. *J. Raman Spectrosc.* **37**, 367.

9. Koljenovic, S., Schut, T. C. B., Wolthuis, R., de Jong, B., Santos, L., Caspers, P. J., Kros, J. M., and Puppels, G. J. (2005) Tissue characterization using high wave number Raman spectroscopy. *J. Biomed. Opt.* **10**, 031116.

10. Krafft, C., Kirsch, M., Beleites, C., Schackert, G., and Salzer, R. (2007) 'Methodology for fiber-optic Raman mapping and FTIR imaging of metastases in mouse brains. *Anal. Bioanal. Chem.* **389**, 1133.

11. Krafft, C., Steiner, G., Beleites, C., and Salzer, R. (2009) 'Disease recognition by infrared and Raman spectroscopy. *J. Biophotonics* **2**, 13.

12. Koljenovic, S., Schut, T. C. B., Vincent, A., Kros, J. M., and Puppels, G. J. (2005) Detection of meningioma in dura mater by Raman spectroscopy. *Anal. Chem.* **77**, 7958.

13. Krafft, C., Sobottka, S. B., Schackert, G., and Salzer, R. (2005) Near infrared Raman spectroscopic mapping of native brain tissue and intracranial tumors. *Analyst* **130**, 1070.

14. Frank, C. J., McCreery, R. L., and Redd, D. C. B. (1995) Raman Spectroscopy of Normal and Diseased Human Breast Tissues. *Anal. Chem.* **67**, 777.

15. Haka, A. S., Shafer-Peltier, K. E., Fitzmaurice, M., Crowe, J., Dasari, R. R., and Feld, M. S. (2005) Diagnosing breast cancer by using Raman spectroscopy. *Proc. Natl. Acad. Sci. USA* **102**, 12371.

16. Chowdary, M. V. P., Kumar, K. K., Kurien, J., Mathew, S., and Krishna, C. M. (2006) Discrimination of normal, benign, and malignant breast tissues by Raman spectroscopy. *Biopolymers* **83**, 556.

17. Kline, N. J. and Treado, P. J. (1997) Raman chemical imaging of breast tissue. *J. Raman Spectrosc.* **28**, 119.

18. Stone, N., Kendall, C., Smith, J., Crow, P., and Barr, H. (2004) Raman spectroscopy for identification of epithelial cancers. *Faraday Discuss.* **126**, 141.

19. Stone, N. and Matousek, P. (2008) Advanced transmission Raman spectroscopy: a promising tool for breast disease diagnosis. *Cancer Res.* **68**, 4424.

20. Brozek-Pluska, B., Placek, I., Kurczewski, K., Morawiec, Z., Tazbir, M., and Abramczyk, H. (2007) 'Breast cancer diagnostics by Raman spectroscopy. *J. Mol. Liq.* **141**, 145.

21. Smith, J., Kendall, C., Sammon, A., Christie-Brown, J., and Stone, N. (2003) 'Raman spectral mapping in the assessment of axillary lymph nodes in breast cancer. *Technol. Cancer Res. Treat.* **2**, 327.

22. Shafer-Peltier, K. E., Haka, A. S., Fitzmaurice, M., Crowe, J., Myles, J., Dasari, R. R., and Feld, M. S. (2002) Raman microspectroscopic model of human breast tissue: implications for breast cancer diagnosis *in vivo. J. Raman Spectrosc.* **33**, 552.

23. Shafer-Peltier, K. E., Haka, A. S., Motz, J. T., Fitzmaurice, M., Dasari, R. R., and Feld, M. S. (2002) Model-based biological Raman spectral imaging. *J. Cell. Biochem.* **S39**, 125.

24. Yu, C. X., Gestl, E., Eckert, K., Allara, D., and Irudayaraj, J. (2006) Characterization of human breast epithelial cells by confocal Raman microspectroscopy. *Cancer Detect. Prev.* **30**, 515.

25. Subramanian, B. and Axelrod, D. E. (2001) Progression of heterogeneous breast tumors. *J. Theor. Biol.* **210**, 107.

26. Gullick, W. J. (2002) A new model for ductal carcinoma *in situ* suggests strategies for treatment. *Breast Cancer Res.* **4**, 176.

27. Subramanian, K., Kendall, C., Stone, N., Brown, J. C., McCarthy, K., Bristol, J., and Chan, Y. H. (2006) Raman spectroscopic analysis of atypical proliferative lesions of the breast. *Proc. SPIE* **6088**, 60880B.

28. Pasteris, J. D., Wopenka, B. Freeman, J. J., Young, V. L., and Brandon, H. J. (1999) Medical mineralogy as a new challenge to the geologist: silicates in human mammary tissue? *Am. Mineral* **84**, 997.

29. Katzin, W. E., Centeno, D. A., Feng, L. J., Kiley, M., and Mullick, F. G. (2005) Pathology of lymph nodes from patients with breast implants: a histologic and spectroscopic evaluation. *Am. J. Surg. Pathol.* **29**, 506.

30. Luke, J. L., Kalasinsky, V. F., Turnicky, R. P., Centeno, J. A., Johnson, F. B., and Mullick, F. G. (1997) Pathological and biophysical findings associated with silicone breast implants: a study of capsular tissues from 86 cases. *Plast. Reconstr. Surg.* **100**, 1558.

31. Schaeberle, M. D., Kalasinsky, V. F., Luke, J. L., Lewis, E. N., Levin, I. W., and Treado, P. J. (1996) Raman chemical imaging: histopathology of inclusions in human breast tissue. *Anal. Chem.* **68**, 1829.

32. Whitley, A. and Adar, F. (2006) 'Confocal spectral imaging in tissue with contrast provided by Raman vibrational signatures. *Cytometry A* **69A**, 880.

33. Wild, C. P. and Hardie, L. J. (2003) Reflux, Barrett's oesophagus and adenocarcinoma: burning questions. *Nat. Rev. Cancer* **3**, 676.

34. Haggitt, R. C. (1994) Barrett's esophagus, dysplasia, and adenocarcinoma. *Hum. Pathol.* **25**, 982.

35. Spechler, S. J. and Goyal, R. K. (1986) Barrett's esophagus. *N. Engl. J. Med.* **315**, 362.

36. Reid, B. J. and Weinstein, W. M. (1987) Barrett's esophagus and adenocarcinoma. *Annu. Rev. Med.* **38**, 477.

37. Quint, L. E., Hepburn, L. M., Francis, I. R., Whyte, R. I., and Orringer, M. B. (1995) Incidence and distribution of distant metastases from newly diagnosed esophageal carcinoma. *Cancer* **76**, 1120.

38. Stein, H. J. and Feith, M. (2005) Surgical strategies for early esophageal adenocarcinoma. *Best Pract. Res. Clin. Gastroenterol.* **19**, 927.

39. Spechler, S. J. (2005) Dysplasia in Barrett's esophagus: limitations of current management strategies. *Am. J. Gastroenterol.* **100**, 927.

40. Bani-Hani, K. E. and Bani-Hani, B. K. (2008) Columnar lined (Barrett's) esophagus: future perspectives. *J. Gastroenterol. Hepatol.* **23**, 178.

41. Reid, B. J., Haggitt, R. C., Rubin, C. E., Roth, G., Surawicz, C. M., Vanbelle, G., Lewin, K., Weinstein, W. M., Antonioli, D. A., Goldman, H., Macdonald, W., and Owen, D. (1988) Observer variation in the diagnosis of dysplasia in Barrett's esophagus. *Hum. Pathol.* **19**, 166.

42. Skacel, M., Petras, R. E., Gramlich, T. L., Sigel, J. E., Richter, J. E., and Goldblum, J. R. (2000) The diagnosis of low-grade dysplasia in Barrett's esophagus and its implications for disease progression. *Am. J. Gastroenterol.* **95**, 3383.

43. Conio, M., Cameron, A. J., Chak, A., Blanchi, S., and Filiberti, R. (2005) Endoscopic treatment of high-grade dysplasia and early cancer in Barrett's oesophagus. *Lancet Oncol.* **6**, 311.

44. Song, L. W. K. and Wang, K. K. (2003) Optical detection and eradication of dysplastic Barrett's esophagus. *Technol. Cancer Res. Treat.* **2**, 289.

45. Wallace, M. B., Sullivan, D., and Rustgi, A. K. (2006) Advanced imaging and technology in gastrointestinal neoplasia: summary of the AGA-NCI Symposium October 4-5. 2004. *Gastroenterology* **130**, 1333.

46. Wilson, B. C. (2007) Detection and treatment of dysplasia in Barrett's esophagus: a pivotal challenge in translating biophotonics from bench to bedside. *J. Biomed. Opt.* **12**, 051401.

47. Kendall, C., Stone, N., Shepherd, N., Geboes, K., Warren, B., Bennett, R., and Barr, H. (2003) Raman spectroscopy, a potential tool for the objective identification and classification of neoplasia in Barrett's oesophagus. *J. Pathol.* **200**, 602.

48. Shetty, G., Kendall, C., Shepherd, N., Stone, N., and Barr, H. (2006) Raman spectroscopy: elucidation of biochemical changes in carcinogenesis of oesophagus. *Br. J. Cancer* **94**, 1460.

49. Hutchings, J., Kendall, C., Shepherd, N., Barr, H., Smith, B., and Stone, N. (2008) Rapid Raman microscopic imaging for potential histological screening. *Proc. SPIE* **6853**, 85305.

50. Evans, G. (2008) Raman analysis speeds into biomedicine. *Biophoton. Int.* **15**, 28.

51. Geibel, J. P. (2005) Secretion and absorption by colonic crypts. *Annu. Rev. Physiol.* **67**, 471.

52. Hidovic-Rowe, D. and Claridge, E. (2005) Modelling and validation of spectral reflectance for the colon. *Phys. Med. Biol.* **50**, 1071.

53. Krafft, C., Codrich, D., Pelizzo, G., and Sergo, V. (2008) Raman and FTIR microscopic imaging of colon tissue: a comparative study. *J. Biophotonics* **1**, 154.

54. Molckovsky, A., Song, L., Shim, M. G., Marcon, N. E., and Wilson, B. C. (2003) Diagnostic potential of near-infrared Raman spectroscopy in the colon: differentiating adenomatous from hyperplastic polyps. *Gastrointest. Endosc.* **57**, 396.

55. Chowdary, M. V. P., Kumar, K. K., Thakur, K., Anand, A., Kurien, J., Krishna, C. M., and Mathew, S. (2007) Discrimination of normal and malignant mucosal tissues of the colon by Raman spectroscopy. *Photomed. Laser Surg.* **25**, 269.

56. Andrade, P. O., Bitar, R. A., Yassoyama, K., Martinho, H., Santo, A. M. E., Bruno, P. M., and Martin, A. A. (2006) Study of normal colorectal tissue by FT-Raman spectroscopy. *Anal. Bioanal. Chem.* **387**, 1643.

57. Haricharan, R. N. and Georgeson, K. E. (2008) Hirschsprung disease. *Semin. Pediatr. Surg.* **17**, 266.

58. Hackam, D. J., Reblock, K. K., Redlinger, R. E., and Barksdale, E. M. (2004) Diagnosis and outcome of Hirschsprung's disease: does age really matter?' *Pediatr. Surg. Int.* **20**, 319.

59. de Jong, B. W. D., Schut, T. C. B., Wolffenbuttel, K. P., Nijman, J. M., Kok, D. J., and Puppels, G. J. (2001) Identification of bladder wall layers by Raman spectroscopy. *J. Urol.* **168**, 1771.

60. Santos, L. F., Wolthuis, R., Koljenovic, S., Almeida, R. M., and Puppels, G. J. (2005) Fiber-optic probes for *in vivo* Raman spectroscopy in the high-wavenumber region. *Anal. Chem.* **77**, 6747.

61. de Jong, B. W. D., Schut, T. C. B., Coppens, J., Wolffenbuttel, K. P., Kok, D. J., and Puppels, G. J. (2002) Raman spectroscopic detection of changes in molecular composition of bladder muscle tissue caused by outlet obstruction. *Vib. Spectrosc.* **32**, 57.

62. de Jong, B. W. D., Schut, T. C. B., Maquelin, K., van der Kwast, T., Bangma, C. H., Kok, D. J., and Puppels, G. J. (2006) Discrimination between nontumor bladder tissue and tumor by Raman spectroscopy' *Anal. Chem.* **78**, 7761.

63. Otite, U., Webb, J. A. W., Oliver, R. T. D., Badenoch, D. F., and Nargund, V. H. (2001) Testicular microlithiasis: is it a benign condition with malignant potential? *Eur. Urol.* **40**, 538.

64. DeCastro, B. J., Peterson, A. C., and Costabile, R. A. (2008) A 5-year followup study of asymptomatic men with testicular microlithiasis. *J. Urol.* **179**, 1420.

65. de Jong, B. W. D., Brazao, C. A. D., Stoop, H., Wolffenbuttel, K. P., Oosterhuis, J. W., Puppels, G. J., Weber, R. F. A., Looijenga, L. H. J., and Kok, D. J. (2004) Raman spectroscopic analysis identifies testicular microlithiasis as intratubular hydroxyapatite. *J. Urol.* **171**, 92.

66. Phillips, C. L., Gattone, V. H., and Bonsib, S. M. (2006) Imaging glomeruli in renal biopsy specimens. *Nephron Physiol.* **103**, 75.

67. Maier, J., Panza, J., Drauch, A., and Stewart, S. (2006) Raman molecular imaging of tissue and cell samples using tunable multiconjugate filter. *Proc. SPIE* **6380**, 638009.

68. Panza, J. L. and Maier, J. S. (2007) Raman spectroscopy and Raman chemical imaging of apoptotic cells. *Proc. SPIE* **6441**, 44108.

69. Taleb, A., Diamond, J., McGarvey, J. J., Beattie, J. R., Toland, C., and Hamilton, P. W. (2006) 'Raman Microscopy for the Chemometric Analysis of Tumor Cells. *J. Phys. Chem.* B **110**, 19625.

70. Roewert-Huber, J., Lange-Asschenfeldt, B., Stockfleth, E., and Kerl, H. (2007) Epidemiology and aetiology of basal cell carcinoma. *Br. J. Dermatol.* **157**, 47.

71. Patel, Y. G., Nehal, K. S., Aranda, I., Li, Y. B., Halpern, A. C., and Rajadhyaksha, M. (2007) Confocal reflectance mosaicing of basal cell carcinomas in Mohs surgical skin excisions. *J. Biomed. Opt.* **12**, 034027.

72. Nijssen, A., Maquelin, K., Santos, L. F., Caspers, P. J., Schut, T. C. B., Hollander, J. C. D., Neumann, M. H. A., and Puppels, G. J. (2007) Discriminating basal cell carcinoma from perilesional skin using high wave-number Raman spectroscopy' *J. Biomed. Opt.* **12**, 034004.

73. Nijssen, A., Schut, T. C. B., Heule, F., Caspers, P. J., Hayes, D. P., Neumann, M. H. A., and Puppels, G. J. (2002) Discriminating basal cell carcinoma from its surrounding tissue by Raman spectroscopy. *J. Invest. Dermatol.* **119**, 64.

74. Short, M. A., Lui, H., McLean, D. I., Zeng, H. S., and Chen, M. X. (2006) Preliminary micro-Raman images of normal and malignant human skin cells. *Proc. SPIE* **6093**, 60930E.

75. Freudiger, C. W., Min, W., Saar, B. G., Lu, S., Holtom, G. R., He, C. W., Tsai, J. C., Kang, J. X., and Xie, X. S. (2008) Label-free biomedical imaging with high sensitivity by stimulated Raman scattering microscopy. *Science* **322**, 1857.

76. Proksch, E., Brandner, J. M., and Jensen, J. -M. (2008) The skin: an indispensable barrier. *Exp. Dermatol.* **17**, 1063.

77. Chan, K. L. A., Zhang, G. J., Tomic-Canic, M., Stojadinovic, O., Lee, B., Flach, C. R., and Mendelsohn, R. (2008) A coordinated approach to cutaneous wound healing: vibrational microscopy and molecular biology. *J. Cell. Mol. Med.* **12**, 2145.

78. Braiman-Wiksman, L., Solomonik, I., Spira, R., and Tennenbaum, T. (2007) Novel insights into wound healing sequence of events. *Toxicol. Pathol.* **35**, 767.

79. Sharifzadeh, M., Zhao, D. Y., Bernstein, P. S., and Gellermann, W. (2008) 'Resonance Raman imaging of macular

pigment distributions in the human retina. *J. Opt. Soc. Am. A* **25**, 947.

80. Leung, I. Y. F. (2008) Macular pigment: new clinical methods of detection and the role of carotenoids in age-related macular degeneration. *Optometry* **79**, 266.

81. Gellermann, W., Ermakov, I. V., McClane, R. W., and Bernstein, P. S. (2002) Raman imaging of human macular pigments. *Opt. Lett.* **27**, 833.

82. Jacob, R. F., Cenedella, R. J., and Mason, R. P. (2001) Evidence for distinct cholesterol domains in fiber cell membranes from cataractous human lenses. *J. Biol. Chem.* **276**, 13573.

83. Sijtsema, N. M., Duindam, J. J., Puppels, G. J., Otto, C., and Greve, J. (1996) Imaging with extrinsic Raman labels. *Appl. Spectrosc.* **50**, 545.

84. Duindam, H. J., Vrensen, G., Otto, C., Puppels, G. J., and Greve, J. (1995) New approach to assess the cholesterol distribution in the eye lens: confocal Raman microspectroscopy and filipin cytochemistry. *J. Lipid Res.* **36**, 1139.

85. Uzunbajakava, N., Lenferink, A., Kraan, Y., Willekens, B., Vrensen, G., Greve, J., and Otto, C. (2003) Nonresonant Raman imaging of protein distribution in single human cells. *Biopolymers* **72**, 1.

86. Jacob, J. T. and Ham, B. (2008) Compositional profiling and biomarker identification of the tear film. *Ocul. Surf.* **6**, 175.

87. Grus, F. H., Joachim, S. C., and Pfeiffer, N. (2007) Proteomics in ocular fluids. *Proteomics Clin. Appl.* **1**, 876.

88. Filik, J. and Stone, N. (2008) Analysis of human tear fluid by Raman spectroscopy. *Anal. Chim. Acta* **616**, 177.

89. Terada, N., Ohno, N., Saitoh, S., and Ohno, S. (2008) Application of "*in vivo* cryotechnique" to detect erythrocyte oxygen saturation in frozen mouse tissues with confocal Raman cryomicroscopy. *J. Struct. Biol.* **163**, 147.

90. MacNeill, B. D., Lowe, H. C., Takano, M., Fuster, V., and Jang, I. K. (2003) Intravascular modalities for detection of vulnerable plaque: current status. *Arterioscler. Thromb. Vasc. Biol.* **23**, 1333.

91. Fayad, Z. A. and Fuster, V. (2001) Clinical Imaging of the High-risk or Vulnerable Atherosclerotic Plaque. *Circ. Res.* **89**, 305.

92. Rudd, J. H. F., Davies, J. R., and Weissberg, P. L. (2005) Imaging of atherosclerosis—can we predict plaque rupture? *Trends Cardiovasc. Med.* **15**, 17.

93. van de Poll, S. W. E., Schut, T. C. B., van den Laarse, A., and Puppels, G. J. (2002) *In situ* investigation of the chemical composition of ceroid in human atherosclerosis by Raman spectroscopy. *J. Raman Spectrosc.* **33**, 544.

94. Adar, F., Jelicks, L., Naudin, C., Rousseau, D., and Yeh, S. R. (2004) Elucidation of the atherosclerotic disease process in apo E and wild type mice by vibrational spectroscopy. *Proc. SPIE* **5321**, 102.

95. Hewko, M. D., Choo-Smith, L. P., Ko, A. C. T., Smith, M. S. D., Kohlenberg, E. M., Bock, E. R., Leonardi, L., and Sowa, M. G. (2006) Atherosclerosis diagnostic imaging by optical

spectroscopy and optical coherence tomography. *Proc. SPIE* **6078**, E782.

96. Romer, T. J., Brennan, J. F., Puppels, G. J., Zwinderman, A. H., van Duinen, S. G., van der Laarse, A., van der Steen, A. F. W., Bom, N. A., and Bruschke, A. V. G. (2000) Intravascular ultrasound combined with Raman spectroscopy to localize and quantify cholesterol and calcium salts in atherosclerotic coronary arteries. *Arterioscler. Thromb. Vasc. Biol.* **20**, 478.

97. van de Poll, S. W. E., Romer, T. J., Puppels, G. J., and van der Laarse, A. (2002) Raman spectroscopy of atherosclerosis. *J. Cardiovasc. Risk* **9**, 255.

98. Burrowes, K. S., Swan, A. J., Warren, N. J., and Tawhai, M. H. (2008) Towards a virtual lung: multi-scale, multi-physics modelling of the pulmonary system. *Philos. Trans. R. Soc. A* **366**, 3247.

99. Koljenovic, S., Schut, T. C. B., van Meerbeeck, J. P., Maat, A. P. W. M., Burgers, S. A., Zondervan, P. E., Kros, J. M., and Puppels, G. J. (2004) Raman microspectroscopic mapping studies of human bronchial tissue. *J. Biomed. Opt.* **9**, 1187.

100. Kajekar, R. (2007) Environmental factors and developmental outcomes in the lung. *Pharmacol. Ther.* **114**, 129.

101. Krafft, C., Codrich, D., Pelizzo, G., and Sergo, V. (2008) Raman mapping and FTIR imaging of lung tissue: congenital cystic adenomatoid malformation. *Analyst* **133**, 361.

102. Adzick, N. S. and Harrison, M. R. (1993) Management of the fetus with a cystic adenomatoid malformation. *World J. Surg.* **17**, 342.

103. Robling, A. G., Castillo, A. B., and Turner, C. H. (2006) Biomechanical and molecular regulation of bone remodeling. *Annu. Rev. Biomed. Eng.* **8**, 455.

104. Sahar, N. D., Hong, S. -I., and Kohn, D. H. (2005) Micro- and nano-structural analyses of damage in bone. *Micron* **36**, 617.

105. Olszta, M. J., Cheng, X. G., Jee, S. S., Kumar, R., Kim, Y. Y., Kaufman, M. J., Douglas, E. P., and Gower, L. B. (2007) Bone structure and formation: a new perspective. *Mater. Sci. Eng. R Rep.* **58**, 77.

106. Timlin, J. A., Carden, A., Morris, M. D., Bonadio, J. F., Hoffler, C. E., Kozloff, K. M., and Goldstein, S. A. (1999) Spatial distribution of phosphate species in mature and newly generated mammalian bone by hyperspectral Raman imaging. *J. Biomed. Opt.* **4**, 28.

107. Yerramshetty, J. S., Lind, C., and Akkus, O. (2006) The compositional and physicochemical homogeneity of male femoral cortex increases after the sixth decade. *Bone* **39**, 1236.

108. Carden, A., Rajachar, R. M., Morris, M. D., and Kohn, D. H. (2003) Ultrastructural changes accompanying the mechanical deformation of bone tissue: a Raman imaging study. *Calcif. Tissue Int.* **72**, 166.

109. McCreadie, B. R., Morris, M. D., Chen, T. -C., Sudhaker Rao, D., Finney, W. F., Widjaja, E., and Goldstein, S. A. (2006) Bone tissue compositional differences in women with and without osteoporotic fracture. *Bone* **39**, 1190.

110. Chen, T. C., Kozloff, K. M., Goldstein, S. A., and Morris, M. D. (2004) Bone tissue ultrastructural defects in a mouse model for osteogenesis imperfecta: a Raman spectroscopy study. *Proc. SPIE* **5321**, 85.

111. Crane, N. J., Gomez, L. E., Ignelzi, M. A. Jr., and Morris, M. D. (2004) Compatibility of histological staining protocols for bone tissue with Raman microspectroscopy and imaging. *Calcif. Tissue Int.* **74**, 86.

112. Widjaja, E., Crane, N., Chen, T., Morris, M. D., Ignelzi, M. A. Jr., and McCreadie, B. (2003) Band-target entropy minimization (BTEM) applied to hyperspectral Raman image data. *Appl. Spectrosc.* **57**, 1353.

113. Golcuk, K., Mandair, G. S., Callender, A. F., Sahar, N., Kohn, D. H., and Morris, M. D. (2006) Is photobleaching necessary for Raman imaging of bone tissue using a green laser? *Biochim. Biophys. Acta* **1758**, 868.

114. Kazanci, M., Roschger, P., Paschalis, E. P., Klaushofer, K., and Fratzl, P. (2006) Bone osteonal tissues by Raman spectral mapping: orientation–composition. *J. Struct. Biol.* **156**, 489.

115. Kazanci, M., Wagner, H. D., Manjubala, N. I., Gupta, H. S., Paschalis, E., Roschger, P., and Fratzl, P. (2007) Raman imaging of two orthogonal planes within cortical bone. *Bone* **41**, 456.

116. Hofmann, T., Heyroth, F., Meinhard, H., Franzel, W., and Raum, K. (2006) Assessment of composition and anisotropic elastic properties of secondary osteon lamellae. *J. Biomech.* **39**, 2282.

117. Gupta, H. S., Stachewicz, U., Wagermaier, W., Roschger, P., Wagner, H. D., and Fratzl, P. (2006) Mechanical modulation at the lamellar level in osteonal bone. *J. Mater. Res.* **21**, 1913.

118. Tarnowski, C. P., Ignelzi, M. A. Jr., Wang, W., Taboas, J. M., Goldstein, S. A., and Morris, M. D. (2004) Earliest mineral and matrix changes in force-induced musculoskeletal disease as revealed by Raman microspectroscopic imaging. *J. Bone Miner. Res.* **19**, 64.

119. Crane, N. J., Morris, M. D., Ignelzi, M. A., and Yu, G. (2005) Raman imaging demonstrates FGF2-induced craniosynostosis in mouse calvaria' *J. Biomed. Opt.* **10**, 031119.

120. Crane, N. J., Popescu, V., Morris, M. D., Steenhuis, P., and Ignelzi, M. A. (2006) Raman spectroscopic evidence for octacalcium phosphate and other transient mineral species deposited during intramembranous mineralization. *Bone* **39**, 434.

121. Chavassieux, P., Seeman, E., and Delmas, P. D. (2007) Insights into material and structural basis of bone fragility from diseases associated with fractures: how determinants of the biomechanical properties of bone are compromised by disease. *Endocr. Rev.* **28**, 151.

122. Ralston, S. H. (2005) Genetic determinants of osteoporosis. *Curr. Opin. Rheumatol.* **17**, 475.

123. Carden, A., Timlin, J. A., Edwards, C. M., Morris, M. D., Hoffler, C. E., Kozloff, K., and Goldstein, S. A. (1999) Raman Imaging of bone mineral and matrix: composition and function. *Proc. SPIE* **3608**, 132.

124. Timlin, J., Carden, A., Morris, M. D., Rajachar, R. M., and Kohn, D. H. (2000) Raman spectroscopic imaging markers for fatigue-related microdamage in bovine bone. *Anal. Chem.* **72**, 2229.

125. Wehrli, F. W., Song, H. K., Saha, P. K., and Wright, A. C. (2006) Quantitative MRI for the assessment of bone structure and function. *NMR Biomed.* **19**, 731.

126. Schulmerich, M. V., Dooley, K. A., Vanasse, T. M., Goldstein, S. A., and Morris, M. D. (2007) Subsurface and transcutaneous Raman spectroscopy and mapping using concentric illumination rings and collection with a circular fiber-optic array. *Appl. Spectrosc.* **61**, 671.

127. Schulmerich, M. V., Cole, J. H., Dooley, K. A., Morris, M. D., Kreider, J. M., Goldstein, S. A., Srinivasan, S., and Pogue, B. W. (2008) Non-invasive Raman tomographic imaging of canine cortical bone tissue. *J. Biomed. Opt.* **13**, 020506.

128. Alfano, M. C., Coulter, I. D., Gerety, M. B., Hart, T. C., Imrey, P. B., LeResche, L., Levy, J., Luepker, R. V., Lurie, A. G., Page, R. C., Rye, L. A., Smith, L., and Walker, C. B. (2001) National Institutes of Health Consensus Development Conference statement: diagnosis and management of dental caries throughout life, March 26-28, 2001. *J. Am. Dent. Assoc.* **132**, 1153.

129. Pretty, I. A. (2006) Caries detection and diagnosis: novel technologies. *J. Dent.* **34**, 727.

130. Ko, A. C. T., Choo-Smith, L. P., Hewko, M., Leonardi, L., Sowa, M. G., Dong, C. C. S., Williams, P., and Cleghorn, B. (2005) *Ex vivo* detection and characterization of early dental caries by optical coherence tomography and Raman spectroscopy. *J. Biomed. Opt.* **10**, 031118.

131. Ko, A. C. T., Choo-Smith, L. P., Hewko, M., Sowa, M. G., Dong, C. C. S., and Cleghorn, B. (2006) Detection of early dental caries using polarized Raman spectroscopy. *Opt. Express* **14**, 203.

132. Sowa, M. G., Popescu, D. P., Werner, J., Hewko, M., Ko, A. C. T., Payette, J., Dong, C. C. S., Cleghorn, B., and Choo-Smith, L. P. (2006) Precision of Raman depolarization and optical attenuation measurements of sound tooth enamel. *Anal. Bioanal. Chem.* **387**, 1613.

133. Ko, A. C. T., Hewko, M., Sowa, M. G., Dong, C. C. S., Cleghorn, B., and Choo-Smith, L. P. (2008) Early dental caries detection using a fibre-optic coupled polarization-resolved Raman spectroscopic system. *Opt. Express* **16**, 6274.

134. Yokoyama, E., Kakino, S., and Matsuura, Y. (2008) Raman imaging of carious lesions using a hollow optical fiber probe. *Appl. Opt.* **47**, 4227.

135. Eick, J. D., Gwinnett, A. J., Pashley, D. H., and Robinson, S. J. (1997) Current concepts on adhesion to dentin. *Crit. Rev. Oral Biol. Med.* **8**, 306.

136. Wang, Y., Spencer, P., and Yao, X. M. (2006) Micro-Raman imaging analysis of monomer/mineral distribution in intertubular region of adhesive/dentin interfaces. *J. Biomed. Opt.* **11**, 024005.

137. Parthasarathy, R., Thiagarajan, G., Yao, X., Wang, Y. P., Spencer, P., and Wang, Y. (2008) Application of multivariate spectral analyses in micro-Raman imaging to unveil structural/chemical features of the adhesive/dentin interface. *J. Biomed. Opt.* **13**, 014020.

138. Santini, A. and Miletic, V. (2008) Comparison of the hybrid layer formed by Silorane adhesive, one-step self-etch and etch and rinse systems using confocal micro-Raman spectroscopy and SEM. *J. Dent.* **36**, 683.

139. Kinoshita, H., Miyoshi, N., Fukunaga, Y., Ogawa, T., Ogasawara, T., and Sano, K. (2008) Functional mapping of carious enamel in human teeth with Raman microspectroscopy. *J. Raman Spectrosc.* **39**, 655.

140. Bulatov, V., Feller, L., Yasman, Y., and Schechter, I. (2008) Dental enamel caries (early) diagnosis and mapping by laser Raman spectral imaging. *Instrum. Sci. Technol.* **36**, 235.

141. Pezzotti, G. (2005) Raman piezo-spectroscopic analysis of natural and synthetic biomaterials. *Anal. Bioanal. Chem.* **381**, 577.

142. Xu, C., Yao, X., Walker, M. P., and Wang, Y. (2009) Chemical/molecular structure of the dentin–enamel junction is dependent on the intratooth location. *Calcif. Tissue Int.* **84**, 221.

143. He, L. H., Carter, E. A., and Swain, M. V. (2007) Characterization of nanoindentation-induced residual stresses in human enamel by Raman microspectroscopy. *Anal. Bioanal. Chem.* **389**, 1185.

7

SKIN PHARMACOLOGY AND COSMETIC SCIENCE APPLICATIONS OF IR SPECTROSCOPY, MICROSCOPY, AND IMAGING

RICHARD MENDELSOHN AND CAROL R. FLACH

Department of Chemistry, Newark College, Rutgers University, Newark, NJ, USA

DAVID J. MOORE AND LAURENCE SENAK

ISP Corporation, Wayne, NJ, USA

7.1 INTRODUCTION

The rapid growth in biomedical applications of vibrational microscopy and imaging is in large part due to the vast amount of information inherent in the complete IR or Raman spectrum acquired from each pixel during the measurement. For characterization of tissues and cells, this spectral information directly monitors elements of molecular and supramolecular structure in the sample constituents. This situation is in sharp contrast to imaging techniques based on electronic spectroscopy, that is, absorption or fluorescence, where the relationship between the spectral properties of the chromophore and the molecular structure of the tissue in which it is located is generally not available. As an example (described later in this chapter) of the relationship between the spectra and the molecular structure information, IR images acquired during the healing of wounded skin permit us to distinguish the spatial distribution of collagen from keratin, and even more important, permit us to distinguish between the various forms of keratin in cells that are activated during reepithelialization of the wounded area [1].

For the past several years, our laboratories have utilized IR spectroscopy, microscopy, and imaging to monitor the biophysics and pharmacology of the skin barrier and the spatial distribution of the major constituents in hair [2–7]. This chapter presents an overview of how to effectively use these technologies to generate useful molecular and supramolecular structure information from tissues and cells. Our general approach is reductionist and predicated upon the reasonable assumption that molecular structure information extracted from IR spectra collected from purified tissue components provides an appropriate basis for interpretation of tissue spectra. To illustrate this approach for skin, we demonstrate the sensitivity of the IR spectra of purified ceramides (a major skin lipid class) to lipid chain conformation and packing. We next utilize this information to interpret a structural transition in intact full thickness stratum corneum (SC). This structural information is used to track the kinetics of the restoration of the skin barrier following its disruption by thermal perturbation. Finally, we move to the imaging mode and demonstrate the feasibility of *imaging* conformational order in exogenous lipid vesicles following their application to skin.

Subsequent to studies of the permeability barrier in skin, we demonstrate the use of IR microscopy to track biochemical alterations in single cells through evaluation of natural moisturizing factor (NMF), an important hydration control mechanism in corneocyte biology. Finally, two applications of IR imaging are presented. The first is a cosmetic science experiment detailing the spatial distribution of the lipid and protein constituents in hair. The second is an application to the healing of skin wounds, in which we track temporal and spatial changes in the distribution of skin proteins that

Raman, Infrared, and Near-Infrared Chemical Imaging Edited by Slobodan Šašić and Yukihiro Ozaki
Copyright © 2010 John Wiley & Sons, Inc.

become activated within the first several days of wound healing in an organ culture (skin explant) model.

7.2 IR SPECTROSCOPY OF SKIN AND CERAMIDE MODELS

7.2.1 Supramolecular Organization of Skin

The spatial region of skin of major importance in cosmetic science and pharmacology research is the outermost layer, the SC. This layer constitutes the main barrier to permeability and in addition maintains water homeostasis. A schematic of the SC supramolecular organization given in Figure 7.1 depicts the two major tissue constituents, cells and lipids. The cellular component consists of corneocytes—anucleated, asymmetric, flattened cells filled with keratin and embedded in a hydrophobic lamellar, lipid-rich matrix comprising equimolar proportions of ceramides, fatty acids, and cholesterol. The dimensions of a typical corneocyte are revealed from an AFM image in Figure 7.1a. Their shape is irregularly hexagonal, with an average thickness of a few hundred nanometers in the "Z" direction perpendicular to the skin surface and ~40–50 μm in the "X, Y" directions parallel to the skin surface. In Figure 7.1b, the corneocytes are depicted embedded in the lipid matrix in what is generally termed the "bricks and mortar" model, with the lipid matrix serving as the mortar holding the corneocyte "bricks" in the necessary geometry.

A schematic of the possible modes of organization of the lipid lamellar layers is depicted in Figure 7.1c. The chain packing and conformation for each of these three common bilayer motifs are depicted in a top view perpendicular to the skin surface and a side view. In the orthorhombic phase, the chains are conformationally ordered (all-*trans* conformation) while each central chain is surrounded in a tightly packed fashion by four others. The hexagonal organization, in which a central molecule is surrounded by six others, is also characterized by highly ordered all-*trans* chains. However, a looser chain packing occurs resulting in substantial rotation about the long axis of the lipid molecules. Finally, in the liquid crystalline state generally induced at high temperatures, the chains are conformationally disordered, that is, possess *gauche* rotations, and assume dynamic, irregular, packing motifs. In some descriptions of the lipid organization in skin, the majority of the SC lipids are thought to be segregated into orthorhombic or hexagonally packed domains separated by grain boundaries [8]. The latter are regions suggested to contain molecules that may be conformationally disordered and which therefore might present a possible pathway for diffusion of hydrophobic species.

7.2.2 IR Spectra–Structure Relationships for Chain Order and Packing in Ceramides

IR spectroscopy provides an ideal method for the study of lipid phase behavior. The basic spectra/structure correlations

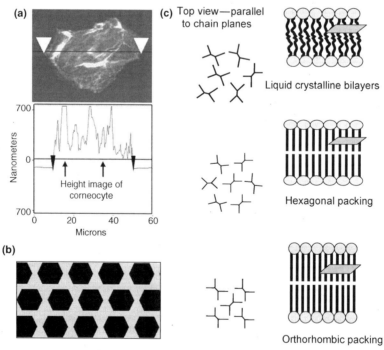

FIGURE 7.1 The major structural components comprising the stratum corneum and the "bricks and mortar" model. (a) An AFM image of an isolated human corneocyte with its height profile along the line drawn across the image shown directly below. (b) The "bricks and mortar" model of the supramolecular organization of the stratum corneum (top view, perpendicular to skin surface). Corneocytes are shown as black hexagons with gray lipid "mortar" between them. (c) A schematic of lipid phases thought to simultaneously exist in the stratum corneum (left: top view, perpendicular to skin surface; right: side view). The degree of order in the lipid phases increases from top to bottom.

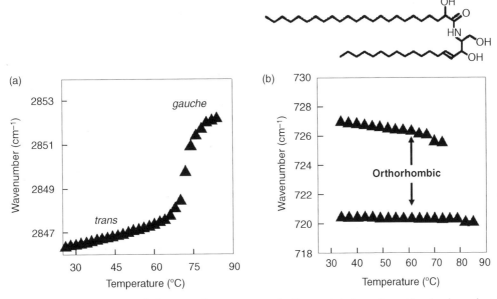

FIGURE 7.2   Temperature-induced changes in IR spectral parameters evaluating acyl chain conformational order and packing in a stratum corneum ceramide, α-hydroxyacid sphingosine (ceramide AS). (Top) Molecular structure of α-hydroxyacid sphingosine. (a) The symmetric CH_2 stretching frequency increases with temperature indicating a loss in acyl chain conformational order (*trans–gauche* isomerization). (b) The CH_2 rocking mode frequencies as a function of temperature highlight the orthorhombic to hexagonal packing transition.

were worked out over a period of 50 years through the seminal studies of alkanes and polyethylene from (amongst others) the laboratories of Snyder (Berkeley) [9–13], Zerbi (Milan) [14–16], and Shimanouchi (Tokyo) [17, 18]. The original correlations have been extended to assemblies of lipids, lipid–protein complexes, and more recently tissues, including skin.

Two IR spectral regions are of interest in the current introduction to the interpretation of lipid spectra. The CH_2 stretching frequencies (2840–2940 cm^{-1}) are well known to be sensitive to chain conformational order (*trans–gauche* isomerization) in the lipid chains. The sensitivity of these frequencies to temperature is depicted in Figure 7.2a, where the major order–disorder transition is shown for a typical skin ceramide, α-hydroxyacid sphingosine (also termed ceramide AS), whose chemical structure is included in the figure. The sigmoid-shaped transition between $\sim$65 and 80°C is accompanied by a $\sim$4–5 cm^{-1} increase in the symmetric CH_2 stretching frequency near 2850 cm^{-1}. This frequency increase reflects the formation of *gauche* rotations in the chains. A comment about precision in the frequency measurement required in the IR measurement is appropriate. Typically, a precision of 0.05 cm^{-1} in this measurement can be readily achieved so that a shift of a few cm^{-1} is easily detected.

Complementary information about chain packing is available from the CH_2 rocking (720–730 cm^{-1}) and scissoring (1460–1474 cm^{-1}) modes. The sensitivity of the CH_2 rocking region to chain packing is depicted in Figure 7.2b. When an orthorhombic perpendicular subcell is present in the supramolecular structure, the rocking modes are split into a doublet with component frequencies at $\sim$720 and 730 cm^{-1}. This is clearly the case for ceramide AS, as is evident from Figure 7.2b in which the temperature dependencies of frequencies of the components are plotted. When the temperature approaches that of the onset of conformational disorder, the orthorhombic doublet collapses to a single band. Equivalent changes (not shown here) are observed for the scissoring contour.

A small change in the chemical structure of the ceramide molecule produces significant changes in the phase state of the molecules. Removal of the α-hydroxy group to form nonhydroxy fatty acid sphingosine (ceramide NS) produces a significantly altered IR response of the spectral parameters considered above. For the symmetric CH_2 stretching frequency ($\nu_{sym}(CH_2)$) that ranges from 2847 to 2855 cm^{-1}, the actual band position can be used to differentiate changes in chain packing from alterations in conformational disorder. Solid–solid phase transitions in which the all-*trans* conformational order persists are revealed through a CH_2 stretching frequency below $\sim$2850 cm^{-1} following the transition. This is the result observed for ceramide NS in Figure 7.3a. The persistence of essentially complete conformational order is indicated by the relatively low frequency above the transition compared with ceramide AS (Figure 7.2a). The presence of a low-temperature orthorhombic phase in the system is confirmed through observation of the characteristic rocking mode doublet depicted in Figure 7.3b. The collapse of the doublet (Figure 7.3b), coupled with the persistence of chain order (Figure 7.3a), reveals the presence of a solid–solid phase transition in

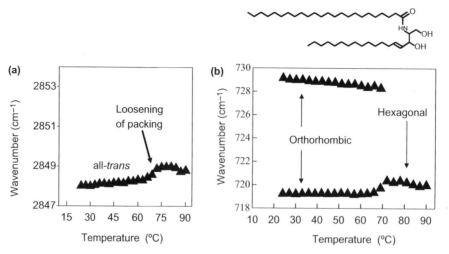

FIGURE 7.3   Temperature-induced changes in IR spectral parameters evaluating acyl chain conformational order and packing in a stratum corneum ceramide, nonhydroxy fatty acid sphingosine (ceramide NS). (Top) Molecular structure of nonhydroxyacid sphingosine. (a) The symmetric CH_2 stretching frequency increases with temperature showing little change over this temperature range indicating a loosening in acyl chain packing, that is, a transition from orthorhombic to hexagonal packing, with essentially no change in conformational order. (b) The temperature dependence of the CH_2 rocking mode frequencies accentuates the orthorhombic to hexagonal packing transition.

which chain packing is altered from orthorhombic to hexagonal.

7.2.3   Phase Transitions in Isolated Stratum Corneum

Perhaps somewhat surprisingly, the above simple framework for tracking conformational order and packing alterations in chain molecules is sufficient to understand a structural transition that we have observed in isolated human SC. In Figure 7.4a, the temperature dependence of $\nu_{sym}(CH_2)$

in human SC is plotted over the range 3–108°C. Two transitions characterized by frequency changes are evident. The transition centered at ~90°C produces the relatively large change in $\nu_{sym}(CH_2)$ (from ~2850.2 to 2853.5 cm^{-1}), which, as noted above, arises from *trans–gauche* isomerization in the chains. In addition, a broad transition from 20 to 40°C is accompanied by a small increase in $\nu_{sym}(CH_2)$ from ~2849.2 to 2850 cm^{-1}.

Spectra of the methylene rocking (711–735 cm^{-1}) region at ~4°C temperature intervals from 6.3 to 79.0°C for an SC

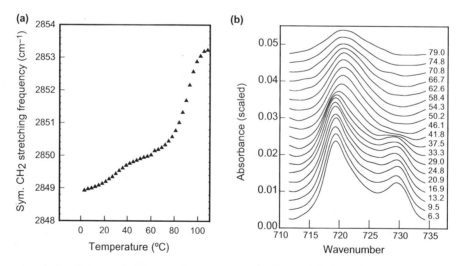

FIGURE 7.4   Temperature-induced changes in IR spectral parameters evaluating acyl chain conformational order and packing in isolated human stratum corneum. (a) The temperature dependence of the symmetric CH_2 stretching frequency displays both the solid–solid phase transition over the 20–40°C temperature range followed by the conformational order to disorder transition from 80 to 100°C. (b) IR spectra (710–735 cm^{-1} region) of the CH_2 rocking modes as a function of temperature as marked. The collapse of the doublet, over the 29.0–41.8°C temperature range, marks the orthorhombic to hexagonal packing transition.

sample are shown in Figure 7.4b. At temperatures <40°C, this mode is split into a doublet with components near 719 and 729 cm^{-1}. As noted previously, the observation of the latter band is reliably diagnostic for orthorhombic perpendicular subcell packing of the lipid chains in the tissue. At temperatures above 40°C, the doublet collapses to a single peak at ~720 cm^{-1}. This spectral change reflects collapse of orthorhombic chain packing. The situation is similar to the transition observed for ceramide NS (Figure 7.3b) and, taken together with the persistence of chain conformational order, is indicative of an orthorhombic → hexagonal packing transition at physiologically relevant temperatures. In the SC, the solid–solid transition is followed by the broad (80–100°C) higher temperature lipid acyl chain order → disorder transition as indicated by the $\nu_{sym}(CH_2)$ increase from ~2850 to 2853 cm^{-1}.

7.3 BARRIER REFORMATION FOLLOWING THERMAL PERTURBATION

The SC has been an obvious target of approaches for transdermal drug delivery based on methods that involve transient modifications of its barrier properties. Traditional functional methods such as *trans*-epidermal water loss utilized to monitor barrier integrity provide no insights into structural modifications that occur subsequent to the application of exogenous molecules to the skin surface. The solid–solid phase transition discussed above provides a useful diagnostic

tool to monitor barrier reformation kinetics [6]. For initial studies, we chose to thermally perturb the barrier.

7.3.1 Experimental Protocols

Dermatomed human cadaver skin samples (500 μm thick) were placed SC side up on a substrate soaked with 0.1% (w/v) trypsin for ~24 h at room temperature. The SC was physically separated, washed with phosphate-buffered saline and distilled water, and dried on a ZnSe IR window. A second window was placed on top of the SC and the sample "sandwich" was contained in a temperature-controlled IR cell. The skin barrier was disrupted by warming to 55°C, restoration was initiated by quenching the sample to either 25 or 30°C, and IR spectra were collected for the desired times.

7.3.2 Results: Kinetics of Barrier Reformation

Spectra of the SC rocking mode contour at various time points following quenching from 55 to 30°C are plotted in Figure 7.5a. The partial reappearance of the 729 cm^{-1} peak unambiguously monitors the reformation of orthorhombic lipid phases. The time dependence of the integrated band intensity at 729 cm^{-1} is plotted in Figure 7.5b. Included in the figure are equivalent data for a sample quenched to 25°C. The rocking data kinetics display exponential growth in the initial stages followed by linear increases at longer times. These data provide strong evidence for partial restoration of the orthorhombic phase component of the SC barrier following

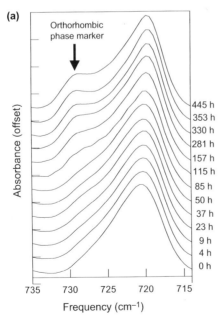

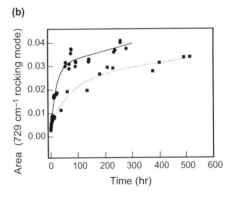

FIGURE 7.5 Kinetics of orthorhombic phase reformation in isolated human stratum corneum after quenching from 55 to 25 or 30°C. (a) Temporal variation in the CH_2 rocking region (710–735 cm^{-1}) after quenching to 30°C. The evolution of the characteristic orthorhombic marker band (729 cm^{-1}) is noted. (b) Time evolution of the area of the 729 cm^{-1} component for a sample quenched to 30°C (●, solid line) and 25°C (■, dotted line). Best-fit lines are depicted assuming that time dependence begins with an exponential phase followed by a linear function.

relatively minor thermal perturbation. The kinetic equations for the data in Figure 7.5 are as follows: (1) for the sample quenched to 30°C, $I_{calc} = -0.0286 \exp(-0.0503T) + 3.12 \times 10^{-5}T + 0.0304$ and (2) for the sample quenched to 25°C, $I_{calc} = -0.0297 \exp(-0.0132T) + 1.69 \times 10^{-5}T + 0.0248$. In these equations, I_{calc} is the peak intensity at $729\,cm^{-1}$ and T is the time (h) following quenching. The half-lives for the exponential part of the growth in each case are $\sim13.8\,h$ (30° quench) and $\sim52.4\,h$ (25° quench). Further analysis of the data reveals that half to three quarters of the orthorhombic phase for a particular quenching temperature was restored during the time course of these experiments.

The importance of this approach lies in the fact that skin barrier damage (or disruption) and reformation are of potential interest in elucidating the mechanism and effects of transdermal drug delivery on skin. Although temperature itself is only rarely used *in vivo* to enhance delivery, it was selected for the current measurements as a convenient laboratory variable to develop a more quantitative IR approach for analyzing barrier recovery. The sensitivity of IR spectral parameters for monitoring lipid reorganization as described is obviously at the core of the approach. Future experiments along these lines are easily envisioned in which the effects of exogenous agents may be tracked.

7.4   IR IMAGING OF ACYL CHAIN CONFORMATIONAL ORDER

The demonstrated sensitivity of the methylene stretching frequencies to chain conformational order provides the opportunity to *image* the conformational order of exogenous lipid-containing formulations upon application to skin. We have successfully demonstrated this approach with liposomes, widely touted as drug delivery vehicles. The method is most convenient if some or all chains of the exogenous lipid molecules are perdeuterated. The symmetric CD_2 stretching frequencies occur between ~2084 and $2092\,cm^{-1}$. This spectral region is free from interference from endogenous skin components, while the frequency position is still sensitive to chain conformational order, that is, a higher frequency reflects the existence of *gauche* bonds in the chains.

A liposome formulation of sn1-perdeuterated, 1-palmitoyl, 2-oleoylphosphatidylcholine (P-d_{31}OPC) was used to demonstrate these ideas. POPC is known to exist in the liquid crystalline (chains conformationally disordered) phase at room temperature. Application of liposomes of P-d_{31}OPC to skin results in their permeation through the SC [2]. The effect of the ordered SC lipid environment on the chain conformation of the P-d_{31}OPC is imaged in Figure 7.6. Figure 7.6a depicts a visible micrograph of the SC. The spatial variation in the CD_2 symmetric stretching frequency and intensity along with the amide II intensity (from skin) for spectra acquired along two lines (at 200 and 50 μm) from the (arbitrary) origin at 0 μm are plotted in Figure 7.6b. The position of the beginning of the SC is indicated both by a decrease in the CD_2 frequency and an increase in the amide II intensity, and is consistent with the optical micrograph. With the CD_2 intensity as a measure of the exogenous lipid concentration, a vesicle reservoir appears to be retained above the skin surface. The majority of lipid has permeated into the epidermal region. Although the CD_2 band intensity is diminished deeper into the skin, it remains above zero at a horizontal position of 120 μm, which corresponds to depths of ~90 and ~70 μm for the lines examined at 200 and 50 μm, respectively. At these depths, the CD_2 stretching frequency increases, possibly reflecting hydration changes at the epidermal/dermal boundary line. In general, the permeated material in the SC epidermal region shows a $1–2\,cm^{-1}$ decrease from the surface layer of liposomes suggesting that the permeated liposomes become more conformationally ordered when they enter the SC. For the individual chains to become ordered, the chains in the exogenous liposomes must have become disrupted and mixed with the SC ceramide lipid chains. The latter as shown previously are highly ordered. This tentative suggestion of liposome disruption may be relevant for studies of release of trapped therapeutic agents from inside.

7.5   IR MICROSCOPY AND IMAGING OF CORNEOCYTES

This section describes the application of IR imaging spectroscopy to a study of isolated human corneocytes. The approach is suitable for monitoring molecular structure, chemical composition, and biological maturation processes occurring in these epidermal cells. The latter are tracked through concentration changes in the relative levels of a substance called the natural moisturizing factor. Our initial experiments describing this method have been published [19]; the current review builds upon this work by describing a new application to follow changes in corneocyte NMF levels as a function of external stress, which in the current case was chosen to be water washing of the skin.

It is beyond the scope of the current work to review in detail the biology of the epidermis and the essential roles of the protein filaggrin and of the NMF in skin barrier biology and hydration. It suffices for our current purposes to note that filaggrin is involved in the generation of the terminally differentiated corneocytes characteristic of the SC. Compaction of keratin is essential to the generation of a robust physical skin permeability barrier that prevents water loss from the body and controls the penetration of external agents. Furthermore, filaggrin proteolysis is the essential step in NMF generation [20]. It is hydrolyzed into hygroscopic components including free amino acids and pyrrolidone

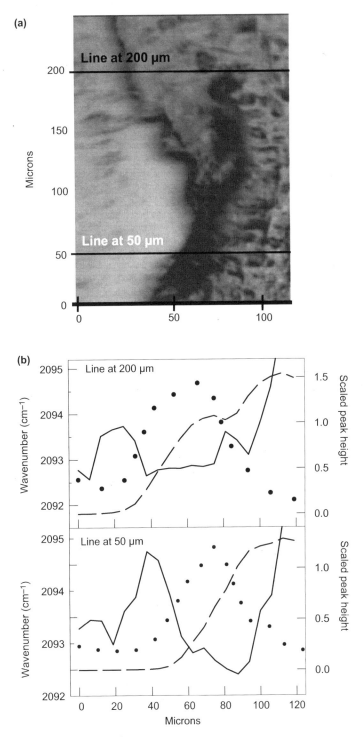

FIGURE 7.6 IR imaging analysis of lipid permeation and acyl chain conformational ordering upon application of 1-palmitoyl-d$_{31}$, 2-oleoylphosphatidylcholine liposomes to pigskin. (a) Visible micrograph of ~5 μm thick pigskin section following application of lipid vesicles to the surface. (b) IR spectral parameters at positions along the line drawn at 200 μm in the visible image (top) and 50 μm (bottom): (i) the wavenumber value of the symmetric CD$_2$ stretching mode arising from the applied liposomes (solid line, left-hand ordinate), (ii) the intensity of the same mode (●, right-hand ordinate), and (iii) the intensity of the amide II mode as a marker for the presence of skin (——, right-hand ordinate).

carboxylic acid, which together account for over 50% of the NMF composition. NMF can contribute up to 30% of the SC's dry weight and is essential in maintaining SC hydration which, in turn, is essential for both enzymatic activity and SC mechanical properties [21–24]. This complex biochemical process is highly regulated and is not yet completely understood; however, recent seminal studies have demonstrated that mutations in the gene encoding for filaggrin play a major role in diseases of the skin barrier, such as ichthyosis vulgaris and atopic dermatitis [25–28].

IR spectroscopic imaging permits the acquisition of an array of spectra within individual corneocytes at a spatial resolution of approximately 10 μm, a dimension convenient for imaging when compared with cell dimensions (Figure 7.1). We have used the approach to track the spatial distribution of various components in skin sections, as well as to image the molecular structure of selected endogenous components [2, 29].

In the current example, corneocyte maturation was studied by acquisition of IR spectra from cells isolated from sequential tape strips collected from volar forearms. Isolation of corneocytes from the tape strips was achieved by flushing the tape with hexane followed, when necessary, by sonication and filtration. Aliquots of corneocytes suspended in hexane were placed onto IR windows and the solvent was evaporated. For the SC maturation studies, spectra were acquired from 20 to 40 individual corneocytes per tape strip and a mean spectrum was generated. As shown in Figure 7.7a, the mean spectra depict differences in NMF levels in cells on tape strips corresponding to different depths beneath the skin surface. The dominant spectral feature at 1404 cm^{-1} arises from carboxylate components in the corneocytes, attributable to the presence of the amino acid breakdown products of filaggrin. These spectral features had previously been correlated with model samples of NMF to confirm assignment of the chemical composition [19]. The protein amide I feature at 1650 cm^{-1}, which occurs in all these spectra (not shown), was used as an internal standard permitting generation of a relative concentration profile of NMF levels. By applying this measure to cells from tape strips 2, 5, and 12, a progressive increase in NMF levels is evident in the outer SC layers moving in from the skin surface (tape 2) to a depth of approximately 6 μm (tape 12). The relative NMF profile is plotted in Figure 7.7b. These data are entirely consistent with our initial results that focused on differences in NMF levels between corneocytes isolated from tapes 3 and 11 [19].

The above results are promising and demonstrate the utility of vibrational spectroscopic approaches to provide molecular and compositional information in research dermatology. Recently, *in vivo* confocal Raman studies have indicated that differences in NMF levels can be measured in patients with atopic dermatitis associated with filaggrin gene mutations [30]. The demonstrated ability of infrared imaging to directly detect changes in the chemical composition of

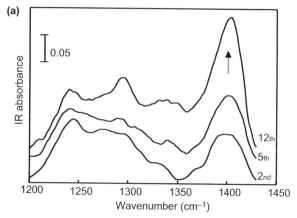

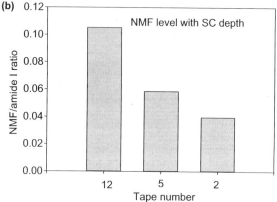

FIGURE 7.7 (a) Mean IR spectra from imaging data of corneocytes collected at different stratum corneum depths by tape stripping. (b) Relative NMF concentration with depth generated from IR imaging spectra as measured by the band area ratio of the carboxylate symmetric stretching mode (∼1404 cm^{-1}) to the amide I band (∼1650 cm^{-1}). A progressive decrease in relative NMF is observed as the cells mature.

corneocytes at different stages of SC maturation suggests that the technology might be of general utility for detecting changes in single cells associated with disease states, environmental stresses, and variations in anatomical sites. The ease of working with tape strips enables the sampling of any anatomic site repeatedly at the same location while providing a high degree (∼0.5 μm) of depth resolution. This, along with the high spectral quality of the IR data, indicates that the protocol may be very useful for clinical applications.

Our initial application of IR microspectroscopy has been focused to track the changes in corneocyte biochemistry induced by a simple stress, namely, the washing and rubbing of skin. Leaching and removal of NMF as a result of cleansing has been suggested to be a source of dry skin [22]. To directly monitor changes in corneocyte NMF levels induced by cleansing, corneocytes were prepared as follows. Two adjacent sites of the inner upper arm were tape stripped 6 times after which one site was treated for 1 min by washing with water (accompanied by physical rubbing) while the

(a)

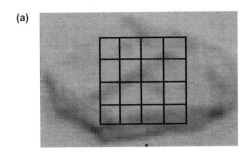

(b)

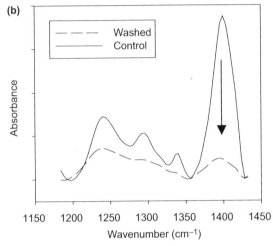

FIGURE 7.8 (a) Visible image of an isolated corneocyte with grid overlay representing the array of 16 IR spectra collected from all corneocytes sampled. Each pixel (spectrum) in the grid samples a $6.25 \, \mu m^2$ area. (b) Mean IR spectra ($1180–1460 \, cm^{-1}$ region) are generated from the 4×4 array of imaging data collected from each of 20 corneocytes obtained from the control and washed sites. Each mean spectrum is generated from 320 IR spectra and a significant decrease in the $1404 \, cm^{-1}$ NMF spectral feature is observed after washing.

other site was untreated. Subsequently, both sites were tape stripped; corneocytes were isolated and collected as previously noted.

Figure 7.8a shows the IR imaging protocol applied to a typical corneocyte visible image. An array of 16 IR spectra (pixel size of $6.25 \, \mu m^2$) is acquired from each corneocyte. Two mean spectra generated from the IR images of 40 corneocytes each, from washed and untreated sites, are displayed in Figure 7.8b. Thus, each mean spectrum is generated from 640 individual IR spectra. It is clear that the mean corneocyte spectrum from the tape strip acquired after washing shows a significant reduction in the $1404 \, cm^{-1}$ peak associated with NMF. In Figure 7.9a and b, spectra from corneocytes have been concatenated and images of correlation coefficients are shown for the 4×4 arrays (16 spectra per cell) from each of the 40 corneocytes isolated from the control site (top in Figure 7.9a and b) and from the treated site (bottom in both Figure 7.9a and b). In Figure 7.9a, spectra in the $1200–1430 \, cm^{-1}$ region have been correlated to the

mean spectrum of the untreated site and similarly to the mean spectrum of the washed site in Figure 7.9b. For the majority of the images, the correlation maps clearly differentiate the washed from the untreated corneocytes, demonstrating that NMF is removed from corneocytes by cleansing skin with water. IR imaging directly detects compositional changes induced by this relatively mild stress. As previously demonstrated in our studies of SC maturation, it is possible to quantitatively image changes in NMF levels in corneocytes by generating images in which the $1404 \, cm^{-1}$ peak characteristic of NMF is normalized to the amide I peak [19]. As illustrated in Figure 7.10, the images of relative NMF concentration in corneocytes from the control and washed sites show marked differences. Washing skin prior to isolation of the cells clearly results in reduced NMF concentrations as directly measured in the current IR imaging studies.

The current results, taken together with the initial studies, demonstrate that IR imaging can track corneocyte maturation in the SC via imaging of NMF levels in cells and suggest experiments focusing on the effects of both environmental stresses and disease-related changes in SC corneocyte biology. Finally, we note that the ease of working with tape strips, the ability to sample repeatedly at the same location (providing a high z-dimension resolution of approximately $0.5 \, \mu m$) and the convenience of sampling any anatomic site, as well as the very high spectral quality of the IR images, all indicate that the application of IR imaging will be of great general utility for clinical studies related to corneocyte biochemistry.

7.6   IR MICROSCOPIC IMAGING OF HAIR

Hair and its constituent keratin fibers have been studied by vibrational spectroscopic methods for decades [31–33]. Such investigations have frequently focused on oxidative damage. More recently, spatially resolved molecular structure information has been acquired with Raman microspectroscopy. As an example, the extensive measurements by Kuzuhara [34–36] elucidated structural changes particularly in the cortex and cuticle induced by a variety of chemical treatments. These studies have been extended to include both white and melanin-enriched black human hair.

FTIR microspectroscopy has also been employed to spatially image cross sections of human hair with both normal imaging instrumentation and with a synchrotron source-based spectrometer [37, 38]. These experiments permit the spatial resolution of cuticle, cortex, and medulla in cross section. To illustrate the utility of IR spectroscopic approaches, some results of imaging experiments from the ISP laboratory are shown below.

Human hair ($\sim$60–80 μm in diameter) is conveniently divided from the outside moving inward into three separate regions, namely, the cuticle, the cortex, and the medulla. The

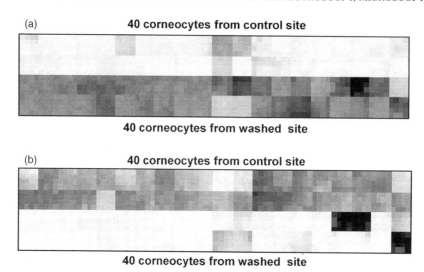

FIGURE 7.9 Correlation coefficient images of multiple corneocytes isolated from untreated and washed sites. (a) Correlation to mean of untreated cells: top two rows are concatenated images of 40 corneocytes isolated from untreated sites and similarly bottom two rows from cells isolated from washed sites. (b) Correlation image to mean of cells from washed sites: concatenation same as in part (a). Correlation coefficient scale: white > gray > black. (See the color version of this figure in Color Plate section with correlation coefficient scale: red > yellow > black.)

outermost region, the cuticle, is comprised of 6–10 layers of flattened, overlapping cells, corresponding to 3–5 μm in thickness, which are largely constructed from amorphous proteins. The cortex contains the bulk of the hair (keratin) fibers that are comprised of two components, the microfibril and the matrix. The microfibrils are highly crystalline regions comprised of α-helical proteins. These proteins, which have a low cysteine disulfide content, are imbedded in the amorphous matrix of high cysteine content. The microfibrils are aligned along the hair fiber axis. Finally, the central medulla (~5–10 μm in diameter) is composed largely of lipids and proteins.

Figure 7.11a–d demonstrates the nature of the information available in the current IR experiment. Captured in this figure are cross sections of bleached human hair that have been embedded in tissue freezing medium, cryogenically microtomed to a thickness of 4 μm, mounted on CaF$_2$ windows and imaged. Figure 11a displays the visible micrograph of the sectioned hair, with clearly resolved cuticle, cortex, and medulla. Figure 7.11b, c, and d depicts IR spatial images of

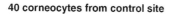

FIGURE 7.10 IR images of relative NMF concentration as measured by band area ratio (1404 cm^{-1}/amide I) in 80 corneocytes, 40 each from the control (top two rows) and washed sites (bottom two rows). The 640 individual spectra collected from the imaging data of the 40 control corneocytes clearly have a higher band area ratio (white > gray > black), which in this image represents a higher concentration of NMF in these cells.

the relative concentrations of protein, lipid, and sulfonate, respectively. Protein levels in Figure 7.11b derived from the amide I intensity are increased in the cortex of the hair fibers compared to the medulla, while clear delineation of the cuticle is not feasible. In contrast, the concentration profiles of lipid chains depicted in Figure 7.11c (as measured from the intensity of the C–H stretching vibration at 2925 cm^{-1}) suggest increased levels of lipids in the medulla compared with the cortex. Finally, Figure 7.11d images sulfonate concentrations in the same set of hair cross sections (measured from the S=O stretching vibration of SO$_3^-$ at 1040 cm^{-1}) and suggests increased levels in the cortex of bleached hair. The limit of spatial resolution available with a standard source is achieved in this experiment. We note that it is feasible to do significantly better with a synchrotron source. The ~10 μm spatial resolution in the current measurement as depicted in the images of Figure 7.11 is surprisingly sufficient to view the medulla region of the fiber. These images clearly demonstrate that simple functional group mapping provides useful contrast for tracking the concentrations of the important chemical species of hair in cross section. The approach is potentially useful for evaluation of the physiological state of hair including assessment of damage from cosmetic treatment.

As an example of the utility of the IR imaging experiment, samples of virgin, bleached, and base-treated hair have been imaged. It is widely known that significant damage to hair is derived from oxidation due to bleaching. The spectroscopic marker suitable for imaging this effect is the sulfonate (S=O) vibration at 1040 cm^{-1}, which is related to oxidized disulfide bond content. Figure 7.12a, b, and c depicts the ratio of the 1040 cm^{-1} band to that of the amide A band found at 3290 cm^{-1} for virgin, bleached, and base-treated hair cross

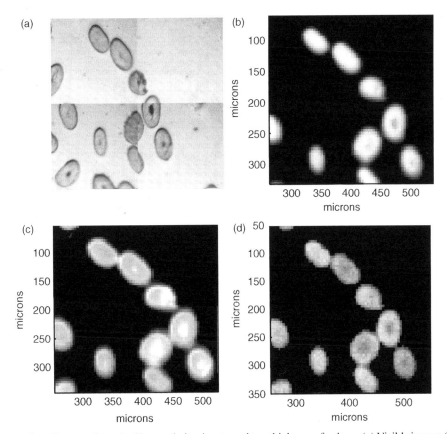

FIGURE 7.11 Cross-sectional image of bleached human hair microtomed to a thickness of ~4 μm. (a) Visible image; (b) IR image of amide I band (~1650 cm^{-1}) intensity representing protein content; (c) IR image of the asymmetric CH$_2$ stretching band (2925 cm^{-1}) intensity depicting primarily lipid content; and (d) IR image of the sulfonate band (~1040 cm^{-1}) intensity potentially useful for evaluating oxidative damage to disulfide bonds predominantly present in the cortex. Grayscale: white > gray > black. (See the color version of this figure in Color Plate section with color scale: red > yellow > blue.)

sections, respectively. This intensity ratio makes possible the assessment of protein structural damage via the relative concentration of the oxidized sulfonate mapped across the hair sections. In common to all three cases, there is elevated sulfonate intensity detected in the cuticle region, most likely due to the exposed nature of the external region of the hair fiber to external oxidative conditions. High scores indicate that the $-SO_3^-$ levels are elevated in the areas of the cortex, and toward the interior of the fiber in general of the bleached hair compared to virgin or base-treated hair. This effect is not quite as pronounced toward the center (medulla) of the fiber, indicating that this region is somewhat shielded from the oxidative effects of the bleaching. From this visualization, it is suggested that structural integrity (S—S bonds) diminished in the matrix portion of the cortex in the bleached hair.

These figures demonstrate that IR imaging can be extended to microscopic visualization of structure and damage to hair as a result of disease or cosmetic treatment. While the work shown in this section does not address the detailed nature of the states of the molecules (conformation, etc.), it clearly demonstrates the feasibility of doing so in a spatially resolved manner.

7.7 VIBRATIONAL MICROSCOPIC IMAGING OF WOUND HEALING

7.7.1 Introduction

In response to injury, skin exhibits a complex temporal and spatial series of cellular processes normally leading to wound healing. The events are considered to occur in three (overlapped) phases: inflammation, proliferation, and maturation/remodeling. Following a cutaneous wound, skin responds very quickly by initiating two main cellular mechanisms geared toward wound closure, namely, reepithelialization and contraction of connective tissue. Within hours following the injury, keratinocytes at the wound edge become activated. Proliferation and migration of keratinocytes commence in the first 2 days via formation of a layer of epidermal cells known as the migrating epithelial tongue (MET) [39, 40]. This process is essentially complete in 7–9 days following the injury at which time the wound bed is covered by a layer of cells. Subsequent to these events, a stratified epidermal layer is reestablished. Following the proliferative phase, granulation tissue forms and the wound begins to contract.

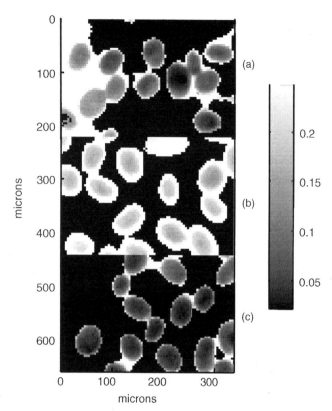

FIGURE 7.12 IR imaging of relative sulfonate content in micro-tomed human hair sections following various treatments. The band area ratio of the $1040\,cm^{-1}$ sulfonate peak over the amide A band ($\sim3200\,cm^{-1}$) is presented for (a) untreated hair, (b) bleached hair, and (c) base-treated hair. (See the color version of this figure in Color Plate section.)

The last stages of the process involve formation of collagen cross-linking both to itself and to other molecules, a process that improves the tissue tensile strength. A schematic of the sequence of events that take place is shown in Figure 7.13 [41]. Over the past two decades, genomic and proteomic approaches have greatly enhanced our under-standing of the spatial and temporal sequence of events.

To date, optical imaging methods to directly characterize the molecular structure and spatial distribution of the newly synthesized tissue constituents have not been available. Knowledge gained at the molecular structure level would be of particular interest for processes such as the evaluation of therapeutic agents and for the study of interfaces between native and artificial skin. There are at least two major impediments to the application of relatively sophisticated spectroscopic methods for the evaluation of wound healing. First is the difficulty in creating reproducible wound models *ex vivo*, which are essential for the systematic tracking of biochemical events over the time course of healing. Second, prior to the development of vibrational microscopy tech-nologies, optical methods were unavailable that could

distinguish the spatial/temporal evolution of closely related protein species (e.g., various forms of keratin) within a single section of wounded tissue or in full thickness (wounded) skin. The current section reviews the efforts of the Rutgers laboratory and their collaborators at the Hospital for Special Surgery (New York City) in demon-strating the utility of IR imaging for monitoring the spatial position of various keratins during the formation and migration of keratinocytes [1].

7.7.2 Methods

7.7.2.1 Human Organ Culture Wound-Healing Model
The model we have used involves creation of acute wounds in human skin acquired during reduction abdominoplasty ("tummy tuck") in accordance with approved institutional protocols. A 3 mm biopsy punch created the wound. Healing processes in skin specimens could be maintained for up to 6 or 7 days in culture in an environmental chamber at 37°C, 5% CO_2, and 95% relative humidity.

7.7.2.2 IR Imaging Experiments
Unwounded and wounded skin samples healing for specific periods of time were selected for IR imaging. Frozen samples were micro-tomed perpendicular to the stratum corneum into 5 μm thick sections and placed on BaF_2 windows. An XY sample stage permitted the collection of IR images (pixel area of 6.25 μm^2) over ~0.5 mm $\times$ 0.5 mm sample areas. The sampling ar-rangement is depicted in Figure 7.14a. Spectral resolution was 8 cm^{-1}.

7.7.3 Results and Discussion

The ability of FT-IR imaging to track molecular structure/composition changes during wound healing is evident from factor analysis of the IR spectral imaging data in the 1185–1475 cm^{-1} region in a spatial area containing a wound edge and a MET 6 days postwounding (Figure 7.15a–f). A visible image of the same skin section used for IR imaging is presented in Figure 7.15a. The MET progresses from the nonwounded area toward the right side of the image as labeled. Four different factors arising from keratin (Figure 7.15f, traces labeled f1–f4) were revealed from factor analysis; their score images are presented in Figure 7.15b–e. The major spectral features in this wave-number region arise from CH_2 and CH_3 bending modes ($\sim1450\,cm^{-1}$), carboxylate symmetric stretch ($\sim1400\,cm^{-1}$), and the amide III mode ($\sim1235\,cm^{-1}$) from keratin. The factor loadings are obviously similar, but nevertheless map to four spatially distinct keratin-rich areas of the tissue. The score image for f1 (Figure 7.15b) highlights the SC and provides a clear marker for the original wound edge (Figure 7.15a). On the left side of this image, high scores carry over into the viable epidermal region. Score images for

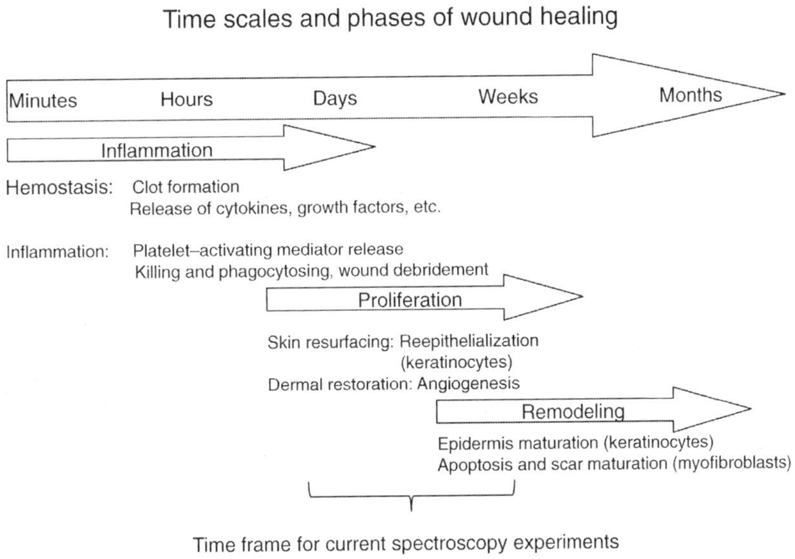

Time scales and phases of wound healing

Minutes   Hours   Days   Weeks   Months

Inflammation

Hemostasis: Clot formation
Release of cytokines, growth factors, etc.

Inflammation: Platelet–activating mediator release
Killing and phagocytosing, wound debridement

Proliferation

Skin resurfacing: Reepithelialization
(keratinocytes)
Dermal restoration: Angiogenesis

Remodeling

Epidermis maturation (keratinocytes)
Apoptosis and scar maturation (myofibroblasts)

Time frame for current spectroscopy experiments

FIGURE 7.13   Flowchart of the complex, overlapping temporal sequence of events that take place following an acute cutaneous wounding of human skin (adapted from Ref. 41).

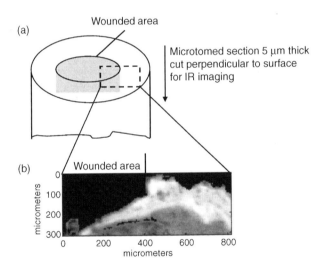

FIGURE 7.14   (a) Schematic depiction of the cutaneous wound-healing model used in the current experiments. A 3 mm punch biopsy is used to generate an acute wound in a human skin specimen. At various time points after wounding, samples are fast frozen and microtomed to 5 μm thick sections for IR imaging (embedding media is not used). (b) An IR image resulting from factor analysis conducted over the amide I and II spectral region (1480–1720 cm^{-1}) on a 6-day postwounded sample is displayed. The score image is shown for a factor loading that strongly resembles keratin revealing the MET and the epidermis (keratin-rich areas) in the nonwounded area. Grayscale coding for scores: white > gray > black. (See the color version of this figure in Color Plate section with color coding for scores: red > yellow > blue.)

the keratin factors f2 (Figure 7.15b) and f3 (Figure 7.15c) are concentrated in the suprabasal and basal epidermal regions, respectively. The keratin imaged in Figure 7.15c encompasses a large region of the MET. The score image for the basal layer keratin (Figure 7.15d) also reveals a diffuse line of keratin-containing cells with elevated scores extending into the lower region of the MET. The final score image (Figure 7.15e) with a keratin spectral signature is spatially restricted to the outer and leading edges of the MET (Figure 15e). Spectra located in the outer edges of the MET at 4 days postwounding (data not shown) share similar characteristics with f4, and reveal a downward frequency shift in the carboxylate symmetric stretch.

At this stage of the experiment, we attempted to correlate the four keratin factors with the particular keratin type corresponding to the factor. We started by correlating our images with information derived from immunofluorescent keratin staining examining the temporal expression of various keratins during healing [1, 42, 43]. In healthy skin, keratinocytes in the basal layer express keratins 5 and 14, while keratins 1 and 10 are expressed in suprabasal keratinocytes. As the wound is reepithelialized, keratinocytes at the wound edge initially become activated; migration and hyperproliferation proceed accompanied by expression of keratins 6, 16, and 17 (K6, K16, and K17) [44]. We have confirmed the presence of activated keratinocytes at the wound edge in our model by staining sections of acute wounds with K17 specific antibody [1]. Twenty-four hours after the wounding, we observed strong suprabasal K17 staining at the wound edge indicating activation of keratinocytes. In addition, keratino-

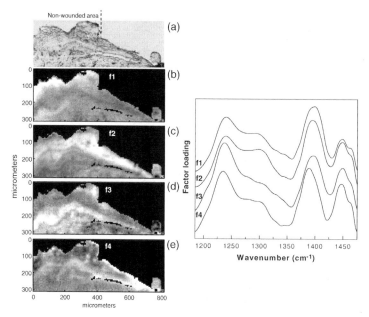

FIGURE 7.15   IR characterization of keratin-rich regions in wounded and nonwounded areas, 6 days postwounding, using factor analysis in the 1185–1480 cm^{-1} spectral region. (a) Visible image of the 5 µm thick skin section used for IR imaging. (b–e) Factor scores for corresponding factor loadings (f1–f4) shown in part (f). In the score images: white > gray > black. See text for details. (See the color version of this figure in Color Plate section with score image scale: red > yellow > blue.)

cytes composing the MET were positive for K17, thus confirming migration of activated suprabasal keratinocytes over the wound bed. As anticipated, we did not observe K17 staining in healthy, unwounded skin.

Since variations in IR spectral bands arise directly from molecular composition and/or structural differences in particular spatial regions of the sample, it is reasonable to assume that our initial hypothesis is correct, namely, that the spectral variations between the various factors (f1–f4) arise from specific types of keratin known to reside in different regions of the epidermis and recovering wound area. Based on similarities between the staining patterns and the IR images, we suggest the following assignments of factors to particular keratins:

1. Factor loading 1 and its corresponding score image (Figure 7.15b) are characteristic of differentiating keratinocytes, rich in K1/10.

2. Factor loading 2, imaged in Figure 7.15c and residing in the suprabasal region proximal to and within the MET, represents regions abundant in K17.

3. High scores for f3 (Figure 15d), found in the basal region extending into the lower MET, depict areas rich in K14.

4. Although we cannot assign f4 to a particular keratin or keratins with certainty, regions with high scores for this factor may represent areas with a higher content of both K14 and K17.

To our knowledge, the current interplay (for classification of very similar proteins) between molecular biology-based methods and vibrational spectroscopic images is a novel approach with many potentially interesting applications. Thus, further development in the spectral analysis of additional wound-healing samples and purified materials under controlled environmental conditions will allow us to exploit the molecular structure information inherent in the data. This in turn may begin to provide a basis for understanding changes in the observed spatial distributions of particular proteins during therapeutic interventions for wound healing.

7.8   CONCLUDING COMMENTS

The point of view we have adopted throughout our development of biomedical applications of IR and Raman imaging is that the molecular structure information inherent in the spectra provides a unique means to interpret the images generated. This is evident in the examples illustrated above. Thus, the approach adds a new dimension to the more traditional means (e.g., absorption, fluorescence, scattering) of evaluating inhomogeneous structures in biological tissues.

It seems fair to note that the techniques as currently applied are limited in two ways. First, the images are of poorer spatial resolution (~10 µm in the IR and ~2 µm in the

Raman) than those generated by traditional means. Second, the weakness of the physical phenomena involved (in particular Raman scattering), and to some extent the vast amounts of spectral information (~100–1000 points per pixel), limits the speed at which images can be generated.

Nevertheless, partial solutions to the above issues are emerging, at least for particular problems. IR spatial resolution may be enhanced through the use of ATR imaging by a factor of 2–4, depending on the particular ATR substrate. More dramatically, for problems not requiring large images, novel approaches such as TERS (tip enhanced Raman spectroscopy), involving the coupling of an atomic force microscope with Raman spectroscopy, combines the sensitivity provided by near-field enhancement of the Raman signal with the nanometer spatial resolution of the AFM.

REFERENCES

1. Chan, K. L. A., Zhang, G., Tomic-Canic, M., Stojadinovic, O., Lee, B., Flach, C. R., and Mendelsohn, R. (2008) A coordinated approach to cutaneous wound healing: vibrational microscopy and molecular biology. *J. Cell. Mol. Med.* **12**, 2145–2154.

2. Xiao, C., Moore, D. J., Rerek, M. E., Flach, C. R., and Mendelsohn, R. (2005) Feasibility of tracking phospholipid permeation into skin using infrared and Raman microscopic imaging. *J. Invest. Dermatol.* **124**, 622–632.

3. Zhang, G., Moore, D. J., Sloan, K. B., Flach, C. R., and Mendelsohn, R. (2007) Imaging the prodrug-to-drug transformation of a 5-fluorouracil derivative in skin by confocal Raman microscopy. *J. Invest. Dermatol.* **127**, 1205–1209.

4. Zhang, G., Flach, C. R., and Mendelsohn, R. (2007) Tracking the dephosphorylation of resveratrol triphosphate in skin by confocal Raman microscopy. *J. Control. Release* **123**, 141–147.

5. Zhang, G., Moore, D. J., Flach, C. R., and Mendelsohn, R. (2007) Vibrational microscopy and imaging of skin: from single cells to intact tissue. *Anal. Bioanal. Chem.* **387**, 1591–1599.

6. Pensack, R. D., Michniak, B. B., Moore, D. J., and Mendelsohn, R. (2006) Infrared kinetic/structural studies of barrier reformation in intact stratum corneum following thermal perturbation. *Appl. Spectrosc.* **60**, 1399–1404.

7. Moore, D. J. and Bi, X. (2007) The application of infrared spectroscopic imaging to skin delivery: visualizing molecular localization in formulations and in skin. In: *Science, Applications of Skin Delivery Systems*, Allured, New York, pp. 49–59.

8. Forslind, B. (1994) A domain mosaic model of the skin barrier. *Acta Derm. Venereol.* **74**, 1–6.

9. Snyder, R. G. (1960) Vibrational spectra of crystalline *n*-paraffins. Part I. Methylene rocking and wagging modes. *J. Mol. Spectrosc.* **4**, 411–434.

10. Snyder, R. G. (1961) Vibrational spectra of crystalline *n*-paraffins. Part II. Intermolecular effects. *J. Mol. Spectrosc.* **7**, 116–144.

11. Snyder, R. G., Hsu, S. L., and Krimm, S. (1978) Vibrational spectra in the C–H stretching region and the structure of the polymethylene chain. *Spectrochim. Acta* **34A**, 395–406.

12. Snyder, R. G., Strauss, H. L., and Elliger, C. A. (1982) C–H stretching modes and the structure of *n*-alkyl chains. 1. Long, disordered chains. *J. Phys. Chem.* **86**, 5145–5150.

13. Snyder, R. G., Liang, G. L., Strauss, H. L., and Mendelsohn, R. (1996) IR spectroscopic study of the structure and phase behavior of long-chain diacylphosphatidylcholines in the gel state. *Biophys. J.* **71**, 3186–3198.

14. Zerbi, G., Magni, R., and Gussoni, M. (1981) Spectroscopic markers of the conformational mobility of chain ends in molecules containing *n*-alkane residues. *J. Mol. Struct.* **73**, 235–237.

15. Zerbi, G., Magni, R., Gussoni, M., Holland Moritz, K., Bigotto, A. B., and Dirlikov, S. (1981) Molecular mechanics for phase transition and melting of *n*-alkanes: a spectroscopic study of molecular mobility of solid *n*-nonadecane. *J. Chem. Phys.* **75**, 3175–3194.

16. Minoni, G. and Zerbi, G. (1982) End effects on longitudinal accordion modes (LAM): fatty acids and layered systems. *J. Phys. Chem.* **86**, 4791–4798.

17. Simanouti, T. and Mizushima, S. -I. (1949) The constant frequency Raman lines of *n*-paraffins. *J. Chem. Phys.* **17**, 1102–1106.

18. Shimanouchi, T. and Tasumi, M. (1971) *Ind. J. Pure Appl. Phys.* **9**, 958–961.

19. Zhang, G., Moore, D. J., Mendelsohn, R., and Flach, C. R. (2006) Vibrational microspectroscopy and imaging of molecular composition and structure during human corneocyte maturation. *J. Invest. Dermatol.* **126**, 1088–1094.

20. Scott, I. R. and Harding, C. R. (1986) Filaggrin breakdown to water binding compounds during development of the rat stratum corneum is controlled by the water activity of the environment. *Dev. Biol.* **115**, 84–92.

21. Rawlings, A. V. (2003) Trends in stratum corneum research and the management of dry skin conditions. *Int. J. Cosmet. Sci.* **25**, 63–95.

22. Rawlings, A. V. and Harding, C. R. (2004) Moisturization and skin barrier function. *Dermatol. Ther.* **17**, 43–48.

23. Rawlings, A. V., Scott, I. R., Harding, C. R., and Bowser, P. A. (1994) Stratum corneum moisturization at the molecular level. *J. Invest. Dermatol.* **103**, 731–740.

24. Harding, C. R., Watkinson, A., Rawlings, A. V., and Scott, I. R. (2000) Dry skin, moisturization and corneodesmolysis. *Int. J. Cosmet. Sci.* **22**, 21–52.

25. Sandilands, A., Smith, F. J., Irvine, A. D., and McLean, W. H. (2007) Filaggrin's fuller figure: a glimpse into the genetic architecture of atopic dermatitis. *J. Invest. Dermatol.* **127**, 1282–1284.

26. Smith, F. J. D., Irvine, A. D., Terron-Kwiatkowski, A., Sandilands, A., Campbell, L. E., Zhao, Y., Liao, H., Evans, A. T., Goudie, D. R., Lewis-Jones, S., Arseculeratne, G., Munro, C. S., Sergeant, A., O'Regan, G., Bale, S. J., Compton, J. G., DiGiovanna, J. J., Presland, R. B., Fleckman, P., and McLean, W. H. I. (2006) Loss of function mutations in the gene

encoding filaggrin cause ichthyosis vulgaris. *Nat. Genet.* **38**, 337–342.

27. Brown, S. J., Relton, C. L., Liao, H., Zhao, Y., Sandilands, A., Wilson, I. J., Burn, J., Reynolds, N. J., and McLean, W. H. I. (2008) Filaggrin null mutations and childhood atopic eczema: a population-based case-control study. *J. Allergy Clin. Immunol.* **121**, 940–946.

28. Irvine, A. D. and McLean, W. H. I. (2006) Breaking the (un) sound barrier: filaggrin is a major gene for atopic dermatitis. *J. Invest. Dermatol.* **126**, 1200–1202.

29. Mendelsohn, R., Chen, H. -C., Rerek, M. E., and Moore, D. J. (2003) Infrared microspectroscopic imaging maps the spatial distribution of exogenous molecules in skin. *J. Biomed. Optics* **8**, 185–190.

30. Kezic, S., Kemperman, P. M. J. J., Koster, E. S., de Jongh, C. M., Thio, H. B., Campbell, L. E., Irvine, A. D., McLean, W. H. I., Puppels, G. J., and Caspers, P. J. (2008) Loss-of-function mutations in the filaggrin gene lead to reduced level of nature moisturizing factor in the stratum corneum. *J. Invest. Dermatol.* **128**, 2117–2119.

31. Frushour, B. G. and Koenig, J. L. (1975) Raman spectroscopy of proteins. In: Clark, R. J. and Hester, R. E. (Eds.), *Advances in Infrared and Raman Spectroscopy*, Vol. 1, Heyden and Son, London, pp. 35–97.

32. Church, J. S., Corino, G. L., Woodhead, A. L. (1997) Analysis of merino wool cuticle and cortical cells. *Biopolymers* **42**, 7–17.

33. Stassburger, J., Breuer, M. M. (1985) Quantitative Fourier transform infrared spectroscopy of oxidized hair. *J. Soc. Cosmet. Chem.* **36**, 61–74.

34. Kuzuhara, A. (2005) Analysis of structural change in keratin fibers resulting from chemical treatments using Raman spectroscopy. *Biopolymers* **77**, 335–344.

35. Kuzuhara, A. (2006) Analysis of structural changes in permanent waved human hair using Raman spectroscopy. *Biopolymers* **85**, 274–283.

36. Kuzuhara, A. (2006) Analysis of structural changes in bleached keratin fibers (black and white human hair) using Raman spectroscopy. *Biopolymers* **81**, 506–514.

37. Bantignies, J. -L., Fochs, G., Carr, G. L., Williams, G. P., Lutz, D., and Marull, S. (1998) Organic reagent interaction with hair spatially characterized by synchrotron IMS. *Int. J. Cosmet. Sci.* **20**, 381–394.

38. Chan, K. L. A., Kazarian, S. G., Mavraki, A., and Williams, D. R. (2005) Fourier transform infrared imaging of human hair with a high spatial resolution without the use of a synchrotron. *Appl. Spectrosc.* **59**, 149–155.

39. Singer, A. J., and Clark, R. A. F. (1999) Cutaneous wound healing. *N. Engl. J. Med.* **341**, 738–746.

40. Coulombe, P. A. (2003) Wound epithelialization: accelerating the pace of discovery. *J. Invest. Dermatol.* **121**, 219–230.

41. Li, J., Chen, J., and Kirsner, R. (2007) Pathology of acute wound healing. *Clin. Dermatol.* **25**, 9–18.

42. Patel, G. K., Wilson, C. H., Harding, K. G., Finlay, A. Y., and Bowden, P. E. (2006) Numerous keratinocyte subtypes involved in wound re-epithelialization. *J. Invest. Dermatol.* **126**, 497–502.

43. Usui, M. L., Underwood, R. A., Mansbridge, J. N., Muffley, L. A., Carter, W. G., and Olerud, J. E. (2005) Morphological evidence for the role of suprabasal keratinocytes in wound reepithelialization. *Wound Rep. Reg.* **13**, 468–479.

44. Freedberg, I. M., Tomic-Canic, M., Komine, M., and Blumenberg, M. (2001) Keratins and the keratinocyte activation cycle. *J. Invest. Dermatol.* **116**, 633–640.

8

NEAR-INFRARED *IN VIVO* SPECTROSCOPIC IMAGING: BIOMEDICAL RESEARCH AND CLINICAL APPLICATIONS

R. Anthony Shaw, Valery V. Kupriyanov, Olga Jilkina, and Michael G. Sowa
Institute for Biodiagnostics, National Research Council of Canada, Winnipeg, Manitoba, Canada

8.1 INTRODUCTION

By virtue of the good transparency of tissue through the range 750–1000 nm, near-IR spectroscopy offers the means to effectively probe certain tissue properties *in vivo*. This possibility was first revealed explicitly by Jöbsis in 1977 [1], with a scientific paper illustrating the ability to monitor hemoglobin saturation (and also suggesting the prospect of monitoring the oxidation state of cytochrome *a,a3*) *in vivo*. This seminal manuscript spawned an *in vivo* near-IR spectroscopy community that continues to flourish today, with ever more adventurous souls devising ever more adventurous measurements. For example, as the fundamental technology has grown more sensitive and less expensive, fiber-optic-based spatial and time-resolved cerebral measurements are now almost routine. The interested reader is referred to Ref. 2 that includes a biographical sketch and Ref. 3 that includes an overview of Professor Jöbsis' career, as well as 11 technical contributions, many of which come from the leading lights active in this research area.

In vivo near-IR spectroscopic measurements provide information regarding tissue blood supply and oxygenation by virtue of the near-IR absorptions of hemoglobin; the near-IR spectrum of oxy-Hb is very dissimilar from that of the deoxygenated counterpart. The near-IR absorptions in the 740–980 nm range arise from electronic transitions, with absorptivities high enough to feature prominently in tissue spectra that also include substantial features from vibrational overtone modes of water. There is a body of work suggesting the possibility that a very weak cytochrome *a,a3* absorption can also be exploited to monitor tissue redox status *in vivo* [4, 5]. This possibility has not yet been explored within imaging studies, however, and will therefore not be discussed further in this chapter.

The majority of *in vivo* spectroscopy measurements have made use of fiber optics as the means both to deliver light to the subject and to convey it back to the detector. By its nature, this approach yields only very limited spatial resolution. Varying the interoptode spacing confers some flexibility; the depth of penetration for light reaching the detector is influenced by the interoptode distance: the greater the separation, the deeper the effective depth of penetration. This factor has been exploited in cerebral spectroscopy as a means to distinguish signals from superficial (close source/detector optode spacing) versus contributions from the target of interest, that is, the brain. While two- or even three-dimensional "imaging" can be achieved by using multiple optodes, and this approach has been exploited for cerebral imaging (and functional imaging) [6, 7], high-resolution two-dimensional imaging requires a novel approach.

This chapter focuses on a technique for macroscopic *in vivo* imaging of superficial (or otherwise exposed) tissue, with imaging targets typically of dimensions ranging from 1×1 to $10 \times 10\,cm^2$. The technique combines very good spatial resolution with spectral resolution that—although modest—is entirely sufficient to recover the essential features of *in vivo* near-IR spectra. Integrating a two-dimensional CCD array with a camera and a wavelength selection device—generally a liquid crystal tunable filter—the imaging system can provide full spectroscopic image data cubes spanning 512×512 pixels (two spatial dimensions) and 41 wavelengths at 10 nm intervals over the range 650–1050 nm (the third, spectroscopic, dimension). This global imaging technique thus provides spectroscopic information simultaneously for the complete field of view.

This chapter begins with a brief overview of the essential features of *in vivo* spectroscopic imaging and of the methodologies used to acquire the data. Distilling useful two-dimensional images from the three-dimensional data sets requires the judicious adoption and/or development of appropriate data processing techniques, which are also surveyed here. We then focus in some detail on two sets of specific applications. We first discuss skin, with applications encompassing dermatological research (skin hydration), burns, and skin flaps in the context of reconstructive surgery. This last application is not only of intrinsic interest, but also of historical note since it was the first strong, clinically relevant application of the 2D imaging technique. We then review a fairly substantial body of work that has illustrated the potential of cardiac imaging both as a research tool and for potential intraoperative use. The chapter closes with a synopsis and overview of future plans and possibilities.

8.2   METHODS

8.2.1   Instrumentation and Measurement Technique

The essential elements of the near-IR imaging technique are a near-IR emitting lamp (or lamps) to illuminate the target of interest, a near-IR camera to capture images, and a tunable filter to permit wavelength selection. A typical experimental arrangement is illustrated by Figure 8.1.

The target is typically illuminated by a pair of quartz halogen floodlights; reflectance images are then acquired with an infrared-sensitive CCD array camera. The majority of studies reported here used a system with 512×512 back-illuminated CCD elements interfaced to a 14/16-bit ST-138 analog to digital converter run in 14-bit mode (Princeton Instruments, Trenton, NJ). The lens is a Nikon Micro AF60 operating at $f/8$. To enhance signal to noise with inconsequential loss in spatial resolution, 2×2 binning is generally performed to produce final images with 256×256 pixels. Each pixel in the final image generally represents a tissue area of approximately 1 mm^2 (depending, of course, on the camera–subject distance).

Because the absorptions of interest are so broad, near-IR spectroscopic imaging does not require high spectral resolution. One wavelength selection device ideally suited for the task is a liquid crystal tunable filter. These devices mount to the camera lens and provide the means to select wavelengths

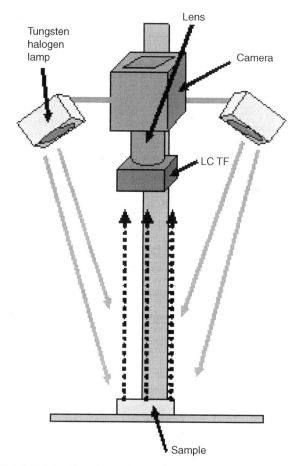

FIGURE 8.1   Experimental setup for near-IR macroscopic imaging measurements. Alternatively, for horizontal line of sight, the lamps are placed on tripod stands with the camera/LCTF mounted on a tripod between them. Typically, the image area lies in the range 1 cm × 1 cm to 10 cm × 10 cm.

with a bandwidth of about 5 nm. In practice, spectroscopic images covering the 650–1050 nm range are typically built from 41 images acquired at 10 nm intervals (co-adding of replicate images—typically five—is commonly done to increase signal to noise). Under this protocol, acquisition of a complete spectroscopic image takes approximately 5 min.

To acquire pseudo-absorbance spectra, raw spectroscopic reflectance imaging data are collected for both for a neutral reflector (e.g., a gray card) and for the sample of interest, providing 65,526 single-beam reference spectra $I_0(\lambda)$ and the same number of sample spectra $I(\lambda)$, respectively. The pseudo-absorbance spectra are derived from reflectance mode measurements by

$$A(\lambda) = -\log[I(\lambda)/I_0(\lambda)]$$

The nature of the spectra measured *in vivo* and their processing to reveal physiological information are discussed in Sections 8.2.2 and 8.2.3.

8.2.2 *In Vivo* Imaging

The basis for *in vivo* near-IR imaging is illustrated by Figure 8.2, which shows the near-IR spectra for deoxy-Hb, oxy-Hb, and water. Pseudo-absorbance spectra measured *in vivo* may be viewed (and reconstructed) as weighted superpositions of these component spectra, with an additional term generally included to represent and account for possible variations in baseline level:

$$\mathbf{A} = (C_{\text{deoxy}}L) \times \varepsilon_{\textbf{deoxy}} + (C_{\text{oxy}}L) \times \varepsilon_{\textbf{oxy}}$$
$$+ (C_{\text{water}}L) \times \varepsilon_{\textbf{water}} + B_{\text{baseline}} \times [\mathbf{1}]$$

Here, the absorptivity spectra are given by the column vectors $\varepsilon_{\textbf{deoxy}}$, $\varepsilon_{\textbf{oxy}}$, and $\varepsilon_{\textbf{water}}$. B_{baseline} represents the baseline shift (which multiplies the unit column vector [**1**] to provide a pseudo-spectrum). The structural tissue components of tissue provide only extraordinarily weak absorptions in the short-wavelength near-IR (SW-NIR) region (750–950 nm) and can be safely neglected in reconstructing spectra.

The three-dimensional nature of the spectroscopic imaging technique is illustrated by Figure 8.3, which highlights the central feature; each of the 256×256 array pixels is associated with a full near-IR spectrum. The absolute concentrations C_{deoxy}, C_{oxy}, and C_{water} are not readily determined since the absolute path length "L" that light travels within tissue before emerging is unknown. However the spectra do provide

- quantitative determination of tissue oxygen saturation parameters, for example, the ratio of oxy- to total hemoglobin $C_{\text{oxy}}/(C_{\text{oxy}} + C_{\text{deoxy}})$,
- qualitative determination of differences in C_{deoxy}, C_{oxy}, and blood volume (as estimated by $C_{\text{deoxy}} + C_{\text{oxy}}$) for

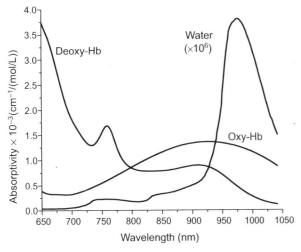

FIGURE 8.2 Molar absorptivity spectra for oxyhemoglobin, deoxyhemoglobin, and water (adapted from data in Refs 59–61). The spectrum of water has been multiplied by 10^6 on the vertical axis.

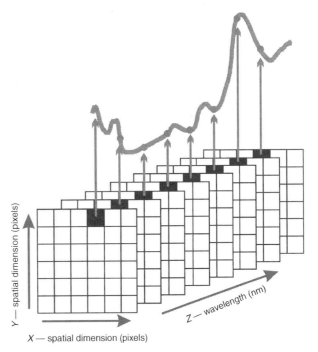

FIGURE 8.3 Schematic diagram to illustrate the three-dimensional nature of the spectroscopic imaging experiment. For *in vivo* studies, images (512×512, binned to 256×256 pixels) are typically acquired at 10 nm intervals over the range 650–1050 nm. The resulting spectroscopic "data cube" may then be processed to provide images reflecting spatial variations in composition.

separate pixels within a single image and changes in the same parameters for individual pixels followed over time.

Physiologically relevant information can therefore be recovered from a simple spectral reconstruction technique (i.e., by determining the fitting coefficients $C_{\text{deoxy}}L$, $C_{\text{oxy}}L$, $C_{\text{water}}L$, and B_{baseline} for each spectrum), as outlined in the following section.

8.2.3 Processing of Biomedical Near-IR Spectroscopic Images

All the standard tools to process images can be exploited in the processing of spectroscopic images and the reader is referred to several excellent textbooks and review articles that outline the current state of the art in image processing [8, 9]. However, it is the ability to exploit the spectral information or the spectral and spatial information concurrently from a scene that conveys the power to spectroscopic imaging compared to techniques that process each image frame separately.

A wide array of techniques have been adopted and adapted for the interpretation of spectroscopic images. These include dimension reduction methods such as principal component [10], independent component [11, 12], and partial least

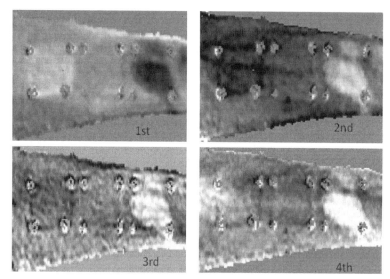

FIGURE 8.4 Principal component decomposition of a spectroscopic reflectance image of a forearm, 30 min after treatment with a moisturizer and acetone. The original spectroscopic image consisted of 75 images taken at 10 nm increments from 960 to 1700 nm. The first four components (score images) capture the variation associated with the effects of the moisturizer (light area in the first principal component image; upper left) and acetone (dark area in the same image) on the skin. To derive these images, the spectroscopic image (which is arranged as a 3D array consisting of *x–y* pixels and wavelength as the three dimensions) was first unwrapped to provide a 2D array containing 65,536 spectra. The 2D array was then subject to a truncated singular value decomposition, determining both the loadings and scores in rank order according to variance. The first vector from the score array was then reshaped to the original image dimension and displayed as the score image; similar steps are done with the second, third, and so on score vectors to generate the remaining score images.

squares analyses [13]. As an example of this general approach, Figure 8.4 illustrates the capability of principal component analysis to reveal useful features in a spectroscopic imaging data set acquired for a forearm. Both unsupervised and supervised classification methods may also be exploited [14], the former generally as any of the various clustering methods and the latter as support vector machines, discriminant partial least squares, and pseudoinverse *k*-nearest neighbor classifiers. These and other data processing techniques are reviewed elsewhere in this book.

Because near-IR spectra measured *in vivo* include contributions from only components whose near-IR spectra are well characterized, by far the most common approach to construct physiologically relevant images is to use a spectral unmixing technique.

8.2.3.1 Spectral Unmixing

Spectral unmixing considers the recorded spectrum at each pixel to be a composite spectrum derived from a number of constituents. Unmixing involves determining the relative contributions of these constituents at each pixel of the image. Techniques such as independent component analysis [15] and multivariate curve resolution [16] may be exploited to this end even if the component spectra are unknown; however, it is generally considered beneficial to make use of other models that do make use of the component spectra if and when they are available.

When the major constituents of the sample and their spectra are known, the component spectra can be used to unmix composite spectra and hence to provide an estimate of the relative abundances of the constituents. A linear mixing model is generally assumed; for *N* known constituents, each composite spectrum is modeled as a linear combination of *N* constituent spectra (see above). To accomplish this, once the data have been collected and converted to pseudoabsorbance spectra, a two-dimensional matrix "**A**" is constructed. Each column represents the spectrum for an individual pixel. The concentrations of deoxy-Hb, oxy-Hb, water (as well as the offset corrections) may then be evaluated as "**C**" using the relationship

$$(\mathbf{CL}) = (\varepsilon'\varepsilon)^{-1}\varepsilon'\mathbf{A}$$

where "ε" is a matrix whose columns comprise the molar absorptivity spectra of the individual components (deoxy-Hb, oxy-Hb, water, and the "offset" spectrum, which is the unit column vector [1]).

For simple multicomponent systems that are completely determined, as is the case for near-IR spectra measured *in vivo*, the above prescription provides reliable estimates of the *relative* constituent concentrations. Since the optical path length "*L*" is generally unknown (but assumed to be constant across the spectral range included in **A**), all concentrations are multiplied by that constant path length. For these reasons, it is often the case that images are constructed as ratios of the CL products (with the unknown "*L*" thus being factored out). The most common approach by far is to construct "oxygenation" or, equivalently, "O_2 saturation" images as the ratio of oxy-hemoglobin to total hemoglobin (and/or myoglobin).

8.3 APPLICATIONS

8.3.1 Skin

Visible and ultraviolet light are strongly absorbed by hemoglobin and skin pigments such as melanin, and as a result light in the UV/visible range penetrates only the uppermost skin layers. In contrast, skin is a relatively weak absorber of near-IR light. Near-IR spectroscopy can therefore be used to interrogate tissues such as the epidermis and dermis that lie beneath the superficial layer. While hemoglobin remains an important chromophore in the near-IR between 700–950 nm, melanin absorbs only very weakly. Variations in skin pigmentation that are often a major confounding factor in visual skin assessment are negligible factors in interpreting near-IR spectroscopic measurements of skin.

Hemoglobin displays distinct absorptions in the near-IR depending on whether the heme is carrying oxygen or is deoxygenated (Figure 8.2). Near-IR reflectance measurements may therefore be processed to reveal noninvasively the relative concentrations and spatial distribution of skin oxy- and deoxy-Hb. Water in tissue is also an important absorber of near-IR light with absorption characteristics distinct from Hb (Figure 8.2). Indeed, most of the strongest features in the near-IR spectrum of tissue between 950 and 2500 nm are the result of absorption bands of water (overtones and combinations of the stretching and bending vibrations of the OH bond). To illustrate, Figure 8.5 graphically demonstrates the correspondence between the absorption features associated with water and their counterparts in the reflectance spectrum of skin. Spectroscopic imaging therefore provides the opportunity to measure the spatial variations in cutaneous water simultaneously with the distribution of hemoglobin concentration and oxygenation.

The visible appearance of skin has long been used to assess the condition and health of the skin. Visual and photographic examinations are relied upon extensively in dermatology, wound management, and surgery and remain the gold standard in many clinical evaluations of skin. However, reliable assessment requires a highly trained evaluator. The question then arises as to the consistency of evaluations made by different observers or by the same observer at different times. The attraction of instrumental methods is that they offer a nonsubjective means to assess skin properties. For example, the parameters provided by near-IR spectroscopic imaging—tissue blood supply, oxygenation, and hydration—are not easily discerned from conventional photographs or by simple visual examination, but they can be crucial to understanding skin health. For example, when confronted with a wound the surgeon is faced

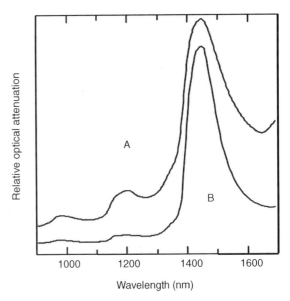

FIGURE 8.5 Near-IR reflectance spectrum of skin (A: top trace) and the transmission spectrum of water (B: lower trace). Overtones and combinations of water OH stretching and bending vibrations dominate the spectra of both pure water and skin.

with the decision of how best to treat the wound. Successful wound management requires an understanding of wound healing, the anatomy in and around the site of the wound, and wound physiology. These are in turn greatly influenced by various factors; age, emotional and nutritional status of the individual, a history of diabetes, smoking, and the mechanical stresses at the site of the wound all affect the healing process and dictate the method of wound treatment.

Objective measurements of the physiological properties of both wound and surrounding tissue provide insight into the nature and the extent of the injury. Near-IR imaging can help fulfill this key role by providing objective indices related both to the health of skin and to its capacity to heal [17]. As a practical matter, the fact that it is closely related to the standard of visual/photographic assessment means that the technique can be integrated readily within current clinical practices associated with dermatology, wound management, and surgical wound closure.

8.3.1.1 Wound Management: Example of Burn Injury Assessment
Burn specialists are often faced with difficult decisions on how best to treat their patients. While many burns will heal spontaneously, others that may appear superficially to be virtually identical in nature will require surgery. There is therefore a need for new prognostic burn characterization techniques to better guide the early decision—whether to perform surgery or wait and allow the wound to heal.

Typically, the burn specialist makes a visual diagnosis based on the surface appearance of the wound. If the wound is very deep or very shallow, that visual diagnosis of the wound shortly after injury is quite accurate. However, burn injuries that are of intermediate depth do not easily fall into pre-defined categories and even experienced clinicians can have difficulty in accurately assessing the capacity of the wound to heal. Adding to the complexity and uncertainty of burn assessment is that the wound evolves and has the capacity to worsen over time. Therefore, patients may be observed for days to weeks before a decision is made on the course of action. Improved wound management stemming from an accurate assessment of wound healing would translate into less morbidity to the patient and shorter periods of hospitalization to effect healing or treatment of the wound.

Near-IR imaging offers a noninvasive means to track physiological changes associated with an injury early in the post-burn period. An adequate supply of oxygenated blood is required for wound healing. Tissue oxygenation and total hemoglobin images derived from near-IR reflectance measurements therefore have a directly interpretable and immediate connection to the healing capacity of the wound [18, 19]. The examples below, from clinical measurements taken on the burn unit at Sunnybrook Hospital (Toronto, Ontario), highlight the particular strengths of the imaging technique.

Figure 8.6 shows both an oxygenation image (left) and a total hemoglobin image (right) of an electric burn on the arm of a patient [19]. The image shows relatively uniform tissue oxygenation with the exception of a poorly oxygenated area that corresponds to the entry point for the electrical current. The total hemoglobin image further shows that no hemoglobin is present at the wound's entry point. This evidence therefore suggests that the area is not being perfused by blood, would therefore likely die without intervention, and that the wound should be treated surgically. Of further interest is a bright ring surrounding the entry point of the wound in the total hemoglobin image, suggesting tissue hyperperfusion. The oxygenation image further indicates that this region of hyperperfusion is well oxygenated, suggesting that the tissue in this area has the capacity to heal from the electrical insult. This illustration suggests that images such as these can delineate the areas of the wound that will need surgical attention.

Because near-IR reflectance imaging is a rapid, nonintrusive, and noncontact method that carries little or no risk to the patient, the technique can be performed routinely to monitor wound behavior over time. To this end, Figure 8.7 documents the changes in tissue oxygenation and total hemoglobin 3, 5, and 8 days after a patient was admitted with partial thickness burn on the shoulder [20]. The oxygenation image displays uniform oxygenation across the shoulder over the 8 days, while the total hemoglobin image indicates that the wound is hyperperfused. This hyperperfusion, most clearly evident on day 5, is waning 8 days after the injury. These images are consistent with a healing wound, and this wound did indeed heal spontaneously.

These two examples illustrate the ability of near-IR imaging to provide the clinician with an objective method to help in the early and reliable assessment of wounds with the potential to improve wound management.

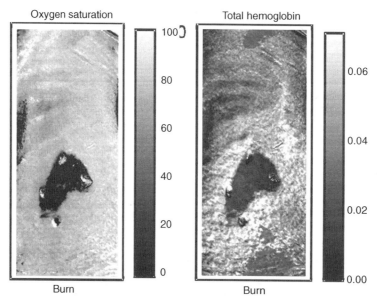

FIGURE 8.6 Near-IR spectroscopic images highlight the spatial variations in tissue oxygenation (left; scale denotes percent oxygenation) and total hemoglobin (right; scale is in mM/cm) that are symptomatic of an electrical burn on the arm. The burned area is distinguished by both poor blood supply and poor tissue oxygenation. The imaged area is approximately $8 \times 4\,cm^2$. Data acquisition time was about 5 min. See Section 8.2.1 for the image acquisition methodology, and Sections 8.2.2 and 8.2.3.1 for the spectral reconstruction technique employed to recover hemoglobin/myoglobin levels. Adapted with permission from Ref. 19.

Oxygen saturation

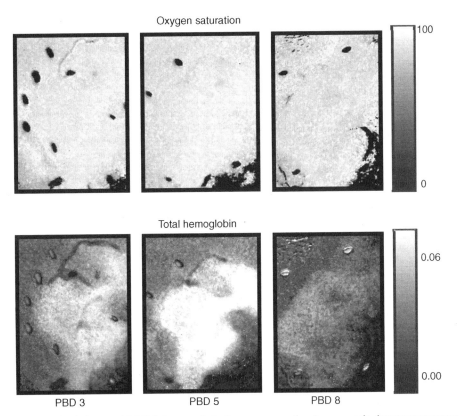

Total hemoglobin

PBD 3   PBD 5   PBD 8

FIGURE 8.7   Near-IR spectroscopic images highlighting trends in tissue oxygenation (upper; scale denotes percent oxygenation) and total hemoglobin (lower; scale is in mM/cm) through the healing process for a burn injury of indeterminate partial thickness on the shoulder of a patient. Images were acquired 3, 5, and 8 days after injury. While the burn injury is as well oxygenated as the surrounding uninjured tissue, the total hemoglobin images show hyperperfusion (elevated hemoglobin levels) at the injury site. Data acquisition time was about 5 min. The imaged area is approximately $2 \times 3 \, cm^2$. See Section 8.2.1 for the image acquisition methodology, and Sections 8.2.2 and 8.2.3.1 for the spectral reconstruction technique employed to recover hemoglobin/myoglobin levels.

8.3.1.2   Surgical Wound Closure: Example of Skin Flap Assessment

Surgical intervention is often the desired course of action for effective treatment of acute and chronic wounds. Intricate skin flap procedures are often relied upon to provide wound closure and coverage for complicated wounds. A wide range of flap procedures is used in reconstructive surgery. Unlike a skin graft that is expected to recruit its own blood supply, flap tissue used in reconstruction retains its vascular attachment and thereby is provided with a supply of blood. The flap must be adequately perfused in order for the procedure to succeed, and the most common complication is compromised arterial supply to or venous drainage from the flap; prolonged periods of little or no blood circulation to the flap will cause the flap tissue to die, requiring the surgery to be repeated.

Salvaging a failing flap requires early detection of perfusion deficits in the flap followed by attempts to reestablish adequate blood circulation to the flap. We have demonstrated that near-IR spectroscopy and imaging can be used at the time of surgery and postoperatively to ensure that the flap is well supplied with oxygenated blood and to check for perfusion (inflow or outflow) abnormalities [21–26].

Figure 8.8 displays a series of near-IR oxygenation images interspersed with photographs of a reverse McFarlane pedicle skin flap on the back of a rat. This is a standard model for the study of skin flap viability, designed to have a predictable failure pattern; since blood is supplied only via a single artery at one end of the flap, the end of the flap distal to the blood supply is fated to die within 72 h. The question addressed by the images shown in Figure 8.8 is whether near-IR images acquired at the time of surgery can reveal the deficit in tissue oxygenation that leads to flap failure. If it is so, then the surgeon can use that immediate feedback to decide whether corrective action is necessary to ensure viability.

The photographs (Figure 8, panels b, d, f, and g) document the fate of the flap over 72 hours. The photograph taken before surgery (panel b) shows healthy skin on the back of the rat. One hour after surgery (panel d) the photograph looks essentially the same. Twelve hours after surgery (panel f) the upper portion of the flap is discoloured (darkened in the grayscale photograph) compared to the lower portion of the flap. After 72 hours the photograph shows that the upper half of the flap has died.

The oxygenation images show a very different time course. Prior to surgery (panel a), the skin on the back of

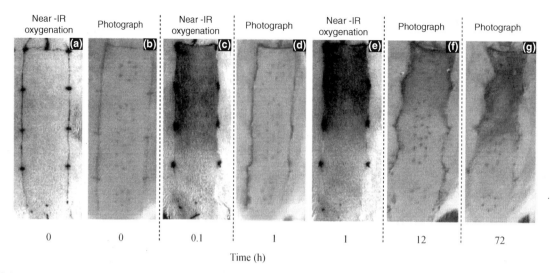

FIGURE 8.8   A series of near-IR oxygenation images and digital photographs of a reverse McFarlane skin flap. Panels a and b compare the oxygenation image and photograph taken before raising the flap. The tissue is well oxygenated and visually appears healthy. Immediately following the flap surgery (panel c), the flap tissue farthest from the blood supply is poorly oxygenated (dark regions denote poor oxygenation). One hour after surgery while the photograph shows an apparently healthy flap, the corresponding oxygenation image clearly shows that the distal end of the flap is in distress. Only after 12 h do visible signs of trouble appear in the 12 h picture (panel f), and by 72 h the upper half of the flap is dead (g). The imaged area is approximately $10 \times 4\,cm^2$. Data acquisition time was about 5 min. See Section 8.2.1 for the image acquisition methodology, and Sections 8.2.2 and 8.2.3.1 for the spectral reconstruction technique employed to recover hemoglobin/myoglobin levels.

the rat is uniformly well oxygenated. However, the oxygenation image acquired immediately following the flap surgery (panel c) reveals that the upper half of the flap, distal to the blood supply, is poorly oxygenated. The oxygenation image taken 1 h after surgery (panel e) is consistent with the oxygenation image acquired immediately after the surgery. By comparison, the 1 h post operative photograph (panel d) shows no signs of the distal portion of the flap being in distress. The oxygenation images clearly show something the eye cannot see until 12 or more hours after the event. Early recognition that the flap is poorly perfused thus dictates a need to restore the circulation then and there, where a repeat operation might otherwise be required.

8.3.1.3   *Dermatological Applications: Example of Skin Hydration*   Near-IR imaging has a number of useful applications in the area of dermatology including mapping tissue vasculature [27, 28], highlighting inflammation [29], and gauging tissue hydration [30–34]. We summarize here an example to demonstrate the latter application.

Skin acts as a barrier both to keep body fluids in and to keep environmental contaminants out. The outermost layer of the epidermis, the stratum corneum, is resistant but not impervious to the passage of water, among many other constituents. As a result, moisture levels in the epidermis, or outer layer of the skin, vary in relation to health, environmental factors, and interventions of many sorts. Epidermal and stratum corneum hydration levels serve as an indication of skin health, and specifically of the effectiveness

of the skin's barrier function. Monitoring hydration can also help in predicting the outcome of treatments for compromised skin.

Water absorption band intensities in the near-IR spectral region have been shown to be good gauges of skin hydration, which may be quantified by simply evaluating the net peak area of the spectral water bands. To illustrate, Figure 8.9 shows band integrations for both the second overtone OH stretching band near 970 nm in the short-wavelength near-IR region and the more intense first overtone OH stretching band near 1450 nm in the long-wavelength near-IR (LW-NIR) region.

To illustrate the capability of the technique, Figure 8.10 shows a hydration image of a forearm for which one area was treated with a moisturizing cream and a second area treated with acetone. For each pixel in the near-IR spectroscopic image, the integrated area of the water band was calculated over the long-wavelength near-IR band (1300–1650 nm) and a linear baseline subtracted. The resulting values were plotted to generate a hydration image that clearly delineates the variations in skin hydration level due to the different treatments.

As illustrated by this example, near-IR spectroscopic imaging offers one of the few methods to determine the spatial distribution of water in skin and tracks its change over time in response to treatment or environmental effects. The technique further provides the means to examine skin barrier function and evaluate perturbations in that function due to chemical and other treatments. As such, the technique offers

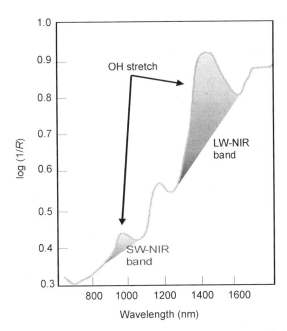

FIGURE 8.9 Skin pseudoabsorbance near-IR spectrum of skin, highlighting the short-wavelength (SW-NIR) and long-wavelength (LW-NIR) water bands used for hydration imaging.

a potentially valuable tool in dermatological applications where the integrity of the barrier function of the stratum corneum is of paramount interest.

8.3.2 Cardiac Imaging

To provide its function of pumping blood through the circulatory system, heart tissue itself requires its own steady and reliable flow of blood. If one of the cardiac arteries supplying this blood is occluded, the risk is local cardiac tissue damage at best, and death at worst. These high stakes motivate a wide

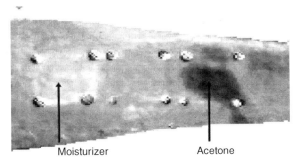

FIGURE 8.10 Near-IR image highlighting spatial variations in skin hydration. Each pixel represents the area of the LW-NIR water absorption (see Figure 8.9), integrated over the wavelength range 1300–1650 nm. Brighter pixels denote higher tissue water content. The dehydrating effects of acetone on the upper layer of skin are evident while the barrier supplied by the moisturizer retains cutaneous water within the treated area.

range of research activities, ranging from fundamental to applied, to characterize cardiac tissue energetics both with normal perfusion and under conditions mimicking those associated with a "heart attack". Cardiac near-IR spectroscopic imaging has therefore been evaluated as a means to delineate regional differences and to track temporal changes in tissue oxygenation and blood supply.

8.3.2.1 *Imaging of Arrested Hearts*   As a first step in evaluating the potential of the technique, near-IR spectroscopic imaging was used to characterize cardiac tissue in arrested porcine hearts [35]. Excised hearts were supplied with a 50:50 mixture of blood and Krebs-Henseleit buffer (KHB, a blood substitute) via a perfusion circuit and were arrested for the duration of the protocol. The heart was suspended in a position suitable to permit imaging of the anterior side, including the anterior wall of the left ventricle.

Once equilibrated, images were acquired for the arrested perfused heart. To produce regional variations in blood (and hence oxygen) delivery, the LAD artery was then occluded for a period of 2 h and images acquired at the 10 min, 1 h, and 2 h time points (this occlusion cuts off the normal blood supply to the left ventricular wall, rendering it ischemic). Blood flow was then restored for 20 min, followed by a period of 10 min of global ischemia—no blood flow to any cardiac vessels—and finally a period of reperfusion.

The three sets of image series in Figure 8.11 illustrate the sensitivity of the near-IR imaging technique in highlighting both regional and temporal variations in deoxy-Hb, oxy-Hb, and blood volume (as estimated by [deoxy-Hb] plus [oxy-Hb]). The images, which reflect the fitting coefficients obtained by spectral reconstruction, clearly highlight the regional increases and decreases in deoxy-Hb and oxy-Hb, respectively, during regional ischemia. Equally striking is the constancy of the blood volume during regional ischemia; although the regional deoxy/oxy balance was clearly and dramatically shifted, regional ischemia did not produce any change in the total blood volume within the affected area—increases in deoxy-Hb were perfectly compensated by decreases in oxy-Hb. On the other hand, global ischemia did precipitate a clear drop in blood volume, as blood was allowed to drain from the heart during the period that supply was halted.

8.3.2.2 *Imaging of Beating Hearts*   Having demonstrated the general feasibility of cardiac imaging on arrested hearts, the next step was to attempt the same experiments on beating hearts [36]. The experiments were essentially the same as those for arrested hearts; porcine hearts were excised and perfused with the same 50:50 blood:KHB mixture, and the imaging setup was identical. The essential distinguishing feature was that the heart was beating. Image acquisition was therefore gated to the cardiac cycle; acquisition was triggered by the peak of the electrocardiogram QRS complex, with a

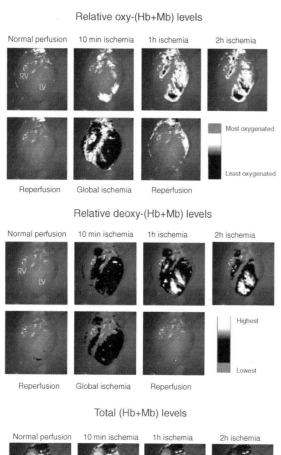

FIGURE 8.11 Near-IR images highlighting regional variations in cardiac oxy-(Hb + Mb), deoxy-(Hb + Mb), and total (Hb + Mb) through a protocol involving both regional and global ischemia in an isolated, arrested heart perfused with 50:50 blood:KHB solution. Note that while regional oxy and deoxy levels were affected by the interruption in local blood flow, the changes were exactly equal and opposite; the total blood supply was unaffected by regional ischemia, as evidenced by the homogeneous distribution in the total (Hb + Mb) image. The imaged area is approximately 12 cm × 12 cm (the heart is approximately 10 cm in height). Data acquisition time was about 5 min. See Section 8.2.1 for the image acquisition methodology, and Sections 8.2.2 and 8.2.3.1 for the spectral reconstruction technique employed to recover hemoglobin/myoglobin levels. Adapted with permission from Ref. 35. (See the color version of this figure in Color Plate section.)

fixed delay time to ensure that all images were acquired at the same point in the cardiac cycle. This point was chosen to be during diastole—the phase during which the heart muscle would normally be relaxed as the heart chambers are filled with blood (these particular experiments made use of isolated isovolumic hearts; within that model, heart motion is moderated somewhat by a balloon that is kept inflated within the left ventricle). This approach to image acquisition permitted averaging of triplicate images (three images of integration time 80 ms were averaged) acquired at each of the 41 wavelengths and ensured that images at each wavelength were registered spatially with those preceding and following.

The success of this approach is illustrated by the image sequence in Figure 8.12, highlighting regional differences and temporal changes in the ratio of deoxy- to oxy-Hb. Again, the regional O_2 deficit triggered by occlusion of the LAD artery is clearly highlighted, as is the global reduction in tissue oxygenation that accompanies cessation of all blood supply. The next step was *in vivo* imaging of porcine hearts exposed in an open chest model [37]—a model mimicking the challenges that would be encountered during cardiac bypass surgery. That investigation highlighted a point that is frequently raised when interpreting the *in vivo* spectra of either skeletal or cardiac muscle, namely, the question of the relative contributions of hemoglobin and myoglobin to the measured spectra. This question was specifically addressed in the next investigation.

8.3.2.3 Relative Contributions of Hemoglobin and Myoglobin

Myoglobin is an oxygen-binding heme protein that is found in the intracellular space of both cardiac and skeletal muscle tissues. Oxygen binding to myoglobin is stronger than it is for hemoglobin, so myoglobin remains completely oxygenated even when blood hemoglobin oxygenation levels are low. Most relevant to spectroscopy and spectroscopic imaging is the fact that the near-IR spectra of myoglobin and hemoglobin (and of the two deoxygenated counterparts) are virtually indistinguishable. As a consequence, "tissue oxygenation" as gauged by near-IR spectroscopy reflects the oxygenation status of both myoglobin and hemoglobin.

In the context of cardiac imaging, it was of interest to determine the extent to which myoglobin contributes to the spectra measured for blood-perfused hearts. To address this question, near-IR spectroscopic images were acquired for eight hearts that were perfused initially with blood-free KHB perfusate [38]. During KHB perfusion, only myoglobin contributed to the heme protein absorption profile. Following spectroscopic image acquisition, the perfusate was switched to the usual 50:50 blood:KHB mixture and a second spectroscopic image was acquired. An ROI (region of interest) was then defined by a patch of tissue that encompassed no superficial blood vessels, and the spectra within that ROI averaged for (1) all hearts during KHB perfusion and (2) all hearts during blood/KHB perfusion. The spectra (Fig-

Deoxy-(Hb+Mb) to oxy-(Hb+Mb) ratios

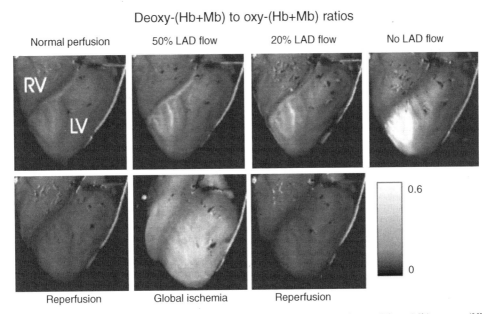

FIGURE 8.12   Near-IR images highlighting regional variations in the ratio of cardiac deoxy-(Hb + Mb) to oxy-(Hb + Mb) through a protocol involving both regional and global ischemia in an isolated, beating heart perfused with 50:50 blood:KHB solution. Image acquisition was gated to the cardiac cycle (see text). As indicated by the scale, lighter areas highlight areas/episodes with relatively poor tissue oxygenation. The imaged area is approximately 8 cm × 8 cm. Data acquisition time was about 5 min. See Section 8.2.1 for the image acquisition methodology, and Sections 8.2.2 and 8.2.3.1 for the spectral reconstruction technique employed to recover hemoglobin/myoglobin levels. Reproduced with permission from Ref. 36.

ure 8.13) were then reconstructed as least squares optimal weighted sums of three component spectra, namely, deoxy (Mb + Hb), oxy(Mb + Hb), and water. Using per-heme absorptivity values (for hemoglobin, the per-heme absorptivity is one-fourth of the molar absorptivity), the path length weighted heme group concentrations were determined to be 0.39 ± 0.05 and 0.62 ± 0.07 mM/cm for KHB- and blood/KHB-perfused hearts, respectively.

The contribution of myoglobin to cardiac spectra and spectroscopic images is therefore substantial. For perfusion with diluted blood, myoglobin contributed over half (63%) of the intensity observed for the combined hemoglobin/myoglobin near-IR features. Extrapolating to perfusion with whole blood, the contribution of myoglobin would still be only marginally less than 50% of the observed near-IR profile. This large contribution of myoglobin to the near-IR spectra (and spectroscopic images) may offer the basis to better characterize the role of myoglobin in cardiac tissue—a role that is not presently well understood.

8.3.2.4 Quantitative Interpretation of Derived Oxygenation Parameters   The preceding section clearly illustrates the utility of near-IR imaging as a qualitative indicator of both spatial and temporal variations in tissue oxygenation and blood supply. However, inspection of the quantitative oxygenation parameters led to a surprising observation; the tissue "oxygenation" never approaches zero, since there is always a significant spectral contribution

from oxy-(Hb + Mb). This is the case both for the local area that is affected when regional blood supply is reduced to zero and for the heart as a whole when the oxygenation of the perfusate is reduced to zero (acute hypoxia).

In light of the above, experiments were designed to better understand the relationship between the oxygenation of blood supplying cardiac tissue and oxygenation parameters derived for that tissue from the near-IR spectra/images [39]. To this end, hearts ($N = 3$) were perfused with a 1:1 blood:KHB mixture, with oxygenation values as measured for samples of the arterial perfusate ranging from 100% to 30%. Spectroscopic images were acquired through this range of blood oxygenation levels, and two ROIs were defined for close scrutiny. One ROI was restricted to the LAD artery (which runs along the surface of the heart), with the rationale that spectra within this region should reflect arterial blood Hb oxygenation with little contribution from Mb. The second ROI encompassed bulk tissue well removed from the major arteries, with the expectation that spectra within this region should broadly reflect both Hb and Mb tissue oxygenation.

Oxygenation values derived spectroscopically for the LAD arterial ROI correlated well with measured arterial blood oxygen saturation values ($R^2 = 0.937$, slope $= 0.72$). For the tissue ROI, which encompasses contributions from myoglobin, arterial blood, and venous blood, the oxygenation parameters derived from the near-IR spectra correlated well ($R^2 = 0.926$) with the mean cardiac arterial/venous saturation. However the slope, 0.40, was much shallower

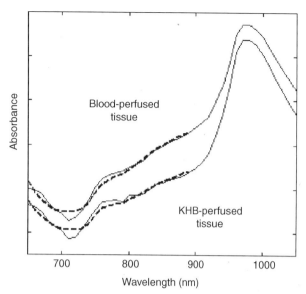

FIGURE 8.13   Near-IR spectra (solid lines) averaged over a region of interest ~0.5 cm × 0.5 cm for an arrested, isolated heart during perfusion with KHB alone (KHB-perfused tissue) and for the same ROI/heart during subsequent perfusion with a 50:50 blood: KHB mixture (blood-perfused tissue). Note the substantial contribution of myoglobin features to the spectrum of KHB-perfused tissue, and the subsequent increase in intensity of the oxy-(Hb + Mb) feature upon perfusion with the blood/KHB mixture. Dotted lines represent the optimal least squares reconstructed spectra, with fitting to optimally reproduce the observed spectra across the region 650–900 nm. The comparison suggested that for hearts perfused with whole blood, myoglobin features account for ~50% of the intensity observed for the oxy- and deoxy-(Hb + Mb) absorptions. See Section 8.2.1 for the image acquisition methodology, and Sections 8.2.2 and 8.2.3.1 for the spectral reconstruction technique employed to recover hemoglobin/myoglobin levels. Reproduced with permission from Ref. 38.

than that for the arterial blood/near-IR saturation relationship (Figure 8.14). This was ascribed to the presence of tissue myoglobin, which raises the "tissue oxygenation" parameter above the value that would be expected in the absence of myoglobin.

8.3.2.5   Application: Cardiac Response to Stress Agents

Having developed the techniques outlined above, more recent investigations have explored modulations of cardiac coronary flow produced *in vivo* by the coronary vasodilator dipyridamole [40] and the β-adrenergic agonist dobutamine [41]. These compounds are of interest since they are both used as stress agents during echocardiographic, nuclear, and MRI stress tests to produce flow maldistribution and myocardial ischemia [42–45].

Although dipyridamole causes coronary vasodilation in normal areas (with no effect on O_2 demand [42]), it does not further dilate vessels that are already dilated within

ischemic tissue. This results in a redistribution of flow favoring normal tissue that is known as the "coronary steal" phenomenon [46–50]. Near-IR imaging provided a striking visualization of this effect within an *in vivo* heart regional ischemia model [40]; dipyridamole further lowered the already reduced oxygenation within a moderately ischemic area due mainly to a considerable decrease in LAD flow, while at the same time the oxygenation of the nonischemic area increased owing, most probably, to increased coronary flow. These oppositely directed changes served to increase the contrast between normal and moderately ischemic areas in the near-IR oxygenation images (Figure 8.15).

The effect of dobutamine is to increase heart rate and blood pressure, as well as dilating coronary blood vessels [42, 43], thereby increasing both oxygen supply and demand. To explore the balance between these effects, a protocol was carried out to image regionally ischemic hearts during dobutamine administration [41]. Near-IR imaging revealed that the oxygen supply accompanying flow augmentation exceeded the excess oxygen demand in the subepicardium and that the effect was most pronounced within the ischemic area; tissue oxygenation was enhanced substantially in the ischemic area and only slightly in normally perfused areas (Figure 8.16). Thus, dobutamine administration increased subepicardial flow and oxygenation both in severely and moderately ischemic areas, hence reducing the contrast to normally perfused regions in the near-IR oxygenation images.

The diametrically opposite effects of dobutamine and dipyridamole on flow and oxygenation distribution in turn promoted decreased and enhanced contrast, respectively, between normal and ischemic areas. These studies clearly show that any intervention that changes the oxygen supply/ demand balance may improve or worsen the delineation of ischemic regions by near-IR imaging.

8.3.2.6   Application: Imaging of Cardiomyopathic Rodent Hearts

While it is well known that diabetic patients have problems with limb microcirculation, it is less widely appreciated that a similar condition also exists in their hearts. Inadequate oxygen supply to the myocardium, combined with the metabolic defect due to diabetes, results in a significant mismatch between energy demand and supply and contributes to the development of diabetic cardiomyopathy.

Laboratory rodents have a well-established genetic background compared to larger animals, thus allowing for bigger and better standardized research projects that may eventually lead to the development of new therapies for human patients. To that end, dynamic, high-resolution optical imaging is a very attractive research tool to monitor oxygenation deficits in rodent hearts. Because mouse hearts are so small, near-IR radiation probes not only epicardial/subepicardial tissue, but also transmural layers. Near-IR imaging therefore provides a

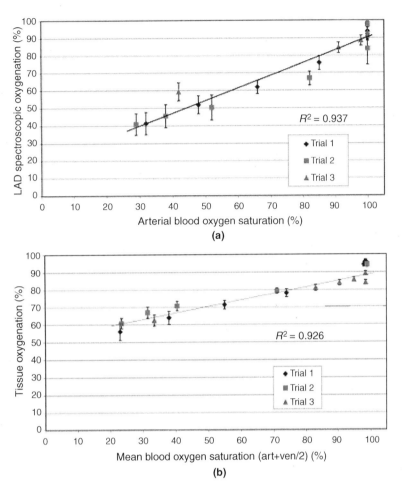

Comparison of calculated oxygenation
with blood oxygen saturation

(a)

(b)

FIGURE 8.14   Relationships between myocardial blood oxygenation (as determined from samples of arterial and venous blood) and blood/ tissue oxygenation as determined by the analysis of near-IR spectroscopic images. "LAD spectroscopic oxygenation" is the ratio of oxy- to total-(Hb + Mb) as determined for a ROI confined within a superficial cardiac artery. "Tissue oxygenation" is the same ratio as determined spectroscopically for a ROI well removed from any superficial artery. Reproduced with permission from Ref. 39.

powerful means to characterize cardiac tissue for rodent models of cardiomyopathy.

Kir6.2$^{-/-}$ mice have a prediabetic type 2 condition and cardiomyopathy [51, 52] that result from a disabling mutation in the potassium conducting subunit, Kir6.2, of the sarcolemmal ATP-sensitive potassium (K_{ATP}) channel [53]. K_{ATP} channels are metabolic sensors located in the cell membrane. In pancreatic β-cells, K_{ATP} channels regulate insulin secretion [54]. Kir6.2$^{-/-}$ mice are lacking glucose-stimulated insulin secretion in the pancreas and develop diabetes mellitus type 2 when they become old and obese [51]. The function of K_{ATP} channels as metabolic sensors is also lost in their hearts and Kir6.2$^{-/-}$ mice develop heart failure under chronic stress conditions and in association with hypertension [52]. Recently, mutations in cardiac K_{ATP}

channels have been identified in human patients with severe diabetic cardiomyopathy of unknown etiology [55].

The response of Kir6.2$^{-/-}$ mouse hearts to stress has been characterized previously by infusion of a mitochondrial uncoupler, 2,4-dinitrophenol (DNP). These hearts suffered a greater decrease in ATP and a higher degree of regional hypoxia in Kir6.2$^{-/-}$ hearts than did control healthy hearts, as demonstrated previously by optical point spectroscopy and NMR methods [56]. By its nature, however, that study could not resolve spatial variations in tissue response.

To test the hypothesis that metabolic stress engenders widespread hypoxic areas in K_{ATP}-deficient hearts, near-IR spectroscopic images were acquired for KHB-perfused control and Kir6.2$^{-/-}$ hearts both at equilibrium and during DNP administration (Figure 8.17). DNP infusion (50 μM, 24 min)

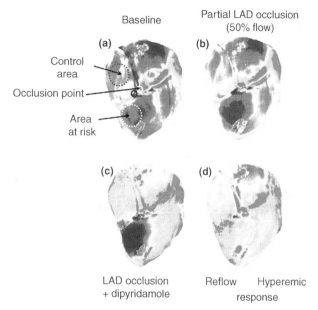

FIGURE 8.15 Effect of dipyridamole infusion on oxygenation in a partial regional ischemia model (dark regions are poorly oxygenated). Partial LAD occlusion produces an ischemic region (b), and dipyridamole infusion with the same occlusion enhances the contrast (c) due to blood flow redistribution from the ischemic to the nonischemic region. The heart is approximately 10 cm in height. Data acquisition time was about 5 min. See Section 8.2.1 for the image acquisition methodology, and Sections 8.2.2 and 8.2.3.1 for the spectral reconstruction technique employed to recover hemoglobin/myoglobin levels. Adapted with permission from Ref. 40. (See the color version of this figure in Color Plate section.)

treatment, in agreement with the data obtained previously by visible range spectroscopy [57].

This work, reported here for the first time [58], shows that near-IR spectroscopic imaging is capable of providing high-resolution oxygenation maps of KHB-perfused mouse hearts. As such, the technique can be used to detect perfusion and metabolism deficits and to characterize microvessel damage in rodent models of cardiac diseases.

8.4 CONCLUSION AND OUTLOOK

Through the work summarized here, near-IR spectroscopic imaging is now well established as a means to probe tissue hydration, blood supply, and oxygenation both for superficial tissue and for internal organs that may be exposed during surgical procedures. The defining features compared to imaging in the visible region are the greater depth of penetration (up to several millimeters) and the insensitivity of the measurement to skin pigmentation, blemishes, scars, and so on—these advantages have been decisive in opening the door to the various applications explored to date.

Translation from the research laboratory to the surgical suite is underway. Collaborative interactions with cardiac and reconstructive surgeons will resolve key questions, addressing both the practical and performance criteria that must be met in order for the methodology to benefit the patient. For example, the regulatory requirements for exploratory measurement in the hospital have been fulfilled by devices recently developed at the NRC Institute for Biodiagnostics, and those devices are being used on hospital wards and in the operating theatre. Growing interest both in the medical community and among clinical instrumentation providers reflects a general recognition of the potential for the technique. It would appear that the only question is which among the many applications will be the first to find acceptance in the form of routine clinical adoption.

uncoupled oxidative phosphorylation in mitochondria and increased oxygen consumption. Therefore, the hearts—perfused under constant flow conditions—became partially hypoxic during DNP infusion. Hypoxia was generally more pronounced in Kir6.2$^{-/-}$ hearts at 10 and 20 min of DNP

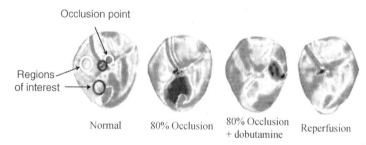

FIGURE 8.16 Effect of dobutamine infusion on oxygenation in a partial regional ischemia model. While 80% occlusion of the LAD produced a clearly ischemic region (dark regions are poorly oxygenated), dobutamine administration increased subepicardial blood flow and oxygenation within that region, hence reducing the contrast to normally perfused regions in the near-IR oxygenation images. The heart is approximately 10 cm in height. Data acquisition time was about 5 min. See Section 8.2.1 for the image acquisition methodology, and Sections 8.2.2 and 8.2.3.1 for the spectral reconstruction technique employed to recover hemoglobin/myoglobin levels. Adapted with permission from Ref. 41. (See the color version of this figure in Color Plate section.)

A control

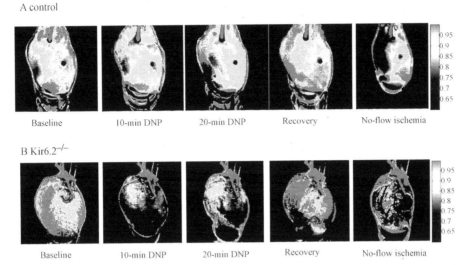

Baseline   10-min DNP   20-min DNP   Recovery   No-flow ischemia

B Kir6.2$^{-/-}$

Baseline   10-min DNP   20-min DNP   Recovery   No-flow ischemia

FIGURE 8.17   Near-IR images highlighting variations in tissue myoglobin oxygenation for KHB-perfused control and Kir6.2$^{-/-}$ mouse hearts following treatment with 2,4-dinitrophenol (dark regions are poorly oxygenated). The Kir6.2$^{-/-}$ mouse is a model that exhibits a prediabetic type 2 condition and cardiomyopathy; due to a mismatch between oxygen delivery and consumption, the Kir6.2$^{-/-}$ heart became more hypoxic than did the control during DNP treatment. The imaged area is approximately 1.5 cm × 1.5 cm (the heart is a little less than 1 cm in height). Data acquisition time was about 5 min. See Section 8.2.1 for the image acquisition methodology, and Sections 8.2.2 and 8.2.3.1 for the spectral reconstruction technique employed to recover hemoglobin/myoglobin levels. (See the color version of this figure in Color Plate section.)

REFERENCES

1. Jöbsis, F. F. (1977).Noninvasive infrared monitoring of cerebral and myocardial oxygen sufficiency and circulatory parameters.*Science* **198**,1264–1267.

2. Delpy, D. T., Ferrari, M., Piantadosi, C. A., and Tamura, M. (2007). Pioneers in biomedical optics: special section honoring Professor Frans F. Jöbsis of Duke University.*J. Biomed. Opt.***12**,062101.

3. Piantadosi, C. A. (2007). Early development of near-infrared spectroscopy at Duke University.*J. Biomed. Opt.***12**,062102.

4. Tachtsidis, I., Tisdall, M., Leung, T. S., Cooper, C. E., Delpy, D. T., Smith, M., and Elwell, C. E. (2007). Investigation of *in? vivo* measurement of cerebral cytochrome-*c*-oxidase redox changes using near-infrared spectroscopy in patients with orthostatic hypotension.*Physiol. Meas.* **28**,199–211.

5. Tisdall, M. M., Tachtsidis, I., Leung, T. S., Elwell, C. E., and Smith, M.(2008). Changes in the attenuation of near infrared spectra by the healthy adult brain during hypoxaemia cannot be accounted for solely by changes in the concentrations of oxy- and deoxy-haemoglobin.*Adv. Exp. Med. Biol.* **614**,217–225.

6. Franceschini, M. A., Joseph, D. K., Huppert, T. J., Diamond, S. G. and Boas, D. A.(2006). Diffuse optical imaging of the whole head.*J. Biomed. Opt.***11**,054007.

7. Nakahachi, T., Ishii, R., Iwase, M., Canuet, L., Takahashi, H., Kurimoto, R., Ikezawa, K., Azechi, M., Sekiyama, R., Honaga, E., Uchiumi, C., Iwakiri, M., Motomura, N., and Takeda, M. (2008). Frontal activity during the digit symbol substitution test determined by multichannel near-infrared spectroscopy.*Neuropsychobiology* **57**,151–158.

8. Gonzalez, R. C. and Woods, R. E. (2007). *Digital Image Processing*, 3rd ed., Prentice Hall.

9. Jähne, B.(2005). *Digital Image Processing*, 6th ed., Springer.

10. Gemperline, P.(2006). *Practical Guide to Chemometrics*, 2nd ed., CRC Press.

11. Roberts, S. and Everson, R.(2001). *Independent Component Analysis: Principles and Practice*, Cambridge University Press.

12. Hyvarinen, A. and Oja, E.(2000). Independent component analysis: algorithms and applications.*Neural Networks* **13**, 411–430.

13. Wold, S., Trygg, J., Berglund, A., and Antti, H.(2001). Some recent developments in PLS modeling.*Chemom. Intell. Lab. Syst.* **58**,131–150.

14. Hastie, T., Tibshirani, R., and Friedman, J. H.(2001). *The Elements of Statistical Learning: Data Mining, Inference, and Prediction*, Springer.

15. Chang, C. -I. (2007). *Hyperspectral Data Exploitation*, Wiley.

16. Rutan, S. C., de Juan, A., and Tauler, R.(2009). Introduction to multivariate curve resolution. In: Brown, S. D., Tauler, R., and Walczak, B. (Eds.), *Comprehensive Chemometrics*, Elsevier, Oxford, pp. 249–259.

17. Khaodhiar, L., Dinh, T., Schomacker, K. T., Panasyuk, S. V., Freeman, J. E., Lew, R., Vo, T., Panasyuk, A. A., Lima, C., Giurini, J. M., Lyons, T. E., and Veves, A. (2007). The use of medical hyperspectral technology to evaluate microcirculatory changes in diabetic foot ulcers and to predict clinical outcomes. *Diabetes Care* **30**,903–910.

18. Sowa, M. G., Leonardi, L., Payette, J. R., Fish, J. S., and Mantsch, H. H.(2001). Near infrared spectroscopic assessment

of hemodynamic changes in the early post-burn period.*Burns* **27**,241–249.

19. Cross, K. M., Leonardi, L., Payette, J. R., Gomez, M., Levasseur,?M. A., Schattka, B. J., Sowa, M. G., and Fish, J. S. (2007). Clinical utilization of near-infrared spectroscopy ?devices for burn depth assessment.*Wound Repair Regen.* **15**,332–340.

20. Cross, K. M., Leonardi, L., Payette, J. R., Gomez, M., Levasseur, M. A., Schattka, B. J., Sowa, M. G., and Fish, J. S.(2009). Unpublished data.

21. Payette, J. R., Kohlenberg, E., Leonardi, L., Pabbies, A., Kerr, P., Liu, K. -Z., and Sowa, M. G. (2005). Assessment of skin flaps using optically based methods for measuring blood flow and oxygenation.*Plast. Reconstr. Surg.***115**, 539–546.

22. Attas, M., Hewko, M. D., Payette, J. R., Posthumus, T., Sowa, M. G., and Mantsch, H. H. (2001). Visualization of cutaneous hemoglobin oxygenation and skin hydration using near infrared spectroscopic imaging.*Skin Res. Technol.* **7**, 238–245.

23. Abdulrauf, B. M., Stranc, M. F., Sowa, M. G., Germscheid, S. L., and Mantsch, H. H.(2000). Novel approach in the evaluation of flap failure using near infrared spectroscopy and imaging.*Can. J. Plast. Surg.* **8**, 68–72.

24. Sowa, M. G., Payette, J. R., Hewko, M. D., and Mantsch, H. H. (1999). Visible-near infrared multispectral imaging of the rat dorsal skin flap.*J. Biomed. Opt.* **4**, 474–481.

25. Mansfield, J. R., Sowa, M. G., Payette, J. R., Abdulrauf, B., Stranc, M. F., and Mantsch, H. H.(1998). Tissue viability by multispectral near infrared imaging: a fuzzy C-means clustering analysis.*IEEE Trans. Med. Imaging* **17**, 1011–1018.

26. Stranc, M. F., Sowa, M. G., Abdulrauf, B., and Mantsch, H. H. (1998). Assessment of tissue viability using near-infrared spectroscopy.*Br. J. Plast. Surg.* **51**, 210–217.

27. Mansfield, J. R., Sowa, M. G., Scarth, G. B., Somorjai, R. L., and Mantsch, H. H.(1997). Fuzzy C-means clustering and principal component analysis of time series from near-infrared imaging of forearm ischemia.*Comput. Med. Imaging Graph.* **21**, 299–308.

28. Vogel, A., Chernomordik, V. V., Riley, J. D., Hassan, M., Amyot, F., Dasgeb, B., Demos, S. G., Pursley, R., Little, R. F., Yarchoan, R., Tao, Y., and Gandjbakhche, A. H. (2007). Using noninvasive multispectral imaging to quantitatively assess tissue vasculature.*J. Biomed. Opt.* **12**, 051604.

29. Stamatas, G. N. and Kollias, N. (2007). *In vivo* documentation of cutaneous inflammation using spectral imaging.*J. Biomed. Opt.* **12**, 051603.

30. Attas, E. M., Sowa, M. G., Posthumus, T. B., Schattka, B. J., Mantsch, H. H., and Zhang, S. L.(2002). Near-IR spectroscopic imaging for skin hydration: the long and the short of it.*Biopolymers* **67**, 96–106.

31. Attas, M., Posthumus, T., Schattka, B. J., Sowa, M. G., Mantsch, H. H., and Zhang, S. L.(2002). Long-wavelength near-infrared spectroscopic imaging for *in-vivo* skin hydration measurements.*Vib. Spectrosc.* **28**, 37–43.

32. Sowa, M. G., Payette, J. R., and Mantsch, H. H.(1999). Near-infrared spectroscopic assessment of tissue hydration following surgery.*J. Surg. Res.* **86**, 626–629.

33. Zhang, S. L., Meyers, C. L., Subramanyan, K., and Hancewicz, T. M.(2005). Near infrared imaging for measuring and visualizing skin hydration. A comparison with visual assessment and electrical methods.*J. Biomed. Opt.* **10**, 031107.

34. Stamatas, G. N., Southall, M., and Kollias, N.(2006). *In Vivo* monitoring of cutaneous edema using spectral imaging in the visible and near infrared.*J. Invest. Dermatol.* **126**, 1753–1760.

35. Nighswander-Rempel, S. P., Shaw, R. A., Mansfield, J. R., Hewko, M., Kupriyanov, V. V., and Mantsch, H. H.(2002). Regional variations in myocardial tissue oxygenation mapped by near-infrared spectroscopic imaging.*J. Mol. Cell. Cardiol.* **34**, 1195–1203.

36. Nighswander-Rempel, S. P., Shaw, R. A., Kupriyanov, V. V., Rendell, J., Xiang, B., and Mantsch, H. H.(2003). Mapping tissue oxygenation in the beating heart with near-infrared spectroscopic imaging.*Vib. Spectrosc.* **32**, 85–94.

37. Kupriyanov, V. V., Nighswander-Rempel, S., and Xiang, B. (2004). Mapping regional oxygenation and flow in pig hearts *in vivo* using near-infrared spectroscopic imaging.*J. Mol. Cell. Cardiol.* **37**, 947–957.

38. Nighswander-Rempel, S. P., Kupriyanov, V. V., and Shaw, R. A. (2005). Relative contributions of hemoglobin and myoglobin to near-infrared spectroscopic images of cardiac tissue.*Appl. Spectrosc.* **59**, 190–193.

39. Nighswander-Rempel, S. P., Shaw, R. A., and Kupriyanov, V. V. (2006). Cardiac tissue oxygenation as a function of blood flow? and pO2: a near-infrared spectroscopic imaging study.*J. Biomed. Opt.* **11**, 054004.

40. Kupriyanov, V. V., Manley, D. M., and Xiang, B.(2008). Detection of moderate regional ischemia in pig hearts *in vivo* by near-infrared and thermal imaging: effects of dipyridamole. *Int. J. Cardiovasc. Imaging* **24**, 113–123.

41. Manley, D. M., Xiang, B., and Kupriyanov, V. V. (2007). Visualization and grading of regional ischemia in pigs *in vivo* using near-infrared and thermal imaging.*Can. J. Physiol. Pharmacol.* **85**, 382–395.

42. Iskandrian, A. S., Verani, M., and Heo, J.(1994). Pharmacologic stress testing: mechanism of action, hemodynamic responses, and results in detection of coronary artery disease.*J. Nucl. Cardiol.* **1**, 94–111.

43. Leppo, J.(1996). Comparison of pharmacologic stress agents.*J. Nucl. Cardiol.* **3**, S22–S26.

44. Gould, L.(1978). Noninvasive assessment of coronary stenosis by myocardial perfusion imaging during pharmacologic coronary vasodilatation. I. Physiologic basis and experimental validation.*Am. J. Cardiol.* **41**, 267–278.

45. Robles, H. B., Lawson, M. A., and Johnson, L. L.(1994). Role of imaging in assessment of ischemic heart disease.*Curr. Opin. Cardiol.* **9**, 435–447.

46. Schaper, W., Lewi, P., Flameng, W., and Gijpen, L.(1973). Myocardial steal produced by coronary vasodilation in chronic coronary artery occlusion.*Basic Res. Cardiol.* **68**, 3–20.

47. Marshall, R. J. and Parrat, J. R.(1973). The effects of dipyridamole on blood flow and oxygen handling in the acutely ischaemic and normal myocardium.*Br. J. Pharmacol.* **49**, 391–399.

48. Cohen, M. V., Sonnenblick, E. H., and Kirk, E. S.(1976). Coronary steal: its role in detrimental effect of isoproterenol

after acute?coronary occlusion in dogs.*Am. J. Cardiol.* **38**, 880–888.

49. Becker, L. C.(1978). Conditions for vasodilator-induced coronary steal in experimental myocardial ischemia.*Circulation* **57**, 1103–1110.

50. Beller, G., Holzgrefe, H. H., and Watson, D. D.(1985). Intrinsic washout rates of thallium-201 in normal and ischemic myocardium after dipyridamole-induced vasodilation.*Circulation* **71**, 378–386.

51. Seino, S., Iwanaga, T., Nagashima, K., and Miki, T.(2000). Diverse roles of K_{ATP} channels learned from Kir6.2 genetically engineered mice.*Diabetes* **49**, 311–318.

52. Kane, G. C., Behfar, A., Dyer, R. B., O'Cochlain, D. F., Liu, X. K., Hodgson, D. M., Reyes, S., Miki, T., Seino, S., and Terzic, A.(2006). KCNJ11 gene knockout of the Kir6.2 K_{ATP} channel causes maladaptive remodeling and heart failure in hypertension.*Hum. Mol. Genet.* **15**, 2285–2297.

53. Miki, T., Nagashima, K., Tashiro, F., Kotake, K., Yoshitomi, H., Tamamoto, A., Gonoi, T., Iwanaga, T., Miyazaki, J., and Seino, S.(1998). Defective insulin secretion and enhanced insulin action in K_{ATP} channel-deficient mice.*Proc. Natl. Acad. Sci. USA* **95**, 10402–10406.

54. Aguilar-Bryan, L. and Bryan, J.(1999). Molecular biology of adenosine triphosphate-sensitive potassium channels.*Endocr. Rev.* **20**, 101–135.

55. Bienengraeber, M., Olson, T. M., Selivanov, V. A., Kathmann, E. C., O'Cochlain, F., Gao, F., Karger, A. B., Ballew, J. D., Hodgson, D. M., Zingman, L. V., Pang, Y. P., Alekseev, A. E., and Terzic, A.(2004). ABCC9 mutations identified in human dilated cardiomyopathy disrupt catalytic KATP channel gating. *Nat. Genet.* **36**, 382–387.

56. Jilkina, O., Kuzio, B., Rendell, J., Xiang, B., and Kupriyanov, V. V.(2006). K^+ transport and energetics in Kir6.2$^{-/-}$ mouse hearts assessed by ^{87}Rb and ^{31}P magnetic resonance and optical spectroscopy.*J. Mol. Cell. Cardiol.* **41**, 893–901.

57. Jilkina, O., Glogowski, M., Kuzio, M. B., Zhilkin, P. A., and Kupriyanov, V. V. (2007). Optical imaging of oxygen saturation in beating K_{ATP}-deficient mouse hearts.*J. Mol. Cell. Cardiol.* **42**, S61.

58. Jilkina, O., Glogowski, M., Kuzio, B., Zhilkin, P. A., Gussakovsky, E., and Kupriyanov, V. V. (2010). Defects in myoglobin oxygenation in KATP-deficient mouse hearts under normal and stress conditions characterized by near infrared spectroscopy and imaging. *Int. J. Cardiol.* [Epub ahead of print] doi:10.1016/j.ijcard.2010.02.009.

59. Prahl, S. Optical absorption of water compendium. Retrieved January 16, 2009 from http://omlc.ogi.edu/spectra/water/abs/index.html.

60. Prahl, S. Tabulated molar extinction coefficient for hemoglobin in water. Retrieved January 16, 2009, from http://omlc.ogi.edu/spectra/hemoglobin/summary.html.

61. Matcher, S. J., Elwell, C. E., Cooper, C. E., Cope, M., and Delpy, D. T.(1995). Performance comparison of several published tissue near-infrared spectroscopy algorithms.*Anal. Biochem.* **227**, 54–68.

PART III

PHARMACEUTICAL APPLICATIONS

9

PHARMACEUTICAL APPLICATIONS OF RAMAN CHEMICAL IMAGING

Slobodan Šašić and Lin Zhang

Analytical Development, Pfizer, Groton, CT, USA

Raman chemical mapping in the field of pharmaceuticals mostly pertains to visualization of the components in tablets. This is understandable because imaging the tablets is relatively easy, and, technically, commercial instrumentation is in most cases quite satisfactory for obtaining reasonably good data. Hence, in the industrial applications (that according to the accessible literature lead the progress in this field), there is no particular need for in-house built instruments with special requirements. In addition, there are a number of manufacturers with state-of-the-art instruments that reflect continuous progress in the available hardware. The application of Raman mapping is clearly growing in the pharmaceutical industry, in particular in the departments of development and manufacturing. Looking-forward perspectives are relatively good because of the mounting evidence that chemical imaging techniques greatly aid in the in-depth understanding of the correlation between final products and manufacturing processes.

The situation is very different when the biochemical materials are concerned. A number of excellent applications of Raman chemical imaging on such samples are already described in previous chapters. To avoid repeats, only applications with active pharmaceutical ingredients (API) are addressed in this chapter. This definition is certainly arguable as it is difficult to draw a line between the biomedical and pharmaceutical fields (indeed, is there a discriminating criterion at all?), but it does facilitate the representation of the chemical imaging works in this book. With this "disclaimer" in mind, there is not so much to report. While mapping tablets is in actual fact effortless, mapping any biological material (e.g., cell) is certainly quite an opposite. Sample preparation,

acquisition details, stability of the sample, quality of the spectra, concentration of the API, and data analysis are all far more complex, by no means straightforward, and compete with the well-established and frequently used techniques (such as fluorescence).

This chapter thus mostly lists applications of Raman mapping to compact materials of industrial interest. Interestingly, this seems to be one of rare fields in which the industry seems to be on the forefront of applications and is behind the drive for improvements in hardware. Additional information may be found in an excellent review paper by Gowen et al. [1] or in another review on similar subject by Šašić [2].

Most of references below list Raman mapping of various tablets. Naturally, in most of them the API is the main target of analysis, but the nature of the mapping data is such that information about all the excipients can be obtained (their spectral traces are retrievable from the convoluted spectra provided their Raman responses are above the limit of detection). Small concentrations of APIs are often detectable. This may seem somewhat surprising because of the inherent inefficiency of the Raman effect (see Chapter 1). However, several factors contribute to the relatively easy detection of the APIs: (1) in most cases, API molecules are (very) strong Raman scatterers, so compared to the excipients their Raman spectra are certainly more intense; (2) tablets are very dense materials, so a large number of scatterers are irradiated in sharp contrast to the liquids where correspondingly low concentrations are much more of a problem; and (3) the laser light is focused by a microscope objective that significantly aids in the efficiency of detection of backscattered Raman light.

The most direct purpose of Raman chemical mapping is to identify and characterize API domains. It is tempting to say "particle sizes" instead of "domains" but in comparison to the methods that detect true particle sizes, chemical mapping described in this book is far coarser and its outcomes probably cannot be considered directly comparable to particle sizing results. Dissipation of irradiation light is significant and the sampling volume across the sample cannot be exactly determined, so the resulting pixels are somewhat tentative and cannot be considered equivalent to true particles. Nevertheless, these images are tremendously useful for monitoring variations in the appearance of the components (API and excipients likewise) in tablets or blends with variations in process parameters, as well as for in-depth understanding of the formulations. Vibrational spectroscopy-based chemical imaging seems to be the key technique for these purposes.

The wealth of information described above is not trivially extricable. The spectral mixing is considerable and sophisticated data analysis tools are frequently, but not necessarily, employed to reach conclusions ([3, 4] and also see chapter on software with references therein). Chemometrics is often indispensable for retrieving signals of minor components and comprehensive evaluation of the data. A number of chemometric approaches and examples from practice are described later in this chapter.

A few typical studies referring to the tablets are now dealt in detail, while some more specific applications regarding the means used or the samples are described in separate paragraphs.

Low-concentration API mapping was described in studies by Šašić [5] and Henson and Zhang [6]. Šašić analyzed alpazolam and xanax tablets from the lots known to experience some issues with particle sizes. The Raman signal of the API in these formulations (respective loadings 0.8% and 0.4% w/w) could not be reliably extracted by a simple univariate analysis. The author points out to misinterpretations that may occur if univariate maps produced at the nonoverlapped API peak are not verified through the raw spectra and reports that multivariate analysis appears to be far more reliable due to the fact that one of the loadings in principal component analysis (PCA) largely overlaps with the spectrum of pure alprazolam (the name of API) (Figure 9.1). Unexpected match between one of PCA loadings and Raman spectrum of API seems to be of tremendous use for visualizing low-loading API in various tablets. As a result, API was imaged in a number of tablets and it was found to be of comparable sizes and homogeneity. In summary, the suspected agglomeration of APIs is not confirmed—if it occurs, it is occasional and a larger study with more tablets is needed for confirmation.

Henson and Zhang [6] used Raman mapping to address the problem of form conversion in the formulation with 0.5% loading. Three forms of the inspected API were known to

exist and the first step of the study was to establish a library with the Raman spectra of those forms and most abundant excipients. In a series of separate experiments, it was confirmed that each of these three forms is retrievable. This was followed by the analysis of the tablet with all the forms present in the ratio of 1:1:8, which is meant to mimic undesired conversion of the API. Using the techniques of partial least squares (PLS) and Euclidean distance (ED) (more details on this work is given below), it was not only confirmed that all these forms are identifiable in the mixed tablets, but also that the ratio of the assigned pixels is close to the relative concentrations of the forms.

Confocal mapping experiments of Ward et al. [7] and Breitenbach et al. [8] also describe very interesting contributions of Raman mapping for better understanding of the solid dosage formulations. Ward et al. [7] described the use of Raman mapping for characterizing amorphous domains of sorbitol (used as a model compound) on the surface of the crystalline sorbitol. Using the 100× objective, these authors were able to collect the spectra from various depths inside the sample. The images obtained from those spectra indicated the volume of the inspected amorphous particle of sorbitol. Breitenbach et al. [8] employed Raman confocal mapping to characterize the physical state of the active, physicochemical stability of the formulation (solid solution of ibuprofen in polyvinylpyrrolidone, PVP) and the homogeneity of the API distribution. The physical state of the API in the polymer matrix was also compared with the solutions of the API in a solvent.

9.1   CORRELATION WITH NIR MAPPING

An interesting attempt to correlate near-infrared (NIR) and Raman mapping was made by Clarke et al. [9]. They proposed "fusion" of the two types of chemical images in cases when the components in a formulation are not all identifiable by one of the imaging methods. By using a microscope slide with reference marks, they were able to collect chemical images of exactly the same area on a tablet with these two mapping platforms. The obtained chemical maps were then overlaid. The analyzed formulations (blend of five components in total) contained two components with strong Raman response and two with strong NIR response. Through analyzing separate chemical images, it was found that four out of five components were detectable by each of the used techniques, with inorganic binder missing from the NIR maps and the disintegrant material missing in the Raman maps. The best combination of the two chemical images was found to contain Raman images of the active and inorganic binder and NIR images of diluent, disintegrant, and lubricant. The combinations of these images provided a comprehensive visual description of the formulation and hinted at the cause of the problem experienced during manufacturing (powder

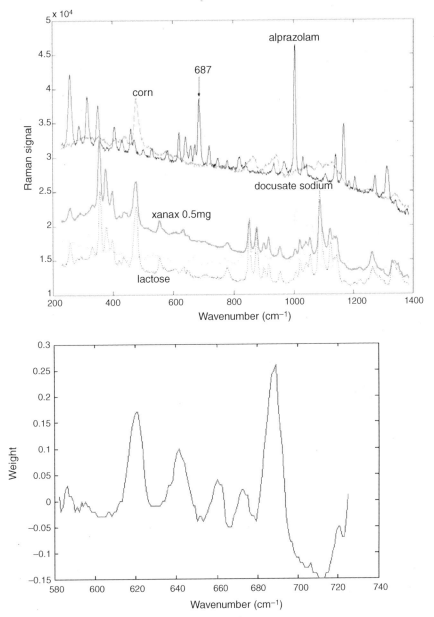

FIGURE 9.1 *Top*: The pure component dispersive Raman spectra of the xanax 0.5 mg formulation together with the spectrum from the tablet. The active component (alprazolam) can clearly be identified through a series of unobstructed peaks centered around the strong band at 687 cm^{-1}. *Bottom*: The PCA of the mapping spectra reveals that one of the loadings strongly resembles the spectrum of alprazolam, thus enabling reliable imaging of the API via the corresponding PC score. Without this (fortunate) overlap, it would have been impossible to image the API due to its weak signal/low concentration [5].

sticking to the tablet tooling), which was the underlying rationale for this study to be conducted.

Somewhat different view is expressed in the in-depth study of Raman and NIR chemical imaging of common pharmaceutical tablets [10]. The key argument in this study is that Raman mapping is superior to NIR mapping with regard to imaging low-concentration components. In this chapter, the excipients explotab and Mg stearate were considered as the low-concentration components and while Raman mapping was capable of retrieving information about

those at least to some extent, NIR mapping failed to provide any information. The explanation was found in the better sensitivity of Raman instruments and much better suitability of Raman spectra for multivariate analysis. NIR bands are broad and indistinguishable and as such very demanding for unraveling. Second derivation, normally applied to improve comprehension of the spectral features in NIR spectroscopy, is of little use here as it produces a very large number of peaks (both positive and negative) that effectively render PCA inapplicable for these data. On the other hand, the sharpness and the easy recognition of the Raman bands (in particular those of Mg stearate in the analyzed formulation) lead to relatively straightforward application of PCA and identification of the bands of those two components (Figure 9.2). PCA was found to be the only way to produce chemical images of the minor components, but significant caution was

advised as this algebraic approach may lead to ambiguous results. PCA did not identify the pure component spectra but rather the linear combinations of the spectra of the present species, and consequently the obtained chemical images may also be seen as linear combinations of these species. For example, the features of explotab may be combined with those of avicel or API, so its final chemical image may be contaminated with distributions of these two components and additional verification was advised to confirm the assignments. Interestingly, the spectral features of Mg stearate were so clear in PCA that its chemical image was quite reliable despite it being only 1% w/w of the formulation.

On the other hand, NIR mapping was much faster. In general, if the component of interest is identifiable by an NIR instrument at the acceptable spatial resolution, it is certainly more recommendable to proceed with this technique as those experiments complete in less than one-fourth of the time needed for Raman mapping.

9.2   BEADS

Beads are very structured and essentially simple formulations, so chemical mapping of them is very different from and less demanding than mapping the tablets. Mapping of beads can provide two types of information: about the thickness and chemical structure of the layers, with the latter being apparently more natural target for chemical imaging. There are two studies [11, 12] on Raman mapping of the beads that discuss advantages of using multivariate data analysis of the mapping spectra in comparison to univariate approaches. If high-quality spectra are obtained from a bead, then univarate analysis is reasonably successful in identifying the chemical structure and spatial characteristics of the layers. However, collecting such spectra is time-consuming and somewhat contrasts the inherent simplicity of the structure of a bead. That is to say, one of the major problems in Raman chemical mapping of the tablets is the spectral interference caused by intimate mixture of the components that requires relatively high-quality spectra to be obtained that are then unraveled by multivariate analysis. Because the components of a bead are spatially separated, the quality of the spectra is only important for the quality of the chemical images that are produced from them, as there is no need for deconvolution. Thus, multivariate approaches can be used solely for denoising. Reference 12 compares Raman images of a bead that are obtained by univariate and multivariate approaches (Figure 9.3). It concludes that significant reduction of acquisition time is achievable if the spectra are denoised by PCA. Raw Raman mapping spectra of low s/n ratio are successfully denoised and images produced from them compare highly to the images obtained from the Raman spectra with high s/n ratio. It is estimated that the acquisition can be lowered to only 10% of that that provides high-quality

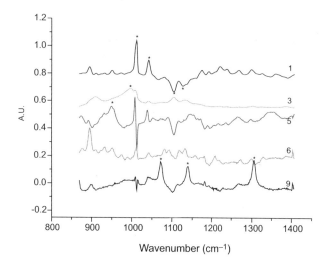

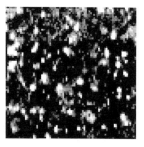

FIGURE 9.2   *Top*: Loadings from the PC analysis of a tablet with 20% API [10]. The stars mark the peaks that can be correlated with the pure component spectra. Of particular interest here are loadings #5 and #9 that can be correlated with explotab and Mg stearate, respectively. This in particular holds for the loading #9 that features three isolated peaks unambiguously assignable to Mg stearate. This allows reliable imaging of Mg stearate (middle, white pixels refer to Mg stearate) despite its low concentration in the formulation and relatively weak Raman scattering ability. Also, an image of explotab (bottom) can be produced although it needs to be additionally verified because of the presence of some peaks in loading #5 that cannot be assigned to explotab only.

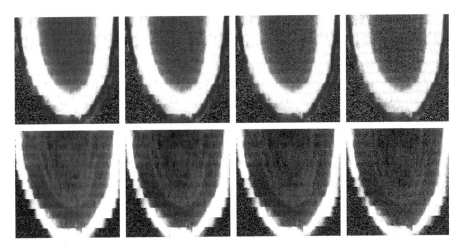

FIGURE 9.3 Raman chemical images of the middle (top) and outer layer (bottom) of the three-layered bead at the respective univariate wavenumbers. The spectra from which the maps are obtained were acquired for 30, 10, 5, and 3 s, left to right, and then denoised by principal component analysis. The size of the maps is approximately 1×0.5 mm^2. The aspect ratio is not maintained in this representation. Binarization of these images reveal that the difference in the number and position of pixels referring to the two shown layers is minimal, which means that 3 s acquisition is practically as effective as that of 30 s.

spectra with minor loss in the information content of the images. This philosophy does not apply to the tablets because of the complexity and heavy overlap of the spectra from them, but it applies to other materials conceptually comparable with beads (i.e., constituents spatially separated).

9.3 GLOBAL ILLUMINATION CHEMICAL IMAGING

Global illumination (GI) Raman chemical imaging is a newer and significantly different approach compared to the more commonly used point or line mapping approaches [13]. As such, it certainly counts fewer applications, and it is yet to establish its niche in the pharmaceutical industry. However, the first few applications that can be seen in the literature are encouraging and reveal that such instruments certainly have significant prospects for widespread use.

Three studies are featured here that address general characteristics of the GI approach in relation to the mapping one. The last one directly refers to pharmaceutical samples while the first two deal with more generic targets.

Markwort et al. [14] first carried out an extensive analysis of the images of polymer samples in which they concluded that the GI approach is indeed faster and of superior visual quality/spatial resolution compared to the point or line mapping. However, they underlined its major drawback originating in photon migration that diminished prospects of unambiguous spectral and spatial identification of the components.

Schlücker et al. [15] studied performances of these two platforms on a highly structured silicon sample that is, as far as the compositions of the sample and the spectra are concerned, significantly simpler compared to all other samples described in this chapter. Their evaluation of the spatial

resolutions achievable on the two platforms was clearly in favor of GI platform. While the separation of about 1 μm was achieved on the grating equipped platforms, submicron/diffraction limited images were obtained with the GI instrument.

Finally, Šašić and Clark [16] discussed the strategy for imaging of pharmaceutical samples on the two platforms. They analyzed the same areas on a common pharmaceutical tablet consisting of five components with a high API loading (25%), and on a formulation blend. The mapping platform was found to be successful in identifying all five components in the tablet. The ability of the mapping platform to produce high-quality spectra that could be analyzed in a multivariate fashion proved to be crucial for successful identification of the low-concentration components such as Mg stearate and explotab with the concentrations of about 1% and 3% w/w, respectively. The applied mathematical methods (PCA and multivariate curve resolution) are not simple and their applications do not proceed in accordance with the theory. In particular, it is worth noting the fortunate situation in which some of the loadings in PCA largely overlapped with the pure component spectra that was vital for the identification of these two components. Mathematically, this has not been expected because the number of informative factors exceeds the nominal number of components. On the other hand, in line with the theory, the spectra obtained on the GI platform were less well defined—the spectra overlapped more strongly and the resolution was rather coarse compared to that on the mapping platform. In part as a consequence of this, the spectral indications and hence images of the minor components could not be retrieved despite the use of the multivariate tools. On the other hand, the visual quality of the much more pixelated GI chemical images of the detectable (more abundant) components was clearly better than that of the mapping

API   Avicel   DCP

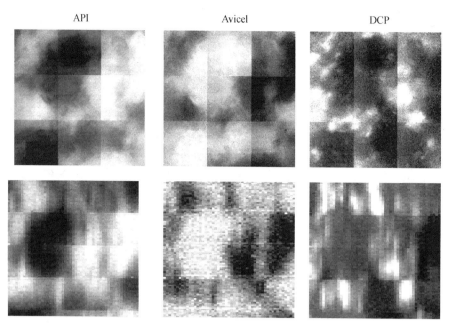

FIGURE 9.4   The univariate images of API, avicel, and DCP (left to right) obtained by the global illumination (top) and line mapping systems (bottom). All corresponding score images are essentially comparable to those shown. The X and Y spatial offsets observable in the images from the two instruments are due to errors when aligning the microscope slides. The size of all shown images is $250 \times 250\,\mu m^2$. This formulation contained high API loading of approximately 25%. The image fidelity is much higher in the global illumination images. Reprinted with permission from Society for Applied Spectroscopy.

platforms and overall the imaging on this platform was found to be faster (Figure 9.4). In this study, no attempt was made to compare the acquisition times on the two platforms because the GI platform was used to collect the spectra that considerably lengthened the acquisition times. It should be borne in mind that that was a suboptimal use of the GI platform, as this platform works best if the bands of interest are identified and imaging conducted strictly across such short regions. The major goal of the cited article, however, was to compare the platforms in as many details as possible, and this was the major reason for collecting the GI spectra. Simple estimates show that GI platform may be much faster to image the components that are imageable on it. More exact estimates are difficult to provide because the experimental details are governed by numerous factors, such as spatial resolution and fidelity of the image, and so only general assessments can be given.

Finally, the images of the blends were also found to be illustrative of the general features of the two platforms. The blends are more demanding in comparison to the tablets as far as Raman chemical imaging is concerned because the density of the sample is significantly smaller, causing much weaker Raman signal to be detected on the charge-coupled device (CCD) camera. In this particular experiment [16], no images were obtained but only the most different spectra in the imaging set were identified. The results showed that the spectral separation was achieved on the mapping platform, that is, the spectra of the pure major excipient and API were detected, while the API spectrum was found to be ubiquitous

and present in all the spectra from the GI platform. This was ascribed to significant photon migration on the GI system. Lesser density of the sample/intensity of Raman signal diminished chances for detecting the components of lower concentrations.

Several very recent applications illustrate effectiveness of the GI chemical imaging for identifying spatially separated particles.

Doub et al. [17] evaluated the feasibility of using GI Raman imaging for identifying chemical identity, particle size, and particle distribution for a corticosteroid in aqueous nasal spray suspension formulations. Several nasal sprays formulated with beclomethasone dipropionate particles ranging from 1.4 to 8.3 µm were imaged by both Raman microscopy and normal microscopy to obtain more exact information about particle sizes.

The Raman spectra of the components of the inspected formulations are shown in Figure 9.5. The strong API band at $1662\,cm^{-1}$ appears very suitable for GI chemical imaging because of its strength and lack of interference. It is worth mentioning that cases such as this are by no means rare in the pharmaceutical environment—the APIs often exhibit strong and nonoverlapped bands that can readily be used for GI chemical imaging. Figure 9.5 also shows the bright-field reflectance and bright-field/Raman overlay images of the nasal spray sample. On the basis of the clear appearance of the Raman peak of the API in the accompanying spectra, one can assign the two color-coded particles to the API. No Raman signal around $1650\,cm^{-1}$ was detected for the

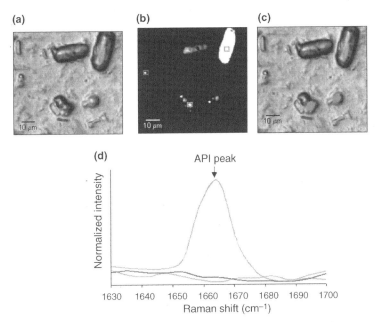

FIGURE 9.5 (a) Bright-field reflectance image, (b) polarized light image, (c) and bright-field/Raman overlay image of Beconase AQ nasal spray sample for a single region of interest with averaged imaging spectrometer-generated Raman spectra, color coded to match indicated regions in the polarized light image (d). Reprinted from [17] with permission from Springer. (See the color version of this figure in Color Plate section.)

placebo sample of similar visual appearance. Such convincing identification of the API led to reviewing much larger areas with the particles from the spray (deposited on Al-coated glass microscope slides) and allowed statistical assessment of the sizes of API particles through binarization of the API images.

Another example is also shown in Figure 9.5 in which an assessment of agglomeration/adherence of the API is given. The green, circled API particles appear adhered to the larger particles of the excipients. The related statistics, although revealing obvious problems with the exactness of the measurements, suggest that relatively small number of relatively large API particles is associated with the excipients.

Conceptually, similar application is described in Ref. 18 although on a very different sample. This time an inhalation formulation was imaged by a GI instrument. The aim of the experiment was simply to determine whether the large particles of lactose carry smaller particles of API. No attempts were made to determine the distribution of the API particles or to characterize them in any other way. Figure 9.6 shows the normal microscopy and Raman chemical image of the API on a particle of lactose of about $100 \times 50 \, \mu m^2$. Raman images as the one in Figure 9.6 can be obtained very quickly. Figure 9.6 was obtained in a few minutes by imaging at only four wavenumbers owing to, again, very favorable situation with the strength of the API Raman band used for imaging and lack of interference. API was clearly visualized on the surface of lactose, which was impossible through the normal microscopy image. It is also worth mentioning that despite the API loading being about 5%, the Raman signal of

the API significantly exceeds that of lactose, so API images are actually of better quality than that of lactose.

Another GI application refers to chemical imaging of granules obtained by the wet granulation process [19]. In fact, granules analyzed in this study were actually the final pharmaceutical product because the granulation was not followed by tabletting but the capsules were simply filled with the granules. The goal of the study was to determine the structure of the granules and hence whether or not the wet granulation proceeds according to the expectation (that the granules are mixtures of the API and the major excipient). In addition to GI chemical mapping, NIR chemical mapping is used in parallel and the results of the methods are compared.

The results revealed that in terms of granulation both platforms identified the majority of granules as mixtures of

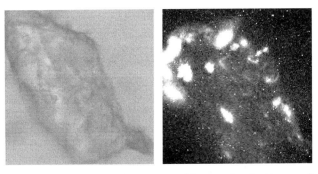

FIGURE 9.6 Normal microscopy and Raman chemical image of an API on the surface of a lactose particle. The sizes of these images are $100 \times 100 \, \mu m^2$.

the two most abundant components (mannitol and API in this case) and that granules of pure materials only occasionally arose. Of those, the pure mannitol ones largely exceeded those of pure API. However, spectral indications of the pure API were much clearer in Raman than in NIR. Incidentally, not only the Raman response of the API was much better than its NIR signal (in relative terms), but also the overlap was significantly smaller leading to a much clearer appearance of the API in the Raman chemical images (Figure 9.7). In addition, the reflection from the granules was better recorded on the normal microscopy camera on the GI platform. As a result, not only chemical images on the GI platform were visually more convincing, but also the normal microscopy images on this platform pointed out to discernible appearance of the pure API particles (as distinctly white pixels). On the other hand, the NIR platform turned out to be much faster despite various combinations of experimental parameters employed on the GI platform aimed at speeding the acquisition. It should be noted here that the NIR measure-ments were additionally troubled with the spatial separation of the granules that led to the complete absence of signal at some pixels. This caused problem with regard to thresholding and contributed to poor identification of pure API granules.

9.4   BIOMEDICAL APPLICATIONS

As mentioned above, following the API in tablets or powders is fairly easy compared to locating it in more demanding biological samples such as tissues or cells. One of the most comprehensive illustrations of the latter (among not so many) may be found in the studies by Ling et al. [20, 21]. Here, a brief summary is given of a quite complex approach (again in comparison to working with tablets or powders) to Raman image the distribution of the anticancer agent paclitaxel in a living tumor cell.

Renishaw 2000 Raman spectroscopic system (Renishaw, Gloucestershire, UK) was used in that study with the laser

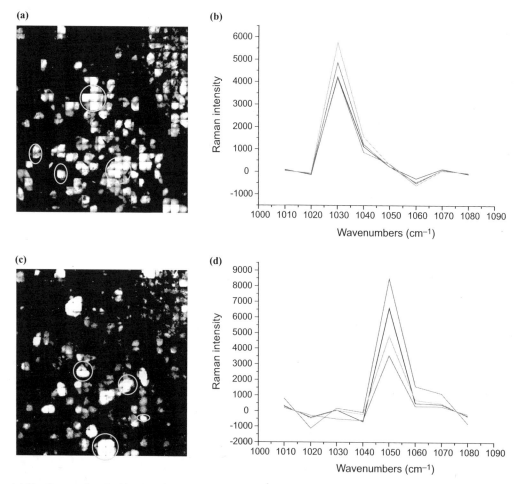

FIGURE 9.7   (a) The Raman chemical image of mannitol at 1030 cm^{-1}. The circled granules are believed to be pure mannitol on the basis of the spectra in (b) that feature only the mannitol band at 1030 cm^{-1}. (c) The Raman chemical image of API at 1050 cm^{-1}. (d) The circled granules are assigned to the pure API on the basis of the presence of only the strong API peak at 1050 cm^{-1}. These images have been produced from the baseline corrected data.

beam being expanded so that, although much better known as a mapping system, this instrument was effectively used here through its global illumination capabilities. The images of paclitaxel were obtained at $1000\,\mathrm{cm}^{-1}$. Fluorescence is an unavoidable problem in bioimaging, so in the preprocessing of the images in Figure 9.8, the key element was found to be subtraction of the images at $1080\,\mathrm{cm}^{-1}$ that are considered to

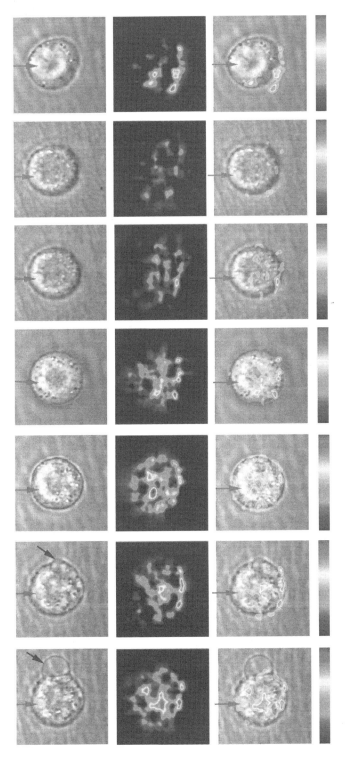

represent fluorescence from the solvent. That way, Raman chemical images were obtained. Before this step, various other processing tools are employed, such as correction for nonuniform illumination, noise reduction, and so on to account for complex optical features of the sample.

The experiment proceeded with exposing the cancer cells to paclitaxel solution for several hours. All the images were obtained with acquisition times of about 5 min per image and with a 60× water immersion objective. The series of produced images illustrated gradual diffusion of paclitaxel into the cell. It was found that paclitaxel concentrated around the center of the cell and the membrane but not in the cell nucleus. This pattern was explained by binding of the agent to the microtubules.

There is another exciting imaging application with the same API. Kang et al. [22] recently reported on coherent anti-Stokes Raman spectroscopy (CARS) imaging of paclitaxel in various polymer films. The study was motivated by the need to determine the distribution of the API in a controlled release drug delivery system and the visualization of the release. Paclitaxel was CARS imaged in three dimensions in matrices of several polymers and its spatial distribution was followed in time and depth during the release from the poly(ethyl-*co*-vinyl acetate) (PEVA, 40% vinyl acetate) film.

The CARS mechanism is described elsewhere in this book. It is a significantly different and far more complex underlying mechanism compared to other mapping/imaging approaches presented here. It is based on collinear overlap of two laser beams that allows selective enhancement of non-overlapped bands of the imaged chemical entity. For this case, it was found that several bands of paclitaxel satisfy conditions for CARS imaging, and thus these were used to produce CARS chemical images.

Figure 9.9 shows an example of *in situ* CARS imaging of paclitaxel in (largely) PEVA film during its release into a phosphate buffer saline medium. The images were obtained by tuning the differences between the two laser beams at $3060\,\mathrm{cm}^{-1}$ to one of the nonoverlapped C–H stretching

FIGURE 9.8 Images before, during, and after an MDA-435 breast cancer cell were exposed to the paclitaxel agent. The first row illustrates the images before the drug treatment. The second and third rows illustrate the images 10 and 45 min during the drug treatment. The fourth to seventh rows illustrate the images 10 min, 1.75 h, 4 h, and 4.5 h after the drug treatment. The left column shows the white light images of the cell that show the cell structure, the center column shows the Raman images of the cell that show the intensity distribution of the $1000\,\mathrm{cm}^{-1}$ Raman band, and the right column shows the overlay of images in the left and center columns. The red arrow points to the cell nucleus region and the blue arrow points to the cell blebbing region. The color bar indicates the relative Raman signal intensity increase from bottom to top. Reprinted with permission from Optical Society of America. (See the color version of this figure in Color Plate section.)

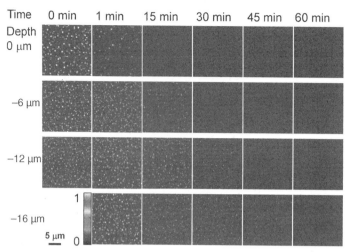

FIGURE 9.9 *In situ* CARS imaging of PTX from a PEVA film. Spectrum color scheme was used to emphasize the change of contrast. The columns were arrayed as time lapse, and rows were arrayed as depth of the film. The CARS images were taken from different depths of the film with $\omega_p - \omega_s$ tuned to 3060 cm^{-1}. The acquisition time for each image was 1.12 s. Reprinted with permission from American Chemical Society. (See the color version of this figure in Color Plate section.)

vibrations of paclitaxel. The speed with which these images were collected was impressive compared to long measurement times normally necessitated in common Raman imaging experiments. The 3D resolution presented is also considered an inherent feature of the technique and does not require any special data analysis approach, which is again very different from the common Raman. The release of the drug seems to be very neatly followed through corresponding CARS images—after 1 h the drug appears to be completely released. The detection limit was found to be 29 mM.

In conclusion, the authors underline several important features of CARS imaging that may propel this technique to be more frequently used for better understanding of mechanism of action of drug delivery systems: (1) there is no need

for labeling that is indispensable in fluorescence microscopy, (2) noninvasiveness, (3) sensitivity (although still inferior to fluorescence), (4) real-time imaging capability, and (5) 3D resolution. However, the overall complexity of the instrumentation may be seen as a considerable obstacle.

9.5   DATA PROCESSING

Data processing is an important part of image analysis. Figure 9.10 illustrates the general role of chemometrics and statistics in this process. In the first step, chemical images of the components are extracted from the unfolded 3D "hyperspectral cubes." Further processing of chemical

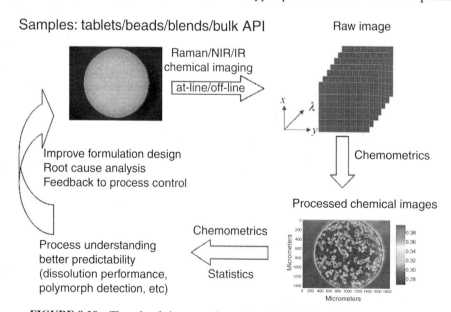

FIGURE 9.10 The role of chemometrics and statistics in chemical image analysis.

images via chemometrics and statistical tools can extract simplified metrics for process understanding, prediction of dissolution, polymorph detection, and so on. This information can then be used to improve formulation design, identify root cause of manufacturing issue, feedback to manufacturing process control, and so on. Designing an appropriate data analysis strategy is often a complex task. This is not only due to the large size of data but also due to the lack of "pure" pixels (pixels containing spectral response from a single chemical species) in most pharmaceutical applications.

Both univariate and multivariate data analysis methods can be applied for producing chemical images. However, multivariate techniques are more comprehensive and much more efficient for imaging components of low concentration. A recent review article by Gendrin et al. gives a nice summary on different chemometric methods for vibrational chemical imaging applications including Raman, NIR, and infrared (IR) mapping and imaging [23]. The discussion below focuses on pharmaceutical application of Raman imaging, but it equally applies to NIR and IR mapping because of the same structure of the data (bearing in mind that differences may occur due to the specificity of noise in Raman data).

As mentioned above, Raman mapping generates data "hypercubes" consisting of four dimensions: x and y spatial dimensions, a wavelength dimension, and spectral response dimension. The numerical format of the data is actually a three-dimensional data cube in which intensity values are stored as a function of spatial dimensions and the wavelength dimension. A typical image data analysis procedure consists of the following steps: (1) data preprocessing to reduce undesired effects (e.g., use of spectral normalization and derivatization to minimize spectral interferences due to surface roughness or particle size effects); (2) generation of chemical maps to visualize spatial distribution of components of interest via appropriate univariate or multivariate methods; and (3) extraction of metrics such as domain size, homogeneity, and so on. Note that depending on the specifics of an application, not all the above steps are needed and some steps could be combined. Two hidden steps unmentioned above are the unfolding of the three-dimensional data into two-way data (pixel × spectral channel) prior to data analysis and the folding of two-way result data to the original image format after data analysis.

Preprocessing is the first step for Raman image analysis. Generally, spectral preprocessing of the Raman spectra data is less demanding than that for NIR and IR spectra except for the situations when cosmic rays or fluorescence background removal is needed.

Cosmic rays are ionizing radiation of extraterrestrial origin that may interact with the CCD to produce random, sharp, and easily recognizable peaks (or rather spikes) in the Raman spectra. These peaks can hamper performance of multivariate chemometric methods. For example, distance-

based classification algorithms can be confused by this kind of noise. Cosmic rays removal by experimental means typically lengthens the experimental time significantly, which is an issue in already quite time-consuming chemical mapping experiments. Zhang and Henson [24] proposed a modified nearest-neighbor comparison algorithm to identify and correct this erroneous signal by mathematical means. The proposed algorithm was based on linear regression of the nearest-neighboring pixel. Via a linear regression, the spectrum from a pixel can be approximated by those from its neighbors. In addition to the use of neighboring pixels, this algorithm allows the incorporation of pure reference spectra of low-concentration components in the regression. This matters for elimination of uncertainty in the assignment of the pixels lacking similar neighbors.

The utility of this algorithm was demonstrated by improvements in multivariate classification analysis. Figure 9.11 shows spectra at several pixels before and after the removal of cosmic rays. These pixels were erroneously classified because of the cosmic spikes overlapping with API bands. Without correction, some API form III and excipient pixels were incorrectly assigned as API forms I or II. A better classification was achieved after applying the proposed algorithm.

Wang et al. proposed the use of wavelets as a processing step before fuzzy c-means clustering analysis of Raman imaging data [25]. Through the testing on Raman images with different noise levels, it is shown that noise removal using wavelets improved the classification accuracy using fuzzy c-means clustering. The preprocessing of spectral data was accomplished by decomposing them in the differential wavelet domain and then a multiscale pointwise product (MPP) criterion was used to discriminate between the true spectral signal and the noise. The approach was tested by classifying Raman images collected from adhesive and dentin interface specimens. When comparing with traditional denoising techniques including spline and Savitzky–Golay filtering, better classification accuracy was achieved with the wavelet filtering. In addition, the localization property of wavelet filters facilitated better visual perspective of images.

After the pretreatment, chemical maps can be generated from the spectra through either univariate or multivariate techniques. Compared to NIR and IR imaging, Raman imaging is more amenable to univariate techniques since Raman spectra typically consist of much sharper and stronger spectral features. Univariate techniques are a straightforward way to generate individual chemical maps in cases when selective bands for the components of interest can be identified. Similar to single point Raman spectroscopy, this approach uses peak height or peak area of those specific bands.

There are numerous multivariate techniques (see Chapter 5). Depending on whether *a priori* information is involved, multivariate methods fall into two categories:

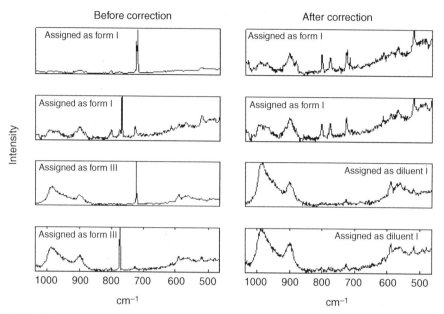

FIGURE 9.11 Spectra of several misassigned pixels before and after cosmic spike removal.

supervised and unsupervised techniques. Unsupervised techniques do not require any prior information about the sample. Examples of these methods are PCA, multivariate curve resolution (MCR), and cluster analysis (CA).

PCA essentially maps each spectrum from a multidimensional space (in terms of wavelength) to a space with reduced number of coordinates. The PCs are constructed to explain as much of the data variance as possible. PC scores can be used to generate chemical maps in lieu of raw spectral responses, while loadings can be used to interpret the physical meaning of each PC.

MCR techniques try to resolve pure component spectra and corresponding distributions directly from the image data by mathematical means. Typically, to ascertain that the estimated solutions do not contain unacceptable values, some constrains are applied, such as nonnegativity of spectral response and concentration values. CA method is a pattern recognition approach that classifies pixels into clusters (groups) without involving training data. CA method assigns pixel memberships by minimizing certain kind of distance-based objective function.

Supervised techniques require reference spectra of pure components and these methods include ordinary least squares (OLS), PLS, and pattern recognition methods such as minimum ED classifier.

A simple side-by-side comparison between the univariate and multivariate methods on the quality of chemical images derived was reported by Šašić [11]. In this study, an artificial model images with different levels of noise and an image from a real bead sample were considered. PCA and OLS were used as multivariate tools. The results revealed that the performance of univariate method deteriorated dramatically

with the increase of the noise level, while the multivariate methods were still able to generate high-quality chemical maps notwithstanding significant noise. Unsurprisingly, OLS performs better than PCA due to the use of the reference spectra. These results clearly demonstrated the advantages of multivariate methods in image analysis.

Zhang et al. compared several multivariate methods using a case study of a model pharmaceutical tablet [26]. The results of that study are summarized in Figure 9.12, which shows RGB composite images for the three major components derived using different multivariate approaches. Red, green, and blue channels were assigned to sodium benzoate, lactose, and avicel, respectively. The results from OLS in Figure 9.12d are deemed to be the most reliable, being based on the reference spectra. For each specific component, respective OLS image contains contrast information. The other results in this figure did not use any *a priori* information about the sample. The result of MCR combined with alternating least squares (ALS) is practically indistinguishable from the OLS result. Although the PCA image in Figure 9.12a somewhat resembles the OLS image, some of the cold pixels present in the OLS image are absent. This is caused by the fact that PCA scores do not necessarily represent the actual distribution of the selected component due to the PC loadings being (more or less of) a mixture of signals from various components. PCs are extracted by finding directions of the highest data variance, and hence the loadings (as mathematical concepts) do not necessarily align with the pure component spectra. The cluster analysis results, KM image in Figure 9.12b, and FCM image in Figure 9.12c show some similarities with the OLS image. However, there is a difference in the boundaries of different

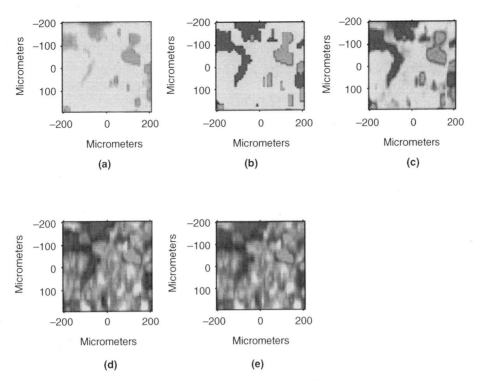

FIGURE 9.12 Composite images of PCA scores (a) KM, (b) FCM, (c) OLS, and (d) MCR/ALS using raw spectra (e). In (a), (d), and (e), images were enhanced to increase the contrast of 95% of the pixels.

components across the images; for example, the domains of red (sodium benzoate) in KM and FCM images are slightly larger than those in DCLS image. This could be caused by the equal population tendency of these clustering algorithms. An advantage of FCM over FM is that FCM is able to preserve the intensity variation within the same cluster, generating a more instructive image.

Widjaja and Seah proposed the use of band-target entropy minimization (BTEM) to recover pure spectra and chemical maps from minor components present in Raman maps of pharmaceutical tablets [27]. BTEM algorithm is a recent variation of MCR techniques. It assumes that the derived pure component spectra have simpler shapes that translate to less "entropy" in comparison to that of the mixture spectra. The algorithm works as follows. First, PCA is applied on the image data to generate a set of eigenvectors. Next, BTEM is performed on selected eigenvectors to derive pure component spectrum, one at a time. The model tablets involved in the study consist of acetaminophen, lactose, avicel, and Mg stearate. Mg stearate was targeted as a minor component with weight percentages ranging 2–0.2%. BTEM was able to resolve the pure spectrum of Mg stearate at levels as low as 0.2 wt%. The derived Mg stearate spectrum was not a perfect match to the reference pure spectrum due to interference from other components. However, this inaccuracy does not significantly affect produced chemical images. On the other hand, the SIMPLISMA resolved [28] Mg stearate

pure spectrum is seriously distorted even at 2 wt% level. It seems that the success of BTEM's detection of minor component relies more on the minor component's distribution variation in the image than on its spectral signal-to-noise ratio. The basis for BTEM detection of minor components is as follows: as long as sufficient variation of the minor component exists across the image, PCA can find a direction aligned closely with that component and this allows subsequent BTEM refinement. Thus, BTEM offers new possibilities for minor component detections, especially in cases when segregation and inhomogeneity occur.

Henson and Zhang [6] reported the use of supervised multivariate classification technique for drug characterization in low dosage pharmaceutical tablets (0.5% w/w of API) using Raman mapping. Multivariate classification was performed to detect the presence of two undesired API polymorphs within tablets. A five-class spectral library was constructed containing Raman spectra of pure components. A target matrix was constructed such that each row of the matrix was a five-element vector consisting of a 1 for the in-class component and 0's for the out-of-class components. Then, a PLS model was constructed between the library spectra and the target matrix. Figure 9.13 shows the PLS score plot using the first three PLS latent variables. Direct use of PLS prediction values for classification is often referred to as PLS discriminant analysis (PLS-DA). However, PLS-DA was found unsuccessful for pixel classification since the

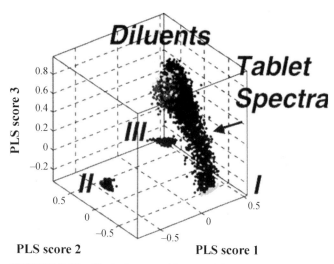

FIGURE 9.13   Pixels from a tablet containing API form I projected onto a PLS score subspace defined by the pure components.

appropriate setting of cutoff values for PLS prediction value is challenging. Thus, classification was accomplished via a minimum ED approach on PLS scores. Basically, this method assigns a pixel's membership according to its closeness to pure component centers in PLS latent space.

Most multivariate image analysis methods in the literature ignore the spatial information during the analysis after unfolding the image data cube into a two-way data matrix. Incorporating spatial information may enhance the cluster analysis performance by improving the robustness and stability of the results. Lin et al. proposed a spatially directed agglomeration clustering method and applied this approach to chloramphenicol palmitate polymorph characterization in tablets [29].

The algorithm is based on the agglomeration clustering analysis approach. The method defines spectral similarity by calculating the norm of the projection residual vector between each pair of spectra. First, the algorithm starts from a single cluster consisting of two nearest pixels. These two pixels are chosen from pixels with Raman spectra belonging to the top 10% most similar ones. Next, the cluster grows by stepwise merging the next pixel that is spatially the nearest to the cluster among unclassified pixels. The constraint is that the spectrum of this pixel should belong to the top 10% most similar ones when comparing with the mean spectrum of the cluster. The algorithm terminates when all the pixels are assigned to respective clusters. It utilizes both spectral differences and spatial closeness of pixels for determining distances in the agglomeration process. The performance of the proposed algorithm is better than that of a classical *K*-means cluster analysis algorithm as it provides more accurate estimates for the contents of different polymorphs.

Raman mapping hardware can be used not only for mapping distributions but also for determining if a component of interest (almost exclusively API) can be determined.

However, because common mapping experiments typically probe only a relatively small area on the sample, a challenging question is how many spectra need to be collected to determine if a sought component (say, undesired API polymorph) of very low concentration is present in the sample. Šašić and Whitlock [30] proposed an optimized, statically based sampling scheme to tackle this problem. The method relies on two simple statistical charts that help predefine the number of spectra to collect that should guarantee the identification of the sought component if it is present in the sample. One of the graphs used determines the probability of detecting a specific event as a function of the total number of probes. In other words, it addresses the question of how many mapping spectra need to be collected to find at least one spectrum with the signal for the component of interest. It is based on binomial distribution with an initial assumption about the concentration of the component of interest. The second graph is used to show the number of experiments needed to confirm a hypothesis. For example, the hypothesis here can be 95% confidence limit for the probability of finding the component of interest in an experiment with strictly determined number of spectra to collect. This study is not exactly a typical chemical mapping experiment because the key goal is in identifying and not necessarily visualizing the component of interest. However, in addition to using the Raman mapping hardware, the data analysis approach heavily relies on identifying the pixels of interest in the relevant chemical image.

REFERENCES

1. Gowen, A. A., O'Donnell, C. P., Cullen, P. J., and Bell, S. E. (2008). Recent applications of chemical imaging to pharmaceutical process monitoring and quality control. *Eur. J. Pharm. Biopharm.* **69**, 10.

2. Šašić, S. (Ed.) (2008). *Pharmaceutical Applications of Raman Spectroscopy*, Wiley, Hoboken, NJ.

3. Andrew, J. J. and Hancewizc, T. M. (1998). Rapid analysis of Raman image data using two-way multivariate curve resolution. *Appl. Spectrosc.* **52**, 797.

4. Malinowski, E. R. (1991). *Factor Analysis in Chemistry*, Wiley, New York.

5. Šašić, S. (2007). Raman mapping of low-content API pharmaceutical formulations. I. Mapping of Alprazolam in Alprazolam/Xanax tablet. *Pharm. Res.* **24**, 58.

6. Henson, M. and Zhang, L. (2006). Drug characterization in low dosage pharmaceutical tablets using Raman microscopic mapping. *Appl. Spectrosc.* **60**, 1247.

7. Ward, S., Perkins, M., Zhang, J., Roberts, C. J., Madden, C. E., Luk, S. Y., Patel, N., and Ebbens, S. J. (2005). Identifying and mapping surface amorphous domains. *Pharm. Res.* **22**, 1195.

8. Breitenbach, J., Schroff, W., and Neumann, J. (1999). Confocal Raman spectroscopy: analytical approach to solid dispersion and mapping of drugs. *Pharm. Res.* **16**, 1109.

9. Clarke, F. C., Jamieson, M. J., Clark, D. A., Hammond, S. V., Jee, R. D., and Moffat, A. C. (2001). Chemical image fusion. The synergy of FT-NIR and Raman mapping microscopy to enable a more complete visualization of pharmaceutical formulations. *Anal. Chem.* **73**, 2369.

10. Šašić, S. (2007). An in-depth analysis of Raman and near-infrared chemical images of common pharmaceutical tablets. *Appl. Spectrosc.* **61**, 239.

11. Šašić, S., Clark, D. A., Mitchell, J. C., and Snowden, M. J. (2004). Univariate versus multivariate Raman imaging: a simulation with an example from pharmaceutical practice. *Analyst* **129**, 1001.

12. Šašić, S., Clark, D. A., Mitchell, J. C., and Snowden, M. J. (2005). Raman line mapping as a fast method for analyzing pharmaceutical bead formulations. *Analyst* **130**, 1530.

13. Treado, P. J. and Nelson, M. P. (2001). Raman imaging. In: Chalmers, J. M. and Griffiths, P.R. (Eds.), *Handbook of Vibrational Spectroscopy*, Wiley, New York.

14. Markwort, L., Kip, B., Da Silva, E., and Roussel, B. (1995). 'Raman imaging of heterogeneous polymers: a comparison of global versus point illumination. *Appl. Spectrosc.* **49**, 1411.

15. Schlücker, S., Schaeberle, M. D., Huffman, S. W., and Levin, I. W. (2003). Raman microspectroscopy: a comparison of point, line, and wide-field imaging methodologies. *Anal. Chem.* **75**, 4312.

16. Šašić, S. and Clark, D. A. (2006). Defining a strategy for chemical imaging of industrial pharmaceutical samples on Raman line-mapping and global illumination instruments. *Appl. Spectrosc.* **60**, 494.

17. Doub, W. H., Wallace, P. A., Spencer, J. A., Buhse, L. F., Nelson, M. P., and Treado, P. J. (2007). Raman chemical imaging for ingredient-specific particle size characterization of aqueous suspension nasal spray formulations: a progress report. *Pharm. Res.* **24**, 934.

18. Šašić, S. (2005). Chemical imaging of pharmaceutical samples by a global illumination Raman imaging instrument, Abstract, XXXII FACSS, Quebec City, Canada.

19. Šašić, S. (2008). Chemical imaging of pharmaceutical granules by Raman global illumination and near-infrared mapping platforms. *Anal. Chim. Acta* **611**, 73.

20. Ling, J. (2008). In: Šašić, S. (Ed.), *Pharmaceutical Applications of Raman Spectroscopy*, Wiley, Hoboken, NJ.

21. Ling, J., Weitman, S. D., Miller, M. A., Moore, R. V., and Bovik, A. C. (2002). Direct Raman imaging techniques for studying the subcellular distribution of a drug. *Appl. Opt.* **41**, 6006.

22. Kang, E., Wang, H., Kwon, I. K., Robinson, J., Park, K., and Cheng, J. (2006). *In situ* visualization of paclitaxel distribution and release by coherent anti-Stokes Raman scattering microscopy. *Anal. Chem.* **78**, 8036.

23. Gendrin, C., Roggo, Y., and Collet, C. (2008). Pharmaceutical applications of vibrational chemical imaging and chemometrics: a review. *J. Pharm. Biomed. Anal.* **48**, 533.

24. Zhang, L. and Henson, M. J. (2007). A practical algorithm to remove cosmic spikes in Raman imaging data for pharmaceutical applications. *Appl. Spectrosc.* **61**, 925.

25. Wang, Y., Wang, Y., and Spencer, P. (2006). Fuzzy clustering of Raman spectral imaging data with a wavelet-based noise-reduction approach. *Appl. Spectrosc.* **60**, 7.

26. Zhang, L., Henson, M. J., and Sekulic, S. S. (2005). Multivariate data analysis for Raman imaging of a model pharmaceutical tablet. *Anal. Chim. Acta* **545**, 262.

27. Widjaja, E. and Seah, R. K. H. (2008). Application of Raman microscopy and band-target entropy minimization to identify minor components in model pharmaceutical tablets. *J. Pharm. Biomed. Anal.* **46**, 274.

28. Windig, W. and Guilment, J. (1991). Interactive self-modeling mixture analysis. *Anal. Chem.* **63**, 1425–32.

29. Lin, W., Jiang, J., Yang, H., Ozaki, Y., Shen, G., and Yu, R. (2006). Characterization of chloramphenicol palmitate drug polymorphs by Raman mapping with multivariate image segmentation using a spatial directed agglomeration clustering method. *Anal. Chem.* **78**, 6003.

30. Šašić, S. and Whitlock, M. (2008). Raman mapping of low-content active-ingredient pharmaceutical formulations. Part II: statistically optimized sampling for detection of less than 1% of an active pharmaceutical ingredient. *Appl. Spectrosc.* **62**, 916.

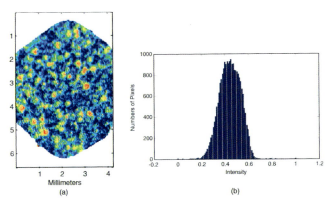

FIGURE 4.2 Presents an image and corresponding histogram derived from a NIRCI data set of a pharmaceutical tablet comprised of ~22,400 NIR spectra. (*See text for full caption.*)

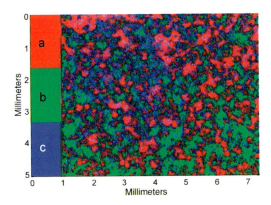

FIGURE 4.7 Chemical image of a three-component mixture: (a), (b), and (c) are the areas holding the pure components and (d) is the area comprising the mixture. (*See text for full caption.*)

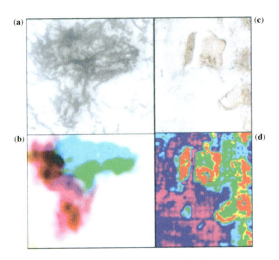

FIGURE 6.1 Raman imaging of a meningioma tumor section. (*See text for full caption.*)

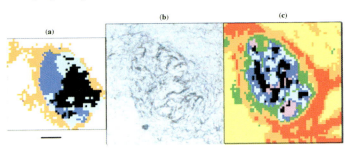

FIGURE 6.6 Factor analysis of a confocal Raman data set delineates skin regions near a wound edge 0.5 days after wounding. (*See text for full caption.*)

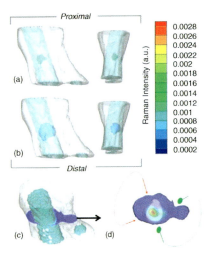

FIGURE 6.10 Raman tomographic images of canine bone tissue. (*See text for full caption.*)

FIGURE 6.4 FTIR microscopic image (a), photomicrograph (b), and Raman microscopic image (c) of ganglia. (*See text for full caption.*)

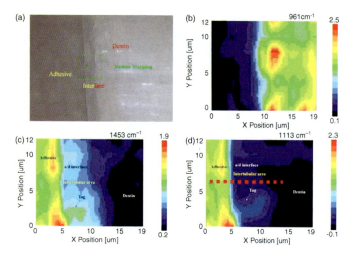

FIGURE 6.12 (a) Visible image of the adhesive/dentin interface with the corresponding micro-Raman spectroscopic images: (*See text for full caption.*)

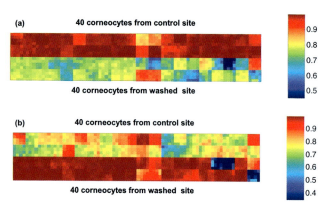

FIGURE 7.9 Correlation coefficient images of multiple corneocytes isolated from untreated and washed sites. (*See text for full caption.*)

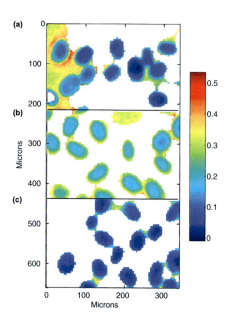

FIGURE 7.12 IR imaging of relative sulfonate content in microtomed human hair sections following various treatments. (*See text for full caption.*)

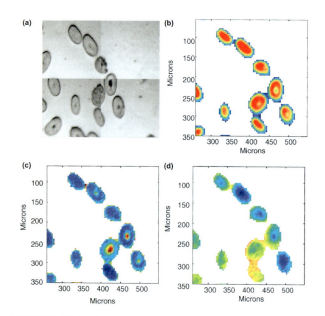

FIGURE 7.11 Cross-sectional image of bleached human hair microtomed to a thickness of ~4 μm. (*See text for full caption.*)

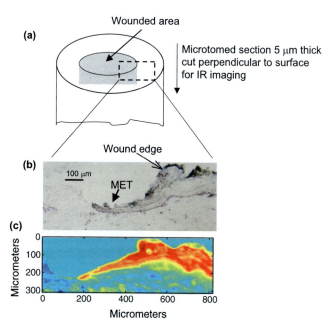

FIGURE 7.14 (a) Schematic depiction of the cutaneous wound-healing model used in the current experiments. A 3 mm punch biopsy is used to generate an acute wound in a human skin specimen. (*See text for full caption.*)

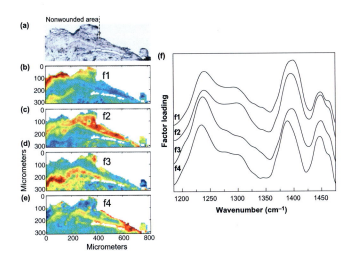

FIGURE 7.15 IR characterization of keratin-rich regions in wounded and nonwounded areas, 6 days postwounding, using factor analysis in the 1185–1480 cm^{-1} spectral region. (*See text for full caption.*)

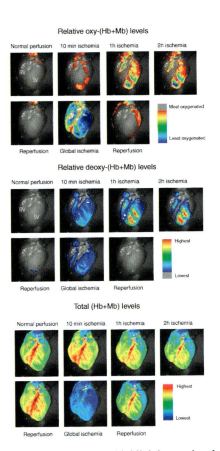

FIGURE 8.11 Near-IR images highlighting regional variations in cardiac oxy-(Hb + Mb), deoxy-(Hb + Mb), and total (Hb + Mb) through a protocol involving both regional and global ischemia in an isolated, arrested heart perfused with 50:50 blood:KHB solution. (*See text for full caption.*)

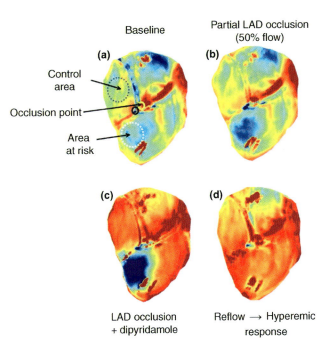

FIGURE 8.15 Effect of dipyridamole infusion on oxygenation in a partial regional ischemia model (dark regions are poorly oxygenated). (*See text for full caption.*)

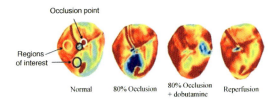

FIGURE 8.16 Effect of dobutamine infusion on oxygenation in a partial regional ischemia model. (*See text for full caption.*)

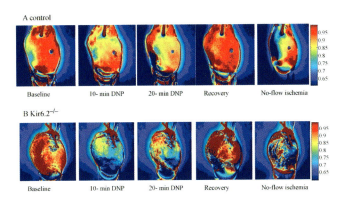

FIGURE 8.17 Near-IR images highlighting variations in tissue myoglobin oxygenation for KHB-perfused control and Kir6.2$^{-/-}$ mouse hearts following treatment with 2,4-dinitrophenol (dark regions are poorly oxygenated). (*See text for full caption.*)

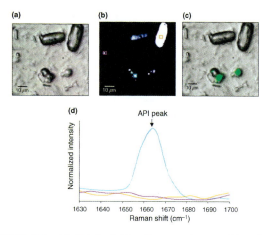

FIGURE 9.5 (a) Bright-field reflectance image, (b) polarized light image, (c) and bright-field/Raman overlay image of Beconase AQ nasal spray sample for a single region of interest with averaged imaging spectrometer-generated Raman spectra, color coded to match indicated regions in the polarized light image (d). Reprinted from [17] with permission from Springer.

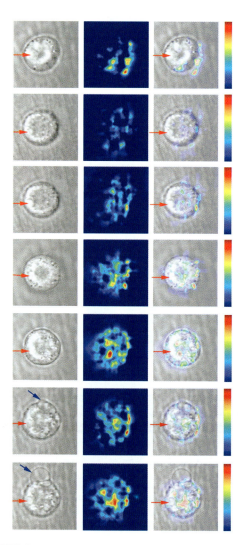

FIGURE 9.8 Images before, during, and after an MDA-435 breast cancer cell were exposed to the paclitaxel agent. (*See text for full caption.*)

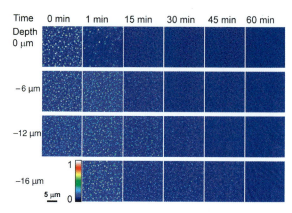

FIGURE 9.9 *In situ* CARS imaging of PTX from a PEVA film. Spectrum color scheme was used to emphasize the change of contrast. (*See text for full caption.*)

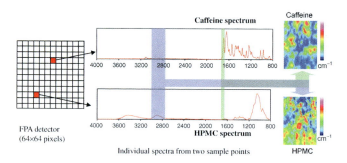

FIGURE 10.1 FPA detector spectra collection.

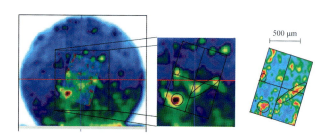

FIGURE 10.8 FTIR spectroscopic images compared with images from X-ray microtomography. Reproduced from Ref. 69 by permission of John Wiley & Sons, Inc.

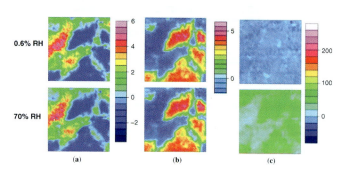

FIGURE 10.11 FTIR images of the PEG-griseofulvin mixture exposed to different relative humidities. (*See text for full caption.*)

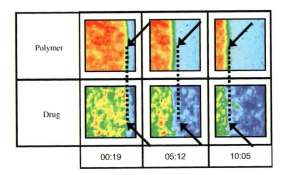

FIGURE 10.13 FTIR imaging of formulation dissolution in transmission. Reprinted from Ref. 83. Copyright 2003, with permission from Elsevier.

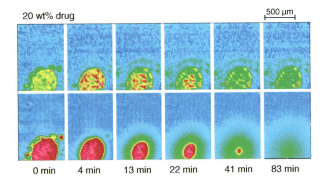

FIGURE 10.14 Dissolution of nifedipine and PEG. (*See text for full caption.*)

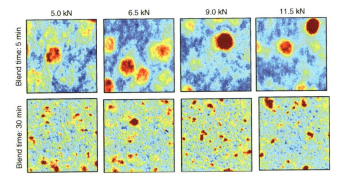

FIGURE 11.11 PLS-DA score contrast images for disintegrant (red = high concentration, blue = low concentration).

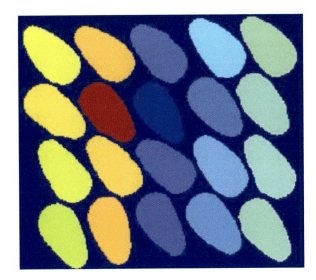

FIGURE 11.16 Intensity map at 1600 nm of 20 tablets with varying content (40–60% API). Reproduced with permission from Ref. 32.

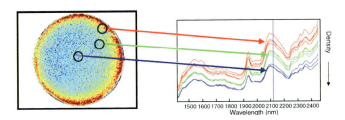

FIGURE 11.21 NIR contrast image of unprocessed spectra (at 2120 nm) of a tablet corresponding to the density within the tablet. Low intensity corresponds to high density. Reproduced with permission from Ref. 37.

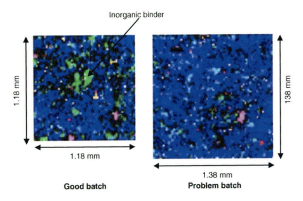

FIGURE 11.24 CIF Images of two Raman and three NIR images of the matrix of a good and a sticking problem blend [40].

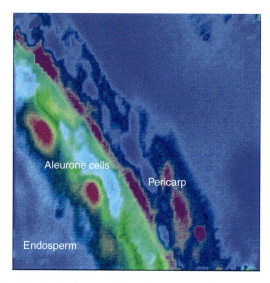

FIGURE 12.2   IR image of the endosperm, the aleurone cells, and the pericarp region of a wheat kernel. (*See text for full caption.*)

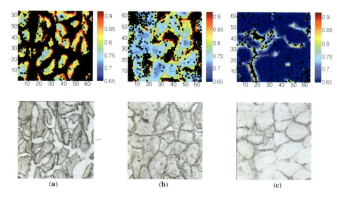

FIGURE 12.5   Chemical images of pork tissue (showing the I_{1630}/I_{1654} band ratio) obtained for three salt concentrations: high (a), medium (b), low (c) (from left to right). Corresponding photomicrographs are shown below the respective IR image.

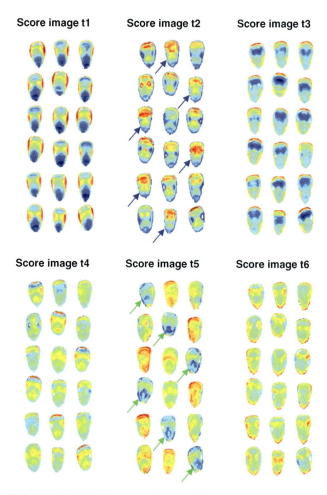

FIGURE 13.4   PCA score images (PC 1–6) after removal of background and other disturbances such as geometrical errors and shadows. Blue arrows indicate soft and green arrows hard maize kernels.

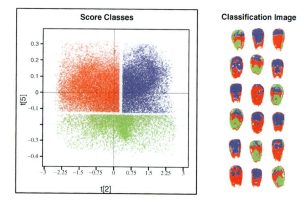

FIGURE 13.6   PCA score plot with classification (green = glassy, red = intermediate, and blue = floury) (left) and the corresponding classification projected onto the score image (right).

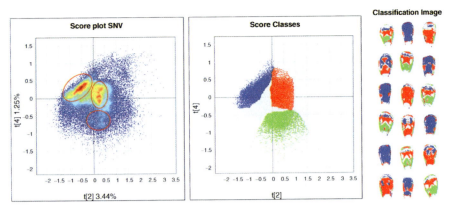

FIGURE 13.7 PCA score plot (PC2 versus PC4) after removal of background and SNV preprocessing (left). (*See text for full caption.*)

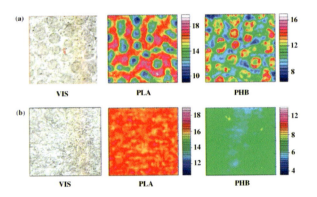

FIGURE 14.3 (a) Visual image (left), PLA-specific FT-IR image (center), and PHB-specific FT-IR image (right) of a PHB/PLA (50:50 wt%) blend. (*See text for full caption.*)

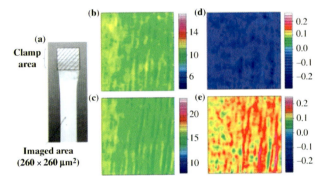

FIGURE 14.7 Optical image (a) and FT-IR images (260×260 μm^2) of A_{0PHB} (b) and A_{0PLA} (c) and the corresponding orientation function (f_∞) images of PHB (d) and PLA (e) of the 200% stretched PHB/PLA (40/60 wt%) blend film (for optimum comparison the f_∞ images (d) and (e) are shown with the same color scale).

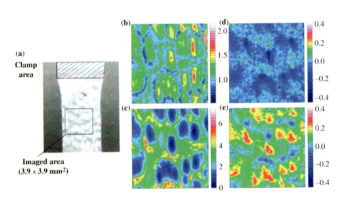

FIGURE 14.5 Optical image (a) and FT-IR images (3.9×3.9 mm^2) of A_{0PHB}/A_{0PLA} (b) and A_{0PLA}/A_{0PHB} (c) and the corresponding orientation function ($f_\perp$) images of PHB (d) and PLA (e) of the 50% stretched PHB/PLA (50/50 wt%) blend film (for optimum comparison the $f_\perp$ images (d) and (e) are shown with the same color scale).

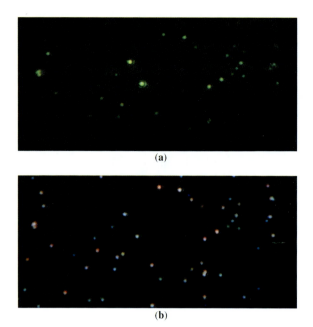

FIGURE 15.4 Microscope images of (a) SERS of TC molecules adsorbed on the Ag nanoaggregates excited at 514 nm and (b) the corresponding LSPR Rayleigh scattering from the Ag nanoaggregates illuminated by the white light through the dark-field condenser lens. The images cover an area of $78 \times 34\ \mu m^2$.

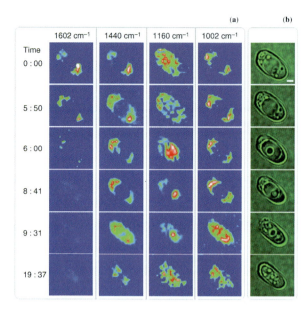

FIGURE 16.4 (a) Time-resolved Raman images and (b) corresponding optical microscope images of a dying *S. cerevisiae* cell.

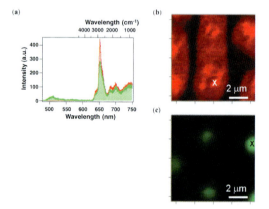

FIGURE 16.9 (a) Spectral profile of the CARS and TPEF signals of a living yeast cell; (b) CARS lateral images of living yeast cells for C−H stretching mode; (c) TPEF lateral images of the same system at 506 nm. (*See text for full caption.*)

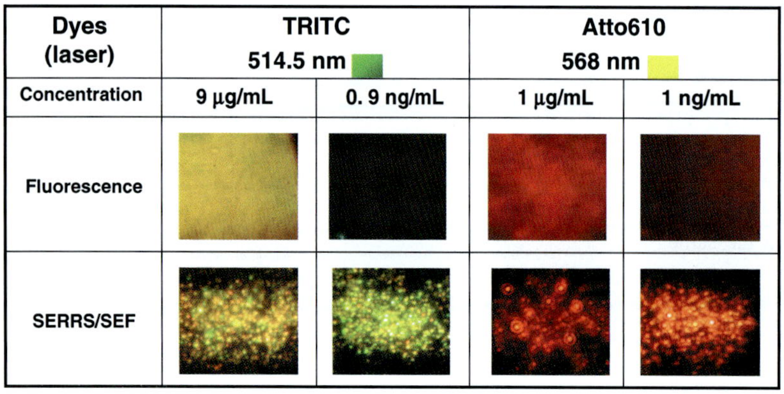

Dyes (laser)	TRITC 514.5 nm		Atto610 568 nm	
Concentration	9 µg/mL	0. 9 ng/mL	1 µg/mL	1 ng/mL
Fluorescence				
SERRS/SEF				

FIGURE 15.9 Fluorescence (the top) and SERRS and SEF (the bottom) microscope images from TRITC and Atto610 at various concentrations.

10

APPLICATIONS OF FTIR SPECTROSCOPIC IMAGING IN PHARMACEUTICAL SCIENCE

SERGEI G. KAZARIAN AND PATRICK S. WRAY

Department of Chemical Engineering, Imperial College London, London, UK

10.1 INTRODUCTION

Spectroscopic imaging using vibrational spectroscopy has significant advantages when studying pharmaceutical samples compared to many other imaging methods, due to the inherent chemical specificity of vibrational spectroscopy and fast acquisition times. FTIR spectroscopic imaging in the mid-IR region has emerged as a very powerful tool in pharmaceutical science and technology.

FTIR imaging in the mid-IR region is particularly attractive for the analysis of pharmaceutical systems because the spectral bands are present in this region (particularly in the "fingerprint" region) not only allow one to easily differentiate the various components in the sample but also provide a wealth of information about the molecular state of a particular component (e.g., amorphous or crystalline), intermolecular interactions, and polymorphic changes. The imaging capability of this method is crucial for obtaining information about the spatial distribution of different components (drug, polymer, excipients) within the tablet. The distribution of these components is often a key property for the performance of the pharmaceutical product. Thus, the spatial distribution of different components has a significant effect on the physical and mechanical properties of the tablet and plays a very important role in the mechanism of drug release (e.g., during dissolution). Content uniformity or a layered structure of the tablet is often required for an efficient or specific type of drug delivery. FTIR imaging is a particularly useful tool for assessing the effects of drug loading and sample preparation methods on drug release.

The key advantage of FTIR imaging, when all spectra are measured simultaneously using a focal plane array (FPA) detector, lies in the application of the technique to samples that change with time. The ability to create spatially resolved chemical snapshots as a function of time offers the possibility of studying dynamic systems such as tablet dissolution via simultaneous measurement of the distribution of polymer, drug and water. The application of FTIR imaging to tablet dissolution overcomes major limitations of the currently used USP dissolution test, which is a rather crude approach since it does not provide any insight into tablet during dissolution. This chapter will not focus on applications of multivariate analysis to imaging of pharmaceutical formulations as there are a number of publications that have already reviewed this [1–4].

This chapter will summarize and discuss the applications of FTIR imaging to pharmaceutical systems, with the main focus on applications of ATR-FTIR imaging, both in micro- and macro-modes [5, 6]. Micro-ATR-FTIR imaging offers imaging with a high spatial resolution by overcoming the diffraction limit of infrared light in air with the use of immersion optics, such as an ATR objective with a germanium (Ge) crystal. Macro-ATR-FTIR imaging without the use of an IR microscope facilitates the imaging of greater fields of view and opens a range of possibilities for studying large areas or whole tablets and the analysis of many samples in a high-throughput manner. The main advantage of ATR-FTIR imaging is its suitability for studying aqueous systems, which is crucial for *in situ* dissolution analysis [7–9]. This chapter will examine many applications of

Raman, Infrared, and Near-Infrared Chemical Imaging   Edited by Slobodan Šašić and Yukihiro Ozaki

macro-ATR-FTIR imaging, mostly pioneered in our group. These applications include the analysis of the spatial distribution of components within tablets, *in situ* imaging of tablet compaction, dissolution and drug release, application to high-throughput analysis, and combining ATR-FTIR imaging with other techniques.

The limitations of ATR-FTIR imaging will be explained as well as methodologies to overcome these limitations or validate the results obtained with ATR-FTIR imaging. The data obtained with macro-ATR-FTIR imaging of model tablets were compared with the results of imaging for the same tablet by X-ray microtomography for validation of the ATR-FTIR imaging approach. Successful application of ATR-FTIR imaging to water-soluble and poor soluble drugs has been demonstrated by combining ATR-FTIR imaging with a flow-through cell and UV/Vis detection of the dissolved drug in water. New developments include combining visible optical imaging with ATR-FTIR imaging of the dissolution of a tablet; this combination allows for the critical assessment of previous studies where conclusions about the mechanism of dissolution were based solely on the visible optical images.

Macro-ATR-FTIR imaging is particularly suitable for the high-throughput analysis of many samples [10]. Thus, the applicability of FTIR imaging to studying many samples simultaneously was successfully achieved using a "drop-on-demand" device by preparing arrays of microdrop samples directly onto the surface of the ATR crystal. Recently, this approach was extended to the use of poly(dimethylsiloxane) (PDMS)-based multichannel devices that allowed the study of the dissolution of several formulations simultaneously. Overall, the ATR-FTIR imaging approach with macro-ATR capability is a very powerful novel tool for high-throughput analysis of pharmaceutical formulations and can provide guidance to the design of pharmaceutical formulations for controlled release.

10.2   MID-INFRARED SPECTROSCOPY

Infrared spectroscopy has long been a useful analytical tool for studying pharmaceutical formulations. It has the potential to reveal much chemical information about the formulations for both the analysis of static samples and samples during dissolution and provides complementary data to that otherwise obtained from the systems studied. For example, the FDA dissolution tests only analyze the dissolution profile under a very controlled environment [11] and do not investigate the formulation itself. Therefore, the opportunity exists to improve the understanding of these chemical formulations. An excellent review on applications of mid-IR spectroscopy to pharmaceuticals was provided recently [12].

10.2.1   ATR and Transmission Approaches

The sampling methodology used in FTIR imaging is based on those developed in conventional FTIR spectroscopy, which needs to be discussed first. There are several approaches that can be used for studying pharmaceutical formulations using IR spectroscopy; these are transmission, diffuse reflectance, and ATR. Diffuse reflectance collects and analyzes scattered IR energy.

Transmission mode is the most well-known form of infrared spectroscopy. A beam of infrared light passes through a sample and examination of the resulting radiation can determine which frequencies of infrared light have been absorbed. The sample must be thin enough to allow the radiation to pass through, but it must also be thick enough such that a reasonable amount of absorbance occurs and so careful sample preparation is required. A suitable thickness is usually between 5 and 20 μm. The form of the chosen sample is also an important factor; powders cause random reflections of the light as it passes through the sample, thus the beam can become too scattered.

A more flexible technique is that of attenuated total reflection spectroscopy. This technique uses an inverted prism-shaped crystal composed of a material with a high refractive index, such as diamond that has a refractive index of 2.4 or zinc selenide that has a similar value of refractive index. The sample is placed on the top surface of the crystal, while the infrared light enters the crystal and approaches the top surface of the crystal at an angle greater than the critical angle. ATR spectroscopy is based on the principle that although total internal reflection occurs at the interface between the crystal and the sample, the radiation does penetrate a few micrometers into the sample, in the form of an evanescent wave, where the absorption by the sample attenuates the beam.

The penetration depth of the evanescent wave into the next medium, which is defined as the distance in which the amplitude of the electric field will fall to $1/e$ of its initial value at the surface, can be calculated using Equation 10.1.

$$DP = \frac{1}{2\pi W N_C (\sin^2\Theta - N_{SC}^2)^{1/2}} \qquad (10.1)$$

where DP is the depth of penetration, W is the wavenumber, N_C is the crystal refractive index, Θ is the angle of incidence, and $N_{SC} = N_{sample}/N_{crystal}$.

The depth of penetration in Equation 10.1 is dependent on several factors listed above. Longer wavelengths have greater depth of penetration. The expected depths of penetration using common polymeric materials with refractive indices of around 1.5 are usually in the range 1–5 μm when working in mid-IR [13]. The equation also states that total internal reflection will only occur if the incidence angle is larger than the critical angle.

To measure a sample in ATR mode, it must be placed on the surface of the crystal; therefore, sample preparation can be very simple. The sample must be pressed onto the crystal with sufficient force to ensure comprehensive contact between the two media because of the shallow penetration depth of the evanescent wave. The crystal must be transparent to infrared radiation and have a high refractive index. Zinc selenide crystals are hard, brittle, and easily scratched but a large, optically consistent crystal can be obtained relatively cheaply. Diamond is another option; however, it is expensive and absorbs certain mid-infrared frequencies of infrared radiation, especially at high temperatures. The main advantages of diamond are that it is able to endure high pressures, its chemical resistance, and it is exceptionally hard making it scratch resistant.

A general perception about ATR spectroscopy is that a high level of force is required to achieve a good contact between sample and ATR crystal, which is often not the case. Adequate contact can be achieved with liquid samples simply by covering the top of the crystal with the sample; however, when working with polymeric materials, some force is required to achieve sufficient contact. The required force can be minimized as flat surfaces help facilitate the contact, while malleable samples require minimum force in compaction. When working with some pharmaceutical polymeric materials in dissolution experiments, the formation of a gel upon contact with water greatly improves the contact of the polymer [14]. When compacting most pharmaceutical samples *in situ* on an ATR crystal, the level of compaction required is an order of magnitude less than in industrial tablet presses, demonstrating that the necessary force is achievable in a lab. Nevertheless, while compaction with a diamond as an ATR crystal is not an issue, it can be harder to achieve a satisfactory compaction on some other crystals, such as zinc selenide, without incurring damage.

10.2.2 FTIR Spectroscopic Imaging

Conventional FTIR spectrometers are single detector devices. One spectrum is obtained that averages information from the total measured volume of the sample, so there is no spatial information in the data acquired. Previously, image production was only possible using mapping techniques, which involved gathering spectral data in a grid pattern from a localized area of the sample to build up a spatially resolved map of the system. This was advanced with the introduction of linear array detectors that gather a line of data at a time, and move across the sample. The major advancement came with the introduction of the FPA detector. The single element detector is replaced with a grid array of detectors (the FPA detector) that collect spatially resolved spectral information from all regions of the sample simultaneously. This means that when acquiring the image, each detector

gathers data during every scan, and therefore every point is measured at the same time. The result is that for a 64×64 detector 4096 spectra are gathered with each scan. The FPA detector can be used in conjunction with a standard interferometer and source. As a 64×64 system records 4096 spectra with each scan, a large amount of data is produced; therefore, in early systems to reduce the computationally intensive nature of this process, step-scan mode was implemented, where the moving mirror changed position in discrete steps. This allowed the computer time to process the data. With modern, more powerful systems and better detectors, continuous-scan mode is used, in which the mirror moves smoothly and continuously. This gathers data much more rapidly.

10.2.3 Introduction to Spectroscopic Imaging

For this explanation, example data acquired with a 64×64 detector will be used. The subject chosen for this image is a model pharmaceutical tablet consisting of caffeine and hydroxypropyl methyl cellulose (HPMC), a cellulose-based polymer. A schematic diagram of the imaging analysis is represented in Figure 10.1.

Each pixel in the grid collects a full infrared spectrum, with a spectral range of 4000–800 cm^{-1}. The concentration of a component is proportional to its absorbance as stated by the Beer–Lambert law. The absorbance of a particular spectral band can be calculated by integrating the area under the spectral band. By plotting the absorbance values of a component spatially, it is possible to generate a relative concentration map. A band must be found for each material, which is well separated from the bands of other materials. For this reason, the band highlighted between 2750 and 3000 cm^{-1} was chosen for water and the band between 1670 and 1730 cm^{-1} was chosen for caffeine (Figure 10.1). By plotting these integral values in a grid matching that of the detector array (taking into account the aspect ratio of the images obtained with macro-ATR using an inverted prism), concentration maps of HPMC and caffeine within the tablet can be constructed as shown in Figure 10.1.

The two images generated in Figure 10.1 represent the same spatial region. One plots the distribution of the absorbance of caffeine, the other plots the distribution of the absorbance of HPMC, and hence they show their corresponding distributions within the imaged area of the sample. Red and pink domains correspond to regions of high concentration, whereas blue domains correspond to regions of low concentration. It can be seen that these images are complementary, that is, where there is a high-concentration region of caffeine there is a relatively lower concentration region of HPMC. There are some regions where low concentration is registered for both; this is most likely a void within the tablet structure or an impurity.

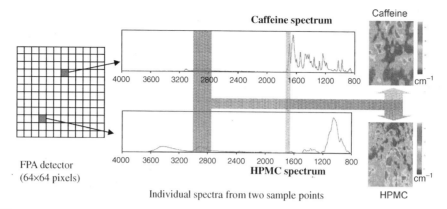

FIGURE 10.1   FPA detector spectra collection. (See the color version of this figure in Color Plate section.)

10.2.4   Imaging Sampling Methodologies

Spectroscopic imaging can be performed in both transmission and ATR modes; however, as explained above, transmission requires significantly more sample preparation. ATR is a surface technique: during imaging the bottom layer of the tablet is measured. Transmission measurements apply through the whole sample: in imaging mode, only an average through the sample is seen [5]. A large particle seen in an image may in fact not be a large particle, but instead consist of two separate particles at different depths that slightly overlap in the path of the beam as shown in Figure 10.2. Nonetheless, transmission imaging has been successfully applied to many materials [15–17].

ATR imaging will not suffer from this "averaging" issue; however, there are some considerations that must be made as the spectroscopic data obtained are sampled from the bottom surface layer of the sample. The probing depth of the image is around several micrometers; however, the individual particles in most powder-based pharmaceutical formulations are much larger than this, which means that a relatively small part of the total volume is imaged and only "the tip of the iceberg" may be seen. This effect is demonstrated in Figure 10.3, where, although the powders had been sieved thoroughly to

a particle diameter of 90 μm, there are some much smaller domains visible. It is probable that each of these domains is actually a small fraction of a much larger particle that lies within the ATR field of view (FOV). Nevertheless, both transmission and ATR mode have uses and an appropriate choice must be made based on the analytical requirements as they differ on many aspects such as spatial resolution, field of view, and the possibility of leaving artifacts [18–20].

The ATR imaging spectrometer is patented by Varian [18]. With modern FPA detectors, an image can be attained in as little as 10 s. Mapping methods that are not strictly imaging are also commonly used. These systems employ a linear array of detectors and a mechanical rastering technique in order to construct a map that will contain the same data as image obtained using an FPA detector [21, 22]. These measurements can achieve a better signal-to-noise ratio than an FPA and can now operate at a speed that rivals that of the older FPA detectors. The use of linear array detectors for the chemical mapping of polymeric materials [23] and pharmaceutical samples [24] has recently been demonstrated.

FIGURE 10.2   Limitations of transmission imaging for thick sample.

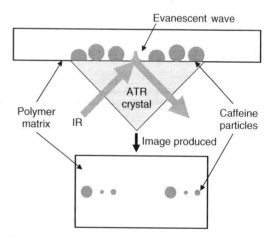

FIGURE 10.3   Effect of probing depth in ATR imaging on imaged particle size.

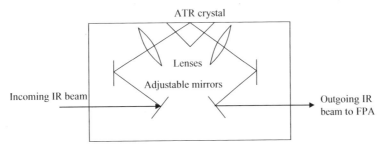

FIGURE 10.4   Internal optics of a diamond Golden Gate™ accessory.

The simultaneous gathering of all the spectra using an FPA detector permits the study of dynamic systems such as the dissolution of pharmaceutical formulations [7, 25]. This principle can then be extended to study many samples at once in high-throughput analysis [26–28]. This technique is also useful for polymers and diffusion processes [5, 29–31].

The most versatile accessory for ATR-FTIR imaging is the Golden Gate™ from Specac; its application for imaging purposes was pioneered in our laboratory. The internal optics for this are shown schematically in Figure 10.4. Shown here working in ATR mode with a small diamond crystal, lenses built into the chamber that houses the ATR accessory focus the beam onto the sample.

As can be seen from the diagram in Figure 10.4, the beam of infrared radiation undergoes total internal reflection at the interface of the ATR crystal with the sample. The angle at which the beam arrives at the interface is about 45°; therefore, as opposed to being imaged with a circular beam, the sample area is imaged by an elliptical beam. As the angle is roughly 45°, the resulting ratio of the height of the image to the width of the image is approximately $1:2^{1/2}$ as shown in Figure 10.5; this aspect ratio must be taken into account when recording and displaying the images. This effect has been minimized by the integration of improved optics into the Golden Gate™ producing images that are closer to square in shape [32].

The spatial resolution of a system is limited by the wavelength of the radiation being used and the numerical aperture (NA). This is described by the Rayleigh criterion,

shown in Equation 10.2, which can be used to calculate the theoretical distance required between two adjacent points in order for them to be just resolved.

$$r = \frac{1.22\lambda}{2\,\text{NA}} \tag{10.2}$$

where r is the distance between two adjacent points that are just resolved (to resolve two points completely the separation distance $2r$ is required), λ is the wavelength of the radiation, and NA is the numerical aperture of the system, defined in Equation 10.3:

$$\text{NA} = n\sin\theta \tag{10.3}$$

where n is the refractive index of the imaging medium between the objective and the sample and θ is the half the angular aperture.

To resolve two points completely from each other, the separation required is a minimum of $2r$. In practice, this cannot be achieved due to practical imperfections such as optical aberrations within the system. A microscope is required to realize the highest spatial resolutions that can be achieved in infrared. In the case of a diamond Golden Gate™ accessory, as shown in Figure 10.4, 13 μm is the demonstrated spatial resolution [33].

As described above, diamond is a particularly useful material for working in ATR mode and has now been fully adapted to work with imaging [5, 33, 34]. The image size for

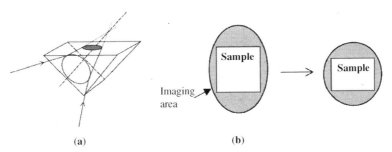

(a)   (b)

FIGURE 10.5   (a) Schematic presentation showing the elliptical imaging area resulting in image stretching. Reprinted with permission from Ref. 33. Copyright 2003 Society of Applied Spectroscopy. (b) Change in imaging aspect ratio. Reprinted with permission from Ref. 33. Copyright 2003 Society of Applied Spectroscopy.

the initial imaging work performed with a diamond was $820\,\mu m \times 1140\,\mu m$, whereas with the advent of the new imaging Golden Gate™ with optimized optics and a modern detector, the imaging area is now $570\,\mu m \times 530\,\mu m$ and thus the aspect ratio has moved from 1:1.4 to 1:1.1.

10.2.5 Imaging FTIR Microspectroscopy

FTIR microspectroscopy employs an imaging or mapping system in conjunction with microscope optics to gather spectra from very localized regions [20]. This has an application in the analysis of heterogeneous samples with small domain sizes, and can be performed in both transmission and ATR modes. In practice, the achievable spatial resolution of microspectroscopic mapping is limited to around 10–15 μm, which can be accounted for by the low throughput of the IR beam with small aperture sizes, though this can be improved by the use of a synchrotron [35–38]. This has allowed the collection of chemical maps of hair showing the medulla region [37, 39]. However, despite the use of a 3 μm aperture, the fingerprint region has a wavelength of 6–11 μm resulting in spectra that would be "contaminated" by the spectral information from the surrounding materials.

As shown by Equation 10.2, the Rayleigh criterion is partially dependent upon the numerical aperture of the system. Therefore, by using an ATR crystal that has a much higher refractive index than air, it is possible to drastically improve the spatial resolution of the system. Chan and Kazarian were able to obtain a resolution of 4 μm using infrared light of wavelength 6 μm [33].

The increased spatial resolution has led to improved detection limits when looking for trace amounts of materials [40]. This is particularly useful when studying pharmaceutical formulations, as heterogeneities are often to be found in the micrometer scale.

10.2.6 ATR-FTIR Expanded Field of View

Attaining the highest spatial resolution is often of significant importance; however, this will come at the expense of the size

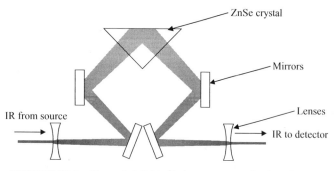

FIGURE 10.6 Expanded field of view accessory for imaging.

of the field of view leading to imaged areas that are rather small (e.g., about $50 \times 50\,\mu m^2$). For studies in which a larger field of view is required, it is possible to use a large ZnSe imaging prism with custom designed expanded optics. The experimental apparatus is analogous to that of a standard Golden Gate™; however, concave lenses are inserted before the imaging optics that expand the beam [41]. The schematic for this system is shown in Figure 10.6.

The area imaged by this accessory is about 15.4 mm × 21.5 mm; this enlarged FOV facilitates the measurement of multiple samples simultaneously, as the samples can be deposited directly on the surface of the crystal within the imaged area. This has been applied to acquire data from more than a hundred static samples in one image, while the dissolution of five samples has been studied simultaneously [41]. A disadvantage of imaging with an expanded FOV is a decrease in the spatial resolution. As the beam is expanded, the projected area for each pixel in the detector increases; the beam expansion also leads to a reduction in the numerical aperture of the system. Thus, expanded field of view imaging cannot be used to obtain data concerning smaller features of the studied samples and so it is only applied to situations in which an enlarged area of interest is necessary.

10.2.7 Quantitative Analysis

ATR is a well-established approach and has long been applied in conjunction with conventional FTIR spectroscopy to obtain quantitative data. ATR-FTIR spectroscopy has been used to determine the diffusion coefficients for liquids in anhydrous lanolin and polyethylene glycol and for urea from polyethylene glycol into medical adhesive [42, 43]. With the introduction of the FPA detector, quantitative analysis can now be performed in imaging mode. This has been demonstrated with pharmaceutical tablets consisting of HPMC as the polymer matrix and niacinamide as a model drug [1]. Concentration profiles were created using the partial least squares (PLS) to quantify the components. From these dissolution profiles, it is possible to obtain a global view of the dissolution and link it to the physical processes occurring in the dissolving tablet for determination of the mechanism of dissolution. FTIR-ATR imaging is also being applied to aid the development of quantitative models for drug dissolution as proposed by Jia and Williams [44]. This is significant since for real systems and those with complex granule structures it is important to have experimental examples as case studies for validation and improving the accuracy of the model, while it is also essential to understand what the effects of digitization may be in terms of introducing errors into the predicted behavior of a system. Imaging can also be applied to quantitatively study the effects of contaminants in formulations [45].

To produce valid quantitative analysis of a system, it is necessary to extract absolute values for the concentrations of the components within the system. This is done using the Beer–Lambert law, shown in Equation 10.4, where A is absorbance, ε is the molar absorption coefficient and $[J]$ is the molar concentration of species J, and l is the sample thickness. This can be written as shown in Equation 10.4.

$$A = \varepsilon[J]l \qquad (10.4)$$

Assuming a constant path length, the absorbance A is proportional to $[J]$, the molar concentration of the component. Therefore, if the molar absorptivity ε is known, or using samples of known concentrations to produce a calibration curve, it is possible to calculate the concentration from the absorbance of a component. In order to do this, it is necessary to know the path length of the radiation in the sample. In transmission, the radiation passes straight through the sample and so the path length is merely the thickness of the sample. In ATR mode, due to the fact that the radiation interacts with the sample via the evanescence wave, instead of the path length, the effective path length is used. For nonpolarized light, this is calculated using Equation 10.5 [10]:

$$\frac{d_e}{\lambda} = \frac{\frac{n_2}{n_1}\cos\theta\left[3\sin^2\theta - 2\left(\frac{n_2}{n_1}\right)^2 + \left(\frac{n_2}{n_1}\right)^2\sin^2\theta\right]}{2\pi\left(1-\left(\frac{n_2}{n_1}\right)^2\right)\left[\left(1+\left(\frac{n_2}{n_1}\right)^2\right)\sin^2\theta - \left(\frac{n_2}{n_1}\right)^2\right]\left(\sin^2\theta - \left(\frac{n_2}{n_1}\right)^2\right)^{0.5}}$$

$$(10.5)$$

where θ is the angle of incidence, n_1 is the refractive index of the crystal, and n_2 is the refractive index of the sample.

When performing quantitative studies it is important to note that even when analyzing a homogeneous sample the absorbance may not always be homogeneous throughout the whole imaged area of the sample. This is because the angle of incidence may not always be uniform across the imaging surface of the crystal [46]. This will have the effect of changing the effective path length and consequently the absorbance. Therefore, it is very important to ensure correct optical alignment of the system for imaging studies. It should also be noted that the mean angle of incidence is heavily dependent upon the alignment of each individual system and so may not match the specification of the manufacturer [47]. More sophisticated methods such as parallel factor analysis (PARAFAC) and multilinear partial least squares (N-PLS) can be used to determine the amount of drug in a formulation as well [48].

10.3 PHARMACEUTICAL INVESTIGATIONS

10.3.1 Polymorphisms

FTIR spectroscopy is well suitable to the study of polymorphism of drugs. The ability to distinguish between polymorphic forms is crucial to the pharmaceutical industry as each polymorphic form can be individually patented; therefore, pharmaceutical companies must find all possible polymorphs of the drugs they have developed otherwise competitors can use the same drug in another crystalline form as in the case of cefdinir [49].

The polymorphic state of the drug can have a large effect on the dissolution properties of a formulation. The amorphous form of a drug typically exhibits a much higher solubility than the crystalline form. Moreover, controlling the crystalline state of the API has significant implications in ensuring the safety and efficacy of the formulation.

Infrared spectra are highly sensitive to polymorphic changes in a compound. Polymorphic transitions will manifest in several forms of spectral change. Upon a transition from an amorphous state to a crystalline structure, the peaks in the spectrum of the compound will become sharper and more defined. A more quantifiable difference comes in the form of a peak shift, for example, amorphous ibuprofen has a carbonyl peak situated at $1730\,\text{cm}^{-1}$, whereas the carbonyl peak in the crystalline form will shift to about $1710\,\text{cm}^{-1}$, because during crystallization hydrogen bonds will form between the drug molecules [7].

This sensitivity has led to the creation of many assays for classifying the crystallinity of a compound [50–54]. For example, sulfamethoxazole has two distinct polymorphic forms and, when studied in diffuse reflectance mode, distinct spectra can be identified that correspond to the different forms. However, in using diffuse reflectance mode there are two parameters for which consistency must be maintained:

1. Production of homogeneous samples for validation and calibration
2. Consistent particle size for all components

Inhomogeneous calibration and validation samples can give incorrect values for IR absorption leading to errors in prediction, while variation in particle size can change the diffuse reflecting properties of a sample. However, an accuracy of 4% is readily achievable when quantifying the crystallinity of a sample.

10.3.2 Supercritical Fluid Investigations

When preparing pharmaceutical formulations there are numerous methods that can be used for embedding the drug within the polymer bulk. Supercritical impregnation has been shown to be practical in preparing samples for which a molecularly dispersed and homogeneous distribution of the drug within the polymer is required. This can be done via particle formation by antisolvent precipitation, aerosolization, and rapid expansion of supercritical fluids [55–57]. Kazarian and Martirosyan [75] have applied ATR-FTIR spectroscopy to study the procedure of supercritical fluid drug impregnation using PVP as the polymer and ibuprofen as a model drug. It was shown that this process is capable of molecularly dispersing the drug in the polymer matrix (the drug is dissolved in polymer matrix at the molecular level). This was revealed by the shift of the carbonyl peak of ibuprofen from 1710 to 1727 cm^{-1} within the PVP, indicating that the intermolecular bonds between drug molecules had been broken. This demonstrated the ability of ATR-FTIR spectroscopy to reveal specific interactions between the C=O groups of the PVP and CO_2. ATR-FTIR spectroscopy using a diamond crystal is particularly suited to this work as it is much stronger than other applicable materials such as zinc selenide, and thus, it can endure the high pressures required to work with supercritical carbon dioxide [58].

10.4   FTIR IMAGING OF PHARMACEUTICALS

Since mid-FTIR imaging was first introduced in 1997, the technique has been applied to study many different systems from all facets of science, from areas such as polymer diffusion and dissolution [29, 59] to the curing rubbers and biological systems [60–62]; it has even found use in forensics [63] and recently in the imaging of live cancer cells [64]. It has been widely applied to the study of pharmaceutical formulations, most commonly to study the controlled release mechanisms of oral dosage formulations [65].

10.4.1   Imaging of Compacted Pharmaceutical Tablets

ATR-FTIR imaging is a valuable tool for studying pharmaceutical tablets, as little or no preparation of the sample is required, and much valuable spectroscopic information can be extracted from imaging studies of these formulations. The lack of sample preparation is a useful property of the ATR methodology as transmission mode requires careful microtoming of the sample into thicknesses of less than 10–20 µm,

while ATR-FTIR imaging allows for the study of many aspects of the tablets. The most important property of imaging is its ability to assess the spatial distribution of different components within the sample [66–68]. By taking several images of a sample, changes in the distribution can be studied; for example, when undergoing compaction, the positions of particles within the tablet rearrange, followed by the particles crumbling, and then the voids within the material collapse resulting in a harder denser material. This process can be studied *in situ* using ATR-FTIR imaging [69, 70].

A custom designed compaction cell has been developed to work in conjunction with a diamond ATR crystal and Golden Gate™ accessory that allows for *in situ* compaction of the pharmaceutical powders into model tablets [68]. The operation of this cell is shown in Figure 10.7.

The brass cell is bolted into place over the diamond plating; the powder mixture is then poured into the hole in the cell, and then the cylindrical punch is placed into the hole on top of the powder before the armature is lowered. A torque screwdriver is then used to wind down a compaction plating from the armature of the Golden Gate™ accessory onto the punch compacting the powder.

Diamond is employed for this approach because when working with compaction only hard ATR crystals can be used, while its feasibility for imaging tablet compaction has already been shown for model tablets consisting of starch and caffeine [19]. This technique has also been used to study the effect that choice of polymer can have on the compaction properties and drug distribution of a formulation using lactose HPMC and Avicel as model excipients and caffeine as a model drug [69]. It was found that the distribution of caffeine is strongly affected by the composition of the polymer matrix used in the tablet. Another important result was also obtained from this work regarding ATR imaging. A general apparent limitation of ATR spectroscopy is that it is a surface technique, so data gathered using this technique would be only valid for the surface of interest and would not

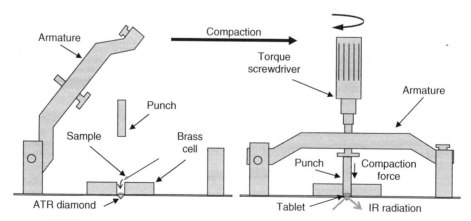

FIGURE 10.7   Schematic presentation of *in situ* imaging of tablet compaction.

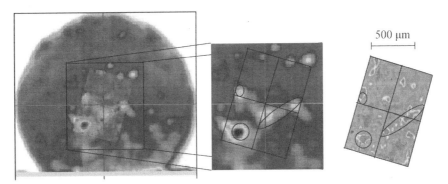

FIGURE 10.8   FTIR spectroscopic images compared with images from X-ray microtomography. Reproduced from Ref. 69 by permission of John Wiley & Sons, Inc. (See the color version of this figure in Color Plate section.)

provide any information from most of the sample. This is true only if there is a significant difference in constituents or the structure of the formulation between the imaged surface and the bulk. Wray et al. [69] used X-ray tomography as a complementary technique with model tablets to compare the ATR-FTIR surface images with the cross-sectional data from the X-ray tomography. They show that there is a similar distribution of drug particles throughout the bulk of the tablet shown in Figure 10.8 and hence that FTIR-ATR imaging is a valid tool for studying pharmaceutical tablets.

10.4.2   Micro-ATR-FTIR Imaging

Macroimaging, such as that with the Golden Gate™, is useful when a large field of view is required. However, sometimes it is necessary to investigate samples on a smaller scale in order to resolve details of smaller features. Micro-ATR imaging can be used to study compacted pharmaceutical tablets; however, it is not possible to perform *in situ* compaction on these samples so they have to be prepared *ex situ*. Micro-ATR imaging will produce an image with a size of approximately $50 \times 50\,\mu m^2$. This has been demonstrated by Chan et al. [19] using model formulations consisting of caffeine, starch, and HPMC as shown in Figure 10.9.

Figure 10.9 shows that the distributions of starch and HPMC are complementary, demonstrating the ability of

micro-ATR-FTIR imaging to spatially separate different chemical domains within a mixture on the micrometer scale. The quality of these images also shows that good contact was made between the micro-ATR crystal and the sample. The large particle seen in the caffeine image is approximately $10\,\mu m$ in diameter; this would not be visible in the macro-ATR imaging demonstrating the effectiveness of being able to image samples on a much smaller scale. The smaller caffeine particles visible have a diameter of the order of $2-3\,\mu m$ only, which is very close to the limits of the spatial resolution of the system; however, they are still clearly visible.

10.4.3   Imaging of Water Sorption in Pharmaceutical Formulations and Human Skin

The manufacture of pharmaceutical solid dosage forms mixes drug compounds that are typically hydrophobic into hydrophilic polymer matrices in order to enhance the dissolution properties of the API. Therefore, during storage and manufacture, these formulations may well absorb water from the atmosphere. This sorption of water can manifest itself in the form of undesirable effects on the dissolution and therapeutic properties of the formulation, hindering the bioavailability of the API. The presence of water can cause the drug to recrystallize, negatively affecting the dissolution

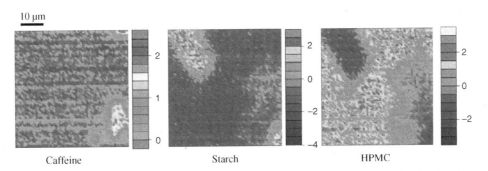

FIGURE 10.9   Micro-ATR-FTIR images showing the distribution of caffeine starch and HPMC in a tablet. Reprinted in part with permission from Ref. 19. Copyright 2003 American Chemical Society.

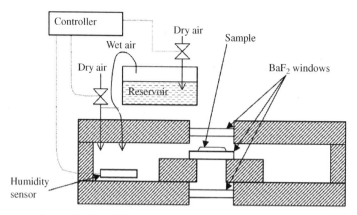

FIGURE 10.10   Schematic diagram of controlled humidity in FTIR transmission imaging mode. Reprinted from Ref. 76. Copyright 2004, with permission from Elsevier.

performance. The sorption of water into the polymer powder can alter its compaction properties, which can then have an effect on the particle morphology within the formed tablet [71]. This has led to a great interest in studying the uptake of water into pharmaceutical formulations [72–74].

The application of conventional FTIR spectroscopy can only give an average value from the sample for the quantity of water absorbed; however, it is unable to display any heterogeneity of water sorption into different domains of the tablet. Previous studies have used solely polymer-based formulations whereas Kazarian and Martirosyan have shown that the composition of the formulation can affect the availability for water sorption [75]. FTIR imaging has been applied to study the heterogeneous distribution of water in pharmaceutical formulations. A controlled humidity cell was combined with FTIR imaging in transmission mode, providing the opportunity of studying water sorption into different domains of the sample *in situ* [76] as shown in Figure 10.10.

The tablets used consisted of polyethylene glycol [49] and griseofulvin. The relative humidity was varied between 0.5% and 90% and the temperature maintained at 25°C. Images were produced to show the distribution of griseofulvin, PEG, and water individually as shown in Figure 10.11.

The work demonstrated that the water preferentially absorbed into the domains of the hydrophilic PEG, rather than the drug. It was also able to reveal that there was a significant increase in the level of water sorption above a relative humidity of 70%, but no effect was seen on the spatial distribution of the components.

As described previously, humidity can affect the compaction properties of tablets. The controlled humidity approach has been applied to study the compaction of pharmaceutical formulations in ATR mode [70]. Tablets consisting of ibuprofen and HPMC were exposed to humidities between 0% and 80% before being compacted into tablets. Results of this work are shown in Figure 10.12.

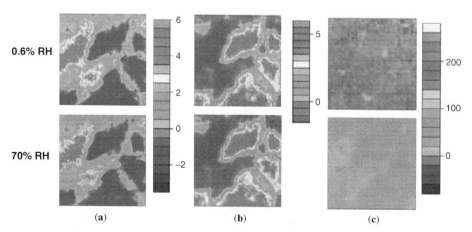

FIGURE 10.11   FTIR images of the PEG-griseofulvin mixture exposed to different relative humidities. The left (a), middle (b), and right (c) columns show the distribution of griseofulvin, PEG, and water, respectively. Reprinted from Ref. 76. Copyright 2004, with permission from Elsevier. (See the color version of this figure in Color Plate section.)

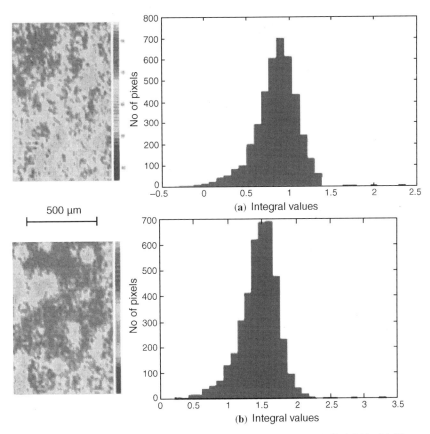

FIGURE 10.12 (a) Image showing FTIR results for compaction of HPMC at 120 MPa and 60% RH with histogram showing the number of pixels at a particular absorbance level. Reproduced from Ref. 70 by permission of John Wiley & Sons, Inc. (b) Image showing FTIR results for compaction of HPMC at 120 MPa and 80% RH with histogram showing the number of pixels at a particular absorbance level. Reproduced from Ref. 70 by permission of John Wiley & Sons, Inc.

These data show that at the same compaction pressure the sample exposed to the higher humidity shows a greater level of infrared absorbance. This is because the density has been increased due to better compaction, showing that the water has had a significant lubricating effect upon the formulation. A higher level of compaction can manifest itself in altered dissolution properties for the formulation. FTIR imaging has demonstrated the ability to reveal information about the properties of compacted tablets. These data have been quantitatively analyzed via the extraction of the absorbance values of the spectral bands, which have then been used to produce histograms showing the range of absorbance values throughout the image.

The combination of macro-ATR-FTIR imaging with a controlled environment accessory allowed analysis of stratum corneum, which is the uppermost layer of the skin, under controlled humidity [77]. The heterogeneous distribution of water in the stratum corneum was analyzed with the aid of a multivariate approach. It has also been shown that ATR-FTIR imaging provides information on the swelling of the stratum corneum as a function of humidity. This approach was also used to image the penetration of liquid ethanol into the skin [77] and showed good potential for studying the transdermal delivery of drugs [78, 79].

10.5 FTIR IMAGING OF TABLET DISSOLUTION

There is an established set of testing procedures for pharmaceutical formulations designed to provide a basis for identification, assay, purity determination, dissolution analysis, and so on; this is published every year in the United States Pharmacopeia (USP) [80]. It is this set of instructions upon which the FDA then carries out enforcement of regulations for pharmaceutical manufacturers.

Unfortunately, despite numerous studies on the dissolution of solid formulations, there is still a lack of understanding of the processes within the formulation (or tablet) upon its contact with dissolution media. The reason for this is conventional dissolution studies, such as USP II, do not investigate the physical processes that may be occurring within the tablet. They are capable only of analyzing the drug concentration in the dissolution medium as a function of time without any insight into the complex processes within the formulation.

Dissolution has a set of USP regulations similar to those of the FDA. The test is a simple repeatable dissolution procedure that makes use of a rotating basket or paddle arrangement inside a chamber filled with the dissolution medium [11].

The formulation is placed inside the basket and the chamber is filled with dissolution medium. The basket is then rotated inside the chamber and the concentration of the drug released into the dissolution medium is measured. The standard dissolution test and consistency of application are required for meaningful comparison; therefore, there are very strict calibration settings [81]. However, this does not provide information about the processes within the tablet that occur during dissolution. It is essential to understand the mechanisms of tablet dissolution in order to produce efficient and reliable tablets.

FTIR imaging has been applied successfully to study many relevant processes, such as pH effects on dissolution [67], polymer behavior under dissolution [14], effects of the initial sample parameters [1, 8], and the occurrence of polymorphic transitions [7, 82].

10.5.1 Transmission Imaging of Dissolution

Koenig and coworkers have used FTIR imaging to analyze the dissolution of drug delivery formulations in transmission mode [83]. Formulations consisting of testosterone as the API and poly(ethylene oxide) (PEO) as the polymer matrix were used. This technique was able to demonstrate the dissolution of the API from the hydrophilic matrix as shown in Figure 10.13.

This work was challenging due to the strong infrared absorption of water in the mid-infrared region. This neces-

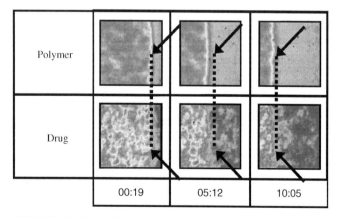

FIGURE 10.13 FTIR imaging of formulation dissolution in transmission. Reprinted from Ref. 83. Copyright 2003, with permission from Elsevier. (See the color version of this figure in Color Plate section.)

sitated the use of deuterated water as well as a very thin spacer, thus restricting investigation to very thin samples (about 10 μm). These samples will not fully represent real tablets as with the application of such thin spacer, the tablets have a very small thickness.

10.5.2 ATR-FTIR Imaging of Dissolution

ATR imaging has the advantages that the path length is independent of the sample thickness and that the depth of penetration of the infrared light into the sample is rather small. Therefore, the dissolution in water-based media of tablets that more closely approximate real-world samples can be studied, as opposed to artificially prepared thin samples.

The possibility of using of macro-ATR-FTIR imaging to examine the dissolution of polymer/drug formulations in contact with water was first demonstrated in a study by Kazarian and Chan [7]. The macro-ATR-FTIR imaging approach, developed in that work, has allowed them to simultaneously study the spatial distribution of both polymer and drug in contact with water as a function of time. The most important finding in that study was that crystallization of the initially molecularly dispersed drug occurred upon contact with the dissolution medium. This was important because crystallization slows overall drug dissolution. These phenomena would not be detected by the conventional dissolution tests and demonstrated that ATR-FTIR imaging can provide important insight into the mechanisms of drug release. That study also demonstrated that the ATR-FTIR imaging approach allows the visualization of the dissolution of inclusion complexes of ibuprofen with cyclodextrines that prevented drug crystallization [7].

Another example of ATR-FTIR imaging of dissolution involved formulations of nifedipine in PEG [82]. Different amounts of crystalline nifedipine were dissolved in molten PEG (MW = 8000) at 70°C to produce samples of 5%, 10%, and 20% drug. The samples were then allowed to cool and solidify before being powdered. The powder was then transferred to the ZnSe ATR crystal that was heated to 60°C to remelt the formulation. The sample was then covered with a glass slide, using a spacer to create a space in which a sample of uniform thickness could form. The sample was then allowed to cool further before the addition of water. FTIR images were then acquired at 5 min intervals. The results are shown in Figure 10.14.

The results show that there is a change in the morphology of the drug as it recrystallizes. This occurs within the polymer matrix upon contact with water, but FTIR imaging shows that this happens in regions in which the water is not yet present. This demonstrates a different trigger for crystallization compared to ibuprofen that exhibits crystallization on direct contact with the dissolution medium [7]. The data also showed that an increase in the drug loading led to much increased crystallization. This work has utilized the great

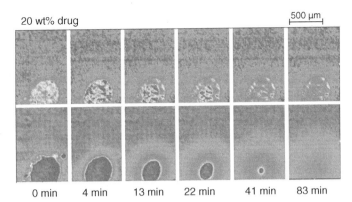

20 wt% drug   500 μm

0 min   4 min   13 min   22 min   41 min   83 min

FIGURE 10.14 Dissolution of nifedipine and PEG. Top row shows drug dissolution and bottom row shows the polymer dissolution. Reprinted in part with permission from Ref. 82. Copyright 2004 American Chemical Society. (See the color version of this figure in Color Plate section.)

potential of FTIR imaging to reveal more information about the dissolution of pharmaceutical formulations [82].

10.5.3 Flow-Through Dissolution Studied with FTIR Imaging

To study flowing dissolution using ATR imaging, a refinement of the simple compaction cell was designed. The cell has been developed such that the tablet can still be compacted *in situ* as shown in Figure 10.15; however, the dissolution medium can then flow through the cell without the need to move the sample. The construction is similar to the standard compaction cell; however, surrounding the punch is retractable metal bolt that is raised after compaction creating a chamber through which the dissolution medium flows as shown in Figure 10.15 [66].

The dissolution cell has a similar experimental setup to that of the compaction cell as shown in Figure 10.15. It is bolted into place over the diamond plating, the only difference being the off-center punch hole and the retractable bolt. There are also flow pipes attached to the side of the block, through which the dissolution medium is pumped.

This equipment is particularly useful for studying the dissolution of model pharmaceutical tablets, as it brings the spatially resolved chemical specificity of FTIR spectroscopic imaging to flow processes. It allows for the study of the ingress of water into the tablet, the formation of polymer gel layers, and ultimately the dissolution of the drug itself. By taking images at regular time intervals, a time-resolved chemical information of the dissolution can be obtained. The dissolution cell is also designed with the punch aligned slightly off center such that the tablet only covers half the face of the diamond. This sets the interface between the tablet and the dissolution medium as the centerline of the image, while also providing space for any potential gelation and expansion or dissolution of the polymer to be observed.

Figure 10.16 reveals the extent of the information that can be obtained using this method. The sample is a model tablet consisting of HPMC and caffeine [66]. The caffeine and HPMC are shown to only cover half the image as expected. It can also be seen that the images are complementary; there are two circular domains of low concentration in the HPMC image, which are matched by two domains of higher concentration in the caffeine image. Water is observed to be filling the empty space on the unoccupied side of the interface; as this is an image taken soon after the water has made contact with the tablet it has not started to ingress into the bulk.

Ibuprofen will tend to crystallize in the presence of acidic media. FTIR imaging can not only determine the presence and distribution of the ibuprofen but will also reveal the crystalline state of the drug. This can be determined through examination of the position of the carbonyl band. This is shown in Figure 10.17a. The peak at 1732 cm^{-1} is from ibuprofen dissolved in PEG, while the peak at 1706 cm^{-1} is crystalline ibuprofen.

Figure 10.17b shows an image taken from the dissolution of an ibuprofen and PEG tablet in acidic media; the image shows the location of the drug within the polymer matrix. The spectrum is extracted from the location indicated by the arrow in which the strong band between 1150 and 970 cm^{-1} is caused by the presence of PEG. The carbonyl peak appears at 1705 cm^{-1}; from this it can be ascertained that the ibuprofen in this system is indeed in the crystalline form. This

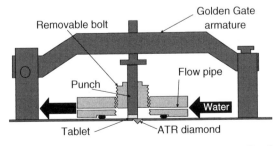

FIGURE 10.15 Schematic representation of the flow dissolution cell combined with ATR accessory.

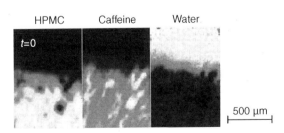

FIGURE 10.16 Images of dissolution of caffeine tablet. Reprinted from Ref. 66. Copyright 2004, with permission from Elsevier.

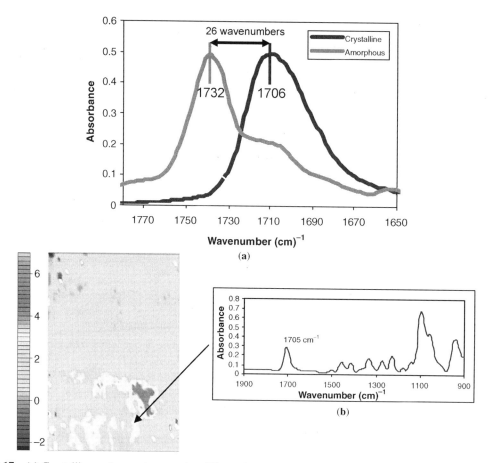

FIGURE 10.17   (a) Crystalline and amorphous peaks of ibuprofen and (b) image of PEG and ibuprofen with extracted spectrum.

demonstrates the power of constructing the images from spectral data.

The *in situ* compaction and dissolution cell uses the diamond crystal meaning that the images produced are 1 mm² or smaller. These images were obtained with a high spatial resolution (about 15 μm), which is useful for studying crystallization and small changes in the structure of the tablet. This FOV only facilitates the study of a relatively small area of small tablets; however, it is often necessary to have a larger field of view [7, 19] to study larger areas of tablets, and the processes of dissolution that occur a larger distance from the original boundary of the tablet. For this a larger crystal must be used. ZnSe is suitable for this purpose; however; it is not as hard as diamond, so compaction cannot be performed *in situ*. Therefore, the tablets must be compacted *ex situ* and then dissolution studied *in situ*. There is the possibility of leakage if care is not taken. In the diamond dissolution cell, as the formulation is compacted onto the diamond, leakage is very unlikely [66]; the swelling of polymers such as HPMC further helps to prevent this ingress [14]. For the ZnSe crystal, the sample can be formed *in situ* if the polymer has a low melting point, and this has been used to study the dissolution of PEG-based

formulations [68]. A schematic diagram of the ZnSe dissolution cell can be seen in Figure 10.18.

One of the most recent examples of applications of ATR-FTIR imaging to dissolution and drug release is the simultaneous FTIR imaging and visible optical photography of an HPMC-based tablet [9]. A custom designed cell was built [9] based on data from previous studies, which showed that a possible leakage of water into the interfacial region between tablet and ATR crystal is not an issue for HPMC-based tablets because of the swelling of the HPMC tablet during water ingress [14]. The new cell, which attached to a diamond ATR accessory, has a visibly transparent window on the top surface. The tablet is compacted *ex situ* and then placed between a diamond crystal and the window. The cell has pipes built into the sides, which allow the dissolution medium to flow through the chamber inside the cell. Thus, visible images were acquired using a CCD camera from the top surface of the tablet simultaneously with ATR-FTIR images measured from the bottom surface during the dissolution of the tablet. This combined approach allowed the study of the moving fronts observed during dissolution. The assignment of the fronts had been a contentious issue as different explanations for the fronts were provided.

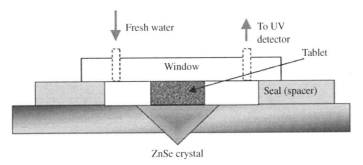

FIGURE 10.18 Schematic diagram showing the ZnSe tablet dissolution cell. Reprinted from Ref. 8. Copyright 2005, with permission from Elsevier.

Consequently, this new imaging approach was applied to a previously studied system that consisted of a colored drug (buflomedil pyridoxal phosphate) and HPMC [9]. Previous assignment of the dissolution fronts for this tablet based on optical photography was not convincing, because photography does not provide a quantitative value for concentrations of the drug, polymer, and water; effects such as changes in the materials' refractive indices, changes in the scattering properties of the medium due to intake of water, and gel formation can all affect the interpretation of the visible imaging data. The ATR-FTIR imaging approach provided reliable interpretation of the fronts and compared them with the appearances of the fronts in visible photography. The three fronts observed in the dissolution of the studied tablets were assigned to true water penetration, total gelification of HPMC, and the erosion front [9]. This assignment of the fronts is crucially important for understanding the mechanism of drug release in HPMC-based tablets. This understanding may help in the designing of new and better drug delivery products.

10.6 ATR-FTIR IMAGING OF COUNTERFEIT TABLETS

FTIR imaging in both micro- and macro-ATR modes can be used to distinguish genuine tablets from counterfeit tablets. The spread of fake tablets presents a significant health threat and counterfeiting of pharmaceutical tablets is a serious crime in both developed and developing countries. Thus, there is often an urgent need to identify the composition of suspect tablets and to use this information in finding the source of these fake medicines. The nondestructive and noninvasive nature of ATR-FTIR imaging is particularly beneficial for the analysis of counterfeit tablets. The distribution of different components can be studied in counterfeit antimalarial tablets in a totally nondestructive way without the need to stain, break, or dissolve the tablets. Following the ATR imaging process, tablets can be analyzed by other complementary techniques.

The combination of micro-ATR-FTIR imaging with desorption electrospray ionization linear ion-trap mass spectrometry (DESI MS) has been applied to analyze counterfeit artesunate antimalarial tablets [84]. Malaria is a potentially fatal disease but the highly active artemisinin derivatives, for example artesunate, are effective for successful treatment. Micro-ATR-FTIR imaging determined the type of drug domains in tablets whereas DESI MS enabled high-sensitivity drug detection.

In these studies, the advantage of the imaging approach to detect domains of highly localized drug with high spatial resolution has been demonstrated. The micro-ATR-FTIR imaging approach collects localized information simultaneously from different areas of the sample in a single measurement, which enables the detection of locally concentrated trace materials due to their heterogeneous distribution. On the other hand, a conventional detector obtains an average signal from the whole sampling area in a single measurement and, therefore, the absorption from the trace materials will be much weaker. Therefore, the imaging approach can, in principle, enhance the sensitivity of the detection of trace materials distributed heterogeneously by many times (depending on the size of the sampling area, particle size, number of pixels in the array detector, and the spatial resolution of the image) in a single measurement compared to conventional spectroscopy with a single detector. This is particularly important in forensic science when one needs to localize and identify a small particle (drug, toxic powders, etc.) on the surface of the tablet or of any other specimen. The sensitivity levels of conventional FTIR spectroscopy and FTIR spectroscopic imaging were compared for a specific system; a model tablet made of polymer and drug, and it was found that for this particular set of samples, the lowest concentration of drug in tablet that could be detected with the conventional MCT detector was 0.35 wt%; whereas while using the ATR imaging approach, the presence of the drug has been detected in a sample containing less than 0.075 wt% of drug [40]. This imaging advantage is crucial in applications such as the screening of counterfeit drugs, samples of which can be highly inhomogeneous because of inadequate manufacturing practices, as was demonstrated in the applications of micro-ATR-FTIR imaging.

In the follow up study, the investigation of counterfeit antimalarial tablets with macro-ATR-FTIR imaging was combined with studies using spatially offset Raman spectroscopy (SORS) [85]. SORS, which is also a nondestructive technique, allows Raman spectra of the tablets to be obtained without removal of their packaging material (e.g., blister packs). In the combined study, SORS was used for the identification of the bulk composition of the tablets through the packaging, while macro-ATR-FTIR spectroscopic imaging using a diamond ATR accessory enabled the study of the spatial distribution of the drug and the excipients over the surface of the tablets. The imaging capability of the macro-ATR-FTIR approach permitted the detection of low-concentration components that were near, or below, the threshold for detection of the SORS technique [85]. Overall, these studies have demonstrated that ATR-FTIR spectroscopic imaging techniques have great potential to help in the forensic characterization of counterfeit tablets.

10.7 ATR-FTIR IMAGING FOR HIGH-THROUGHPUT ANALYSIS

FTIR imaging is an inherently high-throughput technique and is well suited to study many samples simultaneously. Recent studies from our laboratory have demonstrated that the FTIR imaging approach with macro-ATR capability is a very powerful novel tool for high-throughput technology of pharmaceutical formulations and can provide guidance for the design of pharmaceutical formulations for controlled release.

Key ideas of developed methodology [28] have been (i) a combination of macro-ATR with an FPA infrared detector for FTIR imaging; (ii) the use of a microdrop system to deposit a microdroplet directly onto the surface of the ATR crystal; (iii) a combination of the macro-ATR imaging accessory with a controlled humidity cell. This approach allowed imaging of more than 100 samples simultaneously under identical conditions. A schematic diagram of the experimental procedure is shown in Figure 10.19a. Dispenser head 1 was loaded with the sample containing the drug, while dispenser head 2 was loaded with pure PEG (polyethylene glycol). Droplets with different compositions of drug (ibuprofen and nifedipine were used as model drugs) and polymer were prepared by dispensing a different number of drops from each dispenser head onto the same location [28]. The ATR-FTIR image of more than 100 PEG formulations is shown in Figure 10.19b.

This methodology was applied to study PEG/formulatiuons under controlled humidty and to measure the amount of sorbed water in each formulation. The effect of elevated temperature on the stability of all formulations was also studied [86]. This high-throughput approach identified the concentration range for stable formulations and provided evidence that hydrogen bonding between ibuprofen and the polymer is responsible for enhanced stability at higher temperatures [86]. The polymorphic transition of nifedipine was also analyzed using this high-throughput imaging methodology [28]. ATR-FTIR imaging has also been successfully applied to study the permeation of model drugs through human skin in a high-throughput manner [87].

This approach enables the measurement of up to 1024 samples. The introduction of this new accessory with an enhanced field of view provided an opportunity to combine ATR-FTIR spectroscopic imaging with a multichannel grid that allowed the simultaneous imaging of the dissolution of several different formulations [41]. The demonstrated approach was also the first example of the application of spectroscopic imaging to microfluidics and may broaden its future use in miniaturized high-throughput devices [88].

10.8 CONCLUSION

This chapter has covered the important issues concerning the application of FTIR spectroscopy and FTIR imaging to pharmaceutical formulations. The application of conventional spectroscopy to pharmaceuticals was introduced first before moving on to spectroscopic imaging. There was a summary of the hardware necessary for this work and the different techniques applied, which discussed the advantages and disadvantages of both transmission and ATR spectroscopy. The application of these methods was then discussed further with a review of relevant case studies.

The applications shown in this chapter have demonstrated the tremendous analytical potential of FTIR imaging. The most important component of FTIR imaging is the FPA detector, as within a system of coherent optics this provides the very important capabilities of being able to collect both spatial and spectral information simultaneously across a system. Also with modern FPA detectors data acquisition time is very short. This opens up many possibilities for imaging spectroscopy, especially the study of dynamic systems such as the dissolution of solid dosage forms.

The applicability of FTIR imaging is augmented by the versatility of the system in ATR mode. This has led to the development of many customized accessories; the application of these novel techniques to the study of pharmaceutical formulations has been talked about in detail in this chapter. The major utilizations of these accessories have been for compaction, dissolution, controlled humidity, high-throughput analysis, and forensic studies. The use of macro- and micro-ATR imaging allows for investigation of samples at a range of spatial resolutions and fields of view. It is important to remember that ATR imaging does only produce an image of the surface layer of a tablet, though work has been done to show that these data are still relevant. Recent developments

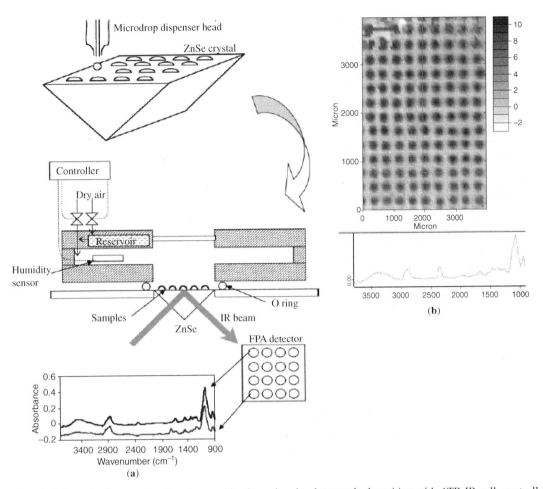

FIGURE 10.19 (a) Schematic diagram showing the combination microdroplet sample deposition with ATR-IR cell, controlled humidity chamber, and infrared array detector, for simultaneous high-throughput analysis of many formulations under controlled environment. (b) The ATR-FTIR image of more than 100 PEG formulations is based on the distribution of the absorbance band of PEG. Reprinted in part with permission from Ref. 28. Copyright 2005 American Chemical Society.

in the area of ATR-FTIR imaging with a variable angle of incidence opens up the opportunity of 3D imaging surface layers of skin in studies of transdermal drug delivery [10] or thin layers of polymeric or pharmaceutical samples [89]. The combination of macro-ATR imaging with mapping [90] offers possibility of obtaining images with large fields of view that may be useful in pharmaceutical analysis. The capabilities of ATR-FTIR imaging are further enhanced by the fact that it is not by necessity a destructive technique, so once a sample has been analyzed *in situ* it can be extracted and reanalyzed elsewhere allowing complementary techniques to be applied to further improve the investigation of a sample. Even when working with dissolution, which is a destructive process, other analysis techniques can be used in conjunction with it, such as UV/Vis or visible optical analysis.

The work in this chapter has demonstrated that FTIR imaging is an invaluable analytical tool for the study of pharmaceutical formulations. FTIR imaging has clear ad-

vantages over the standard pharmaceutical analysis tools such as the USP dissolution test, which produces a dissolution profile, in that it reveals information about what happens within the tablet.

ATR-FTIR imaging has great potential in studies of the interaction of drugs with live cells, which may be useful in the optimization of chemotherapy approaches in cancer treatment [64]. ATR-FTIR imaging can also be a valuable tool for studying protein crystallization in a high-throughput manner under a range of different conditions [91].

REFERENCES

1. van der Weerd, J. and Kazarian, S. G. (2004) Combined approach of FTIR imaging and conventional dissolution tests applied to drug release. *J. Control. Release* **98**(2), 295–305.

2. van der Weerd, J. and Kazarian, S. G. (2007) Multivariate movies and their applications in pharmaceutical and polymer

dissolution studies. In: Grahn, H. F. and Geladi, P. (Eds.), *Techniques and Applications of Hyperspectral Image Analysis*, Wiley, Chichester, pp. 221–260.

3. Šašić, S. (2007) An in depth analysis of Raman and near-infrared chemical images of common pharmaceutical tablets. *Appl. Spectrosc.* **61**(3), 239–250.

4. Gendrina, C., Roggo, Y., and Collet, C. (2008) Pharmaceutical applications of vibrational chemical imaging and chemometrics: a review. *J. Pharm. Biomed. Anal.* **48**(1), 533–553.

5. Kazarian, S. G. and Chan, K. L. A. (2006) Sampling approaches in Fourier transform infrared imaging applied to polymers. *Prog. Colloid Polym. Sci.* **132**(1), 1–6.

6. Kazarian, S. G. and Chan, K. L. A. (2006) Applications of ATR-FTIR spectroscopic imaging to biomedical samples. *Biochim. Biophys. Acta: Biomembr.* **1758**(7), 858–867.

7. Kazarian, S. G. and Chan, K. L. A. (2003) "Chemical photography" of drug release. *Macromolecules* **36**(26), 9866–9872.

8. Kazarian, S. G., Kong, K. W. T., Bajomo, M., Weerd, J. V., and Chan, K. L. A. (2005) Spectroscopic imaging applied to drug release. *Food Bioprod. Process.* **83**(C2), 127–135.

9. Kazarian, S. G. and van der Weerd, J. (2008) Simultaneous FTIR spectroscopic imaging and visible photography to monitor tablet dissolution and drug release. *Pharm. Res.* **25**(4), 853–860.

10. Chan, K. L. A. and Kazarian, S. G. (2007) Attenuated total reflection Fourier transform infrared imaging with variable angles of incidence: a three-dimensional profiling of heterogeneous materials. *Appl. Spectrosc.* **61**(1), 48–54.

11. Cox, D., Douglas, C., Furman, W., Kirchoefer, R., Myrick, J., and Wells, C. (1978) Guidelines for dissolution testing. *Pharm. Technol.* **2**(4), 16–53.

12. Wartewig, S. and Neubert, R. H. H. (2005) Pharmaceutical applications of mid-IR and Raman spectroscopy. *Adv. Drug Deliv. Rev.* **57**(8), 1144–1170.

13. Gupper, A., Wilhelm, P., Schmied, M., Kazarian, S. G., Chan, K. L. A., and Reußner, J. (2002) Combined application of imaging methods for the characterization of a polymer blend. *Appl. Spectrosc.* **56**(12), 1515–1523.

14. van der Weerd, J. and Kazarian, S. G. (2004) Validation of macroscopic ATR-FTIR imaging to study dissolution of swelling pharmaceutical tablets. *Appl. Spectrosc.* **58**(12), 1413–1419.

15. Artyushkova, K., Wall, B., Koenig, J., and Fulghum, J. E. (2001) Direct correlation of X-ray photoelectron spectroscopy and Fourier transform infrared spectra and images from poly (vinyl chloride)/poly (methyl methacrylate) polymer blends. *J. Vac. Sci. Technol. A: Vac. Surf. Films* **19**(6), 2791.

16. Koenig, J. L. and Bobiak, J. P. (2007) Raman and infrared imaging of dynamic polymer systems. *Macromol. Mater. Eng.* **292**(7), 801.

17. Miller-Chou, B. A. and Koenig, J. L. (2003) A review of polymer dissolution. *Prog. Polym. Sci.* **28**(8), 1223–1270.

18. Burka, M. E. and Curbelo, R. (2000) Imaging ATR spectrometer. U.S. Patent 6,141,100.

19. Chan, K. L. A., Hammond, S. V., and Kazarian, S. G. (2003) Applications of attenuated total reflection infrared spectroscopic imaging to pharmaceutical formulations. *Anal. Chem.* **75**(9), 2140–2146.

20. Sommer, A. J., Tisinger, L. G., Marcott, C., and Story, G. M. (2001) Attenuated total internal reflection infrared mapping microspectroscopy using an imaging microscope. *Appl. Spectrosc.* **55**(3), 252–256.

21. Patterson, B. M. and Havrilla, G. J. (2006) Attenuated total internal reflection infrared microspectroscopic imaging using a large-radius germanium internal reflection element and a linear array detector. *Appl. Spectrosc.* **60**(11), 1256–1266.

22. Patterson, B. M., Havrilla, G. J., Marcott, C., and Story, G. M. (2007) Infrared microspectroscopic imaging using a large radius germanium internal reflection element and a focal plane array detector. *Appl. Spectrosc.* **61**(11), 1147–1152.

23. Zhou, X., Zhang, P., Jiang, X., and Rao, G. (2009) Influence of maleic anhydride grafted polypropylene on the miscibility of polypropylene/polyamide-6 blends using ATR-FTIR mapping. *Vib. Spectrosc.* **49**(1), 17–21.

24. Pajander, J., Soikkeli, A. -M., Korhonen, O., Forbes, R. T., and Ketolainen, J. (2008) Drug release phenomena within a hydrophobic starch acetate matrix: FTIR mapping of tablets after *in vitro* dissolution testing. *J. Pharm. Sci.* **97**(8), 3367–3378.

25. Koenig, J. (2002) FTIR imaging of polymer dissolution. *Adv. Mater.* **14**(6), 457–460.

26. Snively, C. M., Oskarsdottir, G., and Lauterbach, J. (2001) Parallel analysis of the reaction products from combinatorial catalyst libraries. *Angew. Chem. Int. Ed. Engl.* **40**(16), 3028–3030.

27. Kubanek, P., Busch, O., Thomson, S., Schmidt, H. W., and Schuth, F. (2004) Imaging reflection IR spectroscopy as a tool to achieve higher integration for high-throughput experimentation in catalysis research. *J. Comb. Chem.* **6**(3), 420–425.

28. Chan, K. L. A., and Kazarian, S. G. (2005) Fourier transform infrared imaging for high-throughput analysis of pharmaceutical formulations. *J. Comb. Chem.* **7**(2), 185–189.

29. Snively, C. M. and Koenig, J. L. (1999) Fast FTIR imaging: a new tool for the study of semicrystalline polymer morphology. *J. Polym. Sci. A, Polym. Chem.* **37**(17), 2353–2359.

30. Ribar, T., Bhargava, R., and Koenig, J. L. (2000) FT-IR imaging of polymer dissolution by solvent mixtures. 1. Solvents. *Macromolecules* **33**(23), 8842–8849.

31. Gupper, A., Chan, K. L. A., and Kazarian, S. G. (2004) FT-IR imaging of solvent-induced crystallization in polymers. *Macromolecules* **37**(17), 6498–6503.

32. Poulter, G. and Thomson, G. (2004) *Spectrometer apparatus.* U.K. Patent GB2420877.

33. Chan, K. L. A. and Kazarian, S. G. (2003) New opportunities in micro- and macro-attenuated total reflection infrared spectroscopic imaging: spatial resolution and sampling versatility. *Appl. Spectrosc.* **57**(4), 381–389.

34. Kazarian, S. G. and Chan, K. L. A. (2004) FTIR imaging of polymeric materials under high-pressure carbon dioxide. *Macromolecules* **37**(2), 579–584.

35. Dumas, P., Jamin, N., Teillaud, J. L., Miller, L. M., and Beccard, B. (2004) Imaging capabilities of synchrotron infrared microspectroscopy. *Faraday Discuss.* **126**(1), 289–302.

36. Carr, G. L. (2001) Resolution limits for infrared microspectroscopy explored with synchrotron radiation. *Rev. Sci. Instrum.* **72**(3), 1613–1619.

37. Briki, F., Busson, B., Kreplak, L., Dumas, P., and Doucet, J. (2000) Exploring a biological tissue from atomic to macroscopic scale using synchrotron radiation: example of hair. *Cell. Mol. Biol.* **46**(5), 1005–1016.

38. Bantignies, J. L., Carr, G. L., Lutz, D., Marull, S., Williams, G. P., and Fuchs, G. (2000) Chemical imaging of hair by infrared microspectroscopy using synchrotron radiation. *J. Cosmet. Sci.* **51**(2), 73–90.

39. Dumas, P. and Miller, L. (2003) The use of synchrotron infrared microspectroscopy in biological and biomedical investigations. *Vib. Spectrosc.* **32**(1), 3–21.

40. Chan, K. L. A. and Kazarian, S. G. (2006) Detection of trace materials with Fourier transform infrared spectroscopy using a multi-channel detector. *Analyst* **131**(1), 126–131.

41. Chan, K. L. A. and Kazarian, S. G. (2006) ATR-FTIR spectroscopic imaging with expanded field of view to study formulations and dissolution. *Lab Chip* **6**(7), 864–870.

42. Wurster, D. E., Buraphacheep, V., and Patel, J. M. (1993) The determination of diffusion coefficients in semisolids by Fourier transform infrared (FT-IR) spectroscopy. *Pharm. Res.* **10**(4), 616–620.

43. Farinas, K. C., Doh, L., Venkatraman, S., and Potts, R. O. (1994) Characterization of solute diffusion in a polymer using ATR-FTIR spectroscopy and bulk transport techniques. *Macromolecules* **27**(18), 5220–5222.

44. Jia, X. and Williams, R. A. (2007) A hybrid mesoscale modelling approach to dissolution of granules and tablets. *Chem. Eng. Res. Design* **85**(7), 1027–1038.

45. Roggo, Y., Edmond, A., Chalus, P., and Ulmschneider, M. (2005) Infrared hyperspectral imaging for qualitative analysis of pharmaceutical solid forms. *Anal. Chim. Acta* **535**(1–2), 79–87.

46. Wessel, E., Heinsohn, G., Kuehne, H. S., Wittern, K., Rapp, C., and Siesler, H. W. (2006) Observation of a penetration depth gradient in attenuated total reflection Fourier transform infrared spectroscopic imaging applications. *Appl. Spectrosc.* **60**(12), 1488–1492.

47. Flichy, N. M. B., Kazarian, S. G., Lawrence, C. J., and Briscoe, B. J. (2002) An ATR-IR study of poly (dimethylsiloxane) under high-pressure carbon dioxide: simultaneous measurement of sorption and swelling. *J. Phys. Chem. B* **106**(4), 754–759.

48. Matero, S., Pajander, J., Soikkeli, A. -M., Reinikainen, S.-P., Lahtela-Kakkonen, M., Korhonen, O., Ketolainen, J., and Poso, A. (2007) Predicting the drug concentration in starch acetate matrix tablets from ATR-FTIR spectra using multi-way methods. *Anal. Chim. Acta* **595**(1–2), 190–197.

49. Cabri, W., Ghetti, P., Pozzi, G., and Alpegiani, M. (2007) Polymorphisms and patent, market and legal battles: cefdinir case study. *Org. Process Res. Dev.* **11**(1), 64–72.

50. Bugay, D. E., Newman, A. W., and Findlay, W. P. (1996) Quantitation of cefepime· 2HCl dihydrate in cefepime· 2HCl monohydrate by diffuse reflectance IR and powder X-ray diffraction techniques. *J. Pharm. Biomed. Anal.* **15**(1), 49–61.

51. Kamat, M. S., Osawa, T., Deangelis, R. J., Koyama, Y., and Deluca, P. P. (1988) Estimation of the degree of crystallinity of cefazolin sodium by X-ray and infrared methods. *Pharm. Res.* **5** (7), 426–429.

52. Sarver, R. W., Meulman, P. A., Bowerman, D. K., and Havens, J. L. (1998) Factor analysis of infrared spectra for solid-state forms of delavirdine mesylate. *Int. J. Pharm.* **167**(1–2), 105–120.

53. Agatonovic-Kustrin, S., Tucker, I. G., and Schmierer, D. (1999) Solid state assay of ranitidine HCl as a bulk drug and as active ingredient in tablets using drift spectroscopy with artificial neural networks. *Pharm. Res.* **16**(9), 1477–1482.

54. Hartauer, K. J., Miller, E. S., and Keith Guillory, J. (1992) Diffuse reflectance infrared Fourier transform spectroscopy for the quantitative analysis of mixtures of polymorphs. *Int. J. Pharm.* **85**(1–3), 163–174.

55. Alessi, P., Cortesi, A., Kikic, I., Foster, N. R., Macnaughton, S. J., and Colombo, I. (1996) Particle production of steroid drugs using supercritical fluid processing. *Ind. Eng. Chem. Res.* **35** (12), 4718–4726.

56. Benedetti, L., Bertucco, A., and Pallado, P. (1997) Production of micronic particles of biocompatible polymer using supercritical carbon dioxide. *Biotechnol. Bioeng.* **53**(2), 232–237.

57. Yeo, S. D. O., Lim, G., Debenedetti, P. G., and Bernstein, H. (1993) Formation of microparticulate protein powders using a supercritical fluid antisolvent. *Biotechnol. Bioeng.* **41**(3), 341–346.

58. Kazarian, S. G., Brantley, N. H., and Eckert, C. A. (1999) Applications of vibrational spectroscopy to characterize poly (ethylene terephthalate) processed with supercritical CO_2. *Vib. Spectrosc.* **19**(2), 277–283.

59. Ribar, T., Koenig, J. L., and Bhargava, R. (2001) FTIR imaging of polymer dissolution. 2. Solvent/nonsolvent mixtures. *Macromolecules* **34**(23), 8340–8346.

60. Oh, S. J. and Koenig, J. L. (1998) Phase and curing behavior of polybutadiene/diallyl phthalate blends monitored by FT-IR imaging using focal-plane array detection. *Anal. Chem.* **70** (9), 1768–1772.

61. Camacho, N. P., West, P., Griffith, M. H., Warren, R. F., and Hidaka, C. (2001) FT-IR imaging spectroscopy of genetically modified bovine chondrocytes. *Mater. Sci. Eng. C* **17**(1–2), 3–9.

62. Colley, C. S., Kazarian, S. G., Weinberg, P. D., and Lever, M. J. (2004) Spectroscopic imaging of arteries and atherosclerotic plaques. *Biopolymers* **74**(4), 328–335.

63. Ricci, C., Phiriyavityopas, P., Curum, N., Chan, K. L. A., Jickells, S., and Kazarian, S. G. (2007) Chemical imaging of latent fingerprint residues. *Appl. Spectrosc.* **61**(5), 514–522.

64. Kuimova, M. K., Chan, K. L. A., and Kazarian, S. G. (2009) Chemical imaging of live cancer cells in the natural aqueous environment. *Appl. Spectrosc.* **63**(2), 164–171.

65. Rafferty, D. W. and Koenig, J. L. (2002) FTIR imaging for the characterization of controlled-release drug delivery applications. *J. Control. Release* **83**(1), 29–39.

66. van der Weerd, J., Chan, K. L. A., and Kazarian, S. G. (2004) An innovative design of compaction cell for *in situ* FT-IR imaging of tablet dissolution. *Vib. Spectrosc.* **35**(1–2), 9–13.

67. van der Weerd, J. and Kazarian, S. G. (2005) Release of poorly soluble drugs from HPMC tablets studied by FTIR imaging and flow-through dissolution tests. *J. Pharm. Sci.* **94**(9), 2096–2109.

68. Chan, K. L. A., Elkhider, N., and Kazarian, S. G. (2005) Spectroscopic imaging of compacted pharmaceutical tablets. *Chem. Eng. Res. Design* **83**(11), 1303–1310.

69. Wray, P. S., Chan, K. L. A., Kimber, J., and Kazarian, S. G. (2008) Compaction of pharmaceutical tablets with different polymer matrices studied by FTIR imaging and X-ray microtomography. *J. Pharm. Sci.* **97**(10), 4269–4277.

70. Elkhider, N., Chan, K. L., and Kazarian, S. G. (2007) Effect of moisture and pressure on tablet compaction studied with FTIR spectroscopic imaging. *J. Pharm. Sci.* **96**(2), 351–360.

71. Yoshinari, T., Forbes, R. T., York, P., and Kawashima, Y. (2003) Crystallisation of Amorphous mannitol is retarded using boric acid. *Int. J. Pharm.* **258**(1–2), 109–120.

72. Aso, Y., Yoshioka, S., Zhang, J., and Zografi, G. (2002) Effect of water on the molecular mobility of sucrose and poly(vinylpyrrolidone) in a colyophilized formulation as measured by 13C-NMR relaxation time. *Chem. Pharm. Bull.* **50**(6), 822–826.

73. Crowley, K. J. and Zografi, G. (2002) Water vapor absorption into amorphous hydrophobic drug/poly (vinylpyrrolidone) dispersions. *J. Pharm. Sci.* **91**(10), 2150–2165.

74. De Brabander, C., Vervaet, C., and Remon, J. P. (2003) Development and evaluation of sustained release mini-matrices prepared via hot melt extrusion. *J. Control. Release* **89**(2), 235–247.

75. Kazarian, S. G. and Martirosyan, G. G. (2002) Spectroscopy of polymer/drug formulations processed with supercritical fluids: *in situ* ATR-IR and Raman study of impregnation of ibuprofen into PVP. *Int. J. Pharm.* **232**(1–2), 81–90.

76. Chan, K. L. A. and Kazarian, S. G. (2004) Visualisation of the heterogeneous water sorption in a pharmaceutical formulation under controlled humidity via FT-IR imaging. *Vib. Spectrosc.* **35**(1–2), 45–49.

77. Chan, K. L. A. and Kazarian, S. G. (2007) Chemical imaging of the stratum corneum under controlled humidity with the attenuated total reflection Fourier transform infrared spectroscopy method. *J. Biomed. Opt.* **12**(4), 044010.

78. Andanson, J. M., Hadgraft, J., and Kazarian, S. G. (2009) *In situ* permeation study of drug through the stratum corneum using ATR-FTIR spectroscopic imaging. *J. Biomed. Opt.* **14**, 034011.

79. Boncheva, M., Tay, F. H., and Kazarian, S. G. (2008) Application of attenuated total reflection Fourier transform infrared imaging and tape-stripping to investigate the three-dimensional distribution of exogenous chemicals and the molecular organization in stratum corneum. *J. Biomed. Opt.* **13**(6), 064009.

80. USP (2008) Reference Standards: An Overview, http//www.usp.org/aboutUSP, viewed July 1, 2008.

81. USP (2007) Dissolution Procedure: Mechanical Calibration and Performance Verification Test.

82. Chan, K. L. A. and Kazarian, S. G. (2004) FTIR spectroscopic imaging of dissolution of a solid dispersion of nifedipine in poly(ethylene glycol). *Mol. Pharm.* **1**(4), 331–335.

83. Coutts-Lendon, C. A., Wright, N. A., Mieso, E. V., and Koenig, J. L. (2003) The use of FT-IR imaging as an analytical tool for the characterization of drug delivery systems. *J. Control. Release* **93**(3), 223–248.

84. Ricci, C., Nyadong, L., Fernandez, F. M., Newton, P. N., and Kazarian, S. G. (2007) Combined Fourier transform infrared imaging and desorption electrospray ionization linear ion trap mass spectrometry for the analysis of counterfeit antimalarial tablets. *Anal. Bioanal. Chem.* **387**(2), 551–559.

85. Ricci, C., Eliasson, C., Macleod, N. A., Newton, P. N., Matousek, P., and Kazarian, S. G. (2007) Characterization of genuine and fake artesunate anti-malarial tablets using Fourier transform infrared imaging and spatially offset Raman spectroscopy through blister packs. *Anal. Bioanal. Chem.* **389**(5), 1525–1532.

86. Chan, K. L. A. and Kazarian, S. G. (2006) High-throughput study of poly(ethylene glycol)/ibuprofen formulations under controlled environment using FTIR imaging. *J. Comb. Chem.* **8** (1), 26–31.

87. Andanson, J. M., Chan, K. L. A., and Kazarian, S. G. (2009) High-throughput spectroscopic imaging applied to permeation through the skin. *Appl. Spectrosc.* **63**, 512–517.

88. Kazarian, S. G. (2007) Enhancing high-throughput technology and microfluidics with FTIR spectroscopic imaging. *Anal. Bioanal. Chem.* **388**(3), 529–532.

89. Chan, K. L. A., Tay, F. H., Poulter, G., and Kazarian, S. G. (2008) Chemical imaging with variable angles of incidence using a diamond attenuated total reflection accessory. *Appl. Spectrosc.* **62**(10), 1102–1107.

90. Chan, K. L. A. and Kazarian, S. G. (2008) Attenuated total reflection–Fourier transform infrared imaging of large areas using inverted prism crystals and combining imaging and mapping. *Appl. Spectrosc.* **62**(10), 1095–1101.

91. Chan, K. L. A., Govada, L., Bill, R. M., Chayen, N. E., and Kazarian, S. G. (2009) ATR-FTIR spectroscopic imaging of protein crystallization. *Anal. Chem.* **81**, 3769–3775.

11

NIR IMAGING APPLICATIONS IN THE PHARMACEUTICAL INDUSTRY

AD GERICH

Schering Plough, Oss, The Netherlands

JANIE DUBOIS AND LINDA H. KIDDER

Malvern Instruments, Columbia, MD, USA

11.1 INTRODUCTION

Although spectroscopic microscopy utilizing other modalities was described as far back as the late 1940s [1], near-infrared spectroscopic imaging, also called NIR chemical imaging (NIRCI), was first conceived relatively late, when implemented by several groups in the early 1990s. The history of NIR spectroscopic imaging instrumental development is presented in Chapter 4 and has also been reviewed in Ref. 2. Its application in the pharmaceutical industry gained momentum almost a decade later with the introduction of commercial instrumentation, and it became one of the most exciting new analytical techniques available within the pharmaceutical industry for formulation development and manufacturing troubleshooting of solid dosage forms. Its value to the industry was derived from the ability to deliver spatially resolved chemical information in solid samples, information that was not readily available using previously existing instrumentation. Today, NIR imaging is employed by most (larger) pharmaceutical companies and contracted out for many intermediate-sized ones, and advantages of this relatively new technique are widely exploited and developed. The increase in the number of application areas within the pharmaceutical industry is due in large part to the unique information that can be accessed. However, it is undoubtedly also encouraged by the simplicity of sample preparation (often limited to placing a tablet in the field of view (FOV)) and the ease with which data can be acquired. Despite the

wealth of new information that is accessed and the realization that it is often critical in formulation development and troubleshooting, deployment of this technique has not yet fully achieved its potential, in part due to the multidisciplinary nature of the technique; NIR imaging combines NIR spectroscopy and advanced mathematical processing of both the spectra (chemometrics) and the resulting chemically segmented images (image processing). The challenge to find personnel with the required balance of skills, coupled with the secrecy of developed methods (necessary to preserve the competitive advantage that they infer), has presented the biggest hurdles for early adoption by the industry at large. The most innovative users today are often part of multidisciplinary teams that combine expertise in different areas (e.g., spectroscopy, chemometrics, image analysis, formulation, and process development), and now the literature abounds with applications of NIRCI from which methods can be extracted. Methods can be implemented into macros that can then be deployed in a turnkey fashion for data processing, making it possible for nonexpert users to successfully implement NIRCI for routine applications.

In this chapter, NIR imaging studies have been grouped into the three most successfully implemented application areas: (1) formulation, process, and quality by design (QbD); (2) quality assurance (QA) and troubleshooting; and (3) the investigation of counterfeit products. There are other types of applications for NRICI, but they tend to rely on the same basic principles of method development presented in this chapter. Although NIRCI is used widely throughout the pharmaceutical development process, extending from formulation development to commercial production, the general consensus remains that the technique is still underutilized.

Product and process understanding have been of particular interest within the pharmaceutical industry since the PAT initiative and QbD were launched by the U.S. FDA [3]. The spatial distribution of active ingredients and excipients is considered to be one of the critical quality attributes of solid dosage forms [4]; NIRCI is most often used to analyze the distribution uniformity of the active ingredients in tablets or powder blends and correlate this with product performance. During the development phase of a pharmaceutical formulation or process, NIR imaging can be extremely helpful in assessing the criticality of distribution of ingredients (either active or excipient) with respect to product quality, a concept at the heart of QbD. At the end of the process, QA of finished solid dose formulations is usually based on assays: purity of the API, dissolution properties, and a number of physical tests. These tests are often destructive, so although they provide an indication that *there is a problem*, the sample is destroyed in the process, making further analysis into *the cause* of the problem impossible. This makes it difficult to remedy the situation. Since these properties are very often related to the spatial distribution of components in a solid dosage form, NIR imaging has a great potential for QA analysis and troubleshooting (i.e., investigation of out-of-specification products). Finally, because poor quality or lack of the active ingredient(s) in counterfeit products can heavily compromise safety, counterfeit analysis, identification, and sourcing are of extreme importance within the pharmaceutical industry and for the health authorities. Counterfeit drugs are often difficult to detect visually, as they have been produced to closely mimic the real drug product. NIRCI has proven to be an invaluable tool for identifying these visually (nearly) identical products in the market.

The objective of this chapter is to describe the various fields of applications and methods used for NIRCI through several illustrative examples. As such, it is not meant to be an exhaustive review of the literature, but rather to provide a few key relevant analytical templates that may be used to solve common problems within the pharmaceutical industry. Many references are provided for review of background information on various topics, and we encourage newcomers in the field to consult these reviews for a more thorough description of the concepts.

11.2 METHODS: STRENGTHS AND LIMITATIONS

Considering the breadth of potential objectives for NIRCI, it should come as no surprise that there is no single analytical method that is optimized for all possible applications. Multiple approaches have been described with respect to sample presentation, data acquisition, spectral preprocessing and processing, and image analysis. In fact, methods may sometimes appear contradictory, depending on the types of application and their objectives; the choices made by the investigator should be relevant to the objectives of the application, even as far as selecting whether NIRCI is the right approach at all! Lyon et al. from the U.S. FDA [5] reported some pharmaceutical applications of NIR technology that described the progression from NIR spectroscopy to imaging as a function of the questions at hand. Once NIRCI has been selected as the instrumental tool, a data acquisition and processing scheme that achieves the relevant analytical goal must be developed, and this is the method development phase, the need for which is universal across analytical techniques. Lewis et al. [6] discussed data acquisition and processing approaches for various applications and LaPlant [7] focused on specific aspects of data acquisition and processing that we also feel are important for proper method development. The general consensus is to try a simple approach first, but that one must be prepared to develop a dedicated (and often more complicated) method to answer very specific questions in complex systems. For example, a quick screening of a powder sample to confirm the presence of component agglomerates can be done very rapidly by acquiring a few spectral data points and calculating a peak height; a more precise characterization of the agglomerates may require data to be acquired over a large spectral range, combined with chemometrics and image analysis. Figure 11.1 shows a block diagram of parameters that should be considered in the development of a NIR imaging application.

Sample preparation for NIR imaging ranges from simply positioning the sample in the field of view to cross-sectioning for the purpose of analyzing the core of a tablet or for certain NIR mapping implementations that require a perfectly flat surface (see Chapter 4 for more details). Sample preparation can have a huge impact on the result and can therefore be quite important. Various aspects of this subject and how it relates to other data acquisition parameters are explained below.

It may sound surprising that very few publications on applications of NIRCI describe any optimization of data acquisition parameters. For example, a vast majority of applications reported in the literature make use of an LCTF and focal plane array-based instrument for which wavelength resolution and range can be adjusted, and it is well understood that these can affect the discrimination between compounds of interest. Yet, very little effort has been put into optimizing these parameters beyond using either the full spectrum or ingredient-specific wavelengths. In other quantitative applications, the signal-to-noise ratio can play an important role that can be influenced simply by acquiring more scans or co-adds. The lack of reports of data acquisition optimization probably stems from two fairly simple facts: most implementations of NIR imaging are so fast that the

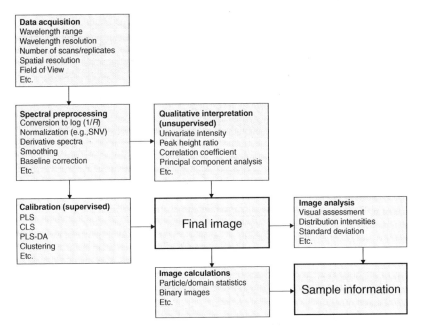

FIGURE 11.1 General routes for NIR imaging applications.

actual gain in time from the selection of an smaller spectral range is not worth the effort, and the signal to noise is so good from single scans with few co-adds that most applications can be solved without the need to increase scan time.

One important consideration in the selection of sample preparation method and spectral range is the relationship between wavelength and depth of penetration in a NIRCI diffuse reflectance measurement. Figure 11.2a shows how the depth of penetration can impact the image produced by a diffuse reflectance measurement. In image a, the diameter of the sphere in the chemical image is smaller than its physical diameter because the radiation penetrates only to depth a. Image b shows a sphere of a diameter equivalent to the actual sphere, but concentric shading represents the mixture spectra that would be measured progressively toward the edge as the beam probes some of the continuous matrix before it hits the sphere itself. In image c, the diameter of the sphere can also be estimated adequately, but all spectra in this circle would be mixture spectra because a volume of the continuous matrix below the sphere is also measured.

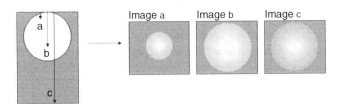

FIGURE 11.2 Schematic representation of the effect of the depth of penetration on the size of the sphere (representing an agglomerate of a single ingredient) measured.

The effect of the depth of penetration can be controlled by selecting the wavelength range. As the depth of penetration increases with the decrease in wavelength, selecting a longer wavelength range effectively reduces the depth and may help detect minor components present close to the surface. The opposite is also possible, where selecting a shorter wavelength may be useful to probe a greater proportion of the sample and thereby "see" agglomeration phenomena that may not be present within a few tens of microns of the surface. Another means of accessing a deeper portion of the tablet is cutting a cross section; this is routinely performed with coated tablets, for which there is no desire to investigate the coating. The tablet can be milled using a bevel-edged blade, cross-sectioned using a sharp straight blade or even sectioned with a microtome.

An interesting aspect of imaging cross sections of tablets was reviewed by Clark and Šašic [8] in Raman mapping experiments. While the technique used is very different in terms of the depth of the measurement (around 2 μm with Raman and around 50–200 μm with NIR), the question raised remains pertinent: How is an image of a cross section able to provide a measurement of the diameter of equal-sized spheres (representing agglomerates of pharmaceutical ingredients) when a microtome will actually not section all of them in the same depth? Figure 11.3 shows a schematic representation of the combined effects of the depth of penetration and two cross-sectional planes on the image acquired from the same sized spheres located at different depths in a continuous matrix.

Two general interpretations can be made from this schematic: (1) a very small depth of penetration will always

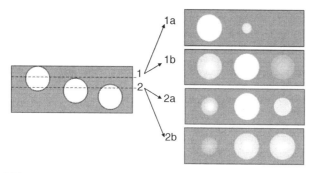

FIGURE 11.3 Schematic representation of the effect of depth of penetration (a represents a short depth of penetration and b represents a long depth) and cross-sectioning plane on the measurement of spheres of identical diameter positioned at different depths in a continuous matrix.

underestimate the size of spheres that have a diameter larger than the depth of penetration if they are not cross-sectioned in the middle of the sphere; (2) a large depth of penetration underestimates the concentration of the sphere ingredients because it measures mixture spectra especially close to the edges (it is, however, better able to estimate the *size* of the sphere). For these identical spheres present at different depths in the sample, size is best estimated using a longer depth of penetration (scenario 1b), but the "purity," or concentration of the ingredient within the sphere, is best estimated in the spheres using a very short depth of penetration (scenario 1a). Overall, the simplified schematic illustrates that there is no "best" data acquisition scheme for all scenarios; the success of a method at answering the question at hand is always directly related to the choices made.

Selecting the FOV or the size of the scanned area is also very important because if the sample size selected is not representative of the distribution of the formulation, no amount of spectral processing is going to give an accurate "picture" of the distribution. For tablets and other samples that can be measured whole, positioning and sometimes fixation of the sample might be the only sample preparation steps. Difficulties arise when parts of powder blends have to be sampled. How does one ensure representative sampling? This is a widespread problem for any analysis from large-scale production and sampling must be considered with the usual statistical considerations in mind. The amount of material sampled is important, sample handling of the thieved powder must not cause segregation within the powder sample, and, finally, a representative area of each sample should be analyzed by imaging. A good example of such a process is described in the section on the analysis of powder blends. It is possible to prepare pressed pellets of a powder sample to stabilize the material, although the high pressure and additional powder handling may compromise the integrity of the sample.

At the sample level, the importance of proper selection of the FOV should never be underestimated. The FOV must be large enough to ensure that the image acquired from a sample is sufficiently representative of the whole sample, even when the sample does not meet expected criteria for distribution of components (i.e., contains very large agglomerates). For example, if it is expected that agglomerations of API will be present in a size range of 10–80 μm in diameter, then imaging a 500 μm FOV could potentially yield enough information to characterize at least a few of these agglomerates. However, if the sample is out of spec, then it may contain agglomerations as large as 300–500 μm, which means a FOV measuring 500 μm simply cannot describe the problem; in reality, the problem may not be seen at all in this small FOV. If the field of view is imposed by instrumentation available and relatively small, then time must be invested in the acquisition of a number of images to be combined into one larger image. While there is no upper limit for the size of the FOV, the largest FOV typically used in the pharmaceutical industry is a few centimeters across for the investigation of tablet identity in a blister pack, where a macro field of view is desired to image the entire blister pack at once. The FOV for global imaging instruments based on refractive optics tends to be more flexible, and large areas are attainable by selecting low magnification optics. Mapping instruments that typically employ mirrored optics may have only one or two magnification options, and large FOVs may only be obtained by moving the x, y stage over larger areas with the standard magnification optics in place.

Related to FOV are the concepts of pixel magnification and spatial resolution. Although they are related, they should not be confused; pixel magnification is determined entirely by the configuration of the optics employed, whereas the spatial resolution is impacted by physics of the type of measurement performed as well as the selection of optics. Most NIRCI pharmaceutical applications reported in the literature are performed in diffuse reflectance. In this mode, the sample is illuminated with NIR radiation and the detector measures the radiation diffusely reflected from the sample. In diffuse reflectance, the diffraction limit at the wavelength of the measurement *and* the depth of penetration of the radiation both affect the actual maximum spatial resolution that can be achieved. These phenomena were explored in detail by Hudak et al. [9], who concluded that the effective spatial resolution in a NIR diffuse reflectance measurement is in the order of 60–90 μm. When the particle size is added to the equation, Clarke et al. [10] reported that the majority of the diffusely reflected radiation may come from the top 50 μm of the sample (with small particles); this led them to conclude that with careful selection of the wavelength range, it is indeed possible to obtain information about the spatial distribution of components similar to that obtained from Raman microscopy, a technique characterized by a much smaller pixel size and consequent longer data acquisition times.

Typical agglomerations (also called particles, aggregates, or domains) of an active ingredient and the vast majority of excipients vary between 5 and 500 μm; it is important to bear the actual spatial resolution in mind when selecting a pixel magnification for a particular experiment to avoid wasting time on "empty" magnification. As the lower limit for spatial resolution has been modeled to be ~50 μm, acquiring data with a magnification of 25 μm per pixel will provide the optimum sampling interval, without empty magnification. In contrast, using a 1 or 10 μm per pixel magnification optic will not provide any additional information regarding the distribution of ingredients. Alternatively, acquiring data at a 100 μm pixel magnification may not provide enough spatial resolution to localize the smaller domains. It is equally important to understand the wavelength-dependant depth of penetration (as reviewed in Figure 11.3). It is possible to take advantage of the repetition of spectral features across different wavelength ranges in the NIR spectrum: choose a longer wavelength range with a shorter depth of penetration to investigate smaller domains, and shorter wavelengths to characterize larger domains.

As real pharmaceutical products are not made of equal-sized spheres, it may be beneficial for the user to control both pixel magnification (through selection of imaging optics) and the depth of penetration (depending on the wavelength of light used to probe the sample).

A circumstance that illustrates this principle is the characterization of magnesium stearate (MgSt). This component is usually present at concentrations around 1%. Used as a lubricant, it does not tend to aggregate, but rather surrounds relatively large particles. The fact that MgSt domains tend to be smaller than the spatial resolution of the technique makes it quite challenging to characterize, as spectra will be thoroughly mixed, with spectral contributions from both MgSt and the particles it surrounds. Fortunately, the NIR spectrum of MgSt contains a strong and sharp band in the long-wavelength range that can be identified even when the component is present in very small amounts. By scanning over (or only selecting) the long-wavelength spectral range, the volume that is being sample can be minimized. Simply put, components present in small amount are easier to locate in smaller sampling volumes, as an agglomerate can represent a more significant fraction of this sampling volume. By applying appropriate data processing, it is possible to pick the distinctive MgSt spectral features from the mixture spectrum and perhaps be able to determine the amount of MgSt measured at each pixel. Interestingly, it is often not necessary to acquire new images to apply the method just described; a small spectral range may simply be extracted from a full spectrum originally acquired. This two-step approach, acquiring the full spectral range but selecting limited wavelength ranges for targeting processing of specific compo-

nents, ensures that most components can be characterized. In conclusion, the spatial resolution, magnification, and field of view should be considered carefully for each application based on the questions to be answered and the knowledge available.

11.2.1 Data Processing

Once spectroscopic data are acquired, the second step concerns the preprocessing of the spectra obtained. A simple rule is to minimize preprocessing as much as possible since preprocessing methods can introduce unwanted and confusing variations in the data set. Reflectance counts are converted into absorbance spectra ($\log(1/R)$) following removal of the detector's dark response and division by an appropriate background response image. Spectral normalization is often applied for pathway corrections of the surface of the sample; normalization such as SNV will often work quite effectively if spectral features of the NIR spectra are more or less alike, which is often the case for NIR spectra because the bulk material contributes strongly to the spectrum. The flip side is that if spectra are relatively different, scaling the absolute responses might compromise the information. At the extreme, normalizing noisy spectra will increase the noise significantly. The effect of normalization should be considered carefully, especially for quantitative data analysis. Beyond normalization, derivative spectra are often used to enhance discrimination of different components. Second derivative spectra are commonly used in NIR spectroscopy because of the easier spectral interpretation; first derivative spectra can be used as well, but interpretation of peak maxima is less intuitive since peak maxima translate to a zero value in the first derivative spectrum. It should be noted that the noise level increases in derivative spectra.

Blend powders and tablets are fairly complex systems, and the key to the successful application of NIRCI has always been a good understanding of what information is desired and what processing of the images is required to achieve it. Most method development strategies involve a progression of data processing, moving from the calculation of a simple single band intensity or integrated area to unsupervised and supervised (calibrated) methods as the sample and the questions increase in complexity.

The simplest and fastest way of getting information out of a NIR image is univariate analysis. When a single wavelength is selected, the contrast in the image is based on the spectral intensity at that wavelength; this may be sufficient to provide the distribution information for components of interest showing distinct spectral features. Visual inspection of the single wavelength image can provide some information and does not need any calibration or complex calculations. For more complex mixtures or components with overlapping spectral features, unsupervised multivariate analysis may be necessary to capture the desired information among the natural variability of the sample and generate chemically relevant image contrast. Principal component analysis (PCA) is widely accepted as a surveying method for chemical imaging data. PCA computes vectors describing the sources of spectral variability and reports scores, or proportions of the importance of these specific sources of spectral variability for each spectrum (pixel) of the image. In samples where the components are segregated into large particles, the PCA loading vectors often resemble the spectra of the pure components, which greatly facilitates interpretation of the variability. When the sample is more homogeneously blended, PCA is much less attractive, as the interpretation of the sources of variability is complicated (Sasic [11] published an interesting discussion on this topic with a comparison of NIR and Raman data) and the scores are generally less strikingly different between pixels. PCA will perform well and be fairly easy to interpret when dealing with a very simple system, consisting of few ingredients that are spatially segregated. More finely blended ingredients produce mixture spectra at more pixels of the tablet that then drive the PCA to look for sources of variability between mixture spectra; the variability then often does not arise from a single ingredient (i.e., a single spectral pattern) and positive and negative bands displaying combined spectral features from multiple ingredients are seen in the loading vectors. When the aim of the analysis is to understand the differences in distribution of ingredients between samples, but not necessarily look to interpret the loading vectors, PCA has proven to be a very powerful tool [12,13]. However, if there is a need to understand the subtle differences between samples that are very similar, have very low content in some ingredients, and consequently display mostly mixture spectra, the use of supervised chemometrics is usually more advisable.

Various supervised chemometrics methods are used to predict the concentration or abundance of ingredients at every pixel of spectral images. Classical least squares (CLS) and partial least squares (PLS) are quite common methods. CLS is a fairly simple method where regression equations are derived from the spectra that link spectral features and concentration. It requires a calibration matrix where the concentration of all ingredients is known and built in the model. It generally performs well with moderately complex systems, but deals relatively poorly with noisy data and very large spectral range selections. PLS is a matrix decomposition method that attempts to describe the concentration of ingredients with as few vectors as possible by taking into account both the spectra and the concentrations. The main advantage of PLS is that one does not necessarily need to know the concentration of all ingredients, although it is important to include all ingredients in the calibration matrix. A PLS model developed to predict a concentration requires a calibration matrix to be built that spans the concentration ranges expected for all ingredients, or at least the main ones. Preparing such a calibration is not an easy task in imaging

because it is practically impossible to prepare a mixture blended finely enough to ensure that all pixels of the image have the same concentration of all ingredients. An average spectrum of a sample image is often calculated to circumvent this difficulty. A more practical approach is the use of PLS discriminant analysis (PLS-DA), or classification PLS, where the model is built with pure ingredient spectra only, each in a separate class, and the model attempts to predict the fit of the pixel spectra in the sample image within the different classes. Some references discussing these different approaches are given in Table 11.1.

As mentioned above, the preparation of a traditional NIR quantitative calibration set should be considered with great care for a NIRCI application. The major drawback of such a calibration set is that it is virtually impossible to prepare homogeneous samples of a specific concentration at the scale of scrutiny used in NIRCI. Simply put, it is almost certain that the powder blend will not be homogeneous at a scale of 40 or even 125 µm per pixel; it is the fundamental reason why this technique is used to assess the quality of blends! Since the preparation of a full calibration set can be quite difficult (or even impossible) and would be very time-consuming, discriminant analysis approaches are more often used in NIR imaging. PLS-DA for example uses a reference library of pure spectra; a model is developed that maximizes the differences between classes (pure ingredients) and can be used to predict the class assignment of pure or mixture spectra. The spectrum acquired at each pixel in an image is then assigned a score for the different components in the library. This approach is probably the most reported supervised analysis in the pharmaceutical NIRCI literature because of the availability of the pure components (both actives and excipients) and simplicity of acquiring their spectral signature. The PLS-DA scores can be related to the abundance (and by extension the concentration) for each library component. The accuracy of the abundance prediction depends on the spectral signatures of the components and the level of optimization of the model to differentiate them. For example, bulk ingredients abundance may be predicted easily with a crude model using a broad wavelength range, but this model may perform poorly for minor components,

TABLE 11.1   Methods Used for Analyzing Hyperspectral Data Cubes in Pharmaceutical Applications

Approach	References
Single wavelength	6, 11, 14–18
Peak–height ratio	16, 18
Correlation coefficient	16
PCA	6, 11–13, 18, 19, 23
CLS	17, 20
PLS-DA (pure spectra)	6, 14, 19, 21–25
PLS (calibration set)	16, 20

Source: C. Ravn et al. (2008) *J. Pharm. Biomed. Anal.* 48.

especially those that have spectral signatures with fewer features or low overall absorbance. Such components may require a separate calibration, which emphasizes their discrimination from the bulk ingredients (as a group) as opposed to the discrimination of the individual bulk ingredients.

It is always useful to explore the impact of the selection of preprocessing and chemometric tools [20, 23]. For example, Gendrin et al. [20] compared the impact of calculating PLS and CLS predictions using normalized absorbance spectra and second derivative spectra. As expected, different processing schemes favor the prediction of different components depending on their abundance, the intensity of the spectral features, the signal-to-noise ratio of the data, and the number of components included in the system. This study emphasizes the need to understand both the system (sample) under investigation and the advantages and disadvantages of the chemometric tools available.

At this point in the analysis, the abundance distribution is often investigated in the form of a histogram plot of the distribution; the x axis of the histogram plot can be a concentration prediction, a peak intensity, a score from a PCA or PLS, a correlation coefficient, and so on. This type of statistical representation of the data emphasizes the presence of low- and high-concentration areas (often called holes and hot spots, respectively). Lewis et al. describe an example of the use of the histogram plot in a blending study in Ref. 26. Figure 11.4 shows histogram plots for a component in three different samples and can serve to understand the correspondence between the histogram and the components distribution. In the first plot, a fairly narrow distribution is observed with no asymmetry in the tails; this is indicative of a homogeneous distribution of the component across the surface of the sample. Such a distribution is characterized by a small standard deviation and skew close to zero. The mean of the distribution is proportional to the abundance (concentration) of that component. The second plot shows significant asymmetry and a broader histogram (higher standard deviation from the mean); the tail to the right of this distribution translates into a positive skew, which is indicative of areas of higher concentration in the sample (i.e., hot spots). Finally, the third plot shows asymmetry on the left side (negative skew), which indicates areas of lower concentration for the particular ingredient (holes, or cold spots).

Finally, image analysis is usually performed on the chemically contrasted images to obtain specific information about spatial distribution of components, preferential colocalization of ingredients, coating thickness, and so on. By converting the score images (or peak height or any other contrast mechanism selected) into a binary image, it is possible to obtain information about the spatial distribution of different components. This information can be extracted directly by visual inspection, but quantification of particle or agglomerate sizes is obtained by measuring the area (i.e., number of pixels) of the continuous regions isolated in the binary image. Binary images, in which pixels are on or off depending on their content of the component of interest, can be used for statistical analysis: mean and variance in diameter, number of particles (domains), nearest neighbor, and other tools to quantify (non)uniformity in the images. This part of the processing will be addressed in some specific applications.

11.3   APPLICATIONS

11.3.1   Formulation, Process, and Quality by Design

11.3.1.1   Blend Uniformity of Powder Samples   Blend uniformity is one of the major pharmaceutical areas in which NIR imaging is applied, in large part due to the push for better process understanding in QbD. Although sampling and handling of powder samples introduces a new challenge compared to tablets or other finished products, analysis of powder blend uniformity by NIR imaging has proven to be quite successful. A uniform blend with respect to active ingredients and excipients is usually desirable to assure a consistent quality of the final product. NIR imaging can be applied to measure blend uniformity from early process development at laboratory scale to full process control in a production environment. In the process development, the choice of a blender type, the influence of mixing time, premixing, humidity, and particle sizes of ingredients are investigated extensively and NIRCI can provide valuable information about the effects of these parameters on blend uniformity and finished products in a relative short time.

One excellent description of blend analysis in a small-scale model system was described in great detail by Ma and Anderson [19], who evaluated the uniformity of the distribution of ingredients in a blend through two types of sampling. First, the top of the blend was imaged in the mini-blender. The content of the mini-blender was then pressed into a compact and the bottom and the cross section of the compact were imaged (Figure 11.5).

In this experiment, the blend was first monitored *in situ* by imaging through a window into the blender, and then compacted to enable an investigation of a cross section of the blender content. Pressing the compacts in the mini-blender should cause very little change to the distribution of actual

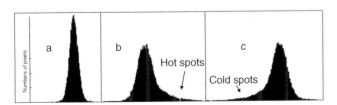

FIGURE 11.4   Histogram plots describing the distribution of a component in a sample.

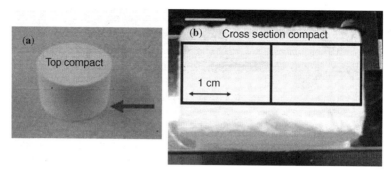

FIGURE 11.5   (a) Compact prepared from blended powder. Images were acquired from the bottom of the compact (indicated by the arrow) and two images were acquired from the cross-sectional areas shown in (b). Reproduced with permission from Ref. 19.

components. Interesting results were obtained that described the progressive mixing of ingredients, but perhaps the most interesting was that the traditional UV analysis of the blend powder to measure API distribution showed an inexplicably large difference between multiple sampling positions at a certain time point. Analysis of the chemical images acquired at the same time point revealed that more API was present near an edge in the top and bottom images of the compact, thereby confirming that there was indeed more API probed by the UV analysis of the measurement made at that edge.

The analysis was continued to evaluate the difference in API concentration at various points in the top, cross section, and bottom of the compacts produced at 10 time points. The bottom showed the largest variability during process, but a relative uniformity at the end of the run (Figure 11.6). Additional statistics were calculated with regard to the size of high-concentration agglomerates that also became more homogeneously distributed as the blend approached the 15 minute end point. A multitude of information about the effect of the order of loading the ingredients, selection of the

blender, blending time, and blend efficiency relative to position within the blender was gathered in this study that can be invaluable in process understanding. A slightly different small-scale blend experiment is shown in Figure 11.7.

In this experiment, nine images (FOV = 13 × 10 mm each, magnification 40 μm pixel size) were concatenated to sample a representative amount of powder, resulting in a sampling area of approximately 4 × 3 cm. A concatenation of smaller images, that is, a mosaic approach, was selected, as opposed to using a lower magnification and thereby a larger FOV because small agglomerations of API are expected; a larger FOV (e.g., magnification 125 μm per pixel) would have enabled the acquisition of this image at once, but each pixel would have probed a larger volume and have revealed only concentration variables arising from the larger agglomerations. In this case, there was an interest in the smaller ones, so a smaller pixel size approximating the maximum spatial resolution of a diffuse reflectance measurement was more appropriate. The pure spectrum of the active ingredient was used to calculate a correlation coefficient image; this simple mathematical processing was selected because the active ingredient had very distinct spectral features (see Figure 11.7b). Once the image contained the desired chemical contrast (in this case the relative abundance of API), image analysis was performed to isolate particles (or domains) of higher abundance. They correspond to the white areas in the binary images. In the binary images, individual pixels are classified as the active ingredient (white) or as bulk material. The threshold for classifying pixels as active ingredient or not was determined based on the contrast of the correlation coefficient. The effect of the blender type for this mixture can be assessed easily by visual inspection. Agglomerates of the active ingredient can be seen in the Turbula-type blender process, while nearly no agglomerates are present in the blend of the Gral-type blender. Quantitative assessment of the detected agglomerates was performed by using a particle statistics function employing binary images in the software ISys 4.0 (Malvern Instruments, UK).

Another small-scale experiment was reported by Li et al. [25], who studied blend uniformity with the objective

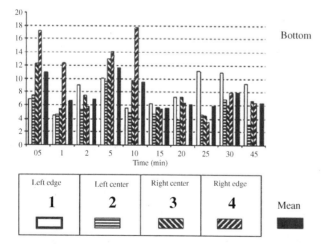

FIGURE 11.6   Predicted concentration of active ingredient (acetaminophen) from different sections of the bottom of the compact. Reproduced with permission from Ref. 19.

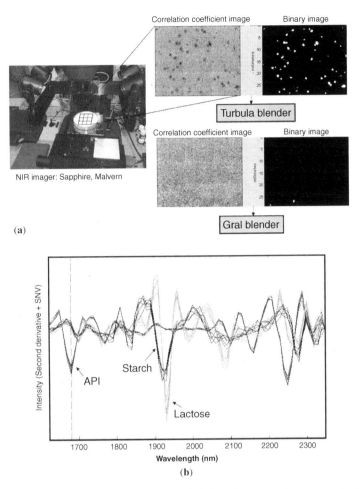

FIGURE 11.7   (a) Correlation coefficient and binary NIR images of API agglomerates in two powder blends processed with different blenders. (b) NIR Imaging spectra of the API and two excipients corresponding to the blend composition as shown in (a).

of measuring the effect of API particle size (i.e., sieve fraction) on blending behavior as a function of time. Small volumes of powder blends were prepared in 20 mL scintillation vials using a benchtop rotating mixer (Glas-Col brand, Terre Haute, IN, USA) and measured using an FPA-based chemical imaging system (Malvern Instruments).

Blend samples containing API of particle sizes 60, 80, 100, 200, and 320 mesh were mixed for 20 min. The NIR images were recorded directly from the top of the vials. SNV normalized and first derivative spectra were used for PLS-2 calibration of the three ingredients: API, HPMC, and microcrystalline cellulose (MCC). On the basis of the predicted concentration of each component, binary images were created to calculate particle statistics in the same manner as described above.

It should be noted that the determination of the threshold in concentration images is subjective and should therefore be set with appropriate reasoning. If concentration differences are relatively large, the threshold can often be determined visually because it corresponds to a change of slope in the histogram plot of the distribution. However, such sharp changes in slope are not always seen. In such cases, a number of standard deviations from the mean or another relevant statistical parameter are usually selected to set the threshold. The most important aspect of this step is to always use the same threshold to process the data so that the binary images obtained are comparable. The total percentage of pixels (i.e., area) multiplied by the average concentration measured in these pixels should correlate with the expected concentration of the ingredient; of course, this will deviate if the sample is not representative of the batch. Comparing the measured abundance and known concentration can help to verify the accuracy of the threshold chosen to create binary images and also establish if the sample is representative. As discussed previously, if large agglomerations are seen, there is an increased likelihood that the abundance measured in the imaged area is not representative of the whole.

Typical particle statistics calculated from the imaged area are shown in Table 11.2, which serves to characterize uniformity further than in terms of concentration variation.

In Table 11.2, a clear relationship between API particle size and the size and number of domains detected exists: As

TABLE 11.2 API Particle/Domain Statistics in Blending Simulation After 20 min

API Size (Mesh)	Total Particles/Domains (>0.001 mm²)		
	Number of Domains	Area %	Diameter (mm)
60	55	4.42	0.28
80	116	5.02	0.22
100	139	4.75	0.18
200	243	4.44	0.13
320	630	4.11	0.08
>320	914	4.30	0.06

Source: Reproduced with permission from Ref. 25.

TABLE 11.3 Particle/Domain Statistics in Blending Simulation with 80-Mesh API

Blending Time (min)	Total Particles/Domains (>0.001 mm²)		
	Number of Domains	Area %	Diameter (mm)
0.17	67	4.89	0.28
0.5	114	6.92	0.25
1.0	108	7.50	0.25
5.0	106	6.40	0.24
10	88	4.95	0.23
20	57	3.28	0.23

Source: Reproduced with permission from Ref. 25.

larger particles are used, the blend yields a much smaller number of larger agglomerations. The proportion of the area of the sample containing higher concentration of API decreases slightly as the particles get smaller, but probably not significantly. Binary images of a 100-mesh and 320-mesh API after 20 min of blending are shown in Figure 11.8. The evolution of API domains in the blend process can be monitored by stopping the blend process and recording a NIR image. In Table 11.3, particle/domain statistics are given for the 80-mesh API as a function of blend time.

Assessing blend uniformity using particle statistics is a very attractive approach to investigate and compare different quality parameters during process development.

Concentration (or abundance) predictions should always be analyzed in conjunction with the agglomerate size measurement and the results presented in Table 11.3 are a good example of the value of this additional piece of information. As the blending time increases, there is a significant change in the number of domains, but the mean diameter of the remaining domains is fairly constant. Since the abundance of the API remains unchanged as blending progresses, it is important to look at the concentration (mean score or other parameter used to quantify API) to understand what is happening: if the number of domains decreases, but the concentration (score) increases in these domains, the API

is likely segregating. On the other hand, if the number of domains decreases and the scores decrease as well, it is an indication that the API is gradually more finely blended.

NIR chemical imaging is used to determine blend quality not only in pharmaceutical formulations involving small molecules, but also for biopharmaceuticals. A common approach to stabilize proteins is to dry them, for example, by freeze drying, spray drying, or supercritical fluid (SCF) drying. However, protein degradation can occur during drying and/or storage of the dried products. Therefore, the addition of sugars or other stabilizing excipients is generally required. Exclusion of protein from sugars could lead to aggregation during storage, and phase separation has been reported to be responsible for failures in protein stabilization. A new SCF procedure was examined and NIRCI was used to study the homogeneity of the SCF samples [16]. NIR images were recorded after preparation of small (thin) pressed compacts using a manual tablet press because the available amount of powder sample was limited. A quantitative PLS calibration model was created and compared with the performance of a correlation coefficient-based model. Binary mixtures were investigated, which makes a correlation coefficient very attractive to use since interpretation is quite easy and reliable, although a quantitative conclusion may be less reliable at very low concentrations. Uniformity of the

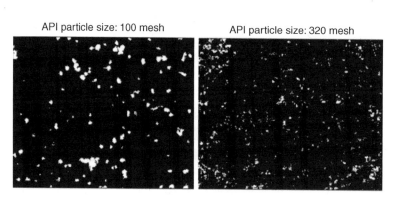

FIGURE 11.8 Binary images of blends prepared from API with a respective particle size of 100 and 320 mesh after 20 min of blending.

SCF samples was assessed by comparing the relative standard deviation (%RSD) of the lysozyme concentration with freeze-dried samples and hand mixed (deliberately nonuniform samples) (Figure 11.9).

11.3.1.2 Blend Uniformity Assessment in Tablets
Uniformity of SCF sample showed comparable or better uniformity than freeze-dried product. No phase separation, agglomeration, was detected due to the SCF process. The model could be used to investigate stability of the formulation as a function of time (and accelerated stress conditions).

Blend uniformity can also be measured in tablets; it is considered a critical quality parameter that can be measured with NIRCI, along with the polymorphic state of API, state of hydration, and content [5]. With respect to product understanding and quality by design, it is now well accepted that formulation developers should know what the desired uniformity of all critical ingredients is and, maybe more im-

portantly, how deviations from this uniformity affect the performance of a tablet. Hilden et al. [13] successfully used NIRCI in a formulation development study and determined the optimal particle size for an excipient to prevent dissolution failure or stability issues in their product. In contrast to blend uniformity of powders, tablets can be analyzed as such, which avoids some of the sample integrity issues related to powders, although representative sampling of tablets (for instance, of a commercial batch) could be difficult as well. To assess blend uniformity in tablets, experimental tablets are often designed to deliberately vary the uniformity of the components by using different mixing times during the blend process [14] or varying API and excipient properties such as particle size. Using this approach, a number of objectives can be studied such as the impact of nonuniformity of specific ingredients, comparison of blend uniformity of commercial products [14, 21], and the impact on uniformity caused by the tabletting process itself (Figure 11.10).

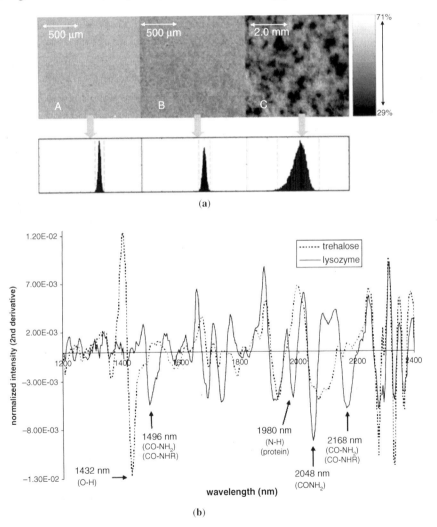

(a)

(b)

FIGURE 11.9 (a) NIR images and histogram plots of the lysozyme concentration distribution of A: supercritical fluid dried sample; B: freeze-dried sample; C: hand-mixed sample. (b) Average NIR imaging spectra of pure trehalose and lysozyme. Reproduced with permission from Ref. 16.

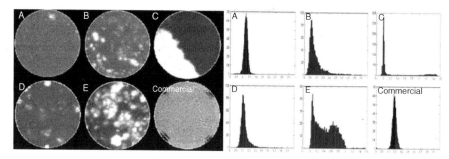

FIGURE 11.10   NIR chemical imaging of the active ingredient distribution in six compositionally identical pharmaceutical model tablets and the corresponding concentration distribution.

The imaging approach should be chosen to achieve as much discriminating power as possible between the tablet ingredients, that is, include both an appropriate spectral range and magnification. If tablet uniformity is compared relatively to a "gold standard" of uniformity, there is often no need to quantify the concentrations of ingredients. PLS-DA or CLS using pure component spectra are often favored for this type of work because the calibration samples need to be only pure ingredients, and the resulting score images are related to the concentration of ingredients at each pixel. The quantification of uniformity can be expressed as %RSD of the predicted individual scores as long as the distribution can be considered to be Gaussian. For severe nonuniformity, the distribution of concentrations/scores is expected to depart significantly from a normal distribution. Statistical values such as skew and kurtosis can be used to assess the distribution with respect to deviation from a normal distribution.

The API is not always the only component investigated; many examples of NIR chemical imaging measurements of distribution of excipients can be found in the literature. The objective is always quite simple: once specific clustering phenomena have been measured, it becomes possible to trace their presence back to certain critical manufacturing variables and possibly correct a problem or modify the process to obtain the desired outcome. A similar approach to that presented above for the API can be followed for studying the uniformity of a specific excipient in tablets. Dissolution properties of an API in a tablet formulation depend strongly on physical distribution of the API and also the excipients. In one particular instance, the distribution of a disintegrant was assessed using NIRCI to investigate the impact of nonuniformity on dissolution behavior [27]. Different blend times were used to incorporate the disintegrant and different compression forces were used to prepare tablets that were then imaged using a Sapphire NIR chemical imaging system (Malvern Instruments) and finally their dissolution profile was measured. Using PLS-DA, a discriminating contrast for the disintegrant was created for a seven-compound tablet formulation after spectral preprocessing consisting in SNV normalization and calculation of the second derivative. In this case, quantification of uniformity was performed using a

measure of regions (domains) of higher abundance of the disintegrant. The domain statistics provide information such as size, shape, and position of each domain and may be combined with the average abundance in each domain for further investigation of heterogeneity.

Results show that the disintegrant is more uniformly distributed after 30 min of blending. PLS-DA score images are shown in Figure 11.11. Dissolution profiles were found to be nearly similar regardless of the distribution of the disintegration agent, which was quite unexpected. It was concluded that the physical properties of the API are probably more critical for dissolution of this formulation and that the distribution of the disintegrant can therefore be considered as noncritical. In quality by design approaches during development, the identification of noncritical attributes is of major importance as well.

In other examples of investigations of excipients with NIRCI, more relationships with process parameters were found to affect distribution of components in the finished product. For example, the compaction force was found to affect dissolution by forming disintegrant clusters that were 40% larger in a particular product; root cause analysis of a processing issue also traced the problem to increased cluster size of a polymer component of the tablets [28].

Quantitative analysis of the clusters in the form of the measurement of their size and number in each tablet provides a statistical means of comparing these samples beyond the pretty picture (procedure described in Ref. 29). This is an example of a product where there probably needs to be manifold differences in cluster size to cause dissolution failure. Other products are more sensitive. In 2005, the FDA Office of Compliance published results indicating a link between API mixing and excipient differences as measured by NIR chemical imaging and dissolution failure in pharmaceutical tablets purchased on the Internet [21]. While there are not really any general rules to predict which formulations are more sensitive to cluster sizes, this measurement is often significantly more relevant to lower dose products because a greater number of larger cluster sizes are indicative of not only possible dissolution problems, but also possible assay failures. Simply put, if the API of a very low dose product is

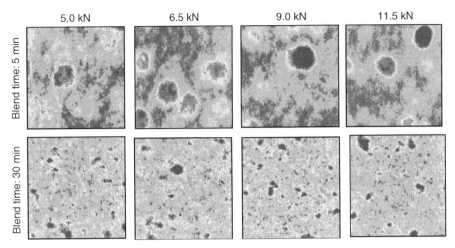

FIGURE 11.11   PLS-DA score contrast images for disintegrant (red = high concentration, blue = low concentration). (See the color version of this figure in Color Plate section.)

present in large clusters, there is an increased chance that some tablets contain a few clusters too many or too little; either case can represent significant deviation from the prescribed dosage. In this case, cluster size measurements using NIR chemical imaging would have been an invaluable tool had it been available during formulation development.

Roggo and Ulmschneider [30] described one such troubleshooting/formulation development application for granules. A PLS classification calculated from images acquired from a broad NIR spectral range allowed the identification and localization of all five components of uncoated granules. The resulting images highlighted an important segregation issue, where the API and the avicel were concentrated in the core of the granules and excipients starch and crospovidone were at the periphery. This information, and the ability to test more samples following trials of process changes, indicated that a premixing step was required to avoid the segregation observed.

11.3.1.3 Process Analytical Technology: Blend Uniformity Monitoring   NIR imaging (Figure 11.12) with all its capabilities for blend uniformity assessment of all critical ingredients is probably one of the most interesting techniques to use as a PAT tool, but its place so far has remained mostly in the quality by design area. Its implementation as a process monitoring tool remains slow, although appropriate measurements are sorted out from the process understanding initiatives. Blend monitoring has been targeted, but final tablets are often discussed. Acquiring imaging at speeds typically encountered in tablet production is a nontrivial exercise, so it is important to understand the real benefits of such endeavors to avoid deploying expensive instrumentation that may in reality be overkill.

The first hurdle to application is that, as stated earlier, representative sampling of production scale blends is tricky. Using a sample thief can already compromise the composi-

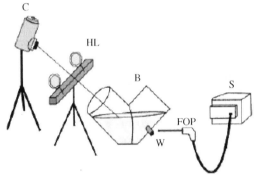

FIGURE 11.12   NIR imaging setup: InSb camera (C), heat lamps (HL), V-blender (B), sapphire window (W), spectrophotometer (S), and fiber-optic probe (FOP). Reproduced with permission from Ref. 15.

tion/uniformity of a sample. Nevertheless, the standard accepted in the industry is currently to perform HPLC on samples that have been removed from the blender at different locations of the blend using a sample thief. This approach is time-consuming (and often operator sensitive) and not considered to be compatible with U.S. FDA process analytical technology (PAT) regulations [31], as it does not provide an understanding of the blending process—it simply gives an assessment of whether or not an end point has been reached with respect to the bulk distribution of the active ingredient.

For direct process analysis (without extracting a sampling from the blend), global imaging systems have been used positioned above or even connected to a blender to measure a small portion of the blend touching a window. Despite the fact that these instrument configurations are far from routinely used, interest is growing and feasibility studies of this type of monitoring have been reported [15, 29]. El-Hagrasy et al. [15] used a liquid nitrogen-cooled InSb imaging camera setup with six discrete bandpass filters encompassing absorption bands of two components of the blend (lactose and salicylic acid) to investigate blend uniformity *in situ*. The main advantage of this type of global imaging approach is the relative large field of view (approximately 15 cm, representing about two-thirds of the top surface of the blend in this case), which is a significant advantage compared to single point NIR blend control (often in the range of 1 cm diameter). The blend process was stopped every 2 min and NIR images were recorded of the top of the blend. The standard deviation of the intensity recorded in one of the six spectral bands was used as a measure of uniformity of the active ingredient. Results showed that the blend was optimally mixed at ~14–16 min, but demixing was clear beyond 20 min. These results were in line with other reference techniques, and in this case the UV and single point NIR spectroscopic blend monitoring. In real life, stopping a (commercial) blend process for process analysis would not be favorable, although this could be acceptable for process development and control of the final blend uniformity (at the mixing time where the blend is expected to be uniform).

The use of an LCTF/FPA NIR imager directly connected to a V-blender via fibre optic (Figure 11.13) was evaluated by Lewis et al. [29]. The basic assumption explored was that, in principle, if high-quality images can be obtained with these complex instrumental configurations, evolution of the blend process could be studied, monitored, or controlled.

The state of uniformity during the blend process was monitored using the histogram plot of the intensity of some significant spectral bands; results are shown for the distribution of the API, as assessed by the intensity at 1632 nm.

The standard deviation values in Figure 11.14 decrease with increased blending time, as expected in a process where the blend is becoming more homogeneous. The narrowing of the histogram distribution indicates that the bulk of pixels are becoming more similar to each other. As explained previously, a positive skew indicates that the tail on the right side of the distribution is longer, which represents a population of pixels with high API content. As the skew decreases with blending time, it means that these high API content domains are mixing with the bulk material and moving toward a mean blending value. The kurtosis is a measure of the "peakedness" of a distribution, or alternatively the size of the tails. A large positive number indicates that the distribution tails are large, implying that a significant population of pixels does not lie close to the mean (indicative of "hot" and "cold" spots in the sample with respect to the component being investigated). In this example, there is no clear trend relating kurtosis and blending time, which implies that although progress of the mixing process can be seen based on the standard deviation and skew values, there are still noticeable domains of API present in the mixture.

The decision on whether imaging measurements in the blender itself will provide a significant benefit over multiple NIR or UV probes has not been made yet, but it is expected that the rapidly increasing body of knowledge gathered with NIRCI in formulation development will help answer this question.

11.3.2   Quality Control

NIR imaging is considered to be particularly useful for quality control of finished products or during different stages of manufacturing. As described earlier, the uniformity of ingredients or the presence of agglomerates of API or excipients can be strongly related to the quality and performance of a finished product. In fact, all applications discussed in this chapter are more or less related to quality control. Existing techniques such as HPLC are widely used to measure the API content, but distribution of ingredients is certainly not applied on a routine basis. Dissolution studies can determine the release characteristics of the API, but like HPLC analysis are destructive and laborious procedures that

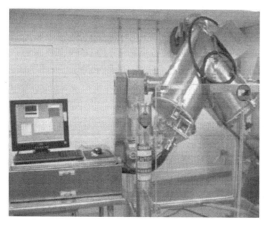

FIGURE 11.13 A prototype NIR imaging blend monitoring instrument mounted to a V-blender. Reproduced with permission from Ref. 29.

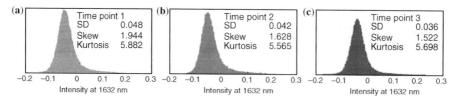

FIGURE 11.14   Distribution of the intensity of 1632 nm corresponding to the API at different blend times. Reproduced with permission from Ref. 29.

cannot provide direct insight into the cause of possible deviations. For more complex formulations, such as multi-layer systems, direct information on the formulation design can be retrieved in a relatively short time by NIR imaging. In this section, a number of application examples are reported that show a wide range of applicability of NIR imaging for quality control.

11.3.2.1 Content Uniformity

Content uniformity of tablets is a common quality parameter describing variance of the API in a product of interest. Although high precision and accuracy make HPLC or UV analysis particularly suitable to obtain reliable API dose results from single tablets, there are generally no equivalent methods to test the excipients, despite the fact that is it now well accepted that excipient uniformity can play a crucial role in the performance of a pharmaceutical formulation. Dissolution properties are of great importance with respect to bioavailability, but stability and other common quality parameters have been shown to be related directly to the uniformity of excipients. The quantitative measurement of the content of API and excipients and their spatial distribution in powder or tablets are referred to as blend uniformity; measurement of only the amounts is referred to as content uniformity.

Since content uniformity is calculated as an intertablet variance, "only" the mean content of single tablets has to be determined. By averaging all spectra/pixels of one tablet, the mean spectrum can be retrieved and used for content predictions [32, 33]. Gendrin et al. performed a feasibility study for two kinds of pharmaceutical samples: simple binary mixtures of API and cellulose and a more complex composition resembling a pharmaceutical formulation containing API, MCC, lactose, and other low-content excipients. PLS-2 (concentration) and CLS calibrations was developed using the mean spectra of each data cube and accurate concentration predictions were obtained. It was shown that optimal spectral preprocessing and calibration settings were dependent on the complexity of the mixture. Both PLS-2 and CLS provided good predictions for the API, while the uniformity of MCC and lactose was better described by PLS-2.

In general, for content uniformity, an average spectrum is calculated from the entire imaged surface and the prediction of concentration made on this unique spectrum. The basic assumption is that the outer layer of the tablet is representative of the whole tablet. This assumption may be confirmed by imaging of multiple subsequent cross sections of a tablet and comparing the individual average spectra. The resulting images can be analyzed individually or reconstructed into a three-dimensional matrix representing the distribution of one or more ingredients (Figure 11.15). As a general rule, if the content (concentration) calculated from the surface corresponds to the expected concentration and aggregates of single components are small, the outer surface is probably a good representation of any layer of the tablet. However, if the concentration calculated deviates from the expected dosage and large aggregates are observed, the outer surface may differ significantly from other layers in the tablet. The good news is that in itself, the measurement of a concentration deviating from the dosage and segregated into discrete domains is an indicator of a problem that probably needs to be addressed! If tablets are pressed manually, there might be an important segregation effect when particle sizes of ingredients differ significantly; recording both sides of a tablet would provide information about possible segregation during the pressing of the tablet. In addition, it should be noted that if the content uniformity of different compounds with large-scale spectral features in different spectral ranges is to be analyzed simultaneously, these wavelength differences may result in different penetration depths within the sample, which could affect the observed compositional information.

An additional advantage of using NIR Imaging for content uniformity analysis is the relative short data acquisition time and the possibility to analyze a large number of tablets simultaneously. A high-throughput content uniformity application was described by Lee and coworkers [32]. In this work, a large FOV (59.5 mm × 47.5 mm) was used, producing relatively large pixel sizes of 186 μm (Matrix NIR, Malvern Instruments) and allowing the simultaneous analysis of 20 tablets. An interesting conclusion of this work is that if the content of the API is relative high and its spectrum

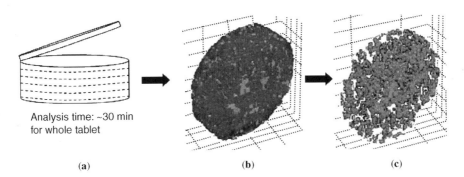

FIGURE 11.15   (a) Schematic of the sample preparation for a three-dimensional analysis of composition, (b) false-color image representing the distribution of API (light) in the excipient matrix (dark), and (c) rendering of the API aggregates without the excipient matrix.

contains characteristic bands, it might be possible to make use of a single or a few wavelengths to obtain the concentration measurement, which reduces data acquisition time from minutes to seconds (for 20 tablets). Compared to traditional HPLC analysis, this can be considered as a huge increase in effectiveness. Figure 11.16 shows an example of this approach, where 20 tablets are positioned in a large FOV and color coded according to the average intensity of a single wavelength at 1600 nm, which in turns corresponds to the concentration of API.

The image provides qualitative information that is visibly accessible; samples colored in red show high intensity at 1600 nm, corresponding to high API concentration. Conversely, samples colored in blue show low intensity at 1600 nm, corresponding to low API concentration. A calibration curve can be developed by plotting the average intensity versus the determined content of a reference meth-

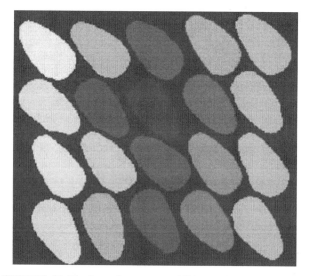

FIGURE 11.16 Intensity map at 1600 nm of 20 tablets with varying content (40–60% API). Reproduced with permission from Ref. 32. (See the color version of this figure in Color Plate section.)

od. The concept of this high-throughput analysis opens clear possibilities for a significant increase in efficiency for content uniformity analysis. Successful implementation strongly depends on the accuracy of the content predictions and the data acquisition time. In low dose products, it should be expected that a single wavelength approach may not provide the accuracy needed and a calibrated model will be necessary to measure content uniformity; data acquisition time would probably go from seconds to a couple of minutes for an array of tablets that translates to <10 s per tablet.

11.3.2.2 Coating Thickness and Multilayer Microspheres
NIR Imaging can be used effectively for the visualization of different layers in complex formulations and determination of coating thickness [4, 33]. A striking example is given in Figure 11.17 in which a multilayer time-release granule is shown. An NIR image of the cross section was recorded and PCA was applied to calculate the different types of variances within the hypercube data. Clearly, three layers can be seen in the granule, without extensive calibration of data processing.

The coating thickness of a single time-release microsphere was investigated by Lewis et al. [33]. Since the magnification of NIR imaging instruments is limited to about 10 μm pixels, relative thin coatings are difficult to quantify in a cross section by NIR imaging. In this case, the microsphere coating thickness was in the order of about 200 μm and the microsphere had been cross-sectioned for data collection (see Figure 11.18). By preparing a binary image based, for instance, on the difference in intensity at 2080 nm, clear differences in intensity exist between the coating and core components. Quantitative calculations can be performed with respect to thickness and regularity of the coating.

If a mixture of two types of microspheres is considered, NIR imaging can be used to identify different types of microspheres based on morphology and/or chemical information. Even using the intensity of a single wavelength can be effective in discriminating between two types of particles.

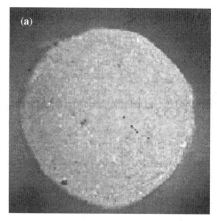
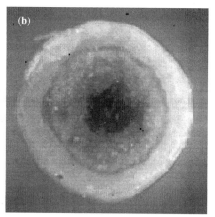

FIGURE 11.17 Visible (a) and NIR principal component images (b) of a pharmaceutical time-release (multi-layer) granule. Reproduced with permission from Ref. 4.

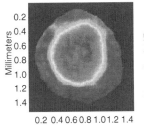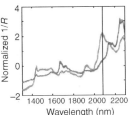

FIGURE 11.18 NIR image of a cross section of a coated microsphere and NIR spectra corresponding to the core and coating of the microsphere. Reproduced with permission from Ref. 33.

If reliable contrast images are obtained, particle statistics can be used for quantification of the microsphere mixture properties such as the ratio between the different types of microspheres and particle size distribution (Figure 11.19).

11.3.2.3 Quality Assessment of Tablets/Capsules

Many of the previously described applications have highlighted the strength of NIRCI for comparing samples; even without any reference spectra of ingredients, the technique can be used to compare chemical composition and structure. This ability was exploited by Westenberger et al. [21] and Veronin and Youan [34], who investigated products commercially available on the Internet. In the first studies, a number of available products on the Internet were examined using traditional and nontraditional methods such as TGA and NIR imaging. Often, products imported through Internet pharmacies are not approved by the FDA, and there are concerns about their safety and effectiveness, which warranted these comparative studies. NIR imaging highlighted differences in formulation between the Internet-bought and the approved U.S. counterpart. Many samples contained different excipients that raised a concern about the shelf life of the product. NIR imaging

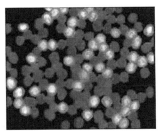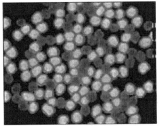

	Red particles	Blue particles	Overall
Number of particles	44	91	135
Mean particle size (mm²)	0.477	0.536	0.517
Std. dev. particle size (mm²)	0.059	0.095	0.089
Mean diameter (mm)	0.779	0.826	0.811
Std. dev. diameter (mm)	0.275	0.348	0.337

FIGURE 11.19 Identification of two types of microspheres based on single wavelength intensity (left: 2050 nm, right 2130 nm). Reproduced with permission from Ref. 33.

was also able to detect manufacturing differences, such as less uniform blending, raising questions about dose uniformity. Most of these formulation and manufacturing issues would not have been evident with traditional testing alone. Veronin and Youan similarly observed very large differences in the structure of tablets purchased over the Internet.

A less common quality control application of NIR imaging is the detection of local contaminations in a solid dose formulation. Local contaminations are often present in relatively small areas of the formulation, and consequently a high spatial resolution is needed for their detection. Raman imaging is considered to be more effective for the investigation of local contaminations in general due to its greater chemical specificity and higher spatial resolution. However, the latter becomes a disadvantage when there is no prior indication of where a small contaminant may be located because a high spatial resolution scan can easily take tens of hours to cover the entire surface of a tablet. It is possible to detect and identify some forms of contaminants, such as degradation products, with NIR imaging when they are localized in distinct areas (domains) of a tablet surface. Detection and even identification are more likely if the impurity covers an area larger than about 30 µm and contains distinct spectral features. PCA analysis or other multivariate analysis can then be used to detect small sources of variance in the hypercube data. In Figure 11.20, an example is shown of the presence of a local degradation product [35, 36].

11.3.2.4 Density of Tablets and Roller Compaction Ribbons

Another application that is of interest for quality assessment of tablets and intermediate products is the examination of density. Ellison et al. [37] investigated the possibility of using NIRCI for tablet density mapping. In pharmaceutical processing, the lubricant Mg stearate can affect compaction efficiency based on blend time and amount of Mg stearate used. Insufficient lubrication produces intra-tablet variations in density. Consistent tablet density profiles and uniform compaction force, as managed by proper lubrication, are important to obtain a predictable performance.

Lactose monohydrate was blended with different amounts of Mg stearate for different mixing times. Compacts were prepared of each blend. NIR chemical images were collected for each tablet, and the density at each image pixel was calculated. Density distribution within compacts was well perceived within the NIR images. Tablets with no Mg stearate or 0.25% Mg stearate were less uniform than tablets with 1.0% lubricant. NIR imaging can be used to nondestructively assess density profiles of tablets and confirm prediction of friction alleviation and improvement in force distribution during tabletting. In this study, the density profiles were both qualitative, showing differences in density profiles between tablets of different blends, and quantitative, providing actual density and tabletting force information within a single tablet (Figure 11.21).

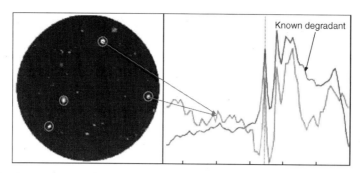

FIGURE 11.20 Tablet showing local contaminations. The NIR spectra of the white areas within the tablet correspond to the NIR spectrum of a known degradant. Reproduced with permission from Ref. 36.

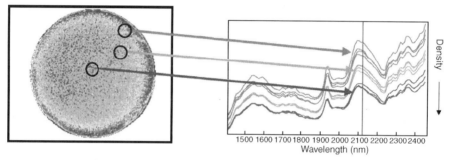

FIGURE 11.21 NIR contrast image of unprocessed spectra (at 2120 nm) of a tablet corresponding to the density within the tablet. Low intensity corresponds to high density. Reproduced with permission from Ref. 37. (See the color version of this figure in Color Plate section.)

A similar investigation of roller compaction ribbons was presented by Lim et al. in 2008 [38], where the team showed differences in the homogeneity of the density across the width of the MCC ribbon as a function of the compaction force. In this particular experiment, the ribbons produced with greater force had a broader gradient of density, as illustrated by the grey scale intensity distribution in Figure 11.22. The density at each point in the images was evaluated by a simple measurement of the baseline shift.

11.3.2.5 Analysis of Tablets in Blister Packaging The possibility to identify and characterize tablets in blister packaging would be of a great potential within the pharmaceutical industry. For quality control of finished products, high-speed cameras could be employed using a large field of

view to include hundreds of tablets within short data acquisition times [39]. The advantage of NIR radiation being able to pass packaging material offers opportunities to identify packaged product without any interference. Malik et al. investigated the feasibility of using NIR imaging not only for identification but also for evaluation of differences in moisture uptake during storage and degradation. An InSb focal plane array video camera with near-IR bandpass cold filter was employed to image a large area (0.5 m) using a tunable filter accessing the spectral range ~1680–2300 nm. With this configuration, the total amount of tablets analyzed simultaneously was ~1300, corresponding to ~16 pixels for each tablet. The measurement was able to identify the active ingredient and follow the moisture uptake in tablets stored in high moisture environment following puncture of the blister. Although the imaging method was not as precise as spectroscopic analysis of single tablets, the method definitely has great potential for possible implementation in the pharmaceutical industry, not in the least because of the enormous gain in efficiency (i.e., 1300 tablets are imaged in a couple of minutes).

Blister packs have also been investigated with NIRCI to detect adulteration (i.e., a "wrong" tablet present in a pack) and empty blisters and in counterfeit analysis (discussed in a separate section). In Figure 11.23, an example is shown of the analysis of tablets in a blister package. One of the tablets was replaced by a tablet containing a different API, which is

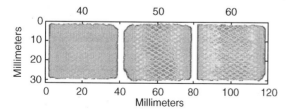

FIGURE 11.22 Roller compaction ribbons showing density variation across the width of the ribbon for compression forces of 40, 50 and 60 kN.

(a) Visible image

(b) NIR PCA score image

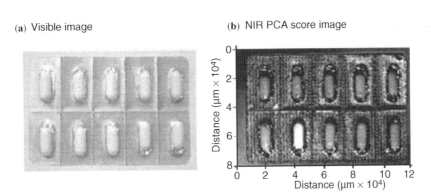

FIGURE 11.23 Visible and NIR PCA score image of a pharmaceutical blister pack. One of the 10 tablets was replaced prior to analysis with a tablet containing a different API. Reproduced with permission from Ref. 15.

clearly seen in the NIR image. The data collection and processing methods described previously for tablets are directly applicable to blister packs; in some cases, single band measurements are sufficient to differentiate products (such as in Figure 11.22), and, in other times, calibrated methods are needed to measure distribution of ingredients.

11.3.2.6 Combining Imaging Techniques

The basic principles of NIR spectroscopy imply that only organic materials display spectral features in this spectral range. When inorganic ingredients are present in a formulation, their domains will appear as black holes in the NIR image because of this absence of a spectrum. Many investigators use the holes as simple indicators of the positioning of inorganic ingredients, but one group (Clarke et al. [40]) pushed the analysis one step further by acquiring both Raman and NIR chemical images of the same area of a tablet surface to investigate the distribution of both organic and inorganic materials in an attempt to understand a "sticky-tablet" problem. The chemical image fusion (CIF) approach required that a microscope slide be calibrated for exact positioning of the sample for both instruments. The data were acquired separately by the two instruments and processed independently to create distribution images of the ingredients. The Raman distribution images of the inorganic binder and API were merged (a Z-concatenation may be applied for this purpose, for example) with the NIR distribution images describing the diluent, disintegrant, and lubricant (Figure 11.24). The compiled CIF images of good and sticky tablets were compared. Unexpectedly, the uniformity of the lubricant (NIR) did not seem to be the main cause for the observed differences although its distribution was not as fine as expected for both batches. The distribution of the inorganic binder in the formulation however seemed to be the main differentiator where, surprisingly, the good batch displayed a less uniform distribution. It was concluded that small particles with an even distribution are not always best, and this example shows the importance of understanding the specifications for various ingredients, an integral part of the QbD framework.

Inorganic binder

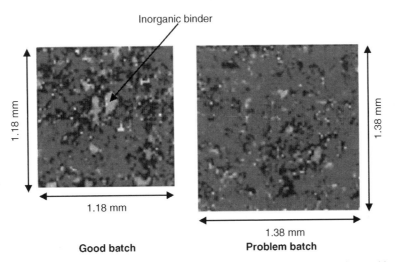

1.18 mm

1.18 mm

1.38 mm

1.38 mm

Good batch

Problem batch

FIGURE 11.24 CIF Images of two Raman and three NIR images of the matrix of a good and a sticking problem blend [40]. (See the color version of this figure in Color Plate section.)

11.3.3 Counterfeit Analysis

Counterfeit pharmaceutical products are a real threat to the health of patients, and the reputation and commercial success of legitimate producers of genuine tablets. Using NIR imaging, multiple samples can be compared simultaneously or detailed compositional information can be obtained and compared from individual samples. The applicability of NIR imaging for detecting counterfeit products was investigated extensively [41–44] and only a few examples are detailed here to explain the main aspects of the problem tackled with NIRCI.

Generally, counterfeits are described as those containing the correct ingredients but having been manipulated in an uncontrolled manner, those containing the wrong active (or inactive) ingredients or any active ingredient at all [41]. In the first instance, a relatively large field of view can be used to screen possible counterfeit products compared to genuine products to compare their chemical makeup (even in the blister pack). In this case, components distribution within a sample is not the main objective, at least initially, but the presence of the right main ingredients is. From a health standpoint, one may be interested in knowing only about the presence of the right active ingredient in the reported concentration. In either case, PCA analysis or peak heights can be used to identify anomalous spectral variations. It is obvious that the spectral features of a possible counterfeit API and the dosage play an important role at high dose, but in relatively low dose products, detecting whether the API is different from the reference can be difficult using a simple PCA, as described previously in the analysis of tablets. However, for many products the lack of any API or major excipients can be detected very easily with this simple approach.

For example, the image on the left-hand side in Figure 11.25 shows that the genuine tablet contains no talc, which has an obvious absorption band in the NIR at 1390 nm. Counterfeit tablet (B3) also does not contain talc, while tablet A2 shows significant absorption representative of talc. On the right-hand side, the wavelength selected is characteristic of

the API that from this image appears not to be present in tablet B3, but present in tablet A2. The histogram characteristics show that the intensity of the API is more widely distributed for tablet A2 than for the genuine tablet, meaning that the API is present but less evenly distributed [42]. When the interest of an evaluation of counterfeit products is restricted to the determination of the presence and concentration of the API, NIR chemical imaging is overkill. Indeed, a single point spectrum acquired from each tablet would provide the same information at lower cost. What drives the use of NIRCI in the analysis of counterfeit products is the additional information provided by the spatial arrangement of the ingredients.

The investigation of counterfeit products using NIRCI generally involves data acquisition and processing very similar to the methods described for quality control, troubleshooting, and formulation development. One example of a more advanced use of NIRCI is an investigation of counterfeit tablets involved in imaging individual tablets and application of a PCA and cluster analysis in an exercise aiming to determine the number of sources for a set of tablets. Counterfeit source identification is useful to determine the extent of the problem by helping understand how many different plants produce counterfeit tablets, regardless of their possible use of common ingredients. Clustering techniques using all spectral information data instead of a single wavelength can be very helpful for classification of unknown products, in order to emphasize the origin of observed variation due to a different API but also due to different excipients. Lopes and Wolff [42] investigated 55 counterfeit Heptodin™ tablets obtained from a market survey and an additional 11 authentic tablets for comparison. PCA and k-means clustering showed both that the counterfeit tablets could be grouped into 13 main groups of which only 18% contained the correct API. Counterfeit tablets also showed differences in excipient composition resulting in a relative large number of clusters. NIR imaging proved to be an excellent tool for counterfeit source analysis in this study.

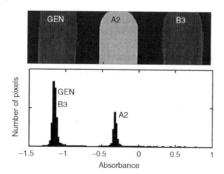
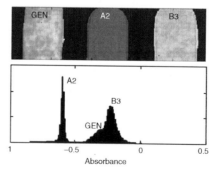

FIGURE 11.25 NIR Images at 1390 nm (left) and 1720 nm (right) and histogram of the image intensities of a genuine Heptodin™ tablet (GEN) and two counterfeit tablets: A2 (containing no API) and B3 (containing API). Reproduced with permission from Ref. 42.

11.4 CONCLUSION

NIR chemical imaging provides invaluable information about pharmaceutical products that is not accessible by traditional techniques. As with most new analytical techniques, a substantial investment in method development has been required. With a variety of experimentally sound approaches to real problems now in the literature, new users can build on early successes and reproduce experiments and apply methods. Once tailored to individual applications, the steps can be documented in reproducible SOPs and macros. Early applications focused on troubleshooting various quality problems, but formulation development (in a QbD approach) and counterfeit analysis are now taking center stage. NIR chemical imaging provides a route to develop better understanding of formulations that should reduce the likelihood of out-of-spec batches. The data produced throughout the formulation development promote a thorough understanding of the product and can therefore be implemented in control of the process, supporting the PAT initiative.

REFERENCES

1. Barer, R., Cole, A. R. H., and Thomson, H. W. (1949). Infrared-microscopy with the reflecting microscope in physics, chemistry and biology. *Nature* **163**(4136), 198–201.

2. Lewis, E. N. and Levin, I. W. (1995). Vibrational spectroscopic microscopy: Raman, near-infrared and mid-infrared imaging techniques. *JMSA* **1**(1), 35–46.

3. www.fda.gov/cder/OPS/PAT.htm.

4. Lewis, E. N., Carroll, J. E., and Clarke, F. M. (2001). A near-infrared view of pharmaceutical formulation analysis. *NIR News* **12**, 16–18.

5. Lyon, R. C., Jefferson, E. H., Ellison, C. D., Buhse, L. F., Spencer, J. A., Nasr, M. M., and Hussain, A. S. (2003). Exploring pharmaceutical applications of near-infrared technology. *Am. Pharm. Rev.* **6**(3), 62–70.

6. Lewis, E. N., Schoppelrei, J., Lee, E., and Kidder, L. H. (2005). Near-infrared chemical imaging as a process analytical tool. In: Bakeev, K. A. (Ed.), *Process Analytical Technology*, Blackwell Publishing Ltd, Oxford, pp. 187–225.

7. LaPlant, F. (2004). Factors affecting NIR chemical images of solid dosage forms. *Am. Pharm. Rev.* **7**(5), 16–24.

8. Clark, D. and Šašic, S. (2006). Chemical images: technical approaches and issues. *Cytometry A* **69A**, 815–824.

9. Hudak, S. J., Haber, K., Sando, G., Kidder, L. H., and Lewis, E. N. (2007). Practical limits of spatial resolution in diffuse reflectance chemical imaging. *NIR News* **18**(6), 4–7.

10. Clarke, F. C., Hammond, S. V., Jee, R. D., and Moffat, A. C. (2002). Determination of the information depth and sample size for the analysis of pharmaceutical materials using reflectance near-infrared microscopy. *Appl. Spectrosc.* **56**(11), 1475–1483.

11. Sasic, S. (2007). An in-depth analysis of Raman and near-infrared chemical images of common pharmaceutical tablets. *Appl. Spectrosc.* **61**(239–250).

12. Clarke, F. (2004). Extracting process-related information from pharmaceutical dosage forms using near infrared microscopy. *Vib. Spectrosc.* **34**, 25–35.

13. Hilden, L. R., Pommier, C. J., Badawy, S. I., and Friedman, E. M. (2008). NIR chemical imaging to guide/support BMS-561389 tablet formulation development. *Int. J. Pharm.* **353**(1-2), 283–290.

14. Lyon, R. C., Lester, D. S., Lewis, E. N., Lee, E., Yu, L. X., Jefferson, E. H., and Hussain, A. S. (2002). Near infrared spectral imaging for quality assurance of pharmaceutical products: analysis of tablets to assess powder blend homogeneity. *AAPS PharmSciTech.* **3**(3), 17.

15. El-Hagrasy, A. S., Morris, H. R., D'Amico, F., Lodder, R. A., and Drennen, J. K. 3rd. (2001). Near-infrared spectroscopy and imaging for the monitoring of powder blend homogeneity. *J. Pharm. Sci.* **90**(9), 1298–1307.

16. Jovanovic, N., Gerich, A., Bouchard, A., and Jiskoot, W. (2006). Near-infrared imaging for studying homogeneity of protein–sugar mixtures. *Pharm. Res.* **23**(9), 2002–2013.

17. Gendrin, C., Roggo, Y., Spiegel, C., and Collet, C. (2008). Monitoring galenical process development by near infrared chemical imaging: one case study. *Eur. J. Pharm. Biopharm.* **68**, 828–837.

18. Roggo, Y., Edmond, A., Chalus, P., and Ulmschneider, M. (2005). Measurement of drug agglomerates in powder blending simulation samples by near infrared chemical imaging. *Anal. Chim. Acta* **535**, 79–87.

19. Ma, H. and Anderson, C. A. (2008). Characterization of pharmaceutical powder blends by NIR chemical imaging. *J. Pharm. Sci.* **97**(8), 3305–3320.

20. Gendrin, C., Roggo, Y., and Collet, C. (2007). Content uniformity of pharmaceutical solid dosage forms by near infrared hyperspectral imaging: a feasibility study. *Talanta* **73**(4), 733–741.

21. Westenberger, B. J., Ellison, C. D., Fussner, A. S., Jenney, S., Kolinski, R. E., Lipe, T. G., Lyon, R. C., Moore, T. W., Revelle, L. K., Smith, A. P., Spencer, J. A., Story, K. D., Toler, D. Y., Wokovich, A. M., and Buhse, L. F. (2005). Quality assessment of internet pharmaceutical products using traditional and non-traditional analytical techniques. *Int. J. Pharm.* **306**(1-2), 56–70.

22. Koehler, F.W., Lee, E., Kidder, L.H., and Lewis, E.N. (2002). Near-infrared spectroscopy: the practical imaging solution. *Spectrosc. Eur.* **14**, 12–19.

23. Burger, J. and Geladi, P. (2007). Spectral pre-treatments of hyperspectral near infrared images: analysis of diffuse reflectance scattering. *J. Near Infrared Spectrosc.* **15**, 29–37.

24. Furukawa, T., Sato, H., Shinzawa, H., Noda, I., and Ochiai S. (2007). Evaluation of homogeneity of binary blends of poly (3-hydroxybutyrate) and poly(L-lactic acid) studied by near infrared chemical imaging (NIRCI). *Anal. Sci.* **23**(7), 871–876.

25. Li, W., Woldu, A., Kelly, R., McCool, J., Bruce, R., Rasmussen, H., Cunningham, J., and Winstead, D. (2008). Measurement of drug agglomerates in powder blending simulation samples by

near infrared chemical imaging. *Int. J. Pharm.* **350**(1-2), 369–373.

26. Lewis, E. N., Kidder, L., and Eunah, L. (2005). NIR chemical imaging as a PAT tool. *Innov. Pharm. Tech.* 107–111.

27. Mellouli, S. and Gerich, A. (2008). Assessment of uniformity of excipients in tablets (internal report: Department of Pharmaceutics, Schering-Plough).

28. Clarke, F. (2004). Extracting process-related information from pharmaceutical dosage forms using near infrared microscopy. *Vib. Spectrosc.* **34**, 25–35.

29. Lewis, E. N., Schoppelrei, J., and Lee, E. (2004). Near Infrared chemical imaging and the PAT initiative. *Spectrosc. Mag.* **04** (26–36).

30. Roggo, Y. and Ulmschneider, M. (2008). Chemical imaging and chemometrics: useful tools for process analytical technology. In: Gad, S. C. (Ed.), *Pharmaceutical Manufacturing Handbook: Regulations and Quality*, Chapter 4.3, Wiley, pp. 411–431.

31. PAT—A Framework for Innovative Pharmaceutical Development, Manufacturing, and Quality Assurance, United States Food and Drug Administration, September 2004.

32. Lee, E., Huang, W., Chen, P., and Vivilecchia, R. (2006). High-throughput analysis of pharmaceutical tablet content uniformity by near-infrared chemical imaging. *Spectroscopy* **21**(11).

33. Lewis, E. N., Kidder, L. H., and Lee, E. (2005). NIR chemical imaging: near infrared spectroscopy on steroids. *NIR News* **16**(2).

34. Veronin, M. A. and Youan, B. B. (2004). Magic bullet gone astray: medications and the Internet. *Science* **305**, 481.

35. Lewis, E. N., Schoppelrei, J., and Lee, E. (2004). Near infrared chemical imaging and the PAT initiative. *Spectrosc. Mag.* **04** (26–36).

36. Lewis, E. N., Lee, E., and Kidder, L. H. (2004). Combining imaging and spectroscopy: solving problems with near-infrared chemical imaging. *Microsc. Today* **12**, (6).

37. Ellison, C. D., Ennis, B. J., Hamad, M. L., and Lyon, R. C. (2008). Measuring the distribution of density and tabletting force in pharmaceutical tablets by chemical imaging. *J. Pharm. Biomed. Anal.* **48**(1), 1–7.

38. Lim, H. -P., Dave, V. S., Kidder, L., Fahmy, R., Bensley, D., O'Brien, C., and Hoag, S. W. (2008). Monitoring variation in porosity of roller compacted ribbons using NIR imaging as a non-destructive analytical tool, Poster presentation, AAPS.

39. Malik, I., Poonacha, M., Moses, J., and Lodder, R. A. (2001). Multispectral imaging of tablets in blister packaging. *AAPS PharmSciTech* **2**(2), E9.

40. Clarke, F. C., Jamieson, M. J., Clark, D. A., Hammond, S. V., Jee, R. D., and Moffat, A. C. (2001). Chemical image fusion. The synergy of FT-NIR and Raman mapping microscopy to enable a more complete visualization of pharmaceutical formulations. *Anal. Chem.* **73**(10), 2213–2220.

41. Dubois, J., Wolff, J. C., Warrack, J. K., Schoppelrei, J., and Lewis, E. N. (2007). NIR chemical imaging for counterfeit pharmaceutical products analysis. *Spectroscopy* **22**, 40–50.

42. Lopes, M. B. and Wolff, J. C. (2009). Investigation into classification/sourcing of suspect counterfeit Heptodin trademark tablets by near infrared chemical imaging. *Anal. Chim. Acta* **633**(1), 149–155.

43. Rodionova, O. Y., Houmøller, L. P., Pomerantsev, A. L., Geladi, P., Burger, J., Dorofeyev, V. L., and Arzamastsev A. P. (2005). NIR Spectrometry for counterfeit drug detection, a feasibility study. *Anal. Chim. Acta* **549** (2005) 151–158.

44. Kidder, L. H., Dubois, J, and Lewis, E.N., Realizing the potential of Near Infrared Chemical Imaging in Pharmaceutical Manufacturing, Pharmaceutical online, 48-52, May 2007; www.pharmaceuticalonline.com.

PART IV

APPLICATIONS IN FOOD RESEARCH

12

RAMAN AND INFRARED IMAGING OF FOODS

Nils Kristian Afseth and Ulrike Böcker

Nofima Mat As, Ås, Norway

12.1 INTRODUCTION

For several decades, vibrational spectroscopy has played an important role in the analysis of foods and food components. On one hand, near-infrared and mid-infrared spectroscopies have been major driving forces for the development of rapid online techniques for industrial applications. On the other, vibrational spectroscopic techniques have been important for the general characterization of foods and food components. In the latter category, IR and in recent years also Raman spectroscopy have made important contributions. The success of vibrational spectroscopy in food analysis is related to their general characteristics: the techniques are rapid, and little or no sample treatment is required, which implies they might also be applied *in vivo*. In many applications, both qualitative and quantitative information from the samples can be obtained. But foods are generally heterogeneous by nature, and traditionally, this has been a challenge mainly overcome by sample homogenization or replicate point measurements to increase reproducibility of sampling. However, a great amount of information is revealed by investigating the heterogeneity of a food matrix, and recently there has been an increased interest in applying Raman and IR imaging for characterization of food samples. This has been forced forward by the continuous development of novel instrumentation of Raman and IR imaging systems.

In general, there are two major ways of obtaining vibrational images of localized regions in foods: mapping or imaging. Mapping involves sequential measurements of spectra of adjacent regions in order to create an image. Imaging, on the other hand, requires an image of the sample to be focused onto an array detector where the intensity of the radiation passing through or being scattered from each region of the sample is measured at each pixel of the detector. The majority of published studies using Raman or IR spectroscopy for mapping or imaging of foods are strictly speaking mapping experiments. But as new imaging instruments are continuously being developed and commercialized, the trend is changing toward more frequent use of "real" imaging systems. In this chapter, however, no attempts will be made to distinguish between these two approaches, as mapping experiments clearly illustrate the potential that lies in using Raman and IR imaging for food analysis.

The aim of this chapter is to review the current status of Raman and IR imaging applications for food analysis. So far, the number of applications is limited, but the potential of using these techniques is far ranging, and the chapter intends to highlight some of these underlying potentials. Studies on agricultural products, studies related to animal breeding, and studies from fields like medical diagnostics and pharmaceutical analysis have also been included when reasonable, as sample similarities often occur, and as the applications provide generic knowledge often directly transferable to the analysis of foods. The chapter is not meant to be an extensive survey of all literature in the field, but intends to provide a state-of-the-art picture and future perspectives regarding Raman and IR imaging within different fields of food analysis.

12.2 VEGETABLES, FRUITS, AND PLANTS

Traditionally, vibrational spectroscopic characterization of plant substances has been a fairly cumbersome approach.

Raman, Infrared, and Near-Infrared Chemical Imaging   Edited by Slobodan Šašić and Yukihiro Ozaki

First, the analyst had to purify the substance to be examined. Various separation techniques, most of them highly intrusive, were used prior to spectroscopic measurements. On top of this, all information regarding spatial distribution was lost. However, the advent of IR and Raman microscopes and imaging techniques introduced a new era of microanalysis of vegetables, fruits, and plants. Now, spatially resolved spectroscopy can be made *in situ*, without the need for stains or chemicals. In recent years, the potential of using IR and Raman imaging for revealing the microstructure of plant tissue has been illustrated through a range of different applications. In the following section, a few of these applications are described.

12.2.1   Revealing the Anatomy of Wheat Grains—A Short Survey of IR Instrumentation

Wheat is among the most-produced grains of the world and constitutes an important part of the human diet. Thus, the microstructure of wheat has been of special interest in regard to its utilization in foods, as the microstructure affects factors ranging from milling quality to breeding considerations. Figure 12.1 shows the main components of the wheat kernel, including the pericarp at the outside of the kernel, the aleurone cells, the endosperm, and the germ.

Raman and IR spectroscopies have been extensively used for the characterization of the microstructure of wheat and its components. Actually, the spectroscopic evaluation of wheat provides an interesting historical survey on the development of Raman and, in particular, IR imaging techniques for the applications on plant tissue. As early as 1993, Wetzel and Reffner published a study that summarizes the early work of using spatially resolved FT-IR microspectroscopy to examine the microstructure of wheat kernels [1]. In their study, an IR microscopy system equipped with a mercury–cadmium–telluride (MCT) detector was used for analyzing sections of wheat kernels. The wheat kernels were soaked overnight in distilled water under refrigeration, and sections of 8 μm thickness were prepared by cryostatic sectioning on a microtome. IR images of tissue areas were made from moving stage experiments, and the authors were able to reveal the spectroscopic and the chemical signatures of the different

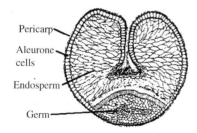

FIGURE 12.1   Wheat cross section revealing the pericarp at the outside of the kernel, the aleurone cells, the endosperm, and the germ.

parts of the wheat kernels. Lipids, carbohydrates, and proteins were observed by moving from the pericarp, through the aleurone layer and the aleurone cell walls, into the endosperm, and further into the germ. Individual differences in the same endosperm and within two parts of the same germ could also be visualized. This study is one of the first to prove the concept of vibrational microscopy for revealing *in situ* information of seed microstructure. This approach and similar studies paved the way for using the information obtained from localized distribution of chemicals within single seeds and link the findings with the seed development process and the quality of seeds for final end use.

As the development of IR microscopy progressed during the 1990s, synchrotron-based IR microscopy was introduced, allowing for improved signal-to-noise performances and increased spatial resolution. In 1998, Wetzel et al. used this technique for advanced characterization of wheat kernels [2]. In their previous study on wheat using a conventional globar IR source, the authors had shown that it was possible to differentiate the aleurone cell layer from the surrounding material of a wheat kernel. But locating the cell wall in between the individual cells was more difficult due to limits in spatial resolution. By introducing synchrotron-based IR, on the other hand, the presence of single aleurone cells could be investigated, and both proteins and carbohydrates of the cell walls provided sufficient contrast to visualize single aleurone cells. Synchrotron-based IR microscopy thus allowed for interrogating single cells and even parts of single cells *in situ*.

Another major development of IR instrumentation involved the introduction of MCT focal plane array (FPA) detection, replacing the more time-consuming point mapping for making IR images and allowing for true IR imaging of samples. Marcott et al. used this technique for the characterization of wheat kernels [3]. IR images were obtained in which each pixel represented 4.5 μm × 4.5 μm areas of the sample, and chemically different regions of the wheat kernel could be clearly visualized. An IR image of wheat kernel is provided in Figure 12.2. The image has been generated by overlaying two images representing the areas under the amide II (protein) band centered at $1550\,cm^{-1}$ and the carbonyl (lipid) band centered at $1735\,cm^{-1}$, thus providing the sufficient contrast for visualization of the endosperm, the aleurone cells, and the pericarp regions.

12.2.2   Wheat Grain Microstructure and Grain Hardness

One of the most important characteristics of wheat kernels used for processing is their hardness. The hardness is related to the texture of the starchy endosperm and severely affects milling performance and the resultant flour quality. It is well known that hard wheat varieties have a mechanical resistant endosperm, whereas the endosperm of soft wheat varieties is

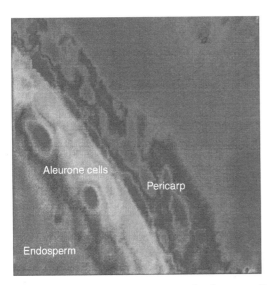

FIGURE 12.2 IR image of the endosperm, the aleurone cells, and the pericarp region of a wheat kernel. The image was generated by overlaying two images representing the areas under the amide II (protein) band centered at 1550 cm^{-1} and the carbonyl (lipid) band centered at 1735 cm^{-1}, respectively. (See the color version of this figure in Color Plate section.)

friable. Nevertheless, the biochemistry of wheat hardness has been a subject for debate over the past years, and elucidating the biochemical factors related to grain hardness has been considered a key for optimization of the wheat milling process.

Piot et al. used confocal Raman microspectroscopy to characterize wheat grains [4, 5]. In their studies, high-resolution spectral mapping was performed on 50 μm thick solid sections with spectra obtained at 2 μm intervals, and a He/Ne laser delivering at 8 mW was used as the excitation source. The researchers were thus able to characterize starch and proteins of the endosperm, as well as the arabinoxylans and ferulic acids of the aleurone–aleurone, the aleurone–pericarp, and the aleurone–endosperm cell walls. The group further focused on the role of the proteins of the endosperm in relation to grain hardness. The secondary structure of the endosperm proteins was estimated by means of decomposing the amide I bands into contributions of α-helical, β-sheet, β-turn, and random coil secondary structures, respectively. Their findings suggested that the α-helical secondary structure of protein could be associated with wheat grain hardness.

The polysaccharides of the endosperm have also been related to grain hardness. Barron et al. used FT-IR microspectroscopy for imaging of wheat endosperm cell walls of four different cultivars of wheat, selected due to their differences in endosperm texture [6]. Images of transverse sections were obtained from which cell contents had been removed by sonication. Based on principal component analysis (PCA) of the images obtained, differences between the structural

heterogeneities of the different varieties could be studied. Hard and soft wheats could not be differentiated using the spectral features derived from entire transverse sections of the endosperm, but when focusing on locations within the grain, the soft and the hard wheat cultivars could clearly be distinguished based on their polysaccharide signatures.

12.2.3 Flax Stems

Natural fibers have always received considerable interest for both textile and industrial purposes, and in order to improve fiber quality, knowledge of chemical components and structure of the cell walls is of vital importance. Flax is among the oldest fiber crops in the world. It is used for the production of linen, and thus efforts have been made to develop noninvasive methods for microstructure characterization of flax stems to improve quality. Himmelsbach et al. used both Raman and IR imaging in order to characterize the distribution of chemical components in flax stem tissue [7, 8]. The FT-IR images of flax stem tissue were obtained by mapping using MCT detection, and from the images they were able to distinguish the primary chemical components such as waxes in the cuticular and epidermal tissues, pectate salts in the epidermal and parenchymal tissues, cellulose in the core and fibers, aromatics in the core, and acetylated polysaccharides in the fibers and core. Corresponding Raman mapping was performed on similar flax stem samples using an FT-Raman microprobe system equipped with a 1064 nm Nd-YAG laser source. As in the IR study, the Raman images gave evidence of wax components in the cuticular and epidermal tissues, and the strongest intensities related to carbohydrates were found in the fiber cells. In addition, the presence of pectins, other noncellulosic polysaccharides, and cellulose was found in the parenchyma tissue, whereas evidence of lignin in core tissue and pigments in epidermal tissue was found.

Retting is one of the postharvest processing steps used to mediate the release of stem fibers from flax. Finding the optimal retting agent is critical for the industry, which makes studying the effect of retting agents an important issue. FT-IR microscopic mapping has been investigated as a tool to study enzymatic retting of flax stems [9]. Using combined IR and visible imaging, it was possible to elucidate the relative loss or changes in the distribution of key chemical components after treatments with enzymes with and without an additional chelator. Also, the IR mapping technique was shown to have advantages over visible imaging alone in that it can detect and locate chemical species present after treatment in relation to the anatomical features of the flax stem.

12.2.4 Exploring the Anatomy of Other Plant Species

Revealing the microstructure of plant tissue is not merely related to improved industrial processing, but might equally well be related to breeding programs for selecting superior varieties for special purposes and prediction of grain quality and nutritive values for humans and animals. Corn and barley constitute examples of varieties important for human consumption, and their microstructure has been extensively investigated. Synchrotron-based FT-IR microspectroscopy has been used to explore the structural–chemical features of corn seed tissue within cellular dimensions [10], and similar approaches have been used to reveal the molecular microstructural–chemical features within tissue in grain barley [11, 12]. The analytical approach in these studies, however, resembles the approaches found in the examination of wheat and flax, and details of these studies will not be discussed further.

Lignin, together with cellulose, is the most abundant biopolymer of the plant cell wall, and understanding the heterogeneity of biopolymers such as lignin and cellulose in the cell wall is essential from the perspectives of both plant biology and commercial utilization. Lignin is produced by condensation reactions of the structural entities coumaryl, coniferyl, and sinapyl alcohols, all aromatic compounds that provide suitable Raman cross sections. Agarwal used Raman imaging to investigate the structure and composition of plant cell walls in black spruce wood, and images showing the lignin and cellulose distribution within the cell walls were obtained [13]. Figure 12.3 shows a Raman image of the spatial distribution of lignin coniferaldehyde and coniferyl alcohol units in a cross section enclosing the cell walls of six adjoining cells. The image was generated using the combined band region of $1519–1712\,cm^{-1}$.

In a similar study, infrared imaging was used to characterize the anatomy of sunflower and maize root [14]. One of the main objectives was to evaluate the potential of using IR for differentiating between two different plant species (i.e., monocotyledons and dicotyledons), using maize and sunflower as representative organisms. PCA enabled the authors to find means for distinct separation of maize and sunflower using IR spectra of the epidermis and xylem, and the infrared band at $1638\,cm^{-1}$, representing hydrocinnamic acid in lignin, provided a conclusive method for distinguishing between maize and sunflower plant tissues.

Another important, though minor, component in plant cell walls is pectin, and Micklander and coworkers investigated the potential of using Raman microscopy for characterizing the distribution of pectin in tissue of potatoes [15]. Measurements were made on raw potato tubers, and spectra were investigated with special emphasis on the unique galacturonic methyl ester peak around $1745\,cm^{-1}$. The authors found that high-quality pectin spectra of potato cell walls could be obtained, but they could not conclude whether these spectra were sufficiently distinctive to be able to distinguish between different qualities of pectin and starch in different potato cultivars.

12.2.5 Pigments and Related Compounds

In Raman spectroscopy, the scattering efficiency of molecular bonds varies widely. The scattering efficiency depends first of all on the polarizability of the electrons constituting functional groups and bonds. Symmetric bonds, like double and triple bonds, are known to provide high Raman scattering efficiencies. But an additional effect, the resonance effect, which arises when the frequency of the laser beam is close to the frequency of an electronic transition, might also enable significant scattering enhancement for particular chemical

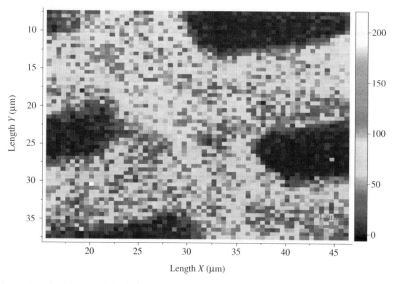

FIGURE 12.3 Raman image showing the spatial distribution of lignin coniferaldehyde and coniferyl alcohol units in a cross section enclosing the cell walls of six adjoining cells in black spruce wood. The image was generated using the combined band region of $1519–1712\,cm^{-1}$.

bonds. In foods, these effects are often observed for colored components and pigments, and the effect might enable very sensitive detection of such components (in the ppm range) [16].

Carotenoids are natural pigments occurring in plants, algae, and other microorganisms. They serve essential functions in plants related to the photosynthesis, and carotenoids like α- and β-carotene are known to transform to vitamin A. The demand for improved cultivars of high nutritional value requires detailed information on the regulation of the carotenoid biosynthesis. Carotenoid molecules consist of a long central chain with a conjugated double bond system, which is the light-absorbing chromophore responsible for the characteristic color of these compounds. Baranska et al. showed that Raman imaging can provide insight into carotenoid accumulation directly in living plant tissue [17]. They focused on the main carotenoids (i.e., β-carotene, α-carotene, lutein, and lycopene) of carrot root. The carotenoids are mainly distinguished due to positional variations of the $-C=C-$ stretching vibration observed within the wavenumber range between 1500 and 1540 cm^{-1}, and Raman images provided insight into the distribution and relative contents of the different carotenoids. From theses results, indications of the developmental regulation of carotenoid genes were obtained.

Polyacetylenes are another group of compounds found in carrots, and their presence in foods has been related, among other things, to bitter off-taste. Polyacetylenes are organic molecules constituting two or several carbon–carbon triple bonds, and these bonds provide unique Raman scattering in the region around 2200 cm^{-1} due to stretching vibrations. Baranska et al. used Raman imaging to provide detailed information regarding these compounds in carrot root, and also found that Raman spectroscopy could be used to distinguish between chemically similar polyacetylenes [18]. Another research group used Raman microscopy to measure the distribution of amygdalin across a bitter almond cotyledon [19]. Amygdalin is an aromatic glycoside incorporating a nitrile group, and besides being related to bitterness, this compound is known to release highly toxic hydrogen cyanide upon hydrolysis. The characteristic Raman bands of amygdalin include the nitrile stretching vibration around 2244 cm^{-1} and the aromatic carbon–hydrogen stretch around 3060 cm^{-1}. In this study, only line maps of Raman spectra were provided, but the study illustrates the potential of using Raman imaging for characterization of cyanogenic glycosides in plant tissue.

Raman imaging has also been used to study and identify secondary metabolites in medicinal and spice plants [20]. Baranska et al. managed to map the distribution of anethole in fennel fruits and curcumin in curcuma roots. Anethole is an aromatic terpenoid used both in food additives and perfumes. Curcumin is a polyphenol and has been related to certain health beneficial effects.

12.2.6 Imaging Techniques for Point Sampling

Microspectroscopic imaging techniques provide means for obtaining spectra with high spatial resolution, enabling high-quality spectra of single components within few micrometer ranges to be obtained. Thus, these techniques are frequently used for point sampling purposes to obtain representative spectra of pure components in heterogeneous matrices. Wetzel and coworkers described one approach for revealing protein spectral details in wheat endosperm [21, 22]. Wheat endosperm is a heterogeneous matrix constituting numerous starch granules (with sizes typically in the range of 5–30 μm) in a network of proteins. The secondary structure of the proteins is in general vital for understanding factors such as digestive behavior, nutritive quality, and texture of plants, and the utilization and availability to humans and animals. In IR and Raman spectroscopies, the amide I band is one of the main bands directly related to protein secondary structures, and deconvolution of this band enables relative quantifications of structures like β-sheet and α-helical conformations, respectively. Wetzel's approach for wheat endosperm analysis involved high-density mapping of sections of wheat endosperm utilizing the high spatial resolution achieved by synchrotron-based IR microspectroscopy. From the spectra of these sections, spectra dominated by protein information could be chosen by eliminating spectra revealing high starch content and high scattering contributions due to the presence of starch granules. By averaging protein spectra from adjacent sections of one sample, representative protein information was obtained allowing estimation of the sample's protein secondary structure. Yu et al. used a similar approach to investigate flax and winterfat (forage) seeds [23, 24].

12.3 ANIMAL TISSUE

Representative sampling is a reoccurring theme in food analysis. For instance, imagine having a large cow muscle on your lab bench. You would like to predict the tenderness of the muscle, but what do you do when you have a NIR probe covering 2 cm^2 of the sample? The sampling area is way too small to represent the entire sample, but too large to say anything about the sample microstructure. Traditionally, these challenges have been faced by grinding the material or making multiple spectroscopic measurements throughout the sample. Recent developments, however, have allowed for imaging of large areas using NIR. At the same time, IR and Raman microspectroscopies have paved the way for detailed characterization of the microstructure of animal tissues. The potential use of these techniques for animal tissue characterization is far ranging, including applications relating lipid deposition in tissue to human health, and linking protein structure to texture and tenderness. In the present section a few applications of IR and Raman microimaging of animal tissue is described.

12.3.1 Heat- and Salt-Induced Changes of Meat

Heating is one of the most important steps in the processing of muscle foods, and heating is often essential in order to achieve a palatable and safe product. Meat undergoes considerable structural changes upon heating both with regard to microstructure and protein structure, and thus the quality of the meat product also changes drastically after cooking. Kirschner et al. used FT-IR microscopic imaging to monitor the denaturation processes in aged beef loin [25], focusing first of all on the denaturation processes in connective tissue and single myofibers, respectively. The analysis comprised raw and heated slices of *Longissimus dorsi* muscles of four Norwegian Red cattle. In the study, the secondary structure changes of the myofibrillar proteins due to heat treatment were monitored in the amide I region of the IR spectra, and evidence for increasing aggregated β-sheet structures and decreasing α-helical structures due to heat treatment was found. The changes found appeared to be more pronounced for the myofibers than for the connective tissue. Images based on the band ratio of two bands in the amide I region related to aggregated β-sheet ($1630\,cm^{-1}$) and α-helical structures ($1654\,cm^{-1}$), respectively, clearly visualized the degree of

protein denaturation due to heat treatment. This is shown in Figure 12.4. The upper image was obtained of a raw muscle section, whereas the lower image was obtained after heat treatment at 70°C. The intensity levels correspond to the band ratios and thus the level of protein denaturation, whereas the black areas correspond to spectra that have been rejected due to failing a signal-to-noise ratio quality test.

Salting is another processing method that affects the secondary structure of proteins and thus the muscle texture. Böcker et al. used IR imaging to study salt-induced chemical changes in pork muscle [26]. Tissue samples from three different degrees of brine salting (1.6%, 7.7%, and 15.4% salt concentrations, respectively) were obtained and analyzed with an FT-IR microscope equipped with an FPA detector. Images made of the band ratios of the amide I bands at 1630 and $1654\,cm^{-1}$, related to aggregated β-sheet and α-helical structures, respectively, are shown in Figure 12.5. The three upper images are chemical images, whereas the three lower are corresponding micrographs. Black domains in the images refer to spectra that have been rejected due to a signal-to-noise ratio quality test. From the images, the effect of decreasing α-helical structures due to salting is clearly seen. In addition, at high salt content the fiber cells shrink to a high degree, increasing the extracellular space. This is visualized by an increase in the black pixel domains in the chemical image.

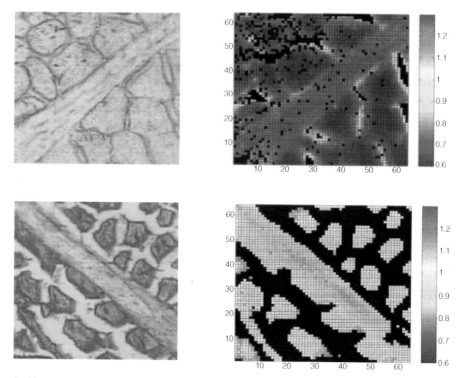

FIGURE 12.4 Chemical images obtained using the I_{1630}/I_{1654} band ratio as a measure for the denaturation level of sections of raw (upper) and heated (70°C, lower) muscle sections of cattle. Corresponding photomicrographs of the IR images are displayed on the left side.

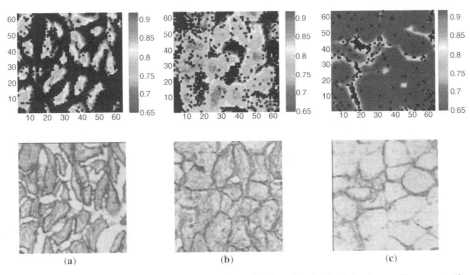

FIGURE 12.5 Chemical images of pork tissue (showing the I_{1630}/I_{1654} band ratio) obtained for three salt concentrations: high (a), medium (b), low (c) (from left to right). Corresponding photomicrographs are shown below the respective IR image. (See the color version of this figure in Color Plate section.)

12.3.2 Imaging Techniques Used for Point Sampling

As discussed in Section 12.2.6, microscopic imaging techniques are frequently used for point sampling purposes to obtain representative spectra of pure components in heterogeneous matrices. This approach has also been used for the collection of single myofiber spectra of meat and pork muscle [27, 28]. Kohler et al., however, introduced an automated multivariate image analysis approach to separate regions of connective tissue and myofibrillar cells in FT-IR images [29]. In this study, a total of 113 IR images of *Longissimus dorsi* muscles aged for four different lengths of time were obtained. The spectra in each pixel of every image were first subjected to a quality test, and spectra of undesirable signal-to-noise ratios or with high absorbances due to water vapor or lipids, were removed. All remaining spectra were then subjected to preprocessing based on extended multiplicative signal correction (EMSC) in order to remove unwanted physical interferences in the spectra. A PCA model was obtained from spectra randomly picked from every image in order to distinguish between connective tissue and myofibrillar cells. This PCA model was then used for the segmentation of myofibrillar cells and connective tissue in all the images. Connective tissue spectra combined with information of their spatial position within a region of connective tissue could then be related to different aging times, and the heterogeneity of the connective tissue could be studied closely.

12.3.3 Bone Tissue

The analysis of bone tissue is most often encountered within biomedical fields, but the characterization of bone tissue might be equally valid in the animal sciences. The overall strength of the bone is clearly one of the factors involved, and Raman spectroscopy has been proposed as one potential method to identify differences in mineral composition of bone. Timlin et al. used Raman imaging to study fatigue-related microdamage in bovine bone [30]. In their study, Raman images of bone tissue with no visible damage, with microcracks, and with diffuse damage were obtained and analyzed. The analysis revealed changes in the Raman phosphate bands in damaged bone that were attributed to the presence of different mineral species, and the study demonstrated the power of spectroscopic imaging for exploring the heterogeneous chemical microstructure of bone. The phosphate v_1 region of the Raman spectra (950–970 cm^{-1}) was of particular interest to the authors. The methodology and the approach should be transferable to studies of bone and bone deformities of other animal species and fish.

12.4 MISCELLANEOUS FOOD PRODUCTS

12.4.1 Biopolymer Blends

Texture and rheological properties are important parameters in food, and the quality of food is often crucially dependent on their microstructure. In many kinds of processed foods, additives like biopolymers are a major factor providing texture, and thus full understanding of the microstructure of the additives is crucial in order to make products with the expected properties. Often, blends of biopolymers are used to provide the appropriate matrix, and imaging techniques based on the chemical contrast between the constituents of a blend, such as Raman and IR imaging, enable unique insights into the microstructure of such systems.

Mousia et al. used FT-IR microscopic imaging to study the spatial variation in the composition of biopolymer blends prepared by extrusion of mixtures of gelatin and maize starch [31]. In their study, FT-IR spectral mapping with a step size of 30 µm was used on thin films of extruded amylopectin–gelatin blends, and maps of the ratio of the area of saccharide bands (1180–953 cm^{-1}) divided by that of the amide I and II bands (1750–1483 cm^{-1}) were used to visualize the spatial variability in the amylopectin-starch ratios. The calculated amylopectin-starch ratios and their dependence on the sampling area were also tested by coaddition of spectra from large aperture measurements and small aperture measurements, respectively, but no significant differences due to the sampling approaches were found. The results suggested a high degree of heterogeneity of the blends, despite the thorough mixing expected due to the extrusion processing. The images also revealed that gelatin constituted the continuous phase in the blends, regardless whether native or pregelatinized starch was used.

Oat products are gaining popularity due to their claimed cholesterol-lowering properties. But unlike most cereal flours, pure oat flour is unsuitable to make stable expanded structures. This is mainly due to the high levels of lipid and soluble gums present. Therefore, oat flour products are usually blended with other cereal flours to make extruded products. Cremer and Kaletunc studied the spatial distribution of starch, protein, and lipid domains in corn and oat flour-based extrudates using FT-IR imaging [32]. From the images, the spatial distribution of lipids (based on the carbonyl band at 1740 cm^{-1}), proteins (based on the amide I and II bands at 1650 and 1550 cm^{-1}), and carbohydrates (based on the region of 1100–1000 cm^{-1}) were obtained, and thus information regarding heterogeneity and interactions between components could be obtained. The results revealed that starch forms a continuous phase in cereal-based extrudates. Among the main components, proteins were the least evenly distributed. The lipid distribution was neither correlated with the starch nor the protein distribution.

Pudney et al. turned their attention to another kind of biopolymer mixture, namely, blends of gellan and κ-carrageenan in water [33, 34]. The authors made Raman images of the gelled soft solid microstructures in various mixtures. The quantification of individual components was performed, and due to the chemical similarity of the two polysaccharides, multivariate curve resolution was employed to provide single component concentration maps. Under certain concentration regimes, the two biopolymers phase separate. This property can be used to produce different microstructures. Thus, two different microstructures were mapped, and quantitative component maps of the two biopolymers were obtained. The resultant concentrations could then be used to produce tie lines for the gellan/κ-carrageenan phase diagram, providing essential knowledge for understanding and manipulating their structures and properties in a systematic way. Raman images of diary spread were also obtained in order to extend the same methodology to investigate even more complex "real-life" samples.

12.4.2 The Microstructure of "Real-Life" Products

Imitation cheese is often used in food products, such as pizzas, as a cheaper alternative to natural cheese. Imitation cheese is a relatively high-fat product (containing around 25% fat), thus consumer awareness related to health aspects forces food manufacturers to develop new imitation cheeses with reduced fat contents. However, a reduction of fat contents is known to affect the texture and melting properties of the product, and knowledge of the microstructure of low-fat imitation cheeses is of great importance. Noronha et al. used FT-IR imaging for examination of the microstructure of starch-containing imitation cheeses [35]. The imitation cheeses contained one of four different types of starch (native, pregelatinized, resistant, or waxy corn), and IR images of the four matrices were obtained. The images provided valuable information regarding the distribution of lipids, proteins, and starch, and the starch type had a clear and visible effect on the distribution of proteins and lipids. Combined with results from electron- and light microscopy, the results suggested that the resistant starch did not gelatinize after proper manufacturing. Pregelatinized and native corn starch, on the other hand, did gelatinize, and thus interacted more with the protein matrix.

In 2007, melamine was identified as the organic compound responsible for the deaths of a significant number of cats and dogs due to adulterated pet food. As it turned out, melamine was intentionally added to pet food in order to boost the apparent protein content of the product, and there is great concern that melamine will enter the food chain again and harm animals or humans. Thus, there is an interest to develop rapid and reliable methods for detection of melamine in pet food. In Raman spectroscopy, marker bands for melamine, in particular the unique triazine ring breathing mode around 670 cm^{-1}, have been identified. Liu et al. performed a small survey to visualize the potential of using Raman imaging for identification of melamine in pet food, and made mixtures of melamine in wheat flour [36]. Figure 12.6 shows images of wheat flour containing 6% of melamine. In the figure, a bright field reflection image is compared with the corresponding Raman image indicating the presence of melamine due to the intensity of the 670 cm^{-1} band. The melamine concentration of the mixture is fairly high, but the authors claim that detection of melamine in concentrations around 0.2 wt% is accomplishable.

12.4.3 Emulsions

A large number of foods exist in the form of emulsions (i.e., mixtures of two immiscible liquids where one liquid is dispersed in the other by the action of emulsifiers like surfactants and/or proteins). Multiple emulsions consist of three phases (water/oil/water or oil/water/oil), and these emulsions are especially interesting in the design of

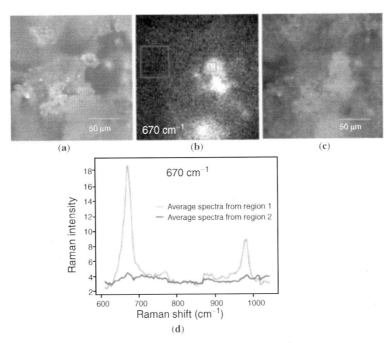

FIGURE 12.6 Bright field reflection image (a), corresponding Raman image at 670 cm^{-1} (b), and fusion of Raman and bright field reflection image of a mixture of whey flour and melamine (6%). The high intensity Raman spectrum of (d) corresponds to the white spots due to melamine in (b).

functional foods due to the possibilities for oxidation protection by the outer layer and for designed slow ingredient release. A rapid technique for the analysis of these mixtures is desirable.

Meyer et al. used coherent anti-Stokes Raman scattering (CARS) microscopy to investigate the composition and molecular distribution in water–oil–water emulsions [37]. CARS is a nonlinear Raman technique that requires excessive instrumentation (i.e., multiple laser sources) compared to conventional Raman spectroscopy, but image acquisition times are often short due to the nonlinear dependence of the signal intensity on the laser intensity. The emulsion components can be imaged separately by choosing the appropriate Raman resonance for the CARS analysis. Meyer made multiple emulsions consisting of inner water droplets and an outer water phase with high sucrose and glucose content. The inner phase was dispersed in MCT-oil droplets with almost the same refractive index (1.449) as the outer water phase (1.445). The similarities of these refractive indices make analysis using microscopic techniques based on transmission or reflection difficult due to lack of contrast. The spectra of the different emulsion components were obtained using conventional Raman spectroscopy to select the Raman resonances specific for the emulsion components, and by choosing these resonances in the CARS analysis, the spatial distribution of the molecules within single emulsion droplets could be imaged. The authors also demonstrated the three-dimensional sectioning capability of CARS microscopy by acquiring images at different focal depths.

12.4.4 Microorganisms

Microbial contamination is a serious issue for the food industry as well as the consumers, and fast, reliable, and nonambiguous methods for characterization and quantification of microorganisms are continuously being developed. Raman and IR spectroscopies have been frequently used for the characterization of microorganisms [38, 39], and when spatial resolution is sufficient, Raman and IR microscopic imaging provides tools for identification, quantification, and investigation of the chemical heterogeneity of microorganisms.

Rösch et al. investigated the spatial heterogeneity of bacteria and bacterial spores commonly encountered in industrial food and pharmaceutical clean rooms using Raman microspectroscopic imaging [40]. The heterogeneity mapping was performed with the overall aim of developing a technique for rapid detection of airborne contaminations within clean rooms. Raman images of nine different microorganisms were obtained, and images were based on the carbon–hydrogen stretch region at ~2900 cm^{-1}, the amide I vibration at ~1650 cm^{-1}, and the methylene deformation at ~1420 cm^{-1}. The heterogeneity, or even lack thereof, within the bacteria and the bacterial spores could be clearly visualized. From these results, the authors could deduce which microorganisms that needed multiple measurement points for representative characterization and which microorganisms that were homogeneous and needed less sampling points.

Escoriza et al. investigated the suitability of using Raman microscopic imaging for the quantification of filtered waterborne bacteria [41]. Chemical images based on the carbon–hydrogen stretching region (2800–3000 cm^{-1}) were obtained using 532 nm laser excitation, and the images were used for enumeration of waterborne bacteria. The intensities of the Raman response correlated well with the number of cells present in drops of sample water on aluminum-coated slides. However, sensitivity might be a limiting factor, thus the development of filters with low Raman scattering background to concentrate the bacteria present is essential. Tripathi et al. also studied waterborne pathogens using Raman imaging [42]. Here, Raman imaging was used to distinguish *Bacillus atrophaeus* and *Escherichia coli* in mixtures at the cellular level. Their results demonstrated the ability of Raman chemical imaging to detect high-quality Raman spectra of individual organisms in the presence of complex biotic backgrounds, by taking advantage of the spatially resolved sampling capability of the technique. An overlay of a Raman image and a bright field image of a mixture of *B. atrophaeus* and *E. coli* bacterial cells is shown in Figure 12.7.

An important step in moving from single point analysis to imaging of microbial strains is the feasibility of detecting localized bacterial colonies. However, the amount of data acquired in a short time frame when using imaging techniques might prevent chemical imaging techniques from being feasible for online or at-line sensing. Gilbert et al. used infrared imaging to distinguish between three different *E. coli* strains [43]. The goal of their investigations was to determine the spatial distribution of the different *E. coli* types. Employing PCA, the feasibility of using infrared imaging for visualization of the spatial distributions of the different *E. coli* types was demonstrated. But the large amount of data obtained required long computational times, thus a wavelet-based data compression method prior to chemometric data analysis was introduced in order to reduce

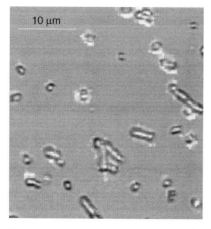

FIGURE 12.7 An overlay of a Raman image and a bright field image of a mixture of *B. atrophaeus* and *E. coli* bacterial cells.

the amounts of data. And with acceptable information losses, computation times could by reduced by more than one order of magnitude.

12.5 CONCLUDING REMARKS AND FUTURE PROSPECTS

From tissue sections to imitation cheeses, the use of Raman and IR imaging of foods spans a wide variety of different applications. In addition, the field of spectroscopic imaging covers a variety of techniques and instrumental configurations, and which technique to choose will mainly depend on the nature of the application. The chemical components to be investigated might favor one of the techniques. Some components (for instance components with highly symmetric bonds, and aqueous solutions) are more appropriately analyzed with Raman spectroscopy, and vice versa, but often various chemical species are interesting making both approaches feasible. Sampling is another issue. Transmission IR microscopic imaging requires thin sections often below 10 µm, whereas considerably thicker samples might be analyzed with Raman techniques. Spatial resolution is no doubt another critical factor. Traditionally, Raman techniques have been able to provide by far the best spatial resolution, but the introduction of synchrotron-based FT-IR and ATR-FT-IR imaging has reduced these differences somewhat [44].

Time is always of the essence, and due to the introduction of FPA or MCT linear array detection, the applications of "real" IR imaging on food samples are "frequently" used. Raman imaging, on the other hand, is still associated with time-consuming mapping or line imaging approaches, even though "real" imaging techniques like wide-field Raman imaging is gaining interest [36, 41, 42]. The feasibility of confocal Raman measurements, i.e. taking the depth of the sample into account, is an additional intriguing possibility for foods. Ideally, one would wish to use the complementary nature of infrared and Raman spectroscopy to fully characterize a sample, but within food analysis the choice of instrumentation is most often a matter of availability and tradition. Naturally, people tend to use what is at hand and what they are familiar with. Instrumentation for vibrational imaging is expensive, and increasing awareness and dropping price-tags are important factors in order to increase the use of Raman and IR imaging in food analysis.

Hyperspectral vibrational images contain huge amounts of data and information. In order to extract this information, the use of multivariate analysis is essential, and features related to component identification and characterization, band resolution, global and local quantitative analysis, and data compression *et cetera* is readily available. However, the main challenge is finding the appropriate robust and automated routines for image analysis, and the role of

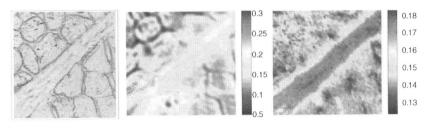

FIGURE 12.8 IR images of a tissue cross section made from the intensity at 1240 cm⁻¹ of original (middle) and spectra subjected to EMSC preprocessing (right). The corresponding light microscopy image is shown to the left.

preprocessing for robust image analysis should not be underestimated. An illustrating example of preprocessing is the use of EMSC to correct IR images of cryosections of *Longissimus dorsi* muscle from cattle subjected to different heat treatments [45]. EMSC is a model based preprocessing method used to separate and characterize physical and chemical information in vibrational spectra [46]. The physical-based features (like scattering effects) might both enhance and obscure the chemical information of spectra. In the cattle-study the authors showed how scatter information in the FT-IR spectra are related to different textural properties of the sample due to heat treatment. The scattering effects might thus misleadingly be interpreted as chemical differences when spectra are not properly preprocessed. However, Figure 12.8 reveals an example of the latter case. In the figure, two chemical images of the same tissue cross section is shown. The images are made from the intensity of the band at 1240 cm⁻¹, which is mainly related to the connective tissue. In the middle image the connective tissue parts are hardly visible, but after EMSC processing (left image) the matrix of connective tissue is clearly revealed.

The fields of medical and pharmaceutical imaging will certainly help to pave the way for new applications within food analysis. The analysis of tissues, cells and even single cells will undoubtedly constitute one central field [47–49]. The qualitative and quantitative characterization and distribution of primary components like proteins, carbohydrates, lipids, water, and even DNA are rich sources of information that relates to fields ranging from food processing, animal welfare, breeding, and quality and heath, to metabolism. For instance, Raman and IR spectroscopies are feasible methods for characterization of lipids in foods and biological systems [50–52], and imaging might be a feasible way of looking at fatty acid metabolism in cells linked to health beneficial aspects. A similar approach for imaging involves the characterization and distribution of secondary metabolites and other minor components like alkaloids, carotenoids, vitamins, or other antioxidants in cells and tissues [53–55].

For many years, Raman and IR spectroscopies have been recognized as feasible tools for detection and characterization of microorganisms and pathogens [38, 39, 56]. Imaging might provide additional aspects both due to rapid screening, sampling and sample heterogeneity, and recent studies have proven the feasibility of rapid IR imaging of microarrays of bacteria, and the detection of pathogens using Raman chemical imaging [57,58]. Another future area involves time-dependent processes. As imaging equipment is continuously being developed allowing for fast imaging of samples, the potential of using Raman and IR imaging for studying time-dependent processes is made possible. Related applications might include a variety of fields from tracking diffusion and component migration in tissue, characterization of phase-changes and crystallizations, to the visualization of water sorption [59–62].

Spectroscopic imaging is a young discipline and has been developing rapidly in recent years. Instrument prices are fairly high, the availability of easy-to-use instrumentation is low, and there is still a lack of awareness and knowledge of instrumentation and their potential in food analysis. Thus, spectroscopic imaging for food analysis will undoubtedly gain increasing interest in the years to come. The techniques provide great possibilities for characterization and understanding of biological systems and processes. As food science is getting increasingly interdisciplinary covering fields like genomics, proteomics, and metabolomics, vibrational imaging will undoubtedly be one of the important tools at the intercepts providing basic and new knowledge and understanding within these and related fields.

REFERENCES

1. Wetzel, D. L., and Reffner, J. A. (1993) Using spatially resolved Fourier-transform infrared microbeam spectroscopy to examine the microstructure of wheat kernels. *Cereal Food World* **38**, 9–20.

2. Wetzel, D. L., Eilert, A. J., Pietrzak, L. N., Miller, S. S., and Sweat, J. A. (1998) Ultraspatially-resolved synchrotron infrared microspectroscopy of plant tissue *in situ. Cell. Mol. Biol.* **44**, 145–168.

3. Marcott, C., Reeder, R. C., Sweat, J. A., Panzer, D. D., and Wetzel, D. L. (1999) FT-IR spectroscopic imaging microscopy of wheat kernels using a mercury-cadmium-telluride focal-plane array detector. *Vib. Spectrosc.* **19**, 123–129.

4. Piot, O., Autran, J. C., and Manfait, M. (2000) Spatial distribution of protein and phenolic constituents in wheat grain as probed by confocal Raman microspectroscopy. *J. Cereal Sci.* **32**, 57–71.

5. Piot, O., Autran, J. C., and Manfait, M. (2002) Assessment of cereal quality by micro-Raman analysis of the grain molecular composition. *Appl. Spectrosc.* **56**, 1132–1138.

6. Barron, C., Parker, M. L., Mills, E. N. C., Rouau, X., and Wilson, R. H. (2005) FT-IR imaging of wheat endosperm cell walls *in situ* reveals compositional and architectural heterogeneity related to grain hardness. *Planta* **220**, 667–677.

7. Himmelsbach, D. S., Khahili, S., and Akin, D. E. (1999) Near-infrared-Fourier-transform-Raman microspectroscopic imaging of flax stems. *Vib. Spectrosc.* **19**, 361–367.

8. Himmelsbach, D. S., Khalili, S., and Akin, D. E. (1998) FT-IR microspectroscopic imaging of flax (*Linum usitatissimum* L.) stems. *Cell. Mol. Biol.* **44**, 99–108.

9. Himmelsbach, D. S., Khalili, S., and Akin, D. E. (2002) The use of FT-IR microspectroscopic mapping to study the effects of enzymatic retting of flax (*Linum usitatissimum* L.) stems. *J. Sci. Food Agric.* **82**, 685–696.

10. Yu, P. Q., McKinnon, J. J., Christensen, C. R., and Christensen, D. A. (2004) Imaging molecular chemistry of pioneer corn. *J. Agric. Food Chem.* **52**, 7345–7352.

11. Yu, P. Q., McKinnon, J. J., Christensen, C. R., and Christensen, D. A. (2004) Using synchrotron transmission FT-IR microspectroscopy as a rapid, direct, and nondestructive analytical technique to reveal molecular microstructural-chemical features within tissue in grain barley. *J. Agric. Food Chem.* **52**, 1484–1494.

12. Yu, P. Q., McKinnon, J. J., Christensen, C. R., Christensen, D. A., Marinkovic, N. S., and Miller, L. M. (2003) Chemical imaging of microstructures of plant tissues within cellular dimension using synchrotron infrared microspectroscopy. *J. Agric. Food Chem.* **51**, 6062–6067.

13. Agarwal, U. P. (2006) Raman imaging to investigate ultrastructure and composition of plant cell walls: distribution of lignin and cellulose in black spruce wood (*Picea mariana*). *Planta* **224**, 1141–1153.

14. Dokken, K. M., and Davis, L. C. (2007) Infrared imaging of sunflower and maize root anatomy. *J. Agric. Food Chem.* **55**, 10517–10530.

15. Thygesen, L. G., Løkke, M. M., Micklander, E., and Engelsen, S. B. (2003) Vibrational microspectroscopy of food. Raman vs. FT-IR. *Trends Food Sci. Technol.* **14**, 50–57.

16. Ozaki, Y., Cho, R., Ikegaya, K., Muraishi, S., and Kawauchi, K. (1992) Potential of near-infrared Fourier transform Raman spectroscopy in food analysis. *Appl. Spectrosc.* **46**, 1503–1507.

17. Baranska, M., Baranski, R., Schulz, H., and Nothnagel, T. (2006) Tissue-specific accumulation of carotenoids in carrot roots. *Planta* **224**, 1028–1037.

18. Baranska, M., and Schulz, H. (2005) Spatial tissue distribution of polyacetylenes in carrot root. *Analyst* **130**, 855–859.

19. Micklander, E., Brimer, L., and Engelsen, S. B. (2002) Noninvasive assay for cyanogenic constituents in plants by Raman spectroscopy: content and distribution of amygdalin in bitter almond (*Prunus amygdalus*). *Appl. Spectrosc.* **56**, 1139–1146.

20. Baranska, M., Schulz, H., Rösch, P., Strehle, M. A., and Popp, J. (2004) Identification of secondary metabolites in medicinal and spice plants by NIR-FT-Raman microspectroscopic mapping. *Analyst* **129**, 926–930.

21. Bonwell, E. S., Fisher, T. L., Fritz, A. K., and Wetzel, D. L. (2008) Determination of endosperm protein secondary structure in hard wheat breeding lines using synchrotron infrared microspectroscopy. *Vib. Spectrosc.* **48**, 76–81.

22. Wetzel, D. L., Srivarin, P., and Finney, J. R. (2003) Revealing protein infrared spectral detail in a heterogeneous matrix dominated by starch. *Vib. Spectrosc.* **31**, 109–114.

23. Yu, P., McKinnon, J. J., Soita, H. W., Christensen, C. R., and Christensen, D. A. (2005) Use of synchrotron-based FTIR microspectroscopy to determine protein secondary structures of raw and heat-treated brown and golden flaxseeds: a novel approach. *Can. J. Anim. Sci.* **85**, 437–448.

24. Yu, P., Wang, R., and Bai, Y. (2005) Reveal protein molecular structural-chemical differences between two types of winterfat (forage) seeds with physiological differences in low temperature tolerance using synchrotron-based Fourier transform infrared microspectroscopy. *J. Agric. Food Chem.* **53**, 9297–9303.

25. Kirschner, C., Ofstad, R., Skarpeid, H. J., Host, V., and Kohler, A. (2004) Monitoring of denaturation processes in aged beef loin by Fourier transform infrared microspectroscopy. *J. Agric. Food Chem.* **52**, 3920–3929.

26. Böcker, U., Ofstad, R., Egelandsdal, B., and Kohler, A. (2004) Study on salt-induced chemical changes in pork muscle by FT-IR imaging. In: *ICoMST 2004*, Helsinki, Finland.

27. Böcker, U., Ofstad, R., Bertram, H. C., Egelandsdal, B., and Kohler, A. (2006) Salt-induced changes in pork myofibrillar tissue investigated by FT-IR microspectroscopy and light microscopy. *J. Agric. Food Chem.* **54**, 6733–6740.

28. Böcker, U., Ofstad, R., Wu, Z. Y., Bertram, H. C., Sockalingum, G. D., Manfait, M., Egelandsdal, B., and Kohler, A. (2007) Revealing covariance structures in Fourier transform infrared and Raman microspectroscopy spectra: a study on pork muscle fiber tissue subjected to different processing parameters. *Appl. Spectrosc.* **61**, 1032–1039.

29. Kohler, A., Bertrand, D., Martens, H., Hannesson, K., Kirschner, C., and Ofstad, R. (2007) Multivariate image analysis of a set of FTIR microspectroscopy images of aged bovine muscle tissue combining image and design information. *Anal. Bioanal. Chem.* **389**, 1143–1153.

30. Timlin, J. A., Garden, A., Morris, M. D., Rajachar, R. M., and Kohn, D. H. (2000) Raman spectroscopic imaging markers for fatigue-related microdamage in bovine bone. *Anal. Chem.* **72**, 2229–2236.

31. Mousia, Z., Farhat, I. A., Pearson, M., Chesters, M. A., and Mitchell, J. R. (2001) FTIR microspectroscopy study of composition fluctuations in extruded amylopectin-gelatin blends. *Biopolymers* **62**, 208–218.

32. Cremer, D. R., Kaletunc, G. (2003) Fourier transform infrared microspectroscopic study of the chemical microstructure of corn and oat flour-based extrudates. *Carbohydr. Polym.* **52**, 53–65.

33. Pudney, P. D. A., Hancewicz, T. M., and Cunningham, D. G. (2002) The use of confocal Raman spectroscopy to characterise the microstructure of complex biomaterials: foods. *Spectroscopy* **16**, 217–225.

34. Pudney, P. D. A., Hancewicz, T. M., Cunningham, D. G., and Brown, M. C. (2004) Quantifying the microstructures of soft

solid materials by confocal Raman spectroscopy. *Vib. Spectrosc.* **34**, 123–135.

35. Noronha, N., Duggan, E., Ziegler, G. R., Stapleton, J. J., O'Riordan, E. D., and O'Sullivan, M. (2008) Comparison of microscopy techniques for the examination of the microstructure of starch-containing imitation cheeses. *Food Res. Int.* **41**, 472–479.

36. Liu, Y., Chao, K., Kim, M. S., Tuschel, D., Olkhovyk, O., and Priore, R. J. (2009) Potential of Raman spectroscopy and imaging methods for rapid and routine screening of the presence of melamine in animal feed and foods. *Appl. Spectrosc.* **63**, 477–480.

37. Meyer, T., Akimov, D., Tarcea, N., Chatzipapadopoulos, S., Muschiolik, G., Kobow, J., Schmitt, M., and Popp, J. (2008) Three-dimensional molecular mapping of a multiple emulsion by means of CARS microscopy. *J. Phys. Chem. B* **112**, 1420–1426.

38. Dalterio, R. A., Nelson, W. H., Britt, D., Sperry, J., and Purcell, F. J. (1986) A resonance Raman microprobe study of chromobacteria in water. *Appl. Spectrosc.* **40**, 271–273.

39. Naumann, D., Helm, D., and Labischinski, H. (1991) Microbiological characterizations by FT-IR spectroscopy. *Nature* **351**, 81–82.

40. Rösch, P., Harz, M., Schmitt, M., Peschke, K. D., Ronneberger, O., Burkhardt, H., Motzkus, H. W., Lankers, M., Hofer, S., Thiele, H., and Popp, J. (2005) Chemotaxonomic identification of single bacteria by micro-Raman spectroscopy: application to clean-room-relevant biological contaminations. *Appl. Environ. Microbiol.* **71**, 1626–1637.

41. Escoriza, M. F., VanBriesen, J. M., Stewart, S., Maier, J., and Treado, P. J. (2006) Raman spectroscopy and chemical imaging for quantification of filtered waterborne bacteria. *J. Microbiol. Methods* **66**, 63–72.

42. Tripathi, A., Jabbour, R. E., Treado, P. J., Neiss, J. H., Nelson, M. P., Jensen, J. L., and Snyder, A. P. (2008) Waterborne pathogen detection using Raman spectroscopy. *Appl. Spectrosc.* **62**, 1–9.

43. Gilbert, M. K., Frick, C., Wodowski, A., and Vogt, F. (2009) Spectroscopic imaging for detection and discrimination of different *E. coli* strains. *Appl. Spectrosc.* **63**, 6–13.

44. Kazarian, S. G., and Chan, K. L. A. (2006) Applications of ATR-FTIR spectroscopic imaging to biomedical samples. *Biochim. Biophys. Acta: Biomembr.* **1758**, 858–867.

45. Kohler, A., Kirschner, C., Oust, A., and Martens, H. (2005) Extended multiplicative signal correction as a tool for separation and characterization of physical and chemical information in Fourier transform infrared microscopy images of cryo-sections of beef loin. *Appl. Spectrosc.* **59**, 707–716.

46. Martens, H., Nielsen, J. P., and Engelsen, S. B. (2003) Light scattering and light absorbance separated by extended multiplicative signal correction. Application to near-infrared transmission analysis of powder mixtures. *Anal. Chem.* **75**, 394–404.

47. Jamin, N., Dumas, P., Moncuit, J., Fridman, W. H., Teillaud, J. L., Carr, G. L., and Williams, G. P. (1998) Highly resolved chemical imaging of living cells by using synchrotron infrared microspectrometry. *Proc. Natl. Acad. Sci. USA* **95**, 4837–4840.

48. Swain, R. J., and Stevens, M. M. (2007) Raman microspectroscopy for non-invasive biochemical analysis of single cells. *Biochem. Soc. Trans.* **35**, 544–549.

49. Uzunbajakava, N., Lenferink, A., Kraan, Y., Volokhina, E., Vrensen, G., Greve, J., and Otto, C. (2003) Nonresonant confocal Raman imaging of DNA and protein distribution in apoptotic cells. *Biophys. J.* **84**, 3968–3981.

50. Beattie, J. R., Bell, S. E. J., and Moss, B. W. (2004) A critical evaluation of Raman spectroscopy for the analysis of lipids: Fatty acid methyl esters. *Lipids* **39**, 407–419.

51. Guillen, M. D., and Cabo, N. (1997) Infrared spectroscopy in the study of edible oils and fats. *J. Sci. Food Agric.* **75**, 1–11.

52. Afseth, N. K., Segtnan, V. H., Marquardt, B. J., and Wold, J. P. (2005) Raman and near-infrared spectroscopy for quantification of fat composition in a complex food model system. *Appl. Spectrosc.* **59**, 1324–1332.

53. Arikan, S., Sands, H. S., Rodway, R. G., and Batchelder, D. N. (2002) Raman spectroscopy and imaging of beta-carotene in live corpus luteum cells. *Anim. Reprod. Sci.* **71**, 249–266.

54. Beattie, J. R., Maguire, C., Gilchrist, S., Barrett, L. J., Cross, C. E., Possmayer, F., Ennis, M., Elborn, J. S., Curry, W. J., McGarvey, J. J., and Schock, B. C. (2007) The use of Raman microscopy to determine and localize vitamin E in biological samples. *FASEB J.* **21**, 766–776.

55. Wold, J. P., Marquardt, B. J., Dable, B. K., Robb, D., and Hatlen, B. (2004) Rapid quantification of carotenoids and fat in Atlantic salmon (*Salmo salar* L.) by Raman spectroscopy and chemometrics. *Appl. Spectrosc.* **58**, 395–403.

56. Yang, H., and Irudayaraj, J. (2003) Rapid detection of foodborne microorganisms on food surface using Fourier transform Raman spectroscopy. *J. Mol. Struct.* **646**, 35–43.

57. Kalasinsky, K. S., Hadfield, T., Shea, A. A., Kalasinsky, V. F., Nelson, M. P., Neiss, J., Drauch, A. J., Vanni, G. S., and Treado, P. J. (2007) Raman chemical imaging spectroscopy reagentless detection and identification of pathogens: signature development and evaluation. *Anal. Chem.* **79**, 2658–2673.

58. Mossoba, M. M., Al-Khaldi, S. F., Kirkwood, J., Fry, F. S., Sedman, J., and Ismail, A. A. (2005) Printing microarrays of bacteria for identification by infrared micro spectroscopy. *Vib. Spectrosc.* **38**, 229–235.

59. Celedon, A., and Aguilera, J. M. (2002) Applications of microprobe Raman spectroscopy in food science. *Food Sci. Technol. Int.* **8**, 101–108.

60. Chan, K. L. A., and Kazarian, S. G. (2004) Visualisation of the heterogeneous water sorption in a pharmaceutical formulation under controlled humidity via FT-IR imaging. *Vib. Spectrosc.* **35**, 45–49.

61. Gupper, A., and Kazarian, S. G. (2005) Study of solvent diffusion and solvent-induced crystallization in syndiotactic polystyrene using FT-IR spectroscopy and imaging. *Macromolecules* **38**, 2327–2332.

62. Zhang, G., Flach, C. R., and Mendelsohn, R. (2007) Tracking the dephosphorylation of resveratrol triphosphate in skin by confocal Raman microscopy. *J. Control. Release* **123**, 141–147.

13

NEAR-INFRARED HYPERSPECTRAL IMAGING IN FOOD RESEARCH

PAUL GELADI

Unit of Biomass Technology and Chemistry, Swedish University of Agricultural Sciences, Umeå, Sweden

MARENA MANLEY

Department of Food Science, Stellenbosch University, Stellenbosch, South Africa

13.1 INTRODUCTION

13.1.1 Bulk Near-Infrared Analysis for Food Products

It has been known for a very long time that food products can be studied by mid-infrared (MIR) (ca. 2500–15,000 nm) and near-infrared (NIR) (780–2500 nm) spectroscopy, as they contain the C–H, O–H, and N–H bonds that have high absorbance in the NIR and MIR wavelength regions. These bonds are present in the major constituents of all biological materials. Already during the 1960s and 1970s the pioneers of NIR spectroscopy were mainly interested in food applications, for example, soybeans [1], meat [2], oilseeds [3], and cereals [3–7]. It was found very early that water, fat, protein, and different carbohydrates could be quantified by NIR calibrations for a wide variety of agricultural products, half-fabricates, and finished consumer products as presented by various authors [8–12]. Later, constituents such as inorganic salts [13], alcohol [14], fatty acids [15], antioxidants [16, 17], and phenolic compounds [17, 18] were also quantified, as were physical parameters such as kernel hardness [19–23], maturity [24], and sensory quality [24–26]. These analyses are usually done on bulk materials, from which a single NIR spectrum is obtained, as the goal is to integrate over an area that is as large as possible in order to avoid sampling errors. Most food products are inhomogeneous by nature, thus requiring integration or homogenization before bulk NIR measurements.

13.1.2 NIR Hyperspectral Imaging

The earliest scientific imaging applications were simple black and white or color images in the visual spectroscopic range, but inspired by satellite imaging, multivariate image analysis [27] was soon becoming useful in the laboratory. Already the first satellite images included wavelengths in the NIR, in addition to visual and MIR wavelengths. Nowadays, hyperspectral imaging [28] is becoming more commonly available, providing complete spectra extending into the long-wave NIR (1100–2500 nm) for each pixel in an image (refer to Chapter 2). The simplest of these images have x and y spatial coordinates and a wavelength coordinate lambda, making a three-way array called a hypercube. This hyperspectral image or hypercube thus allows the description of differences and gradients in the sample under study. Because of this, the samples can be heterogeneous. The spectra in these hyperspectral images show localized spectral features that can be used for exploration and classification. With external information (reference data) also localized, quantitative and qualitative calibrations and predictions are feasible. NIR hyperspectral imaging (NHI) also means that sample preparation different from bulk NIR spectroscopy is needed. Grinding and other homogenization methods are useful for bulk NIR spectroscopic analysis, whereas sampling and sample preparation for NIR imaging have their own peculiarities.

13.1.3 Instrumentation

The NHI instrumentation used for food applications reported in the literature is mainly of three types: (1) Specim-type linescan instruments, (2) Matrix NIR focal plane type and liquid crystal tunable filter (LCTF)-based instruments, and (3) homebuilt instruments often based on Specim PGP (prism grating prism) monochromators. Some authors used instruments based on filters and filter wheels. The sample presentation mode for NHI of food is usually diffuse reflectance, but fluorescence using laser excitation is also used. Only a few applications use transmission.

13.1.4 Sample Preparation and Presentation

For imaging of meat and fish samples, for example, a reasonably flat surface of the right size to fit with the camera and objective (field of view (FOV)) and an optimal illumination system have to be prepared. For cereals, the kernels have to be laid out in the right orientation (embryo or crease up or down) on a suitable background and shadows or kernel overlap should be avoided or minimized. Fruit, vegetables, and other similar food products have the disadvantage of being round and require special care in illumination and imaging, unless an almost flat part of the surface is studied. A more detailed, systematic comparison between bulk NIR spectroscopy and NIR imaging is given in Table 13.1.

13.1.5 Sample Size and Wavelength Range

Two important aspects of NHI of food are magnification and wavelength range. Magnification can accommodate anything between the size of a watermelon (30 cm) and the size of a minuscule bacteria or fungi colony (0.1 mm). The wavelength range can cover from the visible wavelengths with some added wavelengths up to 1100 nm. Alternatively, more advanced cameras use 900–1700 or 1000–2500 nm. Industrial online applications include only a few well selected wavelength regions, as speed is needed for measurement and quality control. For research purposes, a large wavelength range and smaller wavelength spacing can be used because getting the answer in real time is not a requirement.

13.1.6 Localized Properties

Bulk NIR spectroscopy is almost always used to quantify a component (protein, water, fat, carbohydrates) or to describe a physical property, sensory quality, or authentication of a food material. Because of its nature, bulk NIR spectroscopy describes averages. NHI, on the other hand, can be used not only for local quantification of composition but also for localized physical properties or disturbances. The latter is not always translated to concentrations. The localized disturbances may be irregularities in the skin of a fruit or vegetable, the presence of parasites in fish, or beginning of fungal or bacterial infections. Changes in diffuse reflection properties can, for example, be used to identify damage both on and under the skin of fruit and vegetables, even when they are not necessarily associated with chemical changes.

13.1.7 Chapter Details

In this chapter, a brief overview of the current literature covering food applications will be given. Forages and feeds are not included, although some similarities with food applications exist. Applications that use wavelengths below 740 nm only are not mentioned and nor are remote sensing applications. The literature applications are given under the headings cereals, fruit and vegetables, meat and fish, and miscellaneous. This will be followed by an example illustrating the application of NHI in a food product. How to

TABLE 13.1   A Systematic Comparison Between Bulk NIR Spectroscopy and NIR Hyperspectral Imaging

Instrument Setup Attributes	Bulk NIR Spectroscopy	NIR Hyperspectral Imaging
Sample preparation	Grinding, homogenizing	Preferably none
Sample holder	Cup, petri dish	Flat surface, dark or reflecting background
Illumination	May be heterogeneous	As homogeneous as possible
Detector	Si, InGaAs, PbS	Si, InGaAs, HgCdTe, InSb
Wavelength range	Vis-2500 nm	Vis-1000 nm, 900–1700 nm, 1000–2500 nm
Penetration depth	0.1–10 mm	1–2 mm
Mode	Reflection	Reflection
	Transmission	Fluorescence
	Transflection	Transmission
Measurement	Integrating sphere	Focal plane (stare down)
	Rotating–moving cup	Linescan (push broom)
	Flow cell	Point scan (whisk broom)
	Liquid cell	
Desired property	Concentration	Chemical gradients, local damages, local infections, and local parasites

optimally extract useful information from the hypercube, by suitable chemometric techniques, will be shown. The application in this case was the classification of yellow maize kernels of different hardness categories.

13.2 APPLICATIONS IN THE LITERATURE

13.2.1 General Overview

An indication of the number of NHI applications for food reported in peer reviewed journals during the past 15 years is given in Table 13.2. These applications mostly address food quality and/or safety. A number of overview articles have been published on NHI and the food related applications reviewed also mostly include safety and quality [29–31] aspects. The study of Wang and Paliwal [30] in addition also includes bulk NIR applications. Du and Sun [31] also reviewed other imaging techniques such as ultrasound, magnetic resonance imaging (MRI), and various tomographic techniques. Subjects not being addressed until now are adulteration and genetics and these have a future potential in food studies using NHI.

13.2.1.1 Cereals All NIR hyperspectral imaging cereal applications reported thus far have been performed on single kernels [32–39]. Single maize kernels were measured in transmission [32] or diffuse reflection mode [33, 34], and moisture [32], oil, and oleic acid [34] were quantified. Wheat applications were mainly in diffuse reflection mode and encompassed among others the classification of different wheat classes [35, 36] and of the degree of vitreousness in durum wheat [35, 37]. An early detection of sprouting in single wheat kernels was illustrated as having sensitivity greater than either the human eye or wet chemical viscosity testing [38]. Individual wheat grains were classified as blackpoint or fungal affected [39]. More details on instrumentation, wavelength range, imaging mode, image size, and chemometric techniques used are listed in Table 13.3.

13.2.1.2 Fruit and Vegetables Fruit and, to a lesser extent, vegetables are the mainstay of food imaging and are mainly imaged in the visual (400–780 nm) and Herschel (800–1100 nm) ranges. A few applications in the InGaAs

TABLE 13.2   Number of Peer Reviewed Papers Published on NIR Hyperspectral Imaging of Food

	Food Quality	Food Safety
Meat and fish	>10 articles	6–10 articles
Cereals	6–10 articles	6–10 articles
Fruits and vegetables	>10 articles	6–10 articles
Other (e.g., nuts, oilseeds, and mushrooms)	1–5 articles	None

range (900–1700 nm) exist. However, no applications have thus far been reported in the long-wave NIR (1100–2500 nm) range. The majority of the reports on NHI applied to quality of fruit and vegetables had apples as the study subject. These studies mainly focused on determination of starch index [40], bitter pit [41], stem-end/calyx [42], firmness [43], and bruise detection [44–47]. Bruise detection entails the identification of skin and subskin anomalies.

A number of studies addressed important quality aspects of cucumbers, including chilling-induced damage [48, 49], bruise detection [50], and grading [51, 52]. As was the case for most of the fruit and vegetable applications, this was performed in diffuse reflection mode.

Further studies on fruit included determination of fruit firmness for mangoes [53] and peaches [54] as well as acidity in strawberries [55]. One of the few studies performed in transmission was the detection of pits in tart cherries [56]. Quantitative investigations encompassed total soluble solids in mango [53], strawberry [54], apple [44], melon (transmission) [57], and kiwi [58]. The paper reporting the determination of total soluble solids in kiwi is probably the first paper being published on the application of NHI in food. Moisture content was quantified in strawberry [55] and mango [53].

NHI was also applied to fruit safety and mainly constituted the detection of fecal contamination of apples [59–63]. One publication described the fungal infection of mandarins [64].

The authors of the papers on fruit and vegetable quality used a wide variety of multivariate chemometric techniques in order to study the spatial and spectral information. These were principal component analysis (PCA), linear discriminant analysis (LDA), partial least squares (PLS), partial least squares discriminant analysis (PLS-DA), multiple linear regression (MLR), and neural networks (NN). In the studies on food contamination, mainly PCA, PLS-DA, and classification were used. More details on instrumentation, wavelength range, imaging mode, and image size are listed in Table 13.4.

13.2.1.3 Meat and Fish A number of NHI studies have been performed on the evaluation of meat and fish quality. Studies on meat included the prediction of beef tenderness [65] and marbling [66], drip loss, pH, and color for pork [67]. Moisture and fat content were quantified in different fish species [68, 69] and parasites were detected in cod [70, 71].

Fecal and ingesta contamination detection and classification in poultry constitute the majority of safety applications for meat [72–75]. Chickens were also studied for skin tumor detection [76]. The only application using a four-filter instrument studied the detection of chicken heart disease [77]. All meat and fish NHI applications had 1100 nm as the maximum wavelength, meaning that the visible and Herschel

TABLE 13.3 An Overview of NIR Hyperspectral Imaging Applications Covering Quality and Safety Aspects of Maize and Wheat

Product	Application	Wavelength Range (nm)/Number of Wavelengths[a]	Mode/Image Size (Pixels)[a]	Instrumentation	Chemometrics	References
Quality						
Maize	Single kernel/moisture content	700–1090	Transmission, 512 × 512	Homebuilt, CCD camera, LCTF	Preprocessing, PLS, PCR, wavelength selection by GA	[32]
Maize	Effect of grinding on classification accuracy	435–769, 160 wavelengths	Reflection, 640 pixels per line	PIKA linescan camera	Stepwise DA	[33]
Maize	Oil and oleic acid content	950–1700	Reflection, 256 × 320	Matrix NIR	PLS, PC, wavelength selection by GA	[34]
Durum wheat	Classification of rates of vitreousness	650–1100, 90 wavelengths	Reflection 1024 × 1024 and reduced to 512 × 512	Homebuilt, LCTF	DA	[35]
Wheat	Canadian wheat class differentiation	960–1700, 74 wavelengths	Reflection, 640 × 480	Homebuilt, InGaAs camera, LCTF	LDA, QDA, BP-NN	[36]
Durum wheat	Vitreousness and starchiness classification	950–2450	Reflection, 320 × 200	Lextel	Smoothing, visual spectral inspection	[37]
Wheat	Preharvest germination	950–1700	Reflection, 256 × 320	Matrix NIR	Preprocessing, PCA, PLS, visual inspection	[38]
Safety						
Wheat	Classification of sound and stained grains	420–2500	300 × 60	ASD spectrometer, XY scanning	Penalized DA	[39]

[a]As some authors did not report the number of wavelengths or image size, calculations were needed and some numbers may therefore not be correct.

TABLE 13.4 An Overview of NIR Hyperspectral Imaging Applications Covering Quality and Safety Aspects of Fruit and Vegetables

Product	Application	Instrumentation	Wavelength Range (nm)/ Number of Wavelengths[a]	Mode/Image Size (Pixels)[a]	Chemometrics	References
Quality						
Apple	Starch index	Linescan, InGaAs camera, Imspector	900–1700	Reflection, 320×200	Derivatives, PCA, threshold on scores	[40]
Apple	Bitter pit	Linescan, Imspector V9, InGaAs camera	900–1700, reduced to 954–1350	Reflection, 320×240	PLS-DA	[41]
Apple	Stem-end/calyx	Linescan, Imspector V10, CCD camera	400–1000, 1040 wavelengths	Reflection, 800 pixels/ line	PCA and contour levels	[42]
Apple	Firmness, total soluble solids	Linescan, Imspector V9, CCD camera	450–1000	Reflection, 512×512, reduced to 256×256	MLR, wavelength selection	[43]
Apple	Bruise detection	Linescan, Imspector V10, CCD camera	400–1000, reduced to 500–950	Reflection, 800 pixels/ line	PCA and threshold, PLS-DA	[44]
Apple	Bruise detection	Linescan, Imspector V10E, CCD camera	400–1000	Reflection, 400×400	PCA, PLS, wavelength selection	[45]
Apple (Golden delicious)	Bruise detection	Linescan, Imspector V10, CCD camera	400–1000, 1040 wavelengths, reduced to 104	Reflection, 800 pixels/ line, reduced by binning	PCA, classification algorithm	[46]
Apple (Jonagold)	Bruise detection	Linescan, Imspector V10, CCD camera	400–1000, 112 wavelengths	Reflection, 1392×1040	PCA, classification by contour levels	[47]
Cucumber	Chilling damage	Homebuilt linescan, USDA ISL	447–951, 112 wavelengths	Reflection, 146×300	PCA, LDA, weighted combination	[48]
Cucumber	Chilling injury	Homebuilt linescan, USDA ISL	447–951, 112 wavelengths	Reflection, 460×300	PCA, classification on ROI	[49]
Pickling cucumber	Bruise detection	Linescan, Imspector, InGaAs camera	900–1700, 100 wavelengths	Reflection, 320×240	PCA, wavelength selection and ratio on ROI	[50]
Pickling cucumber	Grading (normal versus defective)	Linescan, Imspector V10E, CCD camera	400–1000	Reflection, transmission, 400×400	PLS-DA	[51, 52]
Mango	Total soluble solids, water content, firmness	Linescan, Imspector V10E, CMOS camera	400–1000	Reflection, 512×512	PCA, PLS, NN, MLR	[53]
Peach	Firmness	Linescan, Imspector V9, CCD camera	500–1000	Reflection, 512×512, reduced to 256×256 by binning	MLR, PCR, PLS, wavelength selection	[54]
Strawberry	Moisture content, total soluble solids, acidity	Linescan, Imspector V10, CCD camera	400–1000, 826 wavelengths	Reflection, 400×400	PLS, MLR on reduced wavelength, texture measures.	[55]

(continued)

247

TABLE 13.4 (*Continued*)

Product	Application	Instrumentation	Wavelength Range (nm)/ Number of Wavelengths[a]	Mode/Image Size (Pixels)[a]	Chemometrics	References
Cherry	Presence of pits	Linescan, Imspector, CCD camera	400–1000	Transmission, 512 × 512	NN, PCA	[56]
Melon	Sugar content	Filter wheel, five filters, CCD camera	830, 850, 870, 905, 930	Transmission, 658 × 494, reduced by binning	PLS, stepwise MLR	[57]
Kiwi	Soluble solids	Homebuilt linescan, CCD camera	650–1100, 90 wavelengths	Reflection, 150 × 242	PLS	[58]
Apple Safety	Fecal contamination	Homebuilt linescan, USDA ISL	450–850, 110 wavelengths	Reflection, 408 × 256	PCA, band selection	[59]
Apple	Fecal contamination	Linescan, Imspector V9, CCD camera	425–752, excitation 356	Fluorescence, 512 × 512	PCA	[60]
Apple	Surface defects and contamination	Linescan, Imspector	430–900	Reflection, 256 pixels/ line	PCA, asymmetric second difference	[61]
Apple	Animal fecal contamination	Homebuilt linescan, USDA ISL	425–772, excitation 400	Fluorescence, reflection, 460 pix-els/line	PCA, wavelength selection	[62]
Apple	Fecal contamination	Homebuilt linescan, USDA ISL	447–951	Reflection, 460 pixels/ line	Wavelength ratios and differences, PCA	[63]
Mandarin	Penicillin	Homebuilt, LCTF, CCD camera	460–1020, 56 wavelengths	Reflection, 551 × 551	LDA, CART	[64]

[a]As some authors did not report the number of wavelengths or image size, calculations were needed and some numbers may therefore not be correct.

ranges prevailed. Table 13.5 shows more details on instrumentation and chemometric techniques used for these applications.

13.2.1.4 Miscellaneous

Sweetness according to sucrose, fructose, and glucose concentrations, as well as amino acids, were determined in green vegetable soybeans using PCA and NN [78]. Presence of shell parts in walnut pulp was most effectively determined using the support vector machine (SVM) classifier with Gaussian kernel [79]. Using a sequential process of elimination, the presence or absence of food pathogens could be determined in selected food products [80]. A study of white mushroom used three principal components (PCs) to detect bruises including correction for curvature [81]. Table 13.6 lists more details on instrumentation, wavelength range, imaging mode, image size, and chemometric techniques.

13.3 NIR HYPERSPECTRAL IMAGE ANALYSIS OF A FOOD PRODUCT: MAIZE

13.3.1 Problem Definition and Samples

In this section, a systematic approach is suggested for applying multivariate data analysis on NIR hyperspectral images of food products. This approach follows a sequence of steps that allows the exploration and interpretation of the hypercube. The selected application was the detection and identification of endosperm differences related to hardness of maize (*Zea mays* L.) kernels. Maize kernel hardness is important to the grain and food industry as it influences end-use processing performance. Maize endosperm consists of glassy and floury endosperm that when present in a particular ratio determine whether the kernel is hard or soft.

In this experiment, six maize kernels of each of the three hardness categories, that is, hard, intermediate, and soft, supplied and labeled by experienced breeders, were positioned with the embryo down on a dark background (silicon carbide sandpaper) and an NIR hyperspectral image was acquired. Figure 13.1 shows the digital color image of the samples and the layout of the kernels in terms of hardness categories.

13.3.1.1 Instrumentation

The instrument used for hyperspectral imaging was a sisuChema SWIR (short-wave infrared) hyperspectral, pushbroom imaging system (Specim, Spectral Imaging Ltd, Oulu, Finland). The sisuchema is comprised of an imaging spectrograph PGP coupled with a 2D array mercury–cadmium–telluride (MCT) detector with a spectral range of 1000–2498 nm with ca. 6.5 nm wavelength interval. The use of sisuChema gave maximum 231 pixels per line (*x* coordinate) and 239 wavelengths (lambda coordinate). The scanning stage of the instrument could be programmed to acquire any number of lines (*y* coordinate). Black (0%

reflectance) and white (100% reflectance) reference images were acquired immediately before the sample was imaged. The image acquired with the sisuChema was automatically converted to pseudoabsorbance in the Evince 2.020 hyperspectral image analysis software package (UmBio AB, Umeå, Sweden), taking the internal dark and white references into consideration. Automatic correction for missing pixels was applied. Subsequent data analysis was also performed using Evince 2.020. The original image hypercube was $618 \times 231 \times 239$ (*y* by *x* by lambda), but after cropping to remove background rows and columns at the edges of the original image, a $570 \times 219 \times 239$ image was obtained. The pixel size was approximately $0.2 \times 0.2 \, \text{mm}^2$.

13.3.1.2 Data Reduction Using PCA

The hypercube contains an enormous amount of data, making it necessary to apply data reduction techniques such as PCA. Figure 13.2 shows a typical schematic layout of how PCA can be applied to hypercubes. This is done by reorganizing the hypercube into a very long data matrix ($570 \times 219 = 124{,}830$ rows in this case). The data matrix obtained is usually subjected to spectral preprocessing such as mean centering and/or scaling. This preprocessed data matrix is then decomposed to produce score vectors (the matrix *T*) and loading vectors (the matrix *P*), leaving a residual (the matrix *E*). The score vectors can be reorganized to produce score images. The loading vectors can be shown as line plots to allow spectral interpretation. Scatter plotting of two different score vectors produces a score plot. This plot is very useful for finding outliers, clusters, and gradients.

The first step in multivariate image analysis is thus almost always a simple three-component PCA. This is done to identify background, shading, specular reflection, bad pixels, and edge effects by observing selected score plots and their respective score images interactively. Figure 13.3 shows the first score image after PCA on the mean-centered data (PC1–PC3 component sizes expressed as percentage of sum of squares (SS) = 90.7, 8.3, and 0.64). The image clearly shows background and shading and these have to be removed. This is done more easily by interactive selection between the score image and the score plot. In Figure 13.3, the cluster containing the background pixels is marked. The edge effects, from the cusp of the kernels, that reflects radiation away from the camera lens also need to be removed.

13.3.1.3 Interpretation of Score Images and Score Plots

Figure 13.4 shows the score images for principal components 1–6 after background and shading removal. This figure uses color coding in which low values in the score plot are indicated as blue and high values as red; values in between use the sequence of blue-green-yellow-orange-red. A large area of the same color in a component indicates spectral and spatial similarity in the region (a single kernel in this case). If

TABLE 13.5   An Overview of NIR Hyperspectral Imaging Applications Covering Quality and Safety Aspects of Meat and Fish

Product	Application	Instrumentation	Wavelength Range (nm)/ Number of Wavelengths[a]	Mode/Image Size (Pixels)[a]	Chemometrics	References
Quality						
Beef	Tenderness	Linescan, Imspector V10E, CCD camera	400–1000, 300 wavelengths	Reflection, 800 pixels/ line	PCA, canonical DA	[65]
Pork loin	Quality, marbling	Linescan, Imspector V10E, CMOS camera	400–1000	Reflection	PCA and clustering on ROI	[66]
Pork	Drip loss, pH, color	Linescan, Imspector V10E, CCD camera	400–1000	Reflection	NN	[67]
Coalfish	Moisture content	Titech Visionsort	760–1040, 15 wavelengths	Reflection	PLS	[68]
Fish fillets	Fat and moisture content	Qmonitor	760–1040, 15 wavelengths	Reflection	PLS	[69]
Cod	Detection of nematodes	SpexTubeV	350–950	Reflection, 290 × 290	PLS-DA	[70]
Cod fillets	Detection of parasites	Homebuilt linescan, CCD camera, 15 filters	400–1100	Transmission, 512 × 512, reduced to 256 × 256	PCA classification	[71]
Poultry	Fecal and ingesta contamination	Homebuilt	430–900	Reflection, 1280 × 1024	PCA classification	[72]
Poultry	Detection of contamination on carcasses	Homebuilt linescan	400–1000	Reflection, 1376 × 1040	PLS	[73]
Poultry	Detection of surface fecal contamination	Homebuilt linescan	430–900	Reflection, 1280 × 1024	Wavelength selection, linear regression	[74]
Safety						
Poultry	Contaminant classification	Homebuilt linescan	430–900	Reflection, 1280 × 1024	Spectral angle mapper	[75]
Chicken	Instrument development for skin tumor detection	Homebuilt	420–850	Reflection, 400 pixels/ line	Wavelength reduction	[76]
Chicken	Heart disease detection	Homebuilt	495, 535, 585, 604, 4 filters	Reflection, 640 × 480, reduced to 320 × 240	Univariate	[77]

[a]As some authors did not report the number of wavelengths or image size, calculations were needed and some numbers may therefore not be correct.

TABLE 13.6   An Overview of NIR Hyperspectral Imaging Applications Covering Quality Aspects of Some Miscellaneous Food Products

Product	Application	Instrumentation	Wavelength Range (nm)/ Number of Wavelengths[a]	Mode/Image Size (Pixels)[a]	Chemometrics	References
Quality Soybean	Sucrose, glucose, fructose, amino acid content	Linescan, Imspector, CCD camera	400–1000, 120 wavelengths	Reflection, 484 × 500	PCA, NN, derivatives	[78]
Walnut	Shell and pulp discrimination	Homebuilt, USDA ISL	425–775, 50 wavelengths, UV A excitation	Fluorescence, 512 × 512	SVM, PCA, LDA	[79]
Bacteria	Identification and differentiation	Malvern Sapphire	1200–2350	Reflection, 320 × 256	PLS, classification	[80]
Mushrooms	Bruise damage	Linescan, Imspector V10E, CCD camera	450–1000	Reflection, 580 × 580	Preprocessing, PCA thresholding	[81]

[a]As some authors did not report the number of wavelengths or image size, calculations were needed and some numbers may therefore not be correct.

H	S	I
I	H	S
S	I	H
H	I	S
S	H	I
I	S	H

FIGURE 13.1 Digital image of maize kernels of different hardness (H = hard, I = intermediate, and S = soft).

all kernels have the same color, then distinguishing between them is not possible (e.g., PCs 1, 3, 4, and 6). This entails that no information, in terms of difference in kernel hardness, is available in score images of these components. In the score image of PC2 (Figure 13.4), the soft kernels could be distinguished from the other kernels to some extent. The score image of PC5 shows a clear distinction between the hard, soft, and intermediate kernels. As all colors are present in all kernels, to a greater or lesser extent in PCs 2 and 5, this

distinction is clearly not on a kernel basis, but based on the difference in texture of the endosperm inside the kernels. Principal components 2 and 5 would thus be expected to be most efficient for potential classification of the samples into hard, intermediate, and soft maize kernels. Principal component sizes expressed as %SS (variation explained by each component) are 94.4, 3.8, 1.6, 0.07, 0.05, and 0.02.

13.3.1.4 Detection and Selection of Clusters
Figure 13.5 shows the PCA score plot of PC2 versus PC5 after the removal of background and other disturbances such as geometrical errors and shadows. After trial and error testing of all PC combinations, utilizing interactive comparison between the score plots and score images, this combination of PCs (2 and 5) was confirmed to be the most efficient in discriminating between the different hardness classes. This was already inferred from earlier visual inspection of the score images illustrated in Figure 13.4. Three clusters can be distinguished on the score plot (indicated by ellipses in Figure 13.5). Figure 13.6 shows the respective selected clusters on the PCA score plot and the corresponding classification projected onto the score image. These clusters do not relate to the kernels of different hardness, but are unevenly distributed in all kernels, as can be seen in the classification image (Figure 13.6). Each cluster thus describes physical and chemical similarities within the endosperm of the kernels. Differences between the clusters signify chemical and physical differences between endosperm regions, for example,

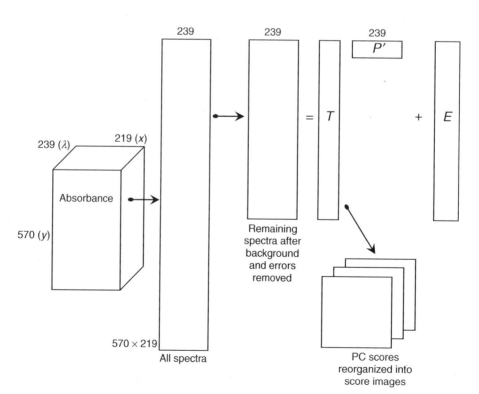

FIGURE 13.2 Schematic layout of PCA to produce principal components (T), loadings (P′), residuals (E), and score images.

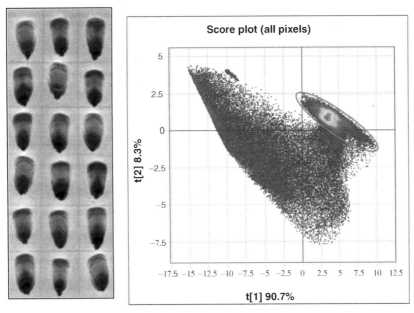

FIGURE 13.3 First PCA score image (left) and corresponding score plot (right) with background included (indicated with ellipse on score plot). The physical size of the image is 11.4 mm × 4.4 mm.

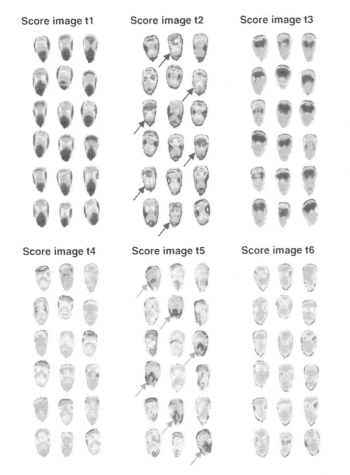

FIGURE 13.4 PCA score images (PC 1–6) after removal of background and other disturbances such as geometrical errors and shadows. Blue arrows indicate soft and green arrows hard maize kernels. (See the color version of this figure in Color Plate section.)

glassy or floury. Glassy and floury endosperm regions are distributed differently in hard, intermediate, and soft maize kernels. Hard kernels have more glassy endosperm (indicated in green) and less floury endosperm (indicated in blue). In soft kernels, this ratio is reversed. Classification was thus found to

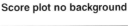
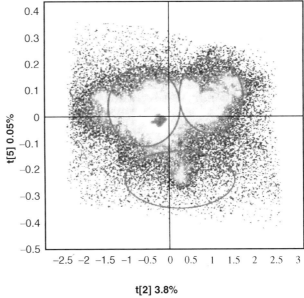

FIGURE 13.5 PCA score plot (PC2 versus PC5) after removal of background and other disturbances such as geometrical errors and shadows. The ellipses indicate clusters that are potentially soft, hard, and intermediate classes.

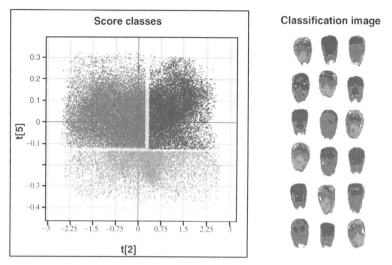

FIGURE 13.6   PCA score plot with classification (green = glassy, red = intermediate, and blue = floury) (left) and the corresponding classification projected onto the score image (right). (See the color version of this figure in Color Plate section.)

be rather in terms of endosperm texture and not in terms of whole hard, intermediate, or soft maize kernels. According to the cereal literature, maize kernels of different hardness contain glassy and floury endosperm in different ratios [82]. From the classification image (Figure 13.6), it is clear that kernels that were labeled as hard by the maize breeders had the maximum amount of glassy endosperm and kernels labeled as soft the maximum amount of floury endosperm. Kernels of intermediate hardness would thus be expected to have varying ratios of glassy and floury endosperm. However, from the score plot and classification image (Figure 13.6), a distinct cluster that could be associated with the larger part of the endosperm present in the maize kernels of intermediate

hardness had been observed. This seems to indicate that maize kernels of intermediate hardness contain not only different ratios of glassy and floury endosperm, but also an intermediate type of endosperm different in physical properties and chemical composition to that of glassy and floury endosperm.

13.3.1.5   Preprocessing of Data   NHI could identify three types of endosperm indicating that a range of endosperm textures could exist. The selection of the classes in Figure 13.6 is not very refined. More refined classes could be found after standard normal variate (SNV) preprocessing and these are shown in Figure 13.7. Standard normal variate is

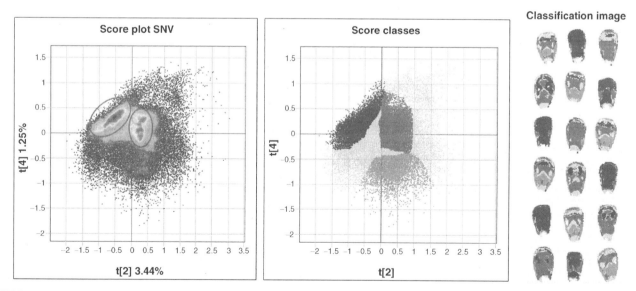

FIGURE 13.7   PCA score plot (PC2 versus PC4) after removal of background and SNV preprocessing (left). The ellipses indicate clusters representing glassy, intermediate, and floury endosperm. PCA score plot with classification (green = glassy, red = intermediate, and blue = floury) (middle) and the corresponding classification projected onto the score image (right). (See the color version of this figure in Color Plate section.)

a row-wise subtraction of the mean and division by standard deviation. Therefore, it entails removal of offset and slope differences. Principal components with %SS of 90.2, 3.44, 2.0, 1.25, and 0.46 were calculated.

Figure 13.7 shows the score plot of PC2 versus PC4 with an indication of possible clusters. In this figure, the clusters were made by interactive polygon selection between the score plot and the score image and are assumed to be more accurate. The final result is the classification image (Figure 13.7). Because of the more accurate description of the clusters, smaller areas of the score plot were selected. The larger unclassified areas are indicated in gray in the classification image. Contrary to the images where no preprocessing was applied, the results where SNV has been applied needed fewer components, indicating that SNV correction indeed is useful. One can make the assumption that the SNV correction removes some scattering effects and therefore needs fewer components for the classification.

13.3.1.6 Interpretation of Spectral Data

To do an interpretation of the endosperm types, it is important to look at spectral information. The average spectra for the glassy and floury endosperm classes are shown in Figure 13.8. It is clearly seen that the floury endosperm has an offset compared

to the glassy endosperm, indicating more scattering. For removing the scattering effect, SNV preprocessing was applied. By comparing the average SNV preprocessed spectra, small chemical differences could now be inferred. However, from the difference plot (Figure 13.8), these chemical differences become more prominent and can be more easily interpreted. Positive differences between glassy and floury endosperm were observed at 1450 (starch and water) and 1929 nm (protein and water). Negative differences between glassy and floury endosperm were observed at 2192 (protein) and 2298 nm (protein).

By combining the two classes and doing PCA on the SNV corrected data, a difference between glassy and floury endosperm was observed in the second PC score. The corresponding second loading line plot (not shown) was similar to the difference plot in Figure 13.8, thus corroborating the interpretation of the difference plot.

13.3.1.7 Partial Least Squares Discriminant Analysis

To test the classification, PLS-DA models were tested on the endosperm classes for the SNV corrected image. PLS-DA makes a regression model between the spectra and a vector of dummy variables. The dummy variable is, for example, 1 for glassy endosperm and 0 for floury endosperm. The quality of

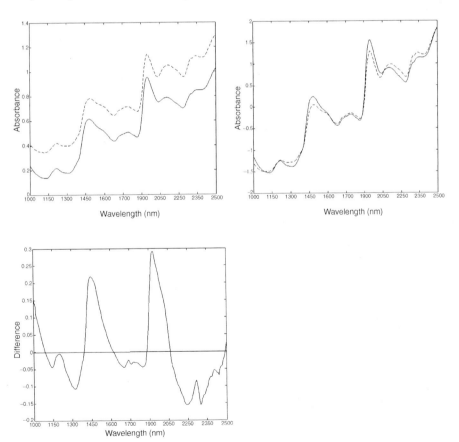

FIGURE 13.8 Average spectra of glassy (solid line) and floury endosperm types. Top left is without preprocessing and top right after SNV applied. Bottom left is the difference of SNV corrected spectra.

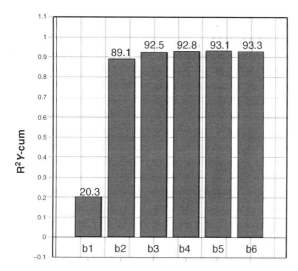

FIGURE 13.9 The evolution of the %SS (indicated as R^2y-cum) of y-variable as a function of the number of PLS components used in PLS-DA modeling.

the PLS-DA model indicated the meaningfulness of the subdivision in glassy and floury endosperm. The results of the PLS-DA model are shown in Figure 13.9, which shows that after three PLS components the explanation of the %SS of the y-variable (referred to as R^2Y-cum in Figure 13.9) was 92.5%. It is also possible to make similar models for training and test sets and to test the accuracy of the prediction of the respective endosperm classes.

13.4 CONSIDERATIONS FOR NIR HYPERSPECTRAL IMAGING AND DATA ANALYSIS OF FOOD PRODUCTS

13.4.1 Sampling and Sample Presentation

Because of the two-dimensional nature of imaging and the limited FOV of camera and lens combinations, the right choice of sample presentation is important. This maize hardness application was made using the maximum amount of kernels to fit the FOV of the camera (Figure 13.1). The shape of the maize kernels in addition requires careful illumination optimization. Still there is no correction for the fact that the kernels are not even or of similar thickness and have a rounded edge (cusp). This problem is often encountered when imaging raw food products.

13.4.2 Image Cleaning

The importance of removing background (Figure 13.3) and any bad pixels, optical, geometrical, or physical disturbance cannot be overstated. High-quality image analyses can be performed only on properly cleaned images. The data analysis presented in this chapter would not have succeeded without efficient removal of irrelevant pixels. By making a simple PCA model of the uncleaned image and using score plots and score images interactively, most irrelevant pixels can be identified and removed sequentially.

13.4.3 Final PCA Model, Cluster Detection, and Selection

After cleaning, a final PCA model encompassing many components can be made. Using score images (Figure 13.4) and score plots (Figure 13.5) interactively, the relevance of the PCs can be evaluated. Using combinations of the most relevant PCs, clusters can be detected based on pixel density (Figure 13.5). The final image is an overlay of selected colored fields on top of a score image, also called a classification image (Figures 13.6 and 13.8). The choice of clusters or classes in the score plots is, however, subjective. Still, the two selections shown (Figures 13.6 and 13.8) led to the same overall conclusions.

Chemometric models change when clusters are chosen differently, but they are still robust if the choices are properly done. It should also be kept in mind that histological background information of the sample should always be taken into account when selecting score plot clusters. It is quite likely that a person without knowledge of the composition of maize kernels could select incorrect classes in a score plot such as the one in Figure 13.5. The PLS-DA models with a high %SS of the y-variable and the accuracy of the predictions confirm that the correct classes for glassy and floury endosperm were chosen.

13.4.4 Penetration Depth

NHI as used in this application is two dimensional, but the images are integrals over a three-dimensional structure, where depth information is also included. It was observed by using stereomicroscopy of cut kernels that some regions are pure glassy endosperm, some pure floury endosperm, but some seemed to be a blend of both. These are the regions that were identified as intermediate.

The application shows additional information about penetration depth of the NIR radiation. Texture differences in the endosperm could be seen through the pericarp (thickness <0.2 mm), but the embryo (1.5–2 mm deep) was never clearly visible. These facts give an indication of the penetration depth in the imaging experiment. As a general rule, it is always advisable to find how deep the radiation penetrates the sample when making hyperspectral images.

13.5 CONCLUSION

The raw materials used in the food industry are heterogeneous and of an organic nature, making NHI an ideal complement to bulk NIR spectroscopy. Many research groups have been using homebuilt and commercial imaging systems to measure different food materials, mainly fruit, vegetables, meat, fish, and cereals. Most applications investigated until now utilized the range of a CCD or CMOS camera (400–1100 nm). Some applications required the InGaAs range (900–1700 nm), while only a few used the extended InGaAs or HgCdTe range (1000–2500 nm). Few applications used transmission or fluorescence with laser or UV lamp excitation, but reflection was the major mode used. The transition from laboratory use to online applications is still big, but it is getting easier as instrumentation improves in versatility, speed, and robustness.

Due to the huge amount of data generated by hyperspectral imaging, the hypercubes constitute extensive sources of chemometric applications, and there are thus infinite opportunities for making local models, both spatially and spectrally. The application of detection and identification of glassy and floury endosperm in maize kernels of different hardness shows how histological knowledge and chemometrics can be combined in a meaningful way.

Future hyperspectral imaging setups will hopefully have a wider wavelength range and better spatial resolution. This will require better detectors (cameras) and better illumination sources. Error-free cameras are also high on the wish list for the future. An interesting idea would be to develop equipment that integrates ultraviolet, visible, NIR, MIR, and Raman spectra with the same spatial resolution and registration.

At present, hyperspectral images are two dimensional. Three-dimensional hyperspectral images made by tomographic techniques would be a huge improvement, but this would require vast improvements in spectroscopic and camera hardware as well as computing power [83].

ACKNOWLEDGMENTS

Julian White (Specim, Spectral Imaging Ltd, Oulu, Finland) for use of the sisuChema instrument, David Nilsson and Oskar Jonsson (Umbio AB, Umeå, Sweden) for use of and help with Evince software, Pioneer Seed (Delmas, South Africa) for supplying samples, The South African–Swedish Research Partnership Programme Bilateral Agreement, National Research Foundation, South Africa (UID 60958 & VR 348-2006-6715) provided funding for exchange of researchers.

ABBREVIATIONS

BP	Back propagation
CART	Classification and regression trees
CCD	Charge-coupled device
CMOS	Complementary metal oxide semiconductor
DA	Discriminant analysis
FOV	Field of view
GA	Genetic algorithm
LCTF	Liquid crystal tunable filter
LDA	Linear discriminant analysis
MCT	Mercury–cadmium–telluride
MIR	Mid-infrared
MLR	Multiple linear regression
MRI	Magnetic resonance imaging
NHI	Hyperspectral imaging
NIR	Near-infrared
NN	Neural network
PC	Principal component
PCA	Principal component analysis
PCR	Principal component regression
PGP	Prism grating prism
PLS	Partial least squares
PLS-DA	Partial least squares discriminant analysis
QDA	Quadratic discriminant analysis
ROI	Region of interest
SNV	Standard normal variate
SS	Sum of squares
SVM	Support vector machine
SWIR	Short-wave infrared
UV	Ultraviolet
VIS	Visible

REFERENCES

1. Ben-Gera, I., and Norris, K. H. (1968a) Determination of moisture content in soybeans by direct spectrophotometry. *Isr. J. Agric. Res.* **18**, 125–132.
2. Ben-Gera, I., and Norris, K. H. (1968b) Direct spectrophotometric determination of fat and moisture in meat products. *J. Food Sci.* **33**, 64–67.
3. Williams, P. C. (1975) Application of near infrared reflectance spectroscopy to analysis of cereal grains and oilseeds. *Cereal Chem.* **52**, 561–576.
4. Williams, P. C., and Thompson, B. N. (1978) Influence of whole meal granularity on analysis of HRS spring wheat for protein and moisture by near infrared reflectance spectroscopy. *Cereal Chem.* **55**, 1014–1037.
5. Williams, P. C., Stevenson, S. G., Starkey, P. M., and Hawtin, G. C. (1978) The application of near infrared reflectance spectroscopy to protein testing in pulse breeding programmes. *J. Sci. Food Agric.* **29**, 285–292.
6. Williams, P. C. (1979) Screening wheat for protein and hardness by near infrared reflectance spectroscopy. *Cereal Chem.* **56**, 169–172.
7. Osborne, B. G. (1981) Application of near infrared reflectance spectroscopy to the analysis of food. *Anal. Proc.* **18**, 488–489.

8. Siesler, H. W., Ozaki, Y., Kawata, S., and Heise, H. M. (Eds.) (2002) *Near-Infrared Spectroscopy: Principles, Instruments, Applications*, Wiley, Weinheim, p. 361.

9. Osborne, B. G., Fearn, T., and Hindle, P. H. (1993) *Practical NIR Spectroscopy with Practical Applications in Food and Beverage Analysis*, 2nd ed., Longman Scientific and Technical, Harlow, p. 227.

10. Burns, D. A.,and Ciurczak, E. W. (Eds.) (2001) *Handbook of Near-Infrared Analysis*, 2nd ed., Marcel Dekker, Inc., New York, p. 814.

11. Williams, P.,and Norris, K. (Eds.) (2001) *Near-Infrared Technology in the Agricultural and Food Industries*, 2nd ed., American Association of Cereal Chemists, St. Paul, MN, p. 296.

12. Ozaki, Y., McClure, W. F.,and Christy, A. A. (Eds.) (2007) *Near-Infrared Spectroscopy in Food Science and Technology*, Wiley–Interscience, Hoboken, NJ, p. 424.

13. Blazquez, C., Downey, G., O'Donnell, C., O'Callaghan, D., and Howard, V. (2004) Prediction of moisture, fat and inorganic salts in processed cheese by near infrared reflectance spectroscopy and multivariate data analysis. *J. Near Infrared Spectrosc.* **12**, 149–157.

14. Baumgarten, G. (1987) The determination of alcohol in wines by means of near infrared technology. *S. Afr. J. Enol. Vitic.* **8**, 75–77.

15. Christopoulou, E., Lazaraki, M., Komaitis, M., and Kaselimis, K. (2004) Effectiveness of determinations of fatty acids and triglycerides for the detection of adulteration of olive oils with vegetable oils. *Food Chem.* **84**, 463–474.

16. Joubert, E., Manley, M., and Botha, M. (2006) The use of NIRS for quantification of mangiferin and hesperidin contents of dried, green honeybush (*Cyclopia genistoides*) plant material. *J. Agric. Food Chem.* **54**, 5279–5283.

17. Manley, M., Joubert, E., and Botha, M. (2006) Quantification of the major phenolic compounds, soluble solid content and total antioxidant activity of green rooibos (*Aspalathus linearis*) by means of near infrared spectroscopy. *J. Near Infrared Spectrosc.* **14**, 213–222.

18. Cozzolino, D., Kwiatkowski, M. J., Parker, M., Cynkar, W. U., Dambergs, R. G., Gishen, M., and Herderich, M. J. (2004) Prediction of phenolic compounds in red wine fermentations by visible and near infrared spectroscopy. *Anal. Chim. Acta* **513**, 73–80.

19. Downey, G., Byrne, S., and Dwyer, E. (1986) Wheat trading in the Republic of Ireland: the utility of a hardness index derived by near infrared reflectance spectroscopy. *J. Sci. Food Agric.* **37**, 762–766.

20. Osborne, B. G. (1991) Measurement of the hardness of wheat endosperm by near infrared spectroscopy. *Postharvest News Inform.* **2**, 331–334.

21. Windham, W. R., Gaines, C. S., and Leffler, R. G. (1993) Effect of wheat moisture content on hardness scores determined by near-infrared reflectance and on hardness score standardization. *Cereal Chem.* **70**, 662–666.

22. Manley, M., Van Zyl, L., and Osborne, B. G. (2002) Using Fourier transform near infrared spectroscopy in determining kernel hardness, protein and moisture content of whole wheat flour. *J. Near Infrared Spectrosc.* **10**, 71–76.

23. Manley, M., Van Zyl, L., and Osborne, B. G. (2001) Deriving a grain hardness calibration for Southern and Western Cape ground wheat samples by means of the particle size index (PSI) method and Fourier transform near infrared (FT-NIR) spectroscopy. *S. Afr. J. Plant Soil* **18**, 69–74.

24. Downey, G., Howard, V., Delahunty, C., O'Callaghan, D., Sheehan, E., and Guinee, T. (2005) Prediction of maturity and sensory attributes of Cheddar cheese using near-infrared spectroscopy. *Int. Dairy J.* **15**, 701–709.

25. Blazquez, C., Downey, G., O'Callaghan, D., Howard, V., Delahunty, C., and Sheehan, E. (2006) Modelling of sensory and instrumental texture parameters in processed cheese by near infrared reflectance spectroscopy. *J. Dairy Res.* **73**, 58–69.

26. Cozzolino, D., Smyth, H. E., Lattey, K. A., Cynkar, W., Janik, L., Damberg, R. G., Francis, I. L., and Gishen, M. (2005) Relationship between sensory analysis and near infrared spectroscopy in Australian Riesling and Chardonnay wines. *Anal. Chim. Acta* **539**, 341–348.

27. Geladi, P. and Grahn, H. F. (1996) *Multivariate Image Analysis*, Wiley, Chichester, UK, p. 316.

28. Grahn, H. F., and Geladi, P. (Eds.) (2007) Techniques and Applications of Hyperspectral Image Analysis, Wiley, Chichester, UK, p. 368.

29. Gowen, A. A., O'Donnell, C. P., Cullen, P. J., Downey, G., and Frias, J. M. (2007) Hyperspectral imaging: an emerging process analytical tool for food quality and safety control. *Trends Food Sci. Technol.* **18**, 590–598.

30. Wang W., and Paliwal, J. (2007) Near-infrared spectroscopy and imaging in food quality and safety. *Sens. Instrum. Food Qual. Saf.* **1**, 193–207.

31. Du, C., and Sun, D. (2004) Recent developments in the applications of image processing techniques for food quality evaluation. *Trends Food Sci. Technol.* **15**, 230–249.

32. Cogdill, R. P., Hurburgh, C. R., and Rippke, G. R. (2004) Single-kernel maize analysis by near-infrared hyperspectral imaging. *Trans. ASAE* **47**, 311–320.

33. Nansen, C., Kolomiets, M., and Gao, Z. (2008) Considerations regarding the use of hyperspectral imaging data in classifications of food products, exemplified by analysis of maize kernels. *J. Agric. Food Chem.* **56**, 2933–2938.

34. Weinstock, B. A., Janni, J., Hagen, L., and Wright, S. (2006) Prediction of oil and oleic acid concentrations in individual corn (*Zea mays* L.) kernels using near-infrared reflectance hyperspectral imaging and multivariate analysis. *Appl. Spectrosc.* **60**, 9–16.

35. Gorretta, N., Roger, J. M., Aubert, M., Bellon-Maurel, V., Campan, F., and Roumet, P. (2006) Determining vitreousness of durum wheat kernels using near infrared hyperspectral imaging. *J. Near Infrared Spectrosc.* **14**, 231–239.

36. Mahesh, S., Manickavasagan, A., Jayas, D. S., Paliwal, J., and White, N. D. G. (2008) Feasibility of near-infrared hyperspectral imaging to differentiate Canadian wheat classes. *Biosystems Eng.* **101**, 50–57.

37. Shahin, M. A., and Symons, S. J. (2008) Detection of hard vitreous and starchy kernels in amber durum wheat samples using hyperspectral imaging (GRL Number M306). *NIR News* **19**, 16–18.

38. Smail, V. W., Fritz, A. K., and Wetzel, D. L. (2006) Chemical imaging of intact seeds with NIR focal plane array assists plant breeding. *Vib. Spectrosc.* **42**, 215–221.

39. Berman, M., Connor, P. M., Whitbourn, L. B., Coward, D. A., Osborne, B. G., and Southan, M. D. (2007) Classification of sound and stained wheat grains using visible and near infrared hyperspectral image analysis. *J. Near Infrared Spectrosc.* **15**, 351–358.

40. Peirs, A., Scheerlinck, N., De Baerdemaeker, J., and Nicolai, B. M. (2003) Starch index determination of apple fruit by means of a hyperspectral near infrared reflectance imaging system. *J. Near Infrared Spectrosc.* **11**, 379–389.

41. Nicolai, B. M., Lotze, E., Peirs, A., Scheerlinck, N., and Theron, K. I. (2006) Non-destructive measurement of bitter pit in apple fruit using NIR hyperspectral imaging. *Postharvest Biol. Technol.* **40**, 1–6.

42. Xing, J., Jancsók, P., and De Baerdemaeker, J. (2007) Stem-end/calyx identification on apples using contour analysis in multispectral images. *Biosystems Eng.* **96**, 231–237.

43. Peng, Y. K., and Lu, R. F. (2008) Analysis of spatially resolved hyperspectral scattering images for assessing apple fruit firmness and soluble solids content. *Postharvest Biol. Technol.* **48**, 52–62.

44. Xing, J., Saeys, W., and De Baerdemaeker, J. (2007) Combination of chemometric tools and image processing for bruise detection on apples. *Comput. Electron. Agric.* **56**, 1–13.

45. ElMasry, G., Wang, N., Vigneault, C., Qiao, J., and ElSayed, A. (2008) Early detection of apple bruises on different background colors using hyperspectral imaging. *LWT Food Sci. Technol.* **41**, 337–345.

46. Xing, J., Bravo, C., Jancsók, P. T., Ramon, H., and De Baerdemaeker, J. (2005) Detecting bruises on Golden Delicious apples using hyperspectral imaging with multiple wavebands. *Biosystems Eng.* **90**, 27–36.

47. Xing, J., Van Linden, V., Vanzeebroeck, M., and De Baerdemaeker, J. (2005) Bruise detection on Jonagold apples by visible and near-infrared spectroscopy. *Food Control* **16**, 357–361.

48. Cheng, X., Chen, Y. R., Tao, Y., Wang, C. Y., Kim, M. S., and Lefcourt, A. M. (2004) A novel integrated PCA and FLD method on hyperspectral image feature extraction for cucumber chilling damage inspection. *Trans. ASAE* **47**, 1313–1320.

49. Liu, Y. L., Chen, Y. R., Wang, C. Y., Chan, D. E., and Kim, M. S. (2005) Development of a simple algorithm for the detection of chilling injury in cucumbers from visible/near-infrared hyperspectral imaging. *Appl. Spectrosc.* **59**, 78–85.

50. Ariana, D. P., Lu, R., and Guyer, D. E. (2006) Near-infrared hyperspectral reflectance imaging for detection of bruises on pickling cucumbers. *Comput. Electron. Agric.* **53**, 60–70.

51. Ariana, D. P., and Lu, R. (2008) Quality evaluation of pickling cucumbers using hyperspectral reflectance and transmittance imaging: part I. Development of a prototype. *Sens. Instrum. Food Qual. Saf.* **2**, 144–151.

52. Ariana, D. P., and Lu, R. (2008) Quality evaluation of pickling cucumbers using hyperspectral reflectance and transmittance imaging: part II. Performance of a prototype. *Sens. Instrum. Food Qual. Saf.* **2**, 152–160.

53. Sivakumar, S. S. (2006) Potential applications of hyperspectral imaging for the determination of total soluble solids, water content and firmness in mango. MSc Thesis, Department of Bioresource Engineering, Macdonald Campus, McGill University, Montreal, Quebec, Canada.

54. Lu, R. F., and Peng, Y. K. (2006) Hyperspectral scattering for assessing peach fruit firmness. *Biosystems Eng.* **93**, 161–171.

55. ElMasry, G., Wang, N., ElSayed, A., and Ngadi, M. (2007) Hyperspectral imaging for nondestructive determination of some quality attributes for strawberry. *J. Food Eng.* **81**, 98–107.

56. Qin, J., and Lu, R. (2005) Detection of pits in tart cherries by hyperspectral transmission imaging. *Trans. ASAE* **48**, 1963–1970.

57. Long, R. L., Walsh, K. B., and Greensill, C. V. (2005) Sugar "imaging" of fruit using a low cost charge-coupled device camera. *J. Near Infrared Spectrosc.* **13**, 177–186.

58. Martinsen, P., and Schaare, P. (1998) Measuring soluble solids distribution in kiwifruit using near-infrared imaging spectroscopy. *Postharvest Biol. Technol.* **14**, 271–281.

59. Kim, M. S., Lefcourt, A. M., Chao, K., Chen, Y. R., Kim, I., and Chan, D. E. (2002) Multispectral detection of fecal contamination on apples based on hyperspectral imagery. Part I. Application of visible and near-infrared reflectance imaging. *Trans. ASAE* **45**, 2027–2037.

60. Kim, M. S., Lefcourt, A. M., Chen, Y. R., Kim, I., Chan, D. E., and Chao, K. (2002) Multispectral detection of fecal contamination on apples based on hyperspectral imagery: part II. Application of hyperspectral fluorescence imaging. *Trans. ASAE* **45**, 2039–2047.

61. Mehl, P. M., Chen, Y.-R., Kim, M. S., and Chan, D. E. (2004) Development of hyperspectral imaging technique for the detection of apple surface defects and contaminations. *J. Food Eng.* **61**, 67–81.

62. Kim, M. S., Lefcourt, A. M., Chen, Y. R., and Kang, S. (2004) Uses of hyperspectral and multispectral laser induced fluorescence imaging techniques for food safety inspection. *Key Eng. Mater.*, **270–273**, 1055–1063.

63. Liu, Y. L., Chen, Y. R., Kim, M. S., Chan, D. E., and Lefcourt, A. M. (2007) Development of simple algorithms for the detection of fecal contaminants on apples from visible/near infrared hyperspectral reflectance imaging. *J. Food Eng.* **81**, 412–418.

64. Gómez-Sanchis, J., Gómez-Chova, L., Aleixos, N., Camps-Valls, G., Montesinos-Herrero, C., Moltó, E., and Blasco, J. (2008) Hyperspectral system for early detection of rottenness caused by *Penicillium digitatum* in mandarins. *J. Food Eng.* **89**, 80–86.

65. Naganathan, G. K., Grimes, L. M., Subbiah, J., Calkins, C. R., Samal, A., and Meyer, G. E. (2008) Visible/near-infrared hyperspectral imaging for beef tenderness prediction. *Comput. Electron. Agric.* **64**, 225–233.

66. Qiao, J., Ngadi, M. O., Wang, N., Gariépy, C., and Prasher, S. O. (2007) Pork quality and marbling level assessment using a hyperspectral imaging system. *J. Food Eng.* **83**, 10–16.

67. Qiao, J., Wang, N., Ngadi, M. O., Gunenc, A., Monroy, M., Gariépy, C., and Prasher, S. O. (2007) Prediction of drip-loss, pH, and color for pork using a hyperspectral imaging technique. *Meat Sci.* **76**, 1–8.

68. Wold, J. P., Johansen, I-R., Haugholt, K. H., Tschudi, J., Thielemann, J., Segtnan, V. H., Narum, B., and Wold, E. (2006) Non-contact transflectance near infrared imaging for representative on-line sampling of dried salted coalfish (bacalao). *J. Near Infrared Spectrosc.* **14**, 59–66.

69. ElMasry, G., and Wold, J. P. (2008) High-speed assessment of fat and water content distribution in fish fillets using online imaging spectroscopy. *J. Agric. Food Chem.* **56**, 7672–7677.

70. Heia, K., Sivertsen, A. H., Stormo, S. K., Elvevoll, E., Wold, J. P., and Nilsen, H. (2007) Detection of nematodes in cod (*Gadus morhua*) fillets by imaging spectroscopy. *J. Food Sci.* **72**, E11–E15.

71. Wold, J. P., Westad, F., and Heia, K. (2001) Detection of parasites in cod filets by using SIMCA classification in multispectral images in the visible and NIR region. *Appl. Spectrosc.* **55**, 1025–1034.

72. Lawrence, K. C., Windham, W. R., Park, B., and Buhr, R. J. (2003) A hyperspectral imaging system for identification of faecal and ingesta contamination on poultry carcasses. *J. Near Infrared Spectrosc.* **11**, 269–281.

73. Lawrence, K. C., Windham, W. R., Park, B., Heitschmidt, G. W., Smith, D. P., and Feldner, P. (2006) Partial least squares regression of hyperspectral images for contaminant detection on poultry carcasses. *J. Near Infrared Spectrosc.* **14**, 223–230.

74. Park, B., Lawrence, K. C., Windham, W. R., and Smith, D. P. (2006) Performance of hyperspectral imaging system for poultry surface fecal contaminant detection. *J. Food Eng.* **75**, 340–348.

75. Park, B., Windham, W. R., Lawrence, K. C., and Smith, D. P. (2007) Contaminant classification of poultry hyperspectral imagery using a spectral angle mapper algorithm. *Biosystems Eng.* **96**, 323–333.

76. Chao, K., Mehl, P. M., and Chen, Y. R. (2002) Use of hyper- and multi-spectral imaging for detection of chicken skin tumors. *Appl. Eng. Agric.* **18**, 113–119.

77. Chao, K., Chen, Y. R., Hruschka, W. R., and Park, B. (2001) Chicken heart disease characterization by multi-spectral imaging. *Appl. Eng. Agric.* **17**, 99–106.

78. Monteiro, S. T., Minekawa, Y., Kosugi, Y., Akazawa, T., and Oda, K. (2007) Prediction of sweetness and amino acid content in soybean crops from hyperspectral imagery. *ISPRS J. Photogramm. Remote Sens.* **62**, 2–12.

79. Jiang, L., Zhu, B., Rao, X., Berney, G., and Tao, Y. (2007) Discrimination of black walnut shell and pulp in hyperspectral fluorescence imagery using Gaussian kernel function approach. *J. Food Eng.* **81**, 108–117.

80. Dubois, J., Lewis, N., Fry, J., Frederick S., and Calvey, E. M. (2005) Bacterial identification by near-infrared chemical imaging of food-specific cards. *Food Microbiol.* **22**, 577–583.

81. Gowen, A. A., O'Donnell, C. P., Taghizadeh, M., Cullen, P. J., Frias, J. M., and Downey, G. (2008) Hyperspectral imaging combined with principal component analysis for bruise damage detection on white mushrooms (*Agaricus bisporus*). *J. Chemom.* **22**, 259–267.

82. Watson, S. A. and Ramstad, P. E. (Eds.) (1987) *Corn: Chemistry and Technology*, American Association of Cereal Chemists, St. Paul, MN, p. 605.

83. Kemsley, E. K., Tapp, H. S., Binns, R., Mackin, R. O., and Peyton, J. A. (2008) Feasibility study of NIR diffuse optical topography on agricultural produce. *Postharvest Biol. Technol.* **48**, 223–230.

PART V

APPLICATIONS IN POLYMER RESEARCH

14

VIBRATIONAL SPECTROSCOPY IMAGING OF POLYMERS

HARUMI SATO AND YUKIHIRO OZAKI

Department of Chemistry, School of Science and Technology, Kwansei Gakuin University, Sanda, Japan

JIANHUI JIANG AND RU-QIN YU

State Key Laboratory of Chemo/Biosensing and Chemometrics, Hunan University, Changsha, China

HIDEYUKI SHINZAWA

Research Institute of Instrumentation Frontier, Advanced Industrial Science and Technology (AIST), Chubu, Nagoya, Japan

14.1 INTRODUCTION

Vibrational spectroscopy is a well-established method to investigate structure and dynamics of polymers [1–5]. One can study constitution, configuration, conformation, and inter- and intramolecular interactions (e.g., hydrogen bondings) of polymers by use of IR and Raman spectroscopy. Both IR and Raman spectroscopy have been used extensively for a variety of polymer researches from basic studies to applications such as structural studies, miscibility, phase transition, crystallization, hydrogen bonding, thermal and mechanical properties, and polymer reactions. IR and Raman spectroscopy are often complementary. IR spectroscopy yields strong bands due to vibrational modes of functional groups with strong polarization, such as OH and C=O stretching modes, while Raman spectroscopy gives intense bands due to vibrational modes of functional groups having large polarizability, such as SS, CX (X = Cl, Br, S), C=C, and C=N stretching modes. Bands due to local vibrational modes such as CH_2 rocking modes appear strongly in IR spectra while those arising from stretching modes of a whole molecule or a large part of molecule, such as accordion modes, emerge strongly in Raman spectra.

For IR spectroscopy of polymers, not only transmittance spectroscopy but also attenuated total reflection (ATR) spectroscopy, reflection spectroscopy, reflection–absorption (RA) spectroscopy, time-resolved spectroscopy, and microspectroscopy are often employed. IR linear dichroism is very important for the orientation measurements of polymers. As for Raman spectroscopy of polymers, not only normal Raman spectroscopy but also resonance Raman spectroscopy, surface-enhanced Raman scattering (SERS), time-resolved-Raman spectroscopy, and Raman microscopy are very useful. One can find a number of good examples of IR and Raman spectroscopic studies of polymers in many references [1–5].

Near-infrared (NIR) spectroscopy has also been utilized in polymer researches and applications [1, 6]. NIR spectroscopy has often been the choice in many practical applications of polymers, such as measurements and predictions of physical properties like density, particle size and crystallinity, online monitoring, and quality control. However, it is also very important to point out that NIR spectroscopy finds its uniqueness in the fundamental polymer research. In fact, it has been used to investigate hydrogen bonding, inter- and intramolecular interactions, polymer reactions, physical

---

Raman, Infrared, and Near-Infrared Chemical Imaging Edited by Slobodan Šašić and Yukihiro Ozaki

properties such as thermal and mechanical properties, and diffusion of solvents in polymers.

14.2 VIBRATIONAL SPECTROSCOPY IMAGING OF POLYMERS

Vibrational spectroscopy imaging of polymers is natural extension of their vibrational spectroscopy mapping [7, 9]. The latter, particularly IR and Raman mapping of polymers, was developed during the past two decades. The development of NIR mapping of polymers was delayed by one decade. In the field of polymer spectroscopy, vibrational spectroscopy mapping has been employed to identify contaminations in polymers and polymeric materials, to determine distributions of components in them, and to investigate structure and morphology of polymers, polymer blends, and polymer composites. The most serious problem in vibrational mapping is that it takes a long time to obtain a mapping. There is no doubt that imaging is much more powerful than mapping. The general history, advantages, and instrumentation of vibrational spectroscopy imaging are described in previous chapters, and thus in this chapter, the advantages are described mainly from the point of polymer applications.

Vibrational spectroscopy imaging has the following advantages in polymer science and technology [7–35]: (i) By a combination of vibrational spectroscopy and digital imaging techniques, one can obtain information about the distribution of components in polymers or the distribution of different morphologies in them. (ii) It is possible to explore dynamic processes such as polymer dissolution on the timescale of the image acquisition process by observing larger sample areas with high spatial resolution.

FT-IR and Raman imaging has often been utilized to investigate phase separation, miscibility, and morphology in polymer blends. For example, Vogel et al. [10] studied phase separation in blends of poly(3-hydroxybutyrate) (PHB) with poly(L-lactic acid) (PLLA) and poly(ε-caprolactone) (PCL) by using FT-IR imaging. Oh et al. [11] reported an FT-IR imaging study on phase-separated morphology of poly(styrene-*co*-allyl alcohol)/polyester blends. Chernev and Wilhelm [12] and Snively and Koenig [13] demonstrated the usefulness of polarization radiation for the production of FT-IR images for the first time. Vogel et al. [14] combined rheo-optical measurement with FT-IR imaging to investigate anisotropic polymer blends.

Wilhelm et al. [15, 16] insisted on the importance of a combination of IR and electron microscopy for characterization of polymer morphologies. They investigated lateral and depth resolution in FT-IR and Raman imaging in comparison with those in SEM. Gupper et al. [18] studied morphology of polymer blends by using Raman imaging.

NIR imaging is relatively new in polymer fields although it has already been used extensively in pharmaceutical applications [19]. Furukawa et al. [20] used this technique to evaluate the homogeneity of binary blends of PHB and PLLA. Shinzawa et al. [35] explored the effect of the grinding on cellulose excipient at the molecular level by NIR imaging.

FT-IR imaging has recently been employed to explore polymer dissolution [21–24]. These sort of attempts are successful particularly for systems evolving rather slowly, because the temporal resolution of the method is low. Koenig and coworkers [21–23] reported investigations on FT-IR imaging of dissolution of polymers. FT-IR imaging allows the detection of initial chemical or physical imperfections in the sample being studied and those generated during the diffusion process. Gupper and Kazarian [24] studied solvent diffusion and solvent-induced crystallization in syndiotactic polystyrene using FT-IR spectroscopy and imaging.

Michaels et al. [25] applied near-field IR imaging and spectroscopy to study a thin film polystyrene/poly(ethyl acrylate) blend. The instrument couples the nanoscale special resolution of scanning probe microscopy with the chemical specificity of vibrational spectroscopy. Its key features include broad tenability and bandwidth, parallel spectral detection for high image acquisition rates, and IR-transparent aperture probes.

Patterson et al. [26] established the ability to collect IR microspectroscopic images of large areas using a large radius hemisphere internal reflection element (IRE) with both a single point and a linear array detector, and used this system for a polymer film research.

14.3 FT-IR IMAGING OF POLYMERS

14.3.1 FT-IR Imaging Study of Phase Separation in Polymer Blends

Recently, Vogel et al. [10] investigated phase separation in blends of PHB with PLLA and PCL as a function of the blend composition by using FT-IR imaging. Chemical structures of PHB, PLLA, and PCL are shown in Figure 14.1. PHB belongs to the poly(hydroxyalkanoates) (PHAs) group, which is synthesized by bacteria from renewable resources [36, 37]. PHB has received much attention as an environment friendly material because of its thermoplasticity coupled with its biodegradability. However, PHB is stiff and rigid because of the perfectly isotactic structure consisting exclusively of the R configuration. To improve its mechanical properties, PHB has to be copolymerized or blended with other polymers: PLLA [38], PCL [39], poly(ethyleneoxide) (PEO) [40], and so on. PHB/PLLA (50/50) blend shows sea–island structure while PHB/PLLA (30/70) blend yields a homogeneous one-phase polymer system.

Figure 14.2 compares FT-IR spectra of the individual blend components, PHB and PLLA (a) with the spectrum

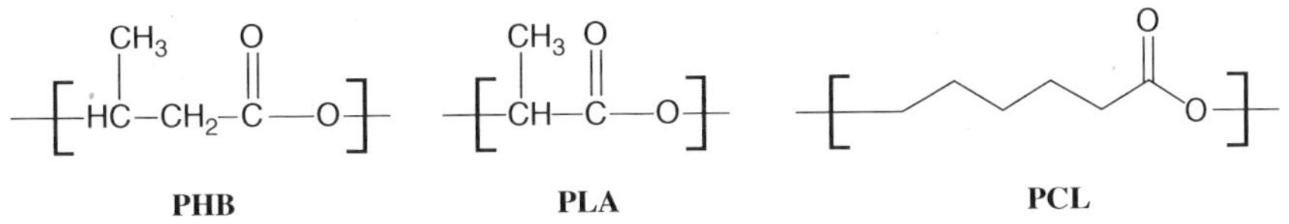

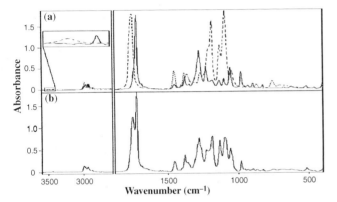

FIGURE 14.1 Chemical structures of PHB, PLA, and PCL. (Reproduced from Ref.10 with permission. Copyright 2008 American Chemical Society.)

of PHB/PLLA (50:50 wt%) blend (b) [10]. In Figure 14.2a, one can find several absorption bands that are specific for the blend components, PHB and PLLA. To compare the PHB/PLLA blends in terms of phase homogeneity, the intensities of the C=O stretching bands at 1723 cm^{-1} for PHB and 1759 cm^{-1} for PLLA were used, respectively. Figure 14.3 depicts visible images and PLLA- and PHB-specific FT-IR images of PHB/PLLA (50:50 wt%) (a) and PHB/PLLA

(30:70 wt%) (b) blends [10]. It can be seen from Figure 14.3 that the 50:50 blend has an island structure with a size of 30–40 μm. Of note is that the PLLA- and PHB-specific FT-IR images are complementary. In contrast to the 50:50 blend, the 30:70 blend does not show phase separation in the visible and FT-IR images. From these results, Vogel et al. [10] concluded that the 50:50 blend is phase separated while the 30:70 blend is a compatible one-phase system. The I_{1723}/I_{1759} intensity ratio was used to compare these blends in terms of homogeneity.

Figure 14.4 plots the ratio versus the concentration of PLLA (wt%) for the PHB-rich and PLLA-rich areas detected in the FT-IR imaging of the blends with different compositions [10]. The plot illustrates that the blends with 15, 30, 60, 70, and 85 wt% PLLA show only small differences in the intensity ratio between the PHB-rich and PLLA-rich areas. This finding indicates that these blends hold homogeneous one-phase polymer systems. On the other hand, the blends with 40 and 50 wt% PLLA show significant differences in the ratio. It is very likely that these blends are separated in two phases with different PHB/PLLA compositions. A miscibility gap was also found from the FT-IR imaging around the 50/50 (w/w) composition for the PHB/PLLA blends [10].

Vogel et al. [14] reports FT-IR imaging obtained with polarized radiation on anisotropic PHB/PLA blends. They

FIGURE 14.2 (a) FT-IR spectra of PHB (—) and PLA (---). (b) FT-IR spectrum of a PHB/PLA (50:50 wt%) blend. (Reproduced from Ref.10 with permission. Copyright 2008 American Chemical Society.)

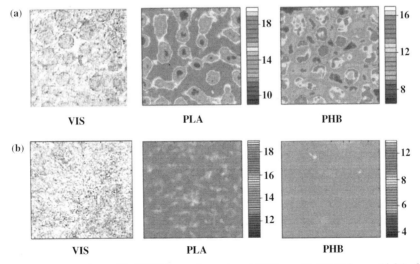

FIGURE 14.3 (a) Visual image (left), PLA-specific FT-IR image (center), and PHB-specific FT-IR image (right) of a PHB/PLA (50:50 wt%) blend. (b) Visual image (left), PLA-specific FT-IR image (center), and PHB-specific FT-IR image (right) of a PHB/PLA (30:70 wt%) blend. (Reproduced from Ref.10 with permission. Copyright 2008 American Chemical Society) (See the color version of this figure in Color Plate section.)

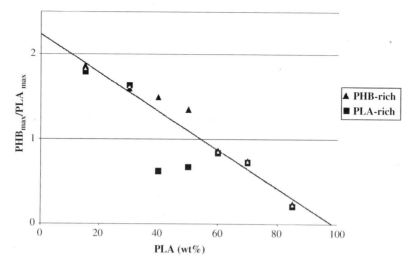

FIGURE 14.4 PHB_{max}/PLA_{max} ratio versus the content of PLA for the PHB- and PLA-rich image areas of the particular blend (see text). (Reproduced from Ref.10 with permission. Copyright 2008 American Chemical Society.)

previously investigated these blends by rheo-optical FT-IR spectroscopy [1, 2], which combines a stress–strain test with *in situ* polarization measurements to detect the structural information on a molecular level simultaneously to the mechanical treatment. They observed interesting orientation phenomena of the PHB and PLA chains in the investigations of the mechanical elongation of PLA-rich (≥ 60 wt% PLA) PHB/PLA blend films (35°C, 10% strain per minute). In these blends, the PLA chains orient in the direction of elongation whereas the PHB chains orient perpendicular to the drawing direction [40]. PHB/PLA blend films with PHB > PLA composition, on the other hand, show mechanical properties similar to PHB homopolymer and could only be oriented by cold drawing in ice water after quenching from the melt [41].

In the FT-IR imaging study, they demonstrated that the FT-IR imaging technique with polarized radiation provides superior details in terms of the characterization of orientation phenomena in anisotropic materials compared to imaging data with unpolarized radiation as well as to dichroic measurements with a single element detector.

To investigate variations in chain orientation induced by the mechanical treatment, the C=O stretching bands of the PHB/PLA blend films were employed to calculate the orientation function $f_\perp$ (assuming a perpendicular transition moment of the C=O absorption bands relative to the polymer chain direction) by

$$f_\perp = -2\frac{R-1}{R+2}$$

where $R = A_\parallel/A_\perp$ is the dichroic ratio of the C=O stretching bands in the polarization spectra. The peak areas under the left wing from 1825 to 1779 cm^{-1} and under the right wing from 1718 to 1691 cm^{-1} were assumed to be characteristic of the PLA and PHB components, respectively. To monitor

changes in the state of order as a function of the mechanical treatment, the structural absorbance A_0, which eliminates the effect of orientation on band intensities, was used:

$$A_0 = \frac{A_\parallel + 2A_\perp}{3}$$

Figure 14.5 shows an optical image (a) and FT-IR images (3.9×3.9 mm^2) of A_{0PHB}/A_{0PLA} (b) and A_{0PHB}/A_{0PLA} (c) and the corresponding orientation function ($f_\perp$) images of PHB (d) and PLA (e) of the 50% stretched PHB/PLA (50/50 wt%) blend film [14]. It can be seen from Figure 14.5b and c that the "islands" are PHB-rich while the matrix has a higher PLA content.

The corresponding orientation function ($f_\perp$) images (Figure 14.5d and e) show that the PHB chains in the "islands" assume a negative orientation ($f_\perp \approx -0.4$) while the PLA chains orient positively in the same domains ($f_\perp \approx 0.3$). In contrast, in the matrix both PHB and PLA orient only very slightly positive ($f_\perp$ between 0 and 0.1). Thus, the two phases of the unstretched PHB/PLA (50/50 wt%) blend film with uniform thickness respond differently to the applied mechanical stress: the PHB-rich phase is extended to higher degrees and lower thickness with opposite orientations of the two polymer components (PHB negative, PLA positive), whereas the PLA-rich phase undergoes only a small elongation with negligible thickness reduction and very low positive orientation for both polymer composites. Figure 14.6 illustrates schematic representation of the orientation mechanism in the PHB-rich domains of a phase-separated PHB/PLA (50/50 wt%) blend film [14].

Vogel et al. [14] measured FT-IR polarization spectra of the whole area, the matrix area, and "island" area of the 50% elongated PHB/PLA (50/50 wt%) blend film. It was found that the polarization spectra measured with a single element

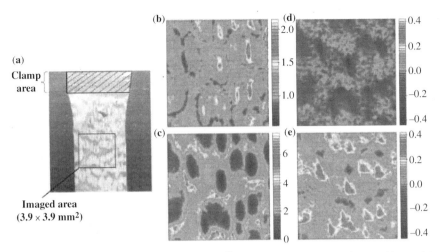

FIGURE 14.5 Optical image (a) and FT-IR images (3.9 × 3.9 mm²) of A_{0PHB}/A_{0PLA} (b) and A_{0PLA}/A_{0PHB} (c) and the corresponding orientation function ($f_\perp$) images of PHB (d) and PLA (e) of the 50% stretched PHB/PLA (50/50 wt%) blend film (for optimum comparison the $f_\perp$ images (d) and (e) are shown with the same color scale). (Reproduced from Ref.14 with permission. Copyright 2008 American Chemical Society) (See the color version of this figure in Color Plate section.)

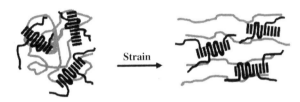

FIGURE 14.6 Schematic representation of the orientation mechanism in the PHB-rich domains of a phase-separated PHB/PLA (50/50 wt%) blend film (black: PHB chains; gray: PLA chains). (Reproduced from Ref.14 with permission. Copyright 2008 American Chemical Society.)

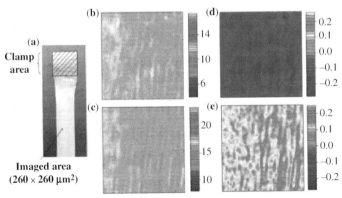

FIGURE 14.7 Optical image (a) and FT-IR images (260 × 260 μm²) of A_{0PHB} (b) and A_{0PLA} (c) and the corresponding orientation function (f_∞) images of PHB (d) and PLA (e) of the 200% stretched PHB/PLA (40/60 wt%) blend film (for optimum comparison the f_∞ images (d) and (e) are shown with the same color scale). (Reproduced from Ref.14 with permission. Copyright 2008 American Chemical Society.) (See the color version of this figure in Color Plate section.)

detector cannot discriminate the different orientation mechanisms in the phase-separated, anisotropic structure of the PHB/PLA blend.

Figure 14.7 depicts an optical image (a) and FT-IR images (260 × 260 μm²) of A_{0PHB} (b) and A_{0PLA} (c) and the corresponding orientation function ($f_\perp$) images of PHB (d) and PLA (e) of the 200% stretched PHB/PLA (40/60 wt%) blend film [14]. In previous studies, PHB/PLA blend films with PLA contents >50 wt% were classified as miscible. The images in Figure 14.7b and c display slightly streaky patterns with reference to the distribution of PHB and PLA over the sampled area. It was found from Figure 14.7d that over the whole area PHB shows a negative orientation function $f_\perp$ in the range from −0.05 to −0.25. The orientation function $f_\perp$ of PLA is positive over the whole image (0.05–0.25) as can be seen in Figure 14.7e.

In this way, in the orientation function ($f_\perp$) images an opposite orientation (PHB negative, PLA positive) was detected over the whole image area. With reference to the streak pattern for the PLA orientation function image, a lower positive chain alignment could be detected for the PLA-rich areas.

14.3.2 FT-IR Imaging Study of Polymer Dissolution by Solvent Mixtures

Miller-Chou and Koenig [21] reported an FT-IR imaging study of the dissolution of entangled poly(α-methylstyrene) (PAMS) in binary solvent mixtures of systematically varied amounts of methyl isobutyl ketone (MIBK) in deuterated cyclohexane (C_6D_{12}). The FT-IR imaging study revealed that, in many of the solvent systems, the solvent did not resolve the polymer uniformly at the polymer–solvent interface, causing cracking and roughening of the polymer edge, which can be directly seen in the images.

Ribar et al. [42] monitored *in situ* dissolution of PAMS in solvent mixtures of MIBK and cyclohexane-*d* by using FT-IR imaging earlier than Miller-Chou and Koenig [21]. The PAMS studied by Ribar et al. [42] was below the entanglement molecular weight. It was found that the dissolution of the entangled PAMS is very different from the

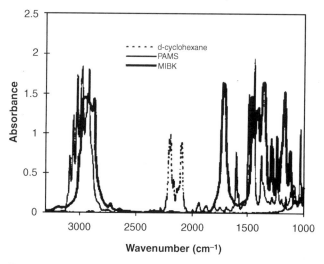

FIGURE 14.8 FT-IR spectra and peaks used to monitor each component (C$_6$D$_{12}$, PAMS, MIBK) in the system. (Reproduced from Ref. 21 with permission. Copyright 2008 American Chemical Society.)

unentangled system. Miller-Chou and Koenig [21] found evidence of solvent segregation, which was not seen in the unentangled PAMS. The entangled PAMS did not dissolve uniformly at the polymer–solvent interface. It is very likely that when the solvents ingress into the polymer, pressure builds up due to limited segmental mobility of the entangled chains and high vapor pressure of cyclohexane-*d*. Large amounts of stress energy are frozen into the polymer in the glass transition. Thus, stress is relieved by cracking, and when the cracks combine, they cause small blocks of the polymer to break away from the bulk polymer.

Figure 14.8 shows FT-IR spectra of PAMS, MIBK, and C$_6$D$_{12}$ [21]. Each component of the system was monitored by a characteristic IR band. A peak at 1600 cm^{-1} due to the ring quadrant stretching mode of PAMS, a band at 1720 cm^{-1} arising from the C=O stretching mode of MIBK, and a peak edge of 2148 cm^{-1} assigned to the CD stretching mode of C$_6$D$_{12}$ were used to characterize each component.

Figure 14.9 shows spectral images showing the concentration of PAMS and MIBK during dissolution of PAMS [21]. Of particular interest is that dissolution did not occur uniformly over the polymer–solvent interface, creating physical peaks and crevices. A similar interface roughening was observed when pure C$_6$D$_{12}$ was used as a solvent.

Figure 14.10 displays spectral images showing the concentration of PAMS, MIBK, and C$_6$D$_{12}$ during dissolution of

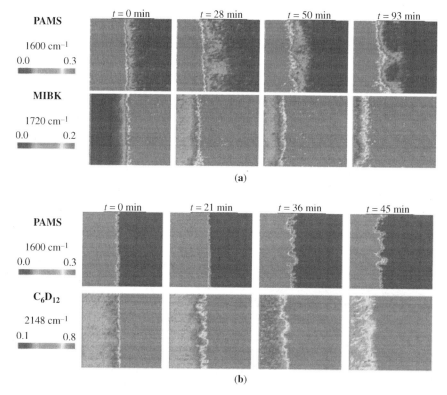

FIGURE 14.9 (a) Spectral images showing the concentration of PAMS and MIBK during dissolution of PAMS. (b) Spectral images showing the concentration of PAMS and cyclohexane-*d* during dissolution of PAMS. (Reproduced from Ref. 21 with permission. Copyright 2008 American Chemical Society.)

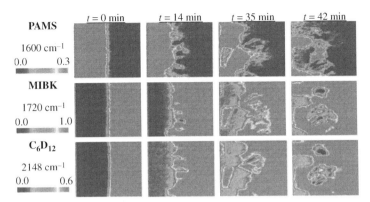

FIGURE 14.10 Spectral images showing the concentration of PAMS, MIBK, and C_6D_{12} during dissolution of PAMS in an 85:15 C_6D_{12}: MIBK solvent solution. (Reproduced from Ref. 21 with permission. Copyright 2008 American Chemical Society.)

PAMS in an 85:15 C_6D_{12}:MIBK solvent solution [21]. It can be seen from these images that the polymer did not dissolve uniformly and blocks of polymer erupted from the bulk polymer interface. However, upon repeating the experiment, the polymer appeared to dissolve uniformly with only a slight roughening of the polymer–solvent interface. Figure 14.11 depicts spectral images showing the concentration of PAMS with corresponding black and white imaging of C_6D_{12} at the same time. When cracks meet, a polymer chunk breaks away from the bulk polymer.

This is a good example of FT-IR imaging studies of polymer dissolution by solvent mixtures.

14.3.3 FT-IR Spectroscopy and Imaging Study of Solvent Diffusion and Solvent-Induced Crystallization in Syndiotactic Polystyrene

Gupper and Kazarian [24] investigated kinetics of solvent diffusion (chloroform) and solvent-induced crystallization in syndiotactic polystyrene (sPS) by using FT-IR transmission imaging and single element detector transmission FT-IR spectroscopy. The appearance of δ crystalline sPS was

monitored as a function of solvent exposure time by spatially resolved information from imaging experiments carried out under controlled environmental conditions (temperature and solvent vapor pressure) and with uniaxial solvent diffusion into the polymer. Polymer crystallization kinetics at various positions in the polymer and solvent diffusion coefficients were determined from a series of time-resolved FT-IR images. It was found from the imaging experiments that solvent diffusion is the limiting factor in the overall crystallization process of an sPS sample.

Solvent-induced crystallization in polymers is fundamental for processing and applications of polymeric materials. Recent rapid development of FT-IR imaging enables one to explore dynamic processes of polymers, such as morphological changes and crystallization in polymer/solvent systems. Gupper and Kazarian [24] introduced a novel application, namely, spatially resolved *in situ* investigations on the kinetics of solvent-induced polymer crystallization by FT-IR transmission imaging and single element detector transmission FT-IR spectroscopy.

The polymer studied was sPS. Compared with the atactic and isotactic forms, sPS has the tendency to crystallize very

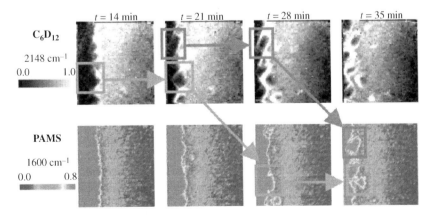

FIGURE 14.11 Spectral images showing the concentration of PAMS with corresponding black and white images of C_6D_{12} at the same time. When cracks meet, a polymer chunk breaks away from the bulk polymer. (Reproduced from Ref. 21 with permission. Copyright 2008 American Chemical Society.)

quickly and to a relatively high extent on exposure to temperatures above T_g or on exposure to certain solvents. It is known that sPS has four main crystalline modifications and several subforms. A principal distinction can be made between the transplanar or TTTT all-*trans* α and β forms with all the phenyl side groups *trans* to each other and the helical or TTGG γ and δ forms, where macromolecules arrange in a helical conformation and the phenyl side groups are *trans–trans–gauche–gauche* to each other along the polymer backbone. To obtain sPS in its helical conformation, a step involving solvents, that is, casting from solution, precipitation from solution, or exposure of an amorphous sample to a suitable solvent, is required. Immirzi et al. [43] studied the use of solvents and their effects on amorphous sPS. It was reported that some organic solvents introduce the γ and some the δ form. The forms can be distinguished by wide-angle X-ray diffraction and vibrational spectroscopy.

Figure 14.12 shows FT-IR spectra of chloroform (a), amorphous (b), and δ crystalline (c) sPS [24]. A band at $1220 \, cm^{-1}$ is assigned to the bending mode of chloroform. This band was used to determine the type of solvent diffusion behavior and the solvent diffusion coefficient. A band at $1275 \, cm^{-1}$ due to sPS was employed to monitor the crystallization process of the polymer. The $1400–900 \, cm^{-1}$ region shown in Figure 14.12 is accessible in imaging and single element detector experiments and contains information about solvent uptake as well as phase transformations of sPS. Figure 14.13 shows position of the front of chloroform as a function of horizontal pixel number of focal plane array (FPA); the gas phase is on the left side, and the polymer is on the right side [24]. In this figure, the gaseous chloroform was supplied from the left side. Profiles are based on the values of the integrated absorbance of the $1220 \, cm^{-1}$ band. The sample was exposed to vapor of chloroform for 52 s in profile (a). For profiles (b–h), solvent exposure time increases in steps of 52 s. The plots in Figure 14.13 show that, with longer solvent

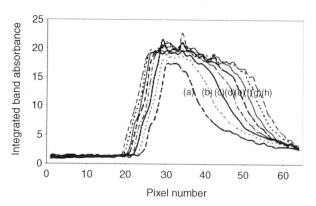

FIGURE 14.13 Position of the front of chloroform as a function of horizontal pixel number of FPA; the gas phase is on the left side, and the polymer is on the right side. Profiles are based on the values of the integrated absorbance of the band of chloroform at $1220 \, cm^{-1}$. (Reproduced from Ref. 24 with permission. Copyright 2005 American Chemical Society.)

exposure time, the solvent front moves to the right and further into the polymer. It can be clearly seen that the polymer film swells on interaction with the solvent. The results in Figure 14.13 reveal the equilibrium concentration of chloroform within the polymer by the plateau of the concentration value around 20, which remains almost constant from the polymer/solvent interface to the solvent diffusion front. It suggests that there is equilibrium between the condensed solvent phase within the polymer and the gaseous supply.

To deduce quantitative information about the diffusion of chloroform, the polymer film edge and the position of the chloroform diffusion front after different solvent exposure times were determined from the plots derived from FT-IR images. Figure 14.14 shows a log–log plot of chloroform diffusion front distance (d in μm) from the polymer/solvent interface versus solvent exposure time [24]. The polymer film edge was found by integration of the sPS bands at 1530 and $1420 \, cm^{-1}$, and the position of the solvent front was determined by integrating the $1220 \, cm^{-1}$ band of chloroform. Both film edge and solvent front were defined as the points at 50% of the maximal integrated absorbance values. The type of solvent diffusion can be determined by the so-called diffusion exponent α. According to Snively and Koenig [44], an α value of 0.5 indicates Fickian and a value of 1.0 case II type diffusion behavior. All experiments with chloroform yielded α values of about 0.5 and indicate Fickian diffusion behavior.

Figure 14.14b plots the diffusion front distance from the polymer/solvent interface (d in cm) versus the square root of solvent exposure time [24]. From the slope of the linear regression, the solvent diffusion coefficient D at 20□ was calculated to be $4 \times 10^{-7} \, cm^2/s$. It is of note that solvent diffusion was directly determined from spatially resolved FT-IR images. The observed value of D is in excellent agreement with the diffusion coefficient of liquid chloroform from mass uptake experiments by Vittoria et al. [45]. It was rather

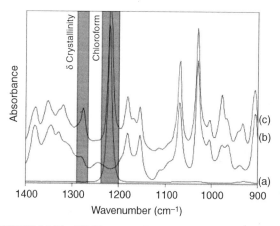

FIGURE 14.12 FT-IR spectra in the $1400–900 \, cm^{-1}$ region of chloroform (a), amorphous (b), and δ crystalline (c) syndiotactic polystyrene. (Reproduced from Ref. 24 with permission. Copyright 2005 American Chemical Society.)

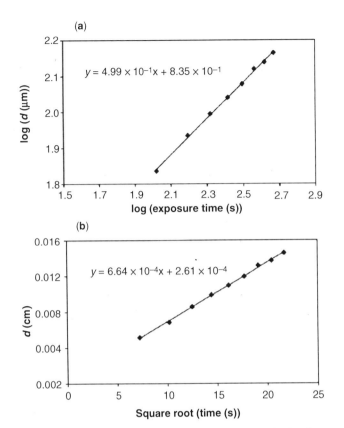

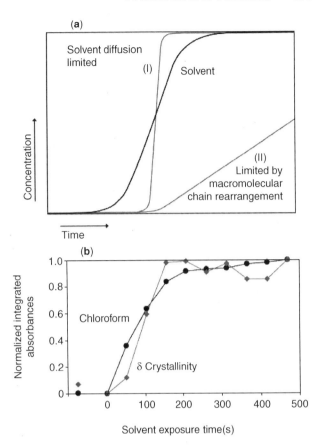

FIGURE 14.14 (a) A log–log plot of chloroform diffusion front distance (*d* in μm) from the polymer/solvent interface versus solvent exposure time. (b) The diffusion front distance from the polymer/solvent interface (*d* in cm) plotted versus the square root of solvent exposure time. (Reproduced from Ref. 24 with permission. Copyright 2005 American Chemical Society.)

FIGURE 14.15 (a) Limiting situations for the overall crystallization process of an sPS sample: (I) the rate of solvent diffusion into the polymer limits the crystallization progress; (II) rearrangement of macromolecular chains is slow compared to the movement of the solvent diffusion front. (b) Observed increase in degree of crystallinity (■) and solvent (●) concentration 50 μm away from the polymer/solvent interface versus solvent exposure time (FPA pixel row 41). (Reproduced from Ref. 24 with permission. Copyright 2005 American Chemical Society.)

remarkable that the diffusion coefficient of chloroform, when supplied in its gaseous form, is about the same as with a supply of liquid chloroform. This indicates that an increase in the amount of solvent present does not affect the plasticization of the polymer and solvent diffusion compared to that of a saturated chloroform atmosphere.

This study also investigated solvent-induced crystallization of sPS. Vittoria et al. [45] revealed that chloroform induces a helical crystalline form of sPS (δ form with solvent molecules forms a complex with the helical macromolecular chains). The fundamental question addressed in their study was whether the overall crystallization process in an sPS sample is limited by the macromolecular chain rearrangements or by the solvent diffusion into the polymer. Figure 14.15a depicts the normalized concentration/time profiles for the two limiting situations [24]. The black solid line represents the solvent concentration profile at a certain position in the polymer and applies to both cases discussed below. The gray lines illustrate the amount of δ crystalline sPS at the same specific location in the sample for two possible scenarios. Curve (I) in Figure 14.15a indicates that

there must be a certain amount of solvent before the macromolecules start to rearrange. As soon as this critical value of solvent concentration is reached, the reorganization into the thermodynamically more stable δ crystalline modification begins. Curve (II) represents the situation where the rearrangement of the macromolecular chains from the randomly coiled amorphous state into the helical conformation is the rate-determining step in the crystallization process.

Figure 14.15b illustrates observed increase in degree of crystallinity (squares) and solvent (circles) concentration 50 μm away from the polymer/solvent interface versus exposure time (FPA pixel row 41) [24]. It can be seen from Figure 14.15b that the pattern reflects curve (I) in Figure 14.15a and suggests a diffusion-limited crystallization process.

Figure 14.16 depicts normalized integrated absorbance profiles for the spectral bands of chloroform and δ crystallinity

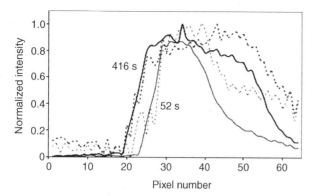

FIGURE 14.16 Normalized integrated absorbance profiles for the spectral bands of chloroform and δ crystallinity as a function of the horizontal pixel number of the FPA detector. (Reproduced from Ref. 24 with permission. Copyright 2005 American Chemical Society.)

as a function of the horizontal pixel number of the FPA detector [24]. The solid lines indicate concentration profiles of chloroform whereas the dashed lines represent the integrated absorbance value of the band at 1275 cm^{-1} due to sPS, or indicator of δ crystallinity. The profiles at both solvent exposure times in Figure 14.16 show that the crystallization process is completed before the solvent equilibrium is achieved in the semicrystalline polymer. This observation led to the conclusion that the overall crystallization process is limited by the solvent diffusivity and plasticization of the polymer rather than by the rearrangement of macromolecular chains.

Gupper and Kazarian [24] also explored how fast the actual crystallization process was occurring and what the lower critical solvent concentration to introduce crystallinity was. Figure 14.17 plots the absorbance of the 1275 cm^{-1} band versus the solvent exposure time [24]. Curves (a–c)

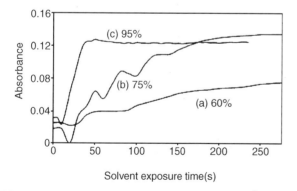

FIGURE 14.17 Absorbance of the band at 1275 cm^{-1} band, an indicator of δ crystallinity, plotted versus chloroform vapor exposure time. Chloroform concentrations were set to levels of 60% (a), 75% (b), and 95% (c) solvent-saturated air entering the sample compartment as indicated in the figure. (Reproduced from Ref. 24 with permission. Copyright 2005 American Chemical Society.)

correspond to polymer films that were subjected to chloroform levels of 60%, 75%, and 95% of the saturated solvent vapor. At the highest solvent vapor concentration, the overall crystallization process is finished in about 60 s. When the solvent vapor pressure is lowered, the crystallization process takes much longer. The solvent molecules move in between single macromolecular chains, increase the distance between them, and allow the polymer to rearrange. The higher the amount of solvent present, the easier it is for the macromolecules to rearrange.

The reported *in situ* FT-IR imaging approach is not limited to sPS and can be applied to diffusion and crystallization studies of a wide range of polymers.

14.4 NIR IMAGING OF POLYMERS

14.4.1 NIR Imaging of Polymer Blends

Furukawa et al. [20] carried out FT-NIR imaging for four kinds of PHB/PLLA polymer blends with the PLLA content ranging from 20 to 80 wt% to elucidate the blend quality. This study highlights the potential of FT-NIR imaging for qualitative and quantitative nondestructive evaluation of blend homogeneity. Furukawa et al. [20] combined FT-NIR data with partial lease squares regression (PLSR).

Figure 14.18a and b shows NIR spectra in the 1200–2400 nm region and their second derivative of neat PHB and PLLA [20]. Bands due to the combination and first overtone modes of the CH vibrations and that arising from the second overtone of the C=O stretching vibration are observed in both PHB and PLLA spectra. It is noted that the bands of PHB and PLLA are significantly overlapping even in the second derivatives spectra. Thus, it was different to use the intensity at a fixed wavelength to monitor the distribution and prediction for the blend components. Consequently, they employed the PLSR.

Figure 14.19 shows the score images of four kinds of PHB/PLLA (80/20, 60/40, 40/60, 20/80) blends derived from PLSR [20]. In the score images, the pixels with higher and low score values indicate PHB- and PLLA-rich phases, respectively. The compositions of polymer blends were estimated by using PLSR method. The predicted concentrations of the components of the polymer blends with averaging over the space are in good agreement with their actual concentrations. The score images directly depict the spatial distributions of the components of PHB/PLLA polymer blends. The standard deviations (STD) of the histograms, indicating the distribution of the score values, show small values for the blends. These results quantitatively and qualitatively show the high kevel of homogeneity of PHB/PLLA blends.

FT-NIR imaging proved to be very useful in the study of binary polymer blends. Even in the homogeneously polymer

(a)

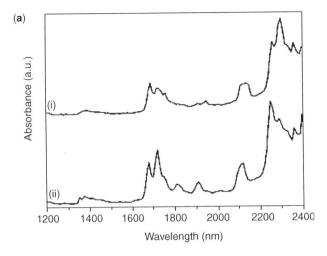

(b)

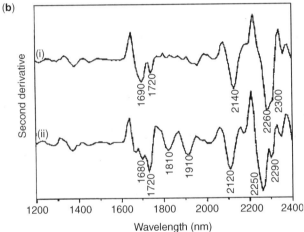

FIGURE 14.18   (a) NIR spectra and (b) their second derivative spectra of neat PHB (i) and PLLA (ii).

blends, FT-NIR imaging can nondestructively investigate a sample over a wide range.

14.4.2   NIR Imaging of Cellulose Tablets

Grinding is a central part of the manufacturing process of pharmaceutical tablets. The main objective of grinding process usually is to simply obtain the uniform distribution of components within tablets, and the uniform distribution enables one to design the well-controlled sustained release of the actives from the tablet. Another important aspect of the grinding process is the expected mechanochemical effect, which induces additional chemical or physical changes in the pharmaceutical ingredients themselves during the process [46]. The control of mechanochemical effect potentially makes it possible to control the desired pharmaceutical property of the final products.

Shinzawa et al. [35] explored the effect of the grinding on cellulose excipient at the molecular level by NIR imaging. Figure 14.20 shows (a) original NIR imaging spectra of the

cellulose tablet ground for 60 min and (b) a typical second derivative spectrum, respectively [35]. Three characteristic peaks are observed in Figure 14.20b. A negative peak observed at 6950 cm^{-1} is assignable to the first overtone of the OH stretching mode of OH groups in the amorphous region with weak hydrogen bonds [35, 47, 48]. A peak at 6780 cm^{-1} is ascribed to hydrogen-bonded OH groups in the semicrystalline region, where the crystalline structure of cellulose is partially disordered. The peak at 6304 cm^{-1} is assigned to hydrogen-bonded OH groups in the crystalline region [35, 47, 48].

In the handling of spectroscopic imaging data, the plot of a peak position can be a useful way to elucidate the structural morphological information of a sample [35, 49]. A substantial change of molecular structure can be detected as a form of band position shift to a higher or lower wavenumber. Band position related to such diatomic OH molecules can be a useful index representing a characteristic feature of molecular structure. For example, peak positions concerning the amorphous band observed in spectral data of the tablet ground for (a) 0 min and (b) 60 min are shown in Figure 14.21 [35]. The amorphous band position in Figure 14.21 shifts to the lower wavenumber with the increase in the grinding time. Since the crystallinity of the cellulose samples clearly decreases with grinding time, now the meaning of the observed shift in Figure 14.21 becomes important. The band position shift to a lower wavenumber direction is obviously correlated with the increased degree of hydrogen bonding in amorphous region along with the grinding time. Thus, it is most likely that the peak position shift is due to the quantitative increase in the amorphous content. Peak positions concerning crystalline band of the tablet ground for (c) 0 min and (d) 60 min are illustrated in Figure 14.21. One can note that the crystalline band in Figure 14.21 shifts to the higher wavenumber with the increase in the grinding time, mostly reflecting the fact that the crystalline content is decreasing. It is also noted that the entire features between amorphous and crystalline peaks become complementary. Such visualization may bring another kind of information on the molecular structure in terms of chemical bond strength, compared to the conventional visualization technique based on spectral intensity that primarily depends on the concentration of component [49]. Averaged positions of the observed three peaks in Figure 14.20b are illustrated in Figure 14.22 [35]. Note that the averaged peak positions are derived from each cellulose tablet. For example, the peaks representing (c) amorphous, (d) semicrystalline, and (e) crystalline structure are shown in Figure 14.22 [35]. As expected, the overall features of three peaks are more or less similar, but the directions where they shift are opposite to each other.

The ability of cellulose to interact with water molecule is important from pharmaceutical point of view. Water molecule usually does not penetrate the crystalline area of cellulose while it is entrapped by the amorphous area. Disordered

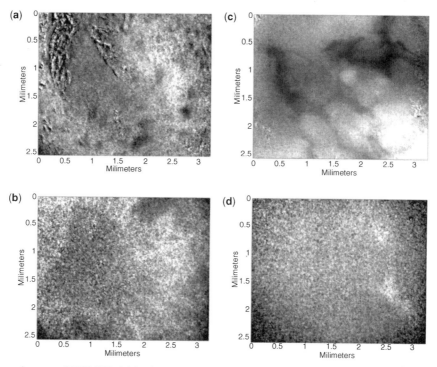

FIGURE 14.19   Score images of PHB/PLLA blends derived from PLSR. PHB/PLLA: (a) 80/20, (b) 60/40, (c) 40/60, (d) 20/80.

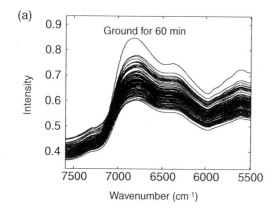

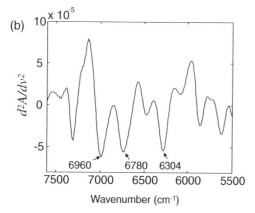

FIGURE 14.20   (a) NIR spectra of cellulose tablet ground for 60 min and (b) a representative second derivative spectrum. (Reproduced from Ref. 35 with permission. Copyright 2009 Society for Applied Spectroscopy.)

alignment of the cellulose polymers is strongly associated with the entrapment of water molecule. For example, the degree of the prolonged retention of the water molecule is related to the amount of the amorphous component [50]. Since Figure 14.21 indicates that the tablet becomes uniformly covered with amorphous structure of the cellulose after sufficient level of grinding, water molecules can be well entrapped in the amorphous region of the tablet and thus eventually leading to the direct contact with active pharmaceutical ingredients [35, 50]. It implies that the tablet ground for 60 min, which is well covered with the amorphous structure, will result in the better solubility of active pharmaceutical ingredients than others.

14.5   RAMAN IMAGING OF POLYMERS

Huan et al. [34] studied phase behavior and compatibility in PET/HDPE polymer blends by using confocal Raman mapping. Maleic anhydride-grafted high-density poly(ethylene) (MAH-HDPE) was prepared by melt mixing HDPE with MAH and peroxide at 220°C [51]. The polymer blends were prepared by compounding the PET with MAH-HDPE or HDPE at 220°C in a twin-screw extruder. Three sets of both types of blends were prepared with the PET/HDPE ratios of 20/80, 50/50, and 80/20 wt%. The spatial resolution of the Raman mapping measurements was about 2 μm. Raman images were recorded by first positioning a polymer sample in the laser focus using a video camera and white light illumination, followed by scanning over the mapping region

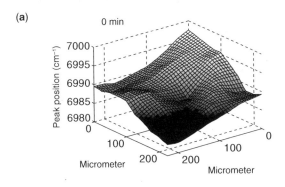

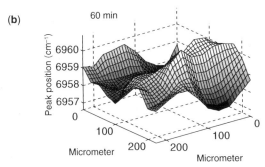

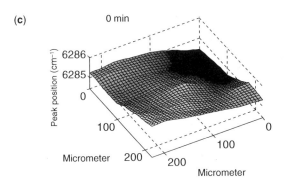

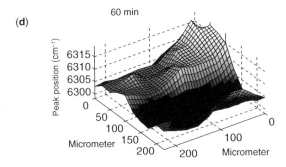

FIGURE 14.21 Peak position of the amorphous band on cellulose tablets ground for (a) 0 min and (b) 60 min and the crystalline band on the cellulose tablets ground for (c) 0 min and (d) 60 min. (Reproduced from Ref. 35 with permission. Copyright 2009 Society for Applied Spectroscopy.)

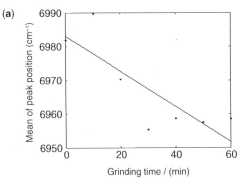

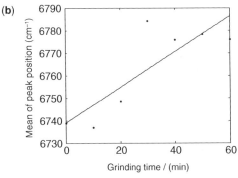

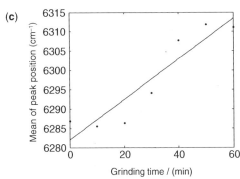

FIGURE 14.22 Averaged peak position for the (a) amorphous, (b) semicrystalline, and (c) crystalline bands. (Reproduced from Ref. 35 with permission. Copyright 2009 Society for Applied Spectroscopy.)

(selected as $60 \times 60 \, \mu m^2$ in step sizes of $2 \, \mu m$) and accumulating a full spectrum at each pixel. A total of 900 Raman spectra (30×30 probe spots) were measured for each sample.

Using polyethylene terephthalate/high-density polyethylene (PET/HDPE) blending system as an example, PET and HDPE are widely used as packaging materials; □HDPE can modify the rheological properties and impact properties of PET and increase the velocity of crystallization, while PET can improve the mechanical properties and thermostability of HDPE. However, they are not thermodynamically miscible. If blended mechanically, the incompatibility between the two polymers may bring poor mechanical properties. If they are blended in the presence of maleic anhydride, a well-known reactive solubilizer, the viscosity status and compatibility between two phases could be improved significantly.

Raman mapping recorded three sets of both types of blends prepared with the PET/HDPE ratios of 20/80, 50/50, and 80/20 wt%. With the aid of the multivariate image segmentation approach based on spatial directed agglomeration clustering [52], the spatial distribution or the degree of

mixing in the polymer blend was studied. The spatial directed agglomeration clustering was implemented for the segmentation of Raman mapping data measured on the 50% HDPE–50% PET incompatible blend.

Figure 14.23a shows an optical image of 50% HDPE–50% PET polymer blend prepared with maleic anhydride [34]. Figure 14.23b illustrates reliability curve for 50% HDPE–50% PET polymer blend prepared with maleic anhydride and Figure 14.23c depicts the dissimilarity curve for

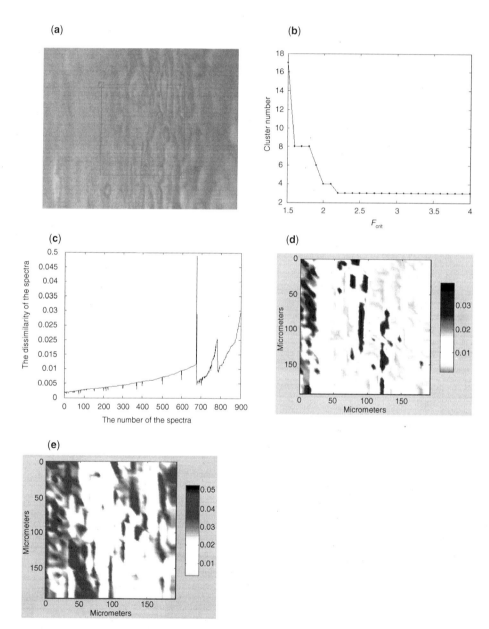

FIGURE 14.23 (a) Optical image of 50% HDPE–50% PET polymer blend prepared with maleic anhydride. (b) Reliability curve for 50% HDPE–50% PET polymer blend prepared with maleic anhydride. (c) Dissimilarity curve for the polymer blend. (d) Correlation map of the polymer blend where the gray scale represents the correlation coefficients with respect to the representative spectra of cluster 1. (e) The correlation map of the polymer blend with respect to the representative spectra of cluster 2.

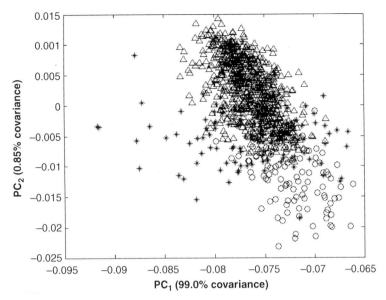

FIGURE 14.24   The projections of the normalized spectra on the first two PCs for 50% HDPE–50% PET polymer blend prepared with maleic anhydride. △: Data points in cluster 1; ○: data points in cluster 2; *:; data points in cluster 3.

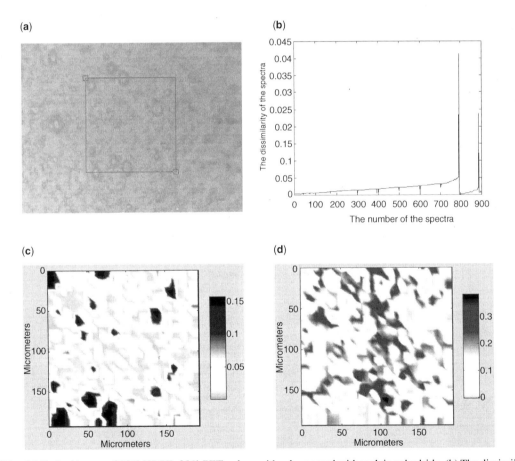

FIGURE 14.25   (a) Optical image of 80% HDPE–20% PET polymer blend prepared with maleic anhydride. (b) The dissimilarity curve for the polymer blend. (c) The correlation map of the polymer blend where the gray scale represents the correlation coefficients with respect to the representative spectra of cluster 1. (d) The correlation map of the polymer blend with respect to the representative spectra of cluster 2.

the polymer blend. Figure 14.23d illustrates the correlation map of the polymer blend where the gray scale represents the correlation coefficients with respect to the representative spectra of cluster 1. The correlation map of the polymer blend with respect to the representative spectra of cluster 2 is shown in Figure 14.23e. Figure 14.24 illustrates the projections of the normalized spectra on the first two PCs for 50% HDPE–50% PET polymer blend prepared with maleic anhydride [34].

Similar analysis was performed on 80% HDPE–20% PET incompatible blend, 20% HDPE–80% PET incompatible blend, 50% HDPE–50% PET semicompatible blend, 80% HDPE–20% PET semicompatible blend, and 20% HDPE–80% PET semicompatible blend. The dissimilarity curves during the corresponding clustering processes are shown in Figures 14.25bFi, 14.26bFi, 14.27bFi, 14.28bFi and 14.29b, respectively [34]. It is observed that there are several leaps in the curves that correspond to the number of clusters in the image data. By inspecting the representative spectrum for each cluster, one could determine the chemical identity of the clusters. Correlation maps of the spectra in the

image with reference to the chemically significant representative spectrum of individual cluster are also shown in Figures 14.25c and d, 14.26c and d, 14.27c and d, 14.28c and d, and 14.29c and d [34].

The pe-abundant areas are present as a clear and broad continuous phase, while pet-abundant areas are dispersed as big islands, indicating the immiscible polymer blend is highly heterogeneous in spatial chemical distribution. The pe-abundant and pet-abundant areas are both present as clear and broad phases that isolated by each other, indicating components in the immiscible polymer blend are mixed poorly on the scale of laser sampling volume. The pet-abundant areas are present as clear and broad continuous phases and pe-abundant areas are dispersed as isolated big islands, indicating the polymer blend is immiscible. The very small pet-abundant areas are dispersed uniformly in the broad continuous phases composed of pe-abundant areas, indicating the heterogeneity in the miscible blend is substantially improved compared to that in the immiscible one. The subphases in miscible blends are much smaller than those in immiscible blends; the vibration intensity ratios for

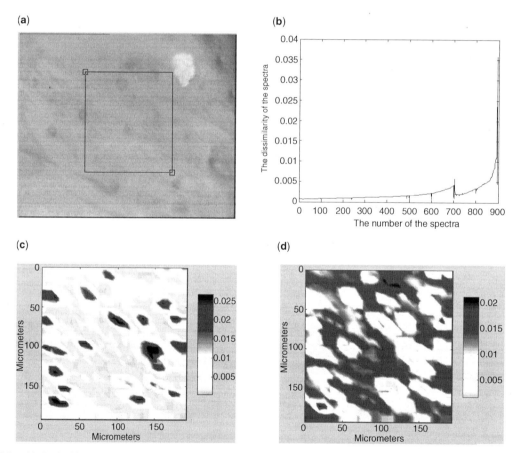

FIGURE 14.26   (a) Optical image of 20% HDPE–80% PET polymer blend prepared with maleic anhydride. (b) The dissimilarity curve for the polymer blend. (c) The correlation map of the polymer blend where the gray scale represents the correlation coefficients with respect to the representative spectra of cluster 1. (d) The correlation map of the polymer blend with respect to the representative spectra of cluster 2.

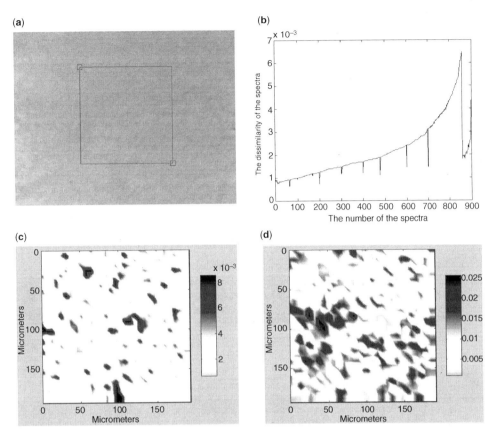

FIGURE 14.27   (a) Optical image of 50% HDPE–50% PET polymer blend prepared without maleic anhydride. (b) The dissimilarity curve for the polymer blend. (c) The correlation map of the polymer blend where the gray scale represents the correlation coefficients with respect to the representative spectra of cluster 1. (d) The correlation map of the polymer blend with respect to the representative spectra of cluster 2.

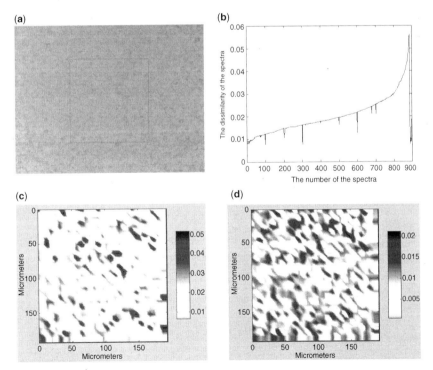

FIGURE 14.28   (a) Optical image of 80% HDPE–20% PET polymer blend prepared without maleic anhydride. (b) The dissimilarity curve for the polymer blend. (c) The correlation map of the polymer blend where the gray scale represents the correlation coefficients with respect to the representative spectra of cluster 1. (d) The correlation map of the polymer blend with respect to the representative spectra of cluster 2.

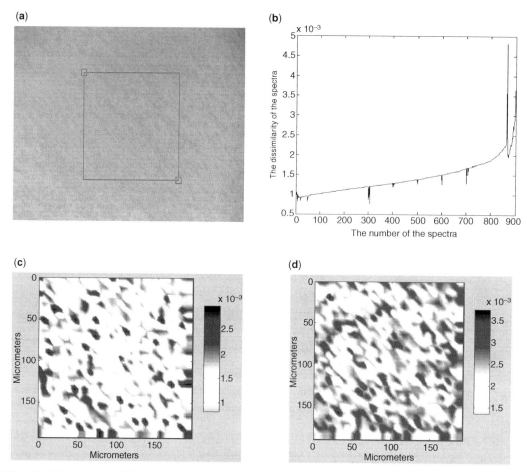

FIGURE 14.29   (a) Optical image of 20% HDPE–80% PET polymer blend prepared without maleic anhydride. (b) The dissimilarity curve for the polymer blend. (c) The correlation map of the polymer blend where the gray scale represents the correlation coefficients with respect to the representative spectra of cluster 1. (d) The correlation map of the polymer blend with respect to the representative spectra of cluster 2.

miscible blends are much smaller than those in immiscible blends. Both miscible blends show improved homogeneity in chemical distributions.

The Raman imaging results were also verified by SEM observation of the polymer blends [34]. The result from the SEM micrographs suggests that when blended mechanically, phase separation could not be avoided between the nonpolar HDPE and the polar PET. However, in the presence of maleic anhydride-grafted HDPE, the reactive solubilization makes it possible to improve the viscosity status and gives marked dispersibility of the polymer blend.

REFERENCES

1. Siesler, H. W. and Holland-Moritz, K. (1980) *Infrared and Raman Spectroscopy*, Marcel Dekker, New York.

2. Christy, A. A., Ozaki, Y., and Gregoriou, V. G. (2001) *Modern Fourier Transform Infrared Spectroscopy. Comprehensive Analytical Chemistry*, Vol. 35, Elsevier.

3. Coleman, M. M., Graf, J. F., and Painter, P. C. (1991) *Specific Interactions and the Miscibility of Polymer Blends*, Technomic Publishing, Lancaster, PA.

4. Everall, N. J., Chalmers, J. M., and Griffiths, P. R. (Eds.) (2007) *Vibrational Spectroscopy of Polymers: Principles and Practice*, Wiley, Chichester, UK.

5. Zerbi, G. (1999) *Modern Polymer Spectroscopy*, Wiley-VCH, Weinheim.

6. Siesler, H. W., Ozaki, Y., Kawata, S., and Heise, H. M. (2002) *Near-Infrared Spectroscopy: Principles, Instruments, Applications*, Wiley-VCH, Weinheim.

7. Bhargava, R., Wang, S.-Q., and Koenig, J. L. (2003) *Adv. Polym. Sci.* **163**, 137.

8. Koenig, J. L. (2005) FTIR imaging of multicomponent polymers. In: Bhargava, R.,and Levin, I. W. (Eds.), *Spectrochemical Analysis Using Infrared Multichannel Detectors*, Blackwell Publishing Ltd, Oxford, UK.

9. Salzer, R., and Siesler, H. W. (Eds.) (2009) *Infrared and Raman Spectroscopic Imaging*, Wiley-VCH, Weinheim.

10. Vogel, C., Wessel, E., and Siesler, H. W. (2008) *Biomacromolecules* **9**, 523.

11. Oh, S. J., Do, J. S., and Ok, J. H. (2003) *Appl. Spectrosc.* **57**, 1058.

12. Chernev, B. and Wilhelm, P. (2006) *Monatsh. Chem.* **137**, 963.

13. Snively, C. M., and Koenig, J. L. (1999) *J. Polym. Sci. B: Polym. Phys.* **37**, 2353.

14. Vogel, C., Wessel, E., and Siesler, H. W., (2008) *Macromolecules* **41**, 2975.

15. Gupper, A., Wilhelm, P., Kothleitner, G., Eichhorn, K.-J., and Pompe, G. (2004) *Macromol. Symp.* **205**, 171.

16. Wilhelm, P., Chernev, B., and Pölt, P. (2005) *Macromol. Symp.* **230**, 105.

17. Wilhelm, P., Chernev, B., Pölt, P., Kothleitner, G., Eichhorn, K.-J., Pompe, G., Johner, N., and Piry, A. (2004) *Spectrosc. Eur.* **16**, 14.

18. Gupper, A., Wilhelm, P., Schmied, M., and Ingolic, E. (2002) *Macromol. Symp.* **184**, 275.

19. Koehler, F. W. IV, Lee, E., Kidder, L. H., and Neil Lewis, E. (2002) *Spectrosc. Eur.* **14**(3), 12.

20. Furukawa, T., Sato, H., Shinzawa, H., Noda, I., and Ochiai, S. (2007) *Anal. Sci.* **23**, 871.

21. Miller-Chou, B. A., and Koenig, J. L. (2002) *Macromolecules* **35**, 440.

22. González-Benito, J. and Koenig, J. L. (2002) *Macromolecules* **35**(19), 7361–7367.

23. Coutls-Lendon, C., and Koenig, J. L. (2005) *Appl. Spectrosc.* **59**, 976.

24. Gupper, A., and Kazarian, S. G., (2005) *Macromolecules* **38**, 2327.

25. Michaels, C. A., Gu, X., Chase, D. B., and Stranick, S. J. (2004) *Appl. Spectrosc.* **58**, 257.

26. Patterson, B. M., Havrilla, G. J., Marcott, C., and Story, G. (2007) *Appl. Spectrosc.* **61**, 1147.

27. Furukawa, T., Sato, H., Kita, Y., Matsukawa, K., Yamaguchi, H., Ochiai, S., Siesler, H. W., and Ozaki, Y. (2006) *Polym. J.* **38**, 1127.

28. Gupper, A., Wilhelm, P., Schmied, M., Kazarian, S. G., Chan, K. L. A., and Reussner, J. (2002) *Appl. Spectrosc.* **56**, 1515.

29. Kazarian, S. G. and Chan, K. L. A., (2004) *Macromolecules* **37**, 579.

30. Wilhelm, P., Chernev, B., and Polt, P. (2005) *Macromol. Symp.* **230**, 105.

31. Vogel, C., Wessel, E., and Siesler, H. W. (2008) *Appl. Spectrosc.* **62**, 599.

32. Wessel, E., Vogel, C., and Siesler, H. W. (2009) *Appl. Spectrosc.* **63**, 1.

33. Kolomiets, O., Hoffmann, U., Geladi, P., and Siesler, H. W. (2008) *Appl. Spectrosc.* **62**, 1200.

34. Huan, S., Lin, W., Sato, H., Yang, H., Jiang, J., Ozaki, Y., Wu, H., Shen, G., and Yu, R. (2007) *J. Raman Spectrosc.* **38**, 260.

35. Shinzawa, H., Awa, K., Ozaki, Y., and Sato, H. (2009) *Appl. Spectrosc.* **63**, 974.

36. Hocking, P. J. and Marchessault, R. H. (1998) Polyhydroxyalkanoates. In: Kaplan, D. L. (Ed.), *Biopolymers from Renewable Resources*, Springer-Verlag, Berlin, p. 220.

37. Doi, Y., and Steinbüchel, A. (Eds.) (2001) *Biopolymers: Polyesters II*, Vol. 3B, Wiley-VCH, Weinheim.

38. Furukawa, T., Sato, H., Murakami, R., Zhang, J., Duan, Y.-X., Noda, I., Ochiai, S., and Ozaki, Y. (2005) *Macromolecules* **38**, 6445.

39. Avella, M. and Martuscelli, E. (1988) *Polymer* **29**, 1731.

40. Vogel, C. (2008) Ph.D. thesis, University of Duisburg-Essen, Essen, Germany.

41. Iwata, T. (2005) *Macromol. Biosci.* **5**, 689.

42. Ribar, T. B., Bhargava, R., and Koenig, J. L. (2000) *Macromolecules*, **33**, 8842.

43. Immirzi, A., De Candia, F., Iannelli, P., Zambelli, A., and Vittoria, V. (1988) *Makromol. Chem., Rapid Commun.* **9**, 761.

44. Snively, C. M. and Koenig, J. L. (1999) *J. Polym. Sci. B: Polym. Phys.* **37**, 2261.

45. Vittoria, V., de Candia, F., Iannelli, P., and Immirzi, A. (1988) *Makromol. Chem., Rapid Commun.* **9**, 765.

46. Awa, K., Okumura, T., Shinzawa, H., Otsuka, M., and Ozaki, Y. (2007) *Anal. Chim. Acta* **619**, 81.

47. Watanabe, A., Morita, S., and Ozaki, Y. (2006) *Appl. Spectrosc.* **60**, 1054.

48. Tsuchikawa, S. and Siesler, H. W. (2003) *Appl. Spectrosc.* **57**, 667.

49. Chan, K. L. A., Kazarian, S. G., Vassou, D., Gionis, V., and Chryssikos, G. D. (2007) *Vib. Spectrosc.* **43**, 221.

50. Shinzawa, H., Morita, S., Awa, K., Okada, M., Noda, I., Ozaki, Y., and Sato, H. (2009) *Appl. Spectrosc.* **63**, 501.

51. Sato, H., Sasao, S., Matsukawa, K., Kita, Y., Ikeda, T., Tashiro, H., and Ozaki, Y. (2002) *Appl. Spectrosc.* **56**, 1038.

52. Lin, W. Q., Jiang, J. H., Yang, H. F., Ozaki, Y., Shen, G. L. and Yu, R. Q. (2006) *Anal. Chem.* **78**, 6003.

PART VI

SPECIAL METHODS

15

SURFACE-ENHANCED RAMAN SCATTERING IMAGING: APPLICATION AND EXPERIMENTAL APPROACH BY FAR-FIELD WITH CONVENTIONAL SETUP

Yasutaka Kitahama, Mohammad Kamal Hossain, and Yukihiro Ozaki

Department of Chemistry, School of Science and Technology, Kwansei Gakuin University, Sanda, Japan

Tamitake Itoh

National Institute of Advanced Industrial Science and Technology, Takamatsu, Kagawa, Japan

Athiyanathil Sujith

National Institute of Technology Calicut, Calicut, Kerala, India

Xiaoxia Han

State Key Laboratory of Supramolecular Structure and Materials, Jilin University, Changchun, China

15.1 INTRODUCTION

In Raman spectroscopy, detailed information about structure of a target molecule is provided by peak positions and intensities of vibrational modes of the functional groups [1]. However, Raman scattering is, in general, weak due to small Raman cross sections, that is, $\sim 10^{-24}$ cm^{-2}. It has widely been known that Raman scattering is enhanced by resonance of a molecular electronic transition with incident light (resonance Raman effect). In 1974, an enhanced Raman spectrum of pyridine adsorbed on an Ag electrode roughened by successive oxidation–reduction cycles was reported [2]. It was originally thought that the enhanced Raman scattering was due to the increase in the surface area of the electrode. In 1977, it was shown that the enormous surface-enhanced Raman scattering (SERS) enhancement factor of 10^5–10^6 for adsorbed pyridine was not proportional to the surface area [3, 4]. It meant that the enhanced Raman scattering depended not only on the surface area but also on the nanostructures responsible for the enhancement of Raman scattering cross section itself. This new phenomenon was termed as SERS.

For SERS studies, various SERS-active systems have been developed—for example, electrodes roughened by oxidation–reduction cycles, island films formed by vapor deposition, lithographically produced nanostructures, and metal colloids prepared by reducing a dissolved metal salt in an aqueous solution [5]. Metal colloids can be easily prepared and have widely been applied to SERS. Metal nanoparticles and nanoaggregates in the colloid solutions have various sizes and structures. In metal nanoparticles, incident light is resonant with plasmon due to dipolar oscillation of conduction band electrons (localized surface plasmon resonance (LSPR)). The LSPR maxima depend on the size and shape of nanoparticles. At a gap of metal nanoparticles, an electromagnetic (EM) field is enormously enhanced. SERS originates in the EM field at the gap with a distance of a few nanometers [6–9]. By conventional ensemble measurements, only an ensemble of SERS spectra of target molecules adsorbed on various nanoparticles and nanoaggregates are measured. Therefore, microscopic imaging is indispensable to investigate correlation between SERS enhancements and individual nanostructures.

Raman, Infrared, and Near-Infrared Chemical Imaging Edited by Slobodan Šašić and Yukihiro Ozaki

Microscopic imaging has been applied to study spatial distribution of target molecules in heterogeneous materials, for example, various biological molecules in a living cell. SERS spectroscopy shows extremely high sensitivity. Indeed, the single molecule spectroscopy by SERS has already been reported [10–14]. In SERS, the vibrational fingerprints provide detailed information about molecular structure, spectral multiplexing is easy, and nonfluorescent molecules can be measured. However, SERS light is emitted from the molecules adsorbed at the junction of noble metal nanostructures [6–9]. In other words, spatial resolution of SERS imaging can be achieved to a few nanometers.

15.2   METHODS AND INSTRUMENTATIONS

A typical Raman spectroscopic setup for SERS imaging is composed of several essential components, namely, an excitation light source (i.e., laser), an appropriate arrangement for Rayleigh scattering rejection (i.e., notch filter), an analyzer (i.e., spectrometer or spectrograph), a detector, a converter and a controller for data acquisition, and so on [15, 16]. Among all these components, the excitation source and the detection system are the centers of interest for imaging purpose.

As for the excitation wavelength for Raman imaging, one has to remember that scattering background and autofluorescence are greatly reduced with longer wavelengths and SERS characteristics become more feasible particularly in biological sectors [17, 18]. Hence, using dyes that adsorb in the red or near-infrared (NIR) wavelength region results in much better signal-to-noise ratios than when using dyes in the blue spectral region [19, 20]. On the other hand, because of insufficient cross section of Raman scattering, great care has to be taken to transfer the scattered light from a sample to a spectrometer. Old-fashioned spectrometers have an entrance slit that has to be kept narrow, and thus optical losses are unavoidable. This issue is elegantly overcome in a confocal Raman microscope by focusing excitation laser beam onto the sample via a microscope objective and collecting scattered signal through another objective lens in transmission configuration. The scattered light enters the spectrometer through a pinhole ensuring an optimal optical throughput. As usual, prior to focusing onto a grating, the scattered light passes a notch filter to remove the reflected and elastically scattered light. Nowadays, compact Raman spectrometer systems are available in which all the components are integrated. They are easy to handle but cannot readily be modified, such that they are difficult to adapt to nonstandard applications. Dedicated and specifically tailored systems may overcome these restrictions. The details of all components of Raman spectrometers can be found in Ref. 16.

Different methods are under development to improve the image quality and to extract as much information as possible [15, 16, 21]. Here, we focus only on two general concepts essential to understand the imaging characteristics.

15.2.1   Point-by-Point Mapping

As explained earlier, in confocal Raman spectroscopy, the excitation laser beam is tightly focused by microscope objectives and the Raman scattering light is collected from a small volume of the sample. The spatial resolution Δd is limited by the Rayleigh criterion as follows:

$$\Delta d = \frac{0.61\lambda}{\text{NA}}$$

where NA is the numerical aperture and λ is the excitation wavelength. As numerical aperture of 1.0 can readily be achieved, Raman spectra can be measured with spatial resolution of ca. $0.5\,\mu\text{m}$ (for excitation of 785 nm). Upon moving the sample through the laser focus (or vice versa) in step widths that correspond to the optical resolution, Raman image of the heterogeneous samples can be obtained, providing molecular structure and microscopic structure information at the same time.

Higher spatial resolution beyond the diffraction limit may be accessible by scanning near-field optical microscopy (SNOM). The sample is irradiated through an optical fiber with an aperture that is smaller than the wavelength or a probe is placed very close to the sample surface and the light interacts with the sample prior to diffraction. The response of the illuminated spot is then detected by the detector system. In analogy to scanning probe microscopies, sequential nanometer-sized spots of the sample are probed and an image of the complete specimen is built up by a raster scan. This technique has been widely used for monitoring by correlated two-photon-induced photoluminescence and Raman scattering imaging. It has been detected that the photoluminescence and SERS are emitted from the gap of a noble metal nanodimer by combination with topographic measurements [17, 22–24]. This is consistent with the calculated results by a finite-difference time-domain (FDTD) calculation [8, 9, 18, 25–28]. In these cases, correlation between SERS and the nanosized morphology is investigated. However, as the sample is probed only in near-field, the number of molecules contributing to the Raman scattering is drastically reduced. In addition, the number of the probed photons passing through the capillary (in reflection configuration) is reduced by several orders of magnitude.

15.2.2   Intensity Mapping

With conventional imaging, upon irradiation the optical emission from the sample is passed to the 2D array of the detector system (e.g., CCD in digital camera). Every pixel of the detector corresponds to the individual real points of the

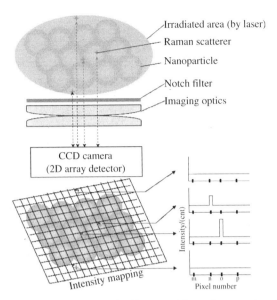

FIGURE 15.1 A schematic diagram of intensity mapping. Different intensities shown in individual pixels correspond to the real points on a sample.

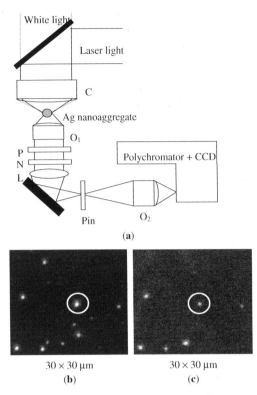

$30 \times 30 \, \mu m$ $30 \times 30 \, \mu m$
      (b)   **(c)**

FIGURE 15.2 (a) The experimental setup that we recently developed to measure LSP resonance and Raman scattering spectra for individual isolated Ag nanoaggregates [20, 27–29, 33, 35, 36]. C, dark- or bright-field condenser; O_1 and O_2, objective lens; P, polarizer; N, notch filter; L, tube lens; pin, pinhole; CCD, charge-coupled device. (b, c) Dark-field and SERS images of selected sample surfaces, respectively [20].

sample and the display shows an intensity mapping of those corresponding points, rather than the individual spectral information. Figure 15.1 shows a schematic diagram for the intensity mapping concept self-explaining the different intensities recorded by individual pixels of 2D array and corresponding to the real points of the sample. For instance, the pixel number "o" represents the maximum intensity scattered from the Raman scatterer, while the pixel numbers "m" and "p" show only the background. Instead of spectral characteristics, only scattering intensity and microscopic structure information can be obtained. To confirm that the bright spots emit SERS or fluorescence, we need to measure the spectral characteristics. To avoid measurement of an ensemble of the bright spots, we limit a measuring area by inserting a pinhole before a polychromator [29], although spectra at various positions are simultaneously measured using a focal plane array detector for a FT-IR imaging [30]. In the next section, this simple and effective method is explained in detail.

15.3 CORRELATION BETWEEN SPR AND SERS IMAGES

We demonstrate here correlation measurements between LSPR Rayleigh scattering and SERS. Use of a microscopic system for the correlated measurements of LSPR Rayleigh scattering and SERS images enabled us to establish relationships between SERS, excitation polarization, excitation energy, and LSPR maxima [18].

Figure 15.2a illustrates the experimental setup for measuring of LSPR Rayleigh scattering spectra of single

Ag nanoaggregates. In an inverted optical microscope, a collimated unpolarized white light beam from a 100 W halogen lamp was introduced into a sample surface through a dark-field condenser lens. Rayleigh scattering light from a single bright spot, likely a single Ag nanoaggregate, was collected using an objective lens and detected using either a digital camera for plasmon and SERS imaging or a polychromator connected to a charge-coupled device for spectral measurements. We detected LSPR Rayleigh scattering light from a single Ag nanoaggregate and minimized background light by selective measurement on a sample area of 1.5 μm diameter (shown in an open circle in Figure 15.2b) using a pinhole (300 μm radius) set in the image plane of the inverted microscope [18].

Figure 15.2a also shows the experimental setup for SERS spectral measurement of single Ag nanoaggregates. The inverted optical microscope was common to detections of Rayleigh scattering and SERS signals. Excitation lasers used were an Ar ion laser, second harmonics (532 nm) of a LD YAG laser, a Kr ion laser, and a He–Ne laser. A set of mirrors (M1–M3) on position controlled ports were used for selecting 457, 488, 514, 532, 567, and 633 nm laser wavelengths for SERS excitations. The laser beams were passed through a polarizer (P1) and a quarter-wave plate (W) and reflected by a half-mirror (HM) into the dark-field condenser lens before focusing at the sample surface. We adjusted the focal point of the laser beams to that of white light by using two convex lenses (L1 and L2). This arrangement was helpful for simple detection of both Rayleigh scattering and SERS signals without additional optics. SERS signal from a single bright spot (Figure 15.2c) was collected using the common objective lens (O$_1$), passed through a holographic notch filter (N) [HNF-(457.8, 488, 514.3, 532, 568.2, and 632.8)], and detected using the CCD cameras in the same way as Rayleigh scattering detection. The excitation laser power was 100 mW/cm^2 at the sample surface. We selected SERS signal from single Ag nanoaggregates and minimized the contribution of background signals to SERS by using a pinhole. The pinhole allowed us selective measurements of SERS signals from a small sample area of 1.5 μm diameter.

15.3.1 Single Nanoparticle and Dimer

SERS enhancement factors have been studied in aspects of both theory and experiments. Theoretical studies based on an electromagnetic model have clarified the correlation among SERS, LSPR, and geometry of isolated Ag nanoaggregates [7]. Experimental studies have revealed that SERS enhancement factors depend on the geometry of Ag nanoaggregates and LSPR Rayleigh scattering spectra [18, 29, 31–35]. These studies have shown that SERS, LSPR Rayleigh scattering, and the geometry of Ag nanoaggregates are related to each other. However, correlation measurements of SERS imaging, LSPR Rayleigh scattering imaging, and SEM imaging of Ag nanoaggregates have not been made because of experimental difficulty. The correlated measurement perfectly enables to evaluate theoretical and experimental results. This evaluation is indispensable for providing prospect of development of geometry that gives rise to high SERS enhancement factors.

Figure 15.3 shows SERS, LSPR Rayleigh scattering, and scanning electron microsccope (SEM) images of the same isolated Ag nanoparticles with adsorbed rhodamine 6G (R6G). SERS and LSPR Rayleigh scattering images of isolated Ag nanoaggregates were observed by using dark-field microscopic system indicated in Figure 15.3. The geometry of the Ag nanoaggregates was observed by FE-SEM. The correlation observations enable us to confirm that dimers of Ag nanoparticles show SERS activity.

Our investigations of LSPR Rayleigh scattering and SERS have directly demonstrated three kinds of relationships between LSPR and SERS. From these investigations, we have identified relationships among excitation polarization, excitation energy, and LSPR energy and have reached to three specific conclusions: (1) SERS bands have the same polarization dependence as that of a longitudinal LSPR band. This indicates that excitation polarization set parallel to the

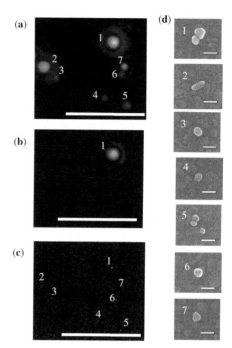

FIGURE 15.3 (a) A LSPR Rayleigh scattering image, (b) a SERS image, (c) and a SEM image of isolated Ag nanoparticles. (d) Enlarged SEM images of each isolated Ag nanoparticles indicated by numbers. Scale bar of (a–c) is 5 mm. Scale bar of (d) is 100 nm.

longitudinal plasmon mode can provide maximum SERS intensity. (2) SERS intensity is dependent on LSPR maximum, suggesting that a larger overlap between molecular absorption band and LSPR band can provide higher SERS intensity. (3) Finally, a spectral shape of SERS is reliant on LSPR energy, indicating that a SERS spectrum can be modulated by a LSPR band shape [18, 29, 31–35].

15.3.2 Nanoaggregate

Figure 15.4a and b, respectively, shows microscope images of SERS and Rayleigh scattering due to LSPR of the Ag nanoaggregates on which 5,5′-dichloro-3,3′-disulfopropyl thiacyanine (TC) was adsorbed from its 0.5 μM aqueous solution excited at 514 nm. One can see various colors of the Ag nanoaggregates shown by their individual LSP resonance in Figure 15.4b. In the case of rhodamines and porphyrins, SERS from Ag nanoaggregates usually exhibits yellow or red color [32, 36]. However, in the case of TC, SERS from Ag nanoaggregates yields various colors, not only red or yellow but also blue. We consider that the reason for showing various colors was due to the absorption bands of TC that appear at short wavelengths (dimer: 408 nm, monomer: 430 nm, J-aggregate: 464 nm) [37].

Figure 15.5a and b shows 514.5 nm excited polarized SERS spectra of a single Ag nanoaggregate on which TC has been adsorbed from its 0.5 μM aqueous solution and the

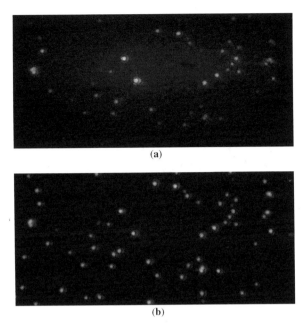

FIGURE 15.4 Microscope images of (a) SERS of TC molecules adsorbed on the Ag nanoaggregates excited at 514 nm and (b) the corresponding LSPR Rayleigh scattering from the Ag nano-aggregates illuminated by the white light through the dark-field condenser lens. The images cover an area of $78 \times 34 \,\mu m^2$. (See the color version of this figure in Color Plate section.)

corresponding polarized scattering spectra of the LSPR, respectively. At the polarization angles of 120° and 210°, the SERS intensity became zero and the maximum, and the LSPR bands appeared at 565 and 710 nm, respectively. This polarization dependence of the SERS spectra is coincident with that of the LSPR band, not at a longer but at a shorter wavelength as shown in Figure 15.5c, and is contrary to those of SERS spectra in our previous works [18, 32, 33]. Note that two LSPR bands emerge at the polarizer angle of 120°, and a calculated LSPR band on the short axis of Ag nanodimer is observed not at 565 nm but at 350 nm [32]. These likely mean that the Ag nanoaggregate excited at 514 nm is not a simple dimer but an aggregate that has a junction on the short axis. We succeeded in reproducing the LSPR bands by FDTD calculation for a rectangle-like nanoaggregate that consists of two-by-three Ag spheres. The appearance of two LSPR bands is due to component nanodimers with different lengths on the short axis of the nanoaggregate as shown in Figure 15.6.

15.3.3 Long-Range Nanostructure

Nowadays, Au nanoparticles have become an indispensable element for the advancement of nanoscience and nanotechnology. Particularly, Au nanostructures and/or clusters are qualified enough for use in various surface-enhanced vibrational (SEV) spectroscopic applications because of their

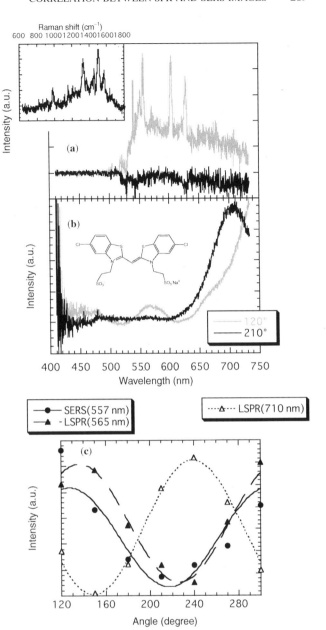

FIGURE 15.5 (a) Polarized SERS spectra of TC molecules adsorbed on a single Ag nanoaggregate excited at 514 nm. Inset: the SERS spectrum in the 1800–600 cm^{-1} region. (b) Polarized reflectance spectra of the same Ag nanoaggregate due to its LSPR. Inset: the chemical structure of 5,5'-dichloro-3,3'-disulfopropyl thiacyanine sodium salt. (c) Polarization dependences of SERS and LSPR.

large spectral enhancement ability, biocompatibility, chemical robustness, and well-established functionalization chemistry [18, 20, 38–40]. In the case of single molecule detection for biomedical applications, the mystery of SERS enhancement at a "hot site" from Au nanoparticles (even from other noble metal nanoparticles) has not yet been extensively studied. One of the reasonable excuses is to

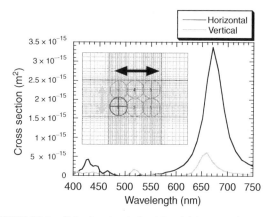

FIGURE 15.6 Calculated polarized Rayleigh scattering spectra of a rectangle-like Ag nanoaggregate. Inset: the nanoaggregate that consists of three-by-two spheres with a diameter of 20 nm. The long and short axes have lengths of 57 nm and 40 (38) nm, respectively.

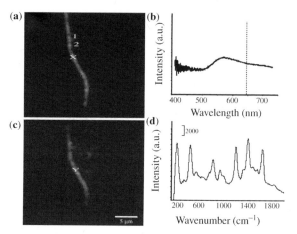

FIGURE 15.7 (a) The SPR image of the anisotropic sample of Au nanoparticle of 50 nm diameter, indicating the intensity variation corresponding to SPR excitations along the line. (b) The SPR spectrum for the same sample obtained at the marked position "X" in Figure 15.7a, indicating a broader SPR peak covering a weak trail of near 700 nm. (c) The SERS image of CV adsorbed on the same sample indicating a good correlation to Figure 15.7a. (d) The SERS spectrum of the CV adsorbed on the same sample obtained at the marked position "Y" in Figure 15.7c. The scale bar shown in Figure 15.7c indicates the size of SPR and SERS images. The dotted vertical line in Figure 15.7b shows the laser excitation position.

prepare a suitable sample and to understand the surface plasmon resonance (SPR)-mediated electromagnetic field distribution. Hence, we have initially developed an innovative, easy to prepare, and less expensive technique [41]. An extensive topographic measurement by atomic force microscope (AFM) and SEM has been adapted. From the initial topographic measurements, it is confirmed that the interparticle gap is ranging from 0 to 5 nm as expected and suitable for a SERS-active substrate. By this proposed technique, it was also possible to prepare a single hot site and 1D or 2D structure of Au nanoparticles. Thus, it opens some other windows of basic research now under sincere consideration. Here, we report only anisotropic Au nanoaggregate and long-range two-dimensional (2D) Au nanoaggregates to elucidate the correlation between SPR excitations and SERS activity [41]. The anisotropic Au sample mostly shows a weak blue-shifted peak near 450 nm in addition to a broadened peak centered at 570 nm covering a trail for another one near 700 nm. In the case of the 2D Au sample, more than one SPR peaks are observed in the longer wavelength region. The SERS observation confirms million times higher enhancement at least in Raman intensity using the Au nanoaggregates adsorbed by dye molecules. The anisotropic sample of Au nanoparticles shows approximately five times higher enhancement in Raman signal compared to that of the 2D sample. A preferentially intense scattering signal in SPR and SERS is observed near the edge in the case of long-range 2D Au nanoaggregates.

Figure 15.7a shows SPR images of the anisotropic sample, that is, elongated aggregate (EA) of Au nanoparticles. The two positions, marked as 1 and 2 in Figure 15.7a, indicates the variation in SPR excitations along the line. In the case of anisotropic assembly of Au nanoparticles, a broader SPR peak centered at 570 nm covering a trail for another peak near 700 nm is observed in addition to a weak peak near the shorter wavelength region that is suppressed in the long-range 2D samples. Figure 15.7b shows an SPR spectrum obtained from

the Au sample at the marked position "X" in Figure 15.7a. The same sample is adsorbed by crystal violet (CV) and investigated with the 647 nm laser excitation. Figure 15.7c shows the SERS image of the same sample of Au nanoparticles. The discontinuous SERS intensity confirms the intensity variation in relation to the SPR excitation along the aggregates as shown in Fig. 15.7a. As mentioned earlier, the SPR solely depends on the variation of local structure in nanometric scale and thus influences the Raman scattering signal. Although the Au nanoparticles are assembled in elongated fashion, the interparticle gaps and individual nanoparticles are not supposed to be similar to each other. It is noteworthy from the SPR observation that the broader peak centered at 570 nm covers well enough the excitation laser wavelength (647 nm) and enhances the SERS signal higher compared to that of the 2D samples as explain below. Figure 15.7d shows the SERS spectrum of CV adsorbed on the EA sample obtained at the marked position "Y" in Figure 15.7c. The peaks observed herewith are consistent with the reported ones [42, 43]. The average SERS intensity seems higher compared to that of the 2D Au samples. A good correlation between the SPR and SERS images is observed herewith for the anisotropic sample of Au nanoparticles.

In the case of long-range self-assembled-like aggregates, the localized SPR gets more freedom and is intuitively influenced by surrounding active hot sites [17, 24, 44]. In turns, the nanoparticles participating in inducing cascaded

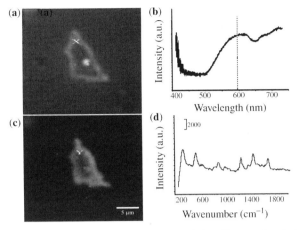

FIGURE 15.8 (a) The SPR image of the 2D sample of Au nanoparticle of 50 nm diameter. (b) The SPR spectrum of the same sample obtained at the marked position "X" in Figure 15.8a, indicating several peaks in the longer wavelength (i.e., ~620, ~670, and ~750 nm). (c) The SERS image of CV adsorbed on the same sample indicating a good correlation to Figure 15.8a. (d) The SERS spectrum of the CV adsorbed on the same sample obtained at the marked position "Y" in Figure 15.8c. The scale bar shown in Figure 15.8c indicates the size of the SPR and the SERS image. The dotted vertical line in Figure 15.8b shows the laser excitation position.

hot sites may behave like nanorods. Indeed, several SPR excitations are observed in this study. Figure 15.8a shows an SPR image of two-dimensional long-ranged assembly (2DLA) self-explaining inhomogeneous distribution of the localized SPR excitations. The plasmon excitation is not so intense like that of the anisotropic samples, elucidating the fact that the SPR excitations on the particular nanoscale position are influenced by surrounding localized energy. Figure 15.8b shows the SPR peaks of the 2DLA sample obtained at the marked position "X" in Figure 15.8a with at least three peaks centered at 620, 670 and 750 nm. Such a spectral discrete behavior is not surprising and is reported for different substrates both theoretically and experimentally [45, 46]. The blue-shifted peak centered at 620 nm could be attributed as the quadruple array resonance and those at 670 and 750 nm might be inferred to discrete peaks of a broad long-wavelength dipolar array resonance termed as longitudinal plasmon resonance. Figure 15.8c shows the SERS image of the same sample and observed inhomogeneous signal distribution. The SERS intensity is found to be higher than that of the RA sample, but not so enhanced like that of the EA sample. Figure 15.8d shows the SERS spectrum of CV adsorbed on the 2DLA sample of Au nanoparticles obtained at the marked position "Y" in Figure 15.8c. The SERS peaks observed herewith are consistent with the reported ones [45, 46].

15.4 APPLICATION OF SERS IMAGING

15.4.1 SERS-Active Substrates for Protein Detections

In conventional SERS-based studies, SERS-active substrates are usually first prepared (e.g., metal colloid, electrodes, or island films), and then analytes are assembled on these substrates for further SERS detections. For most SERS-based protein detections on chips, antibodies and probes are usually linked to metal nanoparticles.

In our studies, we used a contrary way that was based on strong interactions between proteins and Ag nanoparticles [16]. After interactions between proteins and target analytes, we obtained SERS-active substrates by using colloidal Ag staining for total proteins. Figure 15.9 shows fluorescence images of tetramethylrhodamine isothiocyanate (TRITC) and Atto610 molecules after the interactions between the proteins, human immunoglobulin G (IgG), and avidin and their corresponding target analytes, TRITC-antihuman IgG, and Atto610-biotin, respectively. Figure 15.9 also represents their surface-enhanced resonance Raman scattering (SERRS) and surface-enhanced fluorescence (SEF) images after the colloidal Ag staining [47]. Note that the SERRS and SEF images of each fluorescent molecule show significant concentration dependences. For TRITC-antihuman IgG, the color of the SERRS/SEF images changes from yellow, which is similar color to that of the fluorescence image, to green as the concentration decreases. For Atto610-biotin, the color of the SERRS/SEF images change from red, which is similar color to that of the fluorescence image, to orange as the concentration decreases. The blue-shifted color images show the enhancement of SERRS because SERRS occurs near the excitation wavelength. Indeed, the fluorescence maxima of TRITC and Atto610 are located at 580 and 630 nm, respectively. The SERRS peaks of TRITC and Atto610 appear in the 1700–900 and 1400–1000 cm^{-1} regions, corresponding to around 560 and 610 nm, by the excitation at 514 and 568 nm, respectively. From the changes in the SEF and SERRS color images, we find that SEF is reduced remarkably with the decrease in the concentrations of target analytes. Both SERRS and SEF can be observed from the bottom images of Figure 15.9, which are due to distance-dependent enhancement of SERRS and SEF. For the molecules in close proximity to the metallic surface, SERRS is enhanced by an electromagnetic field on the metal surface, but fluorescence is quenched by an energy transfer to the metal surface [48, 49]. Moreover, we found that SERRS images are much more stable than the fluorescent ones that tend toward photobleaching. We can observe stable SERS-active Ag aggregates even at much lower concentration when there are few SEF-active Ag aggregates. This indicates the great potential of SERRS images in ultrasensitive determinations of protein–ligand interactions.

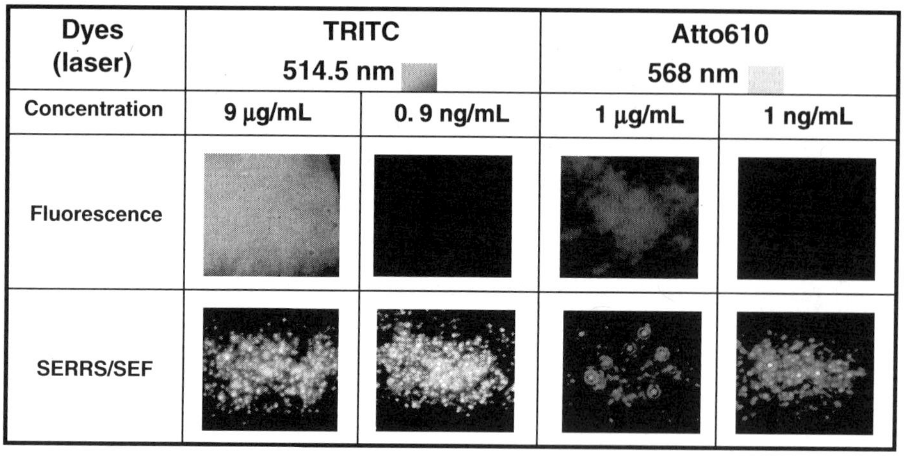

Dyes (laser)	TRITC 514.5 nm		Atto610 568 nm	
Concentration	9 µg/mL	0. 9 ng/mL	1 µg/mL	1 ng/mL
Fluorescence				
SERRS/SEF				

FIGURE 15.9 Fluorescence (the top) and SERRS and SEF (the bottom) microscope images from TRITC and Atto610 at various concentrations. (See the color version of this figure in Color Plate section.)

15.4.2 SERS Imaging for Living Cell Analysis

SERS measurements for biological and medical specimens often use NIR lasers, which can reduce the risk of damaging the specimens by applying high power [50]. The high specificity of vibrational spectra, the reduction of fluorescence, the drop in the detection time (from ~300 to ~1 s compared to normal Raman spectroscopy [50–52]), improvement in spatial detection limit (from ~400 to ~100 nm compared with normal Raman spectroscopy [50–52]), and the insensitivity to the aqueous environment increase the significance of SERS to study complex biological systems like living cells [53–55]. In these experiments, colloidal Ag or Au particles were adsorbed on the cells or incorporated inside the cells, and SERS was applied to monitor the cellular processes and events. Also, SERS-active nanoparticles bring the advantage of detecting or tracking different known biomolecules over fluorescent tags. The use of fluorescent tags suffers from confused overlapping fluorescence spectra broader than SERS spectra and nonuniform photobleaching rates, thus leading us to several potential complications [56].

This section deals with the current scenario of living cell analysis using surface-enhanced Raman spectroscopy. Notable protocols used by many researchers in this regard have been described in this chapter. A great deal of attention has been given to the SERS measurements from inside and outside from the living cells using colloidal noble metal particles with and without probe molecules.

15.4.2.1 Intracellular SERS Measurements · Delivery of nanoparticles into cellular interior, as well as routing of the particles or targeting of cellular compartments, can be achieved in various ways, depending not only on the nature of the experiments but also on the type of cell line and physicochemical particle parameters, such as size, shape, and surface functionalization [57–59]. These methods include fluid-phase uptake from the cultural medium and mechanical methods such as microinjection. The *in vivo* molecular probing of cellular compartments by measuring SERS spectra from endosomes in living individual epithelial cell line IRPT and macrophage cells J774 were reported [60].

SERS signals are highly irreproducible and do not render themselves to acceptable quantification of target intracellular constituents. In addition, delivering SERS inducers (Au/Ag nanoparticles) to points of interest inside cells is another major challenge [61]. In one of their latest works, Kneipp et al. [60] have rightly considered the lysosomal context of nanoparticle aggregates when interpreting the SERS bands. One of the methods to achieve quantitative results is to develop functionalized nanoparticle sensors that employ nanoparticles and have been coated with a molecule that will bind to the analyte of interest. The functionalized nanoparticle probes have several advantages over nonfunctionalized probes: (1) The functional group adds a degree of specificity to the sensor by providing a specific interaction with the target analyte. (2) The analyte molecule does not need to be Raman active or have a particularly large Raman cross section, and (3) the surface is coated with the functional molecule, interfering molecules cannot adsorb to the particle surface, and therefore the background is reduced [62]. Nanoparticle-based pH sensors using 4-mercaptobenzoic acid (4-MBA) was developed by Talley et al. [63]. A Raman spectrum of 4-MBA changes according to the state of the acid group as shown in Figure 15.10. As the pH is reduced and the acid group becomes protonated, the COO^- stretching mode at 1430 cm^{-1} decreases in intensity. The strong ring breathing modes at 1075 and 1590 cm^{-1} are not affected by the

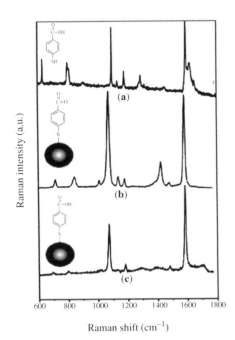

FIGURE 15.10   (a) A Raman spectrum of solid 4-MBA and (b) SERS spectra of 4-MBA attached to silver nanoparticles at pH 12.3 and (c) pH 5.0. The insets to the left of each spectrum illustrate the dominant state of the molecule under the conditions described above [63].

nanoparticles retain their functionality and are not overwhelmed by a large background when they are placed into a biological matrix. The spectrum indicates that the pH surrounding the nanoparticle is below 6, which is consistent with the particles being located inside a lysosome (pH ~5) [64]. Although CHO cells are not normally considered phagocytotic, they have been shown to uptake latex beads as large as 1 mm in diameter [65]. These studies also showed that the phagocytosed latex particles were localized in lysosomes once they were internalized.

Recently, Chourpa et al. [66] proposed an advanced multispectral imaging approach using a combination of SERRS with fluorescence techniques. The fluorescence emission and SERRS spectra of the anticancer drug, mitoxantrone (MTX), were recorded simultaneously over an optical section of the cell (Figure 15.12). The spectral intensity map in Figure 15.12b (average intensity in the spectral region including both fluorescence and SERRS maxima) shows a large zone of high intensity (presumably nuclear and perinuclear drug fluorescence) and at least four particularly bright spots localized in low-intensity zones (cytosol and membrane) of the cell. The spectra corresponding to these spots (Figure 15.12c) confirm the presence of tiny Ag aggregates, since they contain fluorescence background superimposed with SERRS signal of mitoxantrone, noticeable due to the most intense band at $1300 \, cm^{-1}$. The SERRS intensity appears at least as high as that of fluorescence, thus indicating that in terms of sensitivity of subcellular analysis these two techniques are comparable. Figure 15.12c illustrates the semiquantitative analysis of the intracellular spectra since each of them can be deconvoluted into a proportional addition of characteristic fluorescence and SERRS spectra. For instance, the spectrum in Figure 15.12c indicates that the SERRS signal in the given location (inner part of the aggregate 4 located in the membrane of the cell) is

change in pH. The 4-MBA-coated nanoparticle sensors show a pH response in the range of 6–8 that is ideal for biological measurements. To demonstrate the feasibility of utilizing these nanoparticle sensors in living cells, the 4-MBA-functionalized nanoparticles were incorporated into Chinese hamster ovary (CHO) cells by passive uptake. A representative SERS spectrum of the cells incorporated nanoparticles, shown in Figure 15.11, illustrates that the functionalized

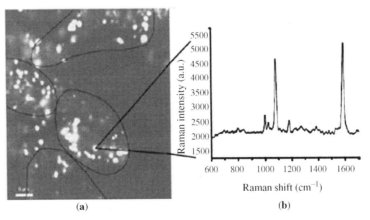

FIGURE 15.11   (a) A confocal image of CHO cells with 4-MBA nanoparticle sensors incorporated into the cells. The cells are outlined in black to facilitate viewing the low contrast of the cells against the bright nanoparticles. (b) A SERS spectrum of one of the nanoparticle sensors (indicating that the pH around the nanoparticle sensor is less than 6).

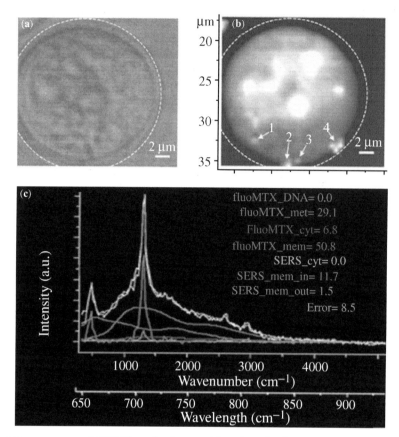

FIGURE 15.12 Combined fluorescence and SERRS confocal spectral imaging of a fluorescent anticancer drug MTX within an MCF-7 cancer cell. (a) White light microscope image of the cell, dashed line shows the outer membrane limits. (b) Spectral intensity distribution map (average intensity in the region of 1200–1500 cm^{-1} or 685–699 nm) reveals four silver colloid aggregates (bright spots 1–4). (c) Intracellular spectrum (inner part of the aggregate 4 located in the membrane of the cell) fitted with a sum of proportional contribution of characteristic fluorescence and SERRS spectra.

colocalized with the fluorescence characteristics of mitoxantrone 33 in a low-polarity environment of membranes (contribution of 50.8%) with that of an oxidative metabolite (contribution of 29.1%). These data, together with 6.8% contribution of the fluorescence characteristic of hydrophilic cytosolic complex of the drug, are in agreement with the assignment of the main SERS pattern to the inner cellular membrane (contribution of 11.7%). Therefore, the combination of SERRS and fluorescence multispectral imaging is a real opportunity to colocalize nanoparticulate SERS substrates with subcellular compartments. Analysis of such SERRS–fluorescence multispectral maps provides multiple information about the drug molecular contacts in a given subcellular compartment.

15.4.2.2 Extracellular SERS Measurements

The SERS measurement of the outer cell membrane of living cells is getting importance because of their sensitivity toward different biological functions of cells [67]. Following are some notable protocols attempted in this regard.

SERS is a promising tool to monitor neurotransmitter release at the single-cell level; it is a sensitive technique that provides structural information about the released compounds and spatial information about their release sites [68]. Furthermore, SERS has recently been successfully applied to differentiating cancerous cells from normal cells. For example, to obtain a highly sensitive cellular image of living normal HEK293 cells and HEK293 cells expressing PLCγ1 using the SERS technique, functional nanoprobes based on Au/Ag core–shell nanoparticles, conjugated with monoclonal antibodies, were used by Lee et al. [69]. PLCγ1 is a protein whose abnormal expression may be associated with tumor development. Schematic illustration of Ag-coated Au nanoprobes with R6G for the SERS imaging of cancer cells is given in Figure 15.13. Functionalized nanoprobes, conjugated with secondary antibodies, were attached only to the markers on cancer cells. On specific binding, nanoprobes were washed using a buffer solution. After washing, the remaining nanoprobes were selectively attached to the cancer markers by antibody–antibody interaction. Each

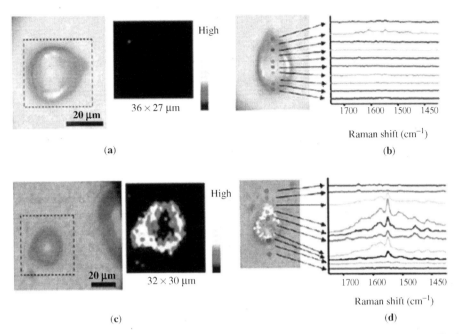

FIGURE 15.13 (a) Normal cell dark-field image and SERS spectroscopy image. (b) Raman measurements. (c) Single cancer cell: bright-field image (left), SERS image (right). (d) Overlay image of bright field and SERS and Raman mapping for single cancer cell. The spots in (b) and (d) indicate the laser spots across the middle of the cell along the y axis.

Raman spectrum was measured by moving down the laser spot from the top of the cell with an interval of 3 mm, but any characteristic SERS signal of R6G was not observed from the normal cells (Figure 15.13a and b). Only a few noise signals caused by fluorescence were observed in the spectra. The SERS image for the same PLCγ1-expressing HEK293 cells is shown in Figure 15.13c. In this figure, the image for the SERS spectra were displayed by a monochrome-decoding method using the strongest Raman peak of R6G at $1650\,cm^{-1}$. The darker the black of the bar, the lower the concentration of SERS nanoprobes attached to the cell. Figure 15.13d shows a combined image of the bright-field and SERS images. It shows the distribution of the PLCγ1 markers on the cell membrane. For a clear understanding, the Raman depth profiling spectra were also measured and are shown in Figure 15.13d. Here, the R6G reporter-labeled SERS spectrum on the cell membrane looks somewhat different from the SERS spectrum of pure R6G. In particular, one strong extra Raman peak was observed at $1542\,cm^{-1}$. However, the other Raman peaks of Raman reporter R6G in HEK cells are well matched with those of pure R6G. In an other work [70], SERS signals of cell components were measured along with SERS signals of Raman reporter indocyanine green (ICG). The relative contributions of ICG and cell components depend on the coadsorption of both kinds of molecules. According to the experimental data, however, the intensity of the strongest Raman peak at $1650\,cm^{-1}$, which was used for Raman mapping, was very much consistent with the intensity of antibody interactions between nanoparticles and biomarkers. Thus, it is believed that SERS imaging with the antibody-

conjugated metal nanoprobes can clearly distinguish between cancerous and noncancerous cells.

Sujith et al. [71] recently attempted to study the cell wall biochemistry of living single yeast by SERS. Figure 15.14a and b show dark-field images of living yeast cells without and with Ag nanoparticles, respectively. Heterogeneously distributed colored spots on the cells in Figure 15.14b correspond to isolated Ag nanoparticles or nanoaggregates. This correspondence is evident from AFM images of a yeast cell surface (Figure 15.14c) and a yeast cell surface adsorbed with Ag nanoparticles (Figure 15.14d). The average diameter of the particles adsorbed on the cell was 96 nm. However, the diameter of individual Ag nanoparticles used for the current work is ~40 nm. Thus, an increase in the observed diameter can be attributed to low resolution of an AFM tip that cannot reach the interface between particle and cell wall, or aggregation of nanoparticles. The height of the particles adsorbed was found in the range 10–23 nm (Figure 15.14e). This small height is possibly due to cell surfaces hollowed by strong adsorption of Ag particles. AFM measurements revealed that about 20% of adsorbed Ag particles form dimer-like aggregates. Figure 15.14f and g shows the dark-field and corresponding SERS image of Ag nanoparticles adsorbed on the cell wall, respectively. The SERS-active spots can clearly be seen in red, green, and yellow colors Figure 15.14g. The distribution of dimer-like aggregates in Figure 15.14d is consistent with that of SERS-active spots in Figure 15.14g. To identify SERS spectra from individual nanoaggregates on single living yeast cell wall shown in Figure 15.15, they measured Raman and SERS spectra of

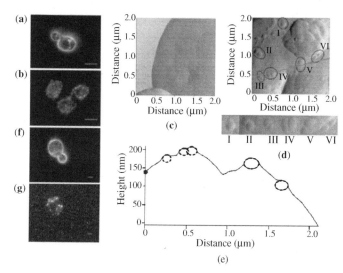

FIGURE 15.14 Dark-field image of (a) yeast cell and (b) Ag adsorbed on yeast cell surfaces (100×) (scale 1 μm). AFM images of (c) an yeast cell and (d) a Ag adsorbed on yeast cell. Dimer nanoaggregates are circled. (e) Height trace along the line drawn in (d) (Ag nanoparticles are schematically represented. The heights of Ag particles from left to right are 10, 10, 10, 23, and 15 nm). (f) Ag nanoaggregates on a yeast cell surface and the corresponding (g) SERS image (60×) (scale 1 μm).

mannan, glucan, and chitin, which are the main components of yeast cell wall. Among the cell wall components, the SERS spectra of mannan from only *Saccharomyces cerevisiae* were similar to SERS spectra in Figure 15.15 and did not show any similarity to the Raman spectra of mannan. Ahern and Garell [72] reported that affinity between mannan and Ag nanoparticles is intrinsically weak. Thus, mannan itself cannot show SERS signal. It is widely known that mannan of yeast cell walls covalently bonded with proteins (mannoproteins) [73]. Moreover, nitrogen atoms in proteins have strong

affinity for Ag atoms. Thus, they attributed the SERS spectra to mannoproteins. The amino acids fenilaranin, tyrosine, tryptophanes, histidine, and so on are present in mannoproteins that have chromophoric groups. So, there is a chance of electronic absorption in the spectral region of excitation, and thereby resonance enhancement. The circles (red spots) in Figure 15.15 indicate similar mannoprotein peaks. However, there are also peaks other than the observed mannoprotein peaks (squares (green spots) in Figure 15.15), which are amide, C–N, protein backbone, and amino acid vibrations [60, 72, 74, 75] related to the bioactivity of living cells such as protein secretion, movement, and so on. The nanoaggregate-by-nanoaggregate spectral variation in Figure 15.15 may arise from yeast cell wall heterogeneity due to clustering or asymmetric cell surface distribution of mannoproteins [76]. The heterogeneity will provide multiple interactions of mannoprotein with Ag nanoaggregates by amino and carboxyl groups.

15.5 BLINKING IN SERS IMAGING

SERS spectroscopy is sensitive enough to measure a Raman spectrum of a single molecule adsorbed on a noble metal nanoaggregate [10, 11]. Recently, several reviews of single molecule SERS spectroscopy were published [6, 12–14]. Some phenomena are considered as evidences for the single molecule detection. First, a statistical distribution of SERS signals does not show Gaussian distribution but Poisson distribution [10, 12]. The former is characteristic of an ensemble of many molecules, and the latter means there is a single or a few molecules adsorbed on a noble metal nanoaggregate. Second, a SERS spectrum from a mixture of two different kinds of analyte molecules is attributed to either analyte molecule [6, 12, 13]. At sufficiently low concentrations of both analytes, a single analyte molecule is adsorbed on a noble metal nanoaggregate. Finally, the blinking SERS emission and the spectral fluctuation are shown in Refs 11, 13, and 14. Blinking is considered as an evidence for single molecule detection also in fluorescence spectroscopy [77].

Figure 15.16 shows time-resolved SE(R)RS spectra of 5,5′-dichloro-3,3′-disulfopropyl thiacyanine, whose structure is shown in an inset of Figure 15.5b, adsorbed on a single Ag nanoaggregate in water [78]. Initially, the SERS spectrum attributed to the J-aggregates with background emission changed to the temporally fluctuated SERRS spectra. This spectral fluctuation coincides with the dissolution of the J-aggregates into the monomer or the dimer. We assume that the anionic thiacyanine molecules are adsorbed on negatively charged citrate-reduced Ag surfaces through their positively charged nitrogen atoms ($=N^+<$) such as rhodamine 6G (R6G) [79]. However, they have two negatively charged $-SO_3^-$ groups unlike R6G. The spectral fluctuation of SERS of the thiacyanine is likely attributed to thermally

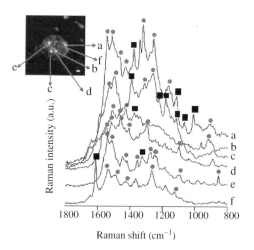

FIGURE 15.15 SERS spectra from different Ag nanoaggregates on a single yeast cell wall. Inset: the SERS image of a single yeast cell (60×) (scale 1 μm). The spectra from "a" to "f" were collected from Ag nanoaggregates a–f in the image.

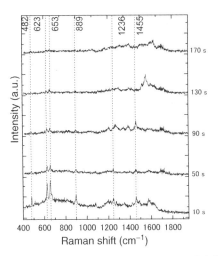

FIGURE 15.16 Time-resolved SE(R)RS spectra of 5,5′-dichloro-3,3′-disulfopropyl thiacyanine adsorbed on a single Ag nanoaggregate in water.

similar to that of temperature dependence of blinking SERS of R6G [28], and it has been reported that the blinking is a thermal activated mechanism. We need to investigate the blinking SERS quantitatively.

We took a video of the blinking SERS from Ag nanoaggregates adsorbed by 5,5′-dichloro-3,3′-disulfopropyl thiacyanine in water using an inverted microscope (Olympus, IX-70) coupling with a cooled digital CCD camera (Hamamatsu, ORCA-AG) whose time resolution was ~60 ms. The video images are shown in Figure 15.17a. Figure 15.17b shows a time profile of the blinking SERS intensity made by the analysis of the video. In this method, many time profiles of SERS intensities can be measured all at once. The spectral fluctuation strongly relates to background emission intensity that composes SERS intensity. The blinking can be quantitatively characterized by some methods. One is the autocorrelation function given by

$$C(\tau) = \sum_{\nu} \left(\langle I_{\nu}(t) \cdot I_{\nu}(t+\tau) \rangle - \langle I_{\nu}(t) \rangle^2 \right)$$

driven movement of the molecules at the small junctions of Ag nanoaggregates by the Coulomb repulsion. Indeed, both of temporally fluctuated and temporally stable SERS spectra were observed from a single Ag nanoaggregate adsorbed by 3,3′-diethyl thiacyanine, cationic dye, that does not form J-aggregates [80]. However, this charge dependence of the blinking SERS was qualitatively investigated in a way

in which $I_{\nu}(t)$ is the intensity at wavenumber ν and time t [81, 82]. It has been reported that the autocorrelation functions, which indicate some periodicity, from each Ag nanoaggregate adsorbed by R6G were not reproduced by any simple function [82]. This suggests a complex process in the blinking. Laser power dependence of blinking was investigated using the rate calculated integrating averaged

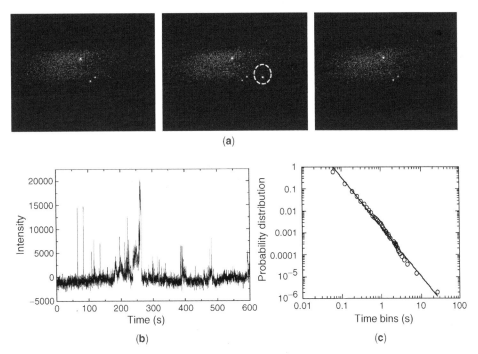

FIGURE 15.17 (a) Microscope images of SERS from single Ag nanoaggregates adsorbed by thiacyanine at various times. (b) Time profiles of blinking SERRS intensity. (c) Distribution of bright events against time bins. The line shows the best fitting result by power law.

autocorrelation function at various laser power [81]:

$$k^{-1} = \int (C(\tau)/C(0)) d\tau$$

On the other hand, power-law statistics are useful for long-range nonexponential behavior of the blinking [83]. Figure 15.17c shows normalized bright time probability distribution of the time profile of SERS intensity (Figure 15.17a) against each bright time bins. The line in Figure 15.17c is given by

$$P(t) = A t^{-\alpha}$$

in which α is the power-law exponent. This may be indicative of the blinking SERS.

15.6 CONCLUSION

In this chapter, we outlined recent studies on the mechanism of SERS imaging and its biological applications. In the introduced SERS imaging, noble metal colloids and a conventional dark-field microscope were used. Thus, these studies on the mechanism of SERS were investigated in terms of correlation with LSPR. SERS has been used for bioapplications from the first stage of its invention. In recent years, biomedical applications of SERS have become a subject with a variety of challenges and opportunities for practitioners of chemistry, biology, and medicine. We reported recent advances in the applications of SERS spectroscopy in protein detection and living cell analysis. Special attention has been given to different protocols used by researchers for intra- and extracellular analysis with and without probe molecules.

REFERENCES

1. Grasselli, J. G. and Bulkin, B. J. (1991) *Analytical Raman Spectroscopy*, Wiley–Interscience, New York.
2. Fleischmann, M., Hendra, P. J., and McQuillan, A. J. (1974) *Chem. Phys. Lett.* **26**, 163.
3. Jeanmaire, D. L. and Van Duyne, R. P. (1977) *J. Electroanal. Chem.* **84**, 1.
4. Albrecht, M. G. and Creighton, J. A. (1977) *J. Am. Chem. Soc.* **99**, 5215.
5. Moskovits, M. (1985) *Rev. Mod. Phys.* **57**, 783.
6. Etchegoin, P. G. and Le Ru, E. C. (2008) *Phys. Chem. Chem. Phys.* **10**, 6079.
7. Xu, H., Bjerneld, E. J., Käll, M., and Börjesson, L. (1999) *Phys. Rev. Lett.* **83**, 4357.
8. Xu, H., Aizpurua, J., Käll, M., and Apell, P. (2000) *Phys. Rev. E* **62**, 4318.
9. García-Vidal, F. J. and Pendry, J. B. (1996) *Phys, Rev. Lett.* **77**, 1163.
10. Kneipp, K., Wang, Y., Kneipp, H., Perelman, L. T., Itzkan, I., Dasari, R. R., and Feld, M. S. (1997) *Phys. Rev. Lett.* **78**, 1667.
11. Nie, S. and Emory, S. R. (1997) *Science* **275**, 1102.
12. Kneipp, J., Kneipp, H., and Kneipp, K. (2008) *Chem. Soc. Rev.* **37**, 1052.
13. Pieczonka, N. P. W., and Aroca, R. F. (2008) *Chem. Soc. Rev.* **37**, 946.
14. Qian,X. -M. and Nie, S. M. (2008) *Chem. Soc. Rev.* **37**, 912.
15. Zander, Ch., Enderlein, J., and Keller, R. A. (2002) *Single Molecule Detection in Solution: Methods and Applications*, Wiley-VCH Verlag GmbH, Berlin.
16. Laserna, J. J. (1996) *Modern Techniques in Raman Spectroscopy*, Wiley, Chichester.
17. Hossain, M. K., Shimada, T., Kitajima, M., Imura, K., and Okamoto, H. (2006) *J. Microsc.* **229**, 327.
18. Itoh, T., Kikkawa, Y., Yoshida, K., Hashimoto, K., Biju, V., Ishikawa, M., and Ozaki, Y. (2006) *J. Photochem. Photobiol. A* **183**, 322.
19. Soper, A. and Legendre, B. L. (1998) *Appl. Spectrosc.* **52**, 1.
20. Bulte, J. W. M. and Modo, M. M. J. (2008) *Nanoparticles in Biomedical Imaging: Emerging Technologies and Applications*, Springer Science + Business Media, New York.
21. Dickinson, M. E., Bearman, G., Tille1, S., Lansford, R., and Fraser, S. E. (2001) *BioTechniques*, **32**, 1272.
22. Imura, K., Okamoto, H.; Hossain, M. K., and Kitajima, M. (2006) *Chem. Lett.* **35**, 78.
23. Imura, K., Okamoto, H.; Hossain, M. K., and Kitajima, M. (2006) *Nano Lett.* **6**, 2173.
24. Hossain, M. K., Shimada, T., Kitajima, M., Imura, K., and Okamoto, H. (2008) *Langmuir* **24**, 9241.
25. Hao, E. and Schatz, G. C. (2004) *J. Chem. Phys.* **120**, 357.
26. Futamata, M., Maruyama, Y., and Ishikawa, M. (2002) *Vib. Spectrosc.* **30**, 17.
27. Futamata, M., Maruyama, Y., and Ishikawa, M. (2003) *J. Phys. Chem. B* **107**, 7607.
28. Maruyama, Y., Ishikawa, M., and Futamata, M. (2003) *J. Phys. Chem. B* **108**, 673.
29. Itoh, T., Hashimoto, K., and Ozaki, Y. (2003) *Appl. Phys. Lett.* **83**, 2274.
30. Petibois, C. and Déléris, G. (2006) *Trends Biotechnol.* **24**, 455.
31. Itoh, T., Hashimoto, K., Ikehata, A., and Ozaki, Y. (2003) *Appl. Phys. Lett.* **83**, 5557.
32. Itoh, T., Hashimoto, K., Ikehata, A., and Ozaki, Y. (2004) *Chem. Phys. Lett.* **389**, 225.
33. Itoh, T., Biju, V., Ishikawa, M., Kikkawa, Y., Hashimoto, K., Ikehata, A., and Ozaki, Y. (2006) *J. Chem. Phys.* **124**, 134708.
34. Itoh, T., Kikkawa, Y., Biju, V., Ishikawa, M., Ikehata, A., and Ozaki, Y. (2006) *J. Phys. Chem. B* **110**, 21536.
35. Itoh, T., Yoshida, K., Biju, V., Kikkawa, Y., Ishikawa, M., and Ozaki, Y. (2007) *Phys. Rev. B* **76**, 085405.
36. Itoh, T., Hashimoto, K., Biju, V., Ishikawa, M., Wood, B. R., and Ozaki, Y. (2006) *J. Phys. Chem. B* **110**, 9579.

37. Yao, H., Kitamura, S., and Kimura, K. (2001) *Phys. Chem. Chem. Phys.* **3**, 4560.

38. Baker, G. A. and Moore, D. S. (2005) *Anal. Bioanal. Chem.* **382**, 1751.

39. Maier, S. A., Brongersma, M. L., Kik, P. G., Meltzer, S., Requicha, A. A. G., and Atwater, H. A. (2001) *Adv. Mater.* **13**, 1501.

40. Schatz, G. C. and Van Duyne, R. P. (2002) *Handbook of Vibrational Spectroscopy*, Wiley, Chichester.

41. Hossain, M. K., Kitahama, Y., Huang, G. G., Kaneko, T., and Ozaki, Y. (2008) *Appl. Phys. B*, **95**, 165.

42. Liang, E. J., Ye, X. L., and Kiefer, W. (1997) *J. Phys. Chem. A.* **101**, 7330.

43. Watanabe, T. and Pettinger, B. (1982) *Chem. Phys. Lett.* **89**, 501.

44. Shimada, T., Imura, K., Hossain, M. K., Kitajima, M., and Okamoto, H. (2008) *J. Phys. Chem. C* **112**, 4033.

45. Le, F., Brandl, D. W., Urzhumov, Y. A., Wang, H., Kundu, J., Halas, N. J., Aizpurua, J., and Nordlander, P. (2008) *ACS Nano* **2**, 708.

46. Hossain, M. K., Kitahama, Y., Biju, V. P., Kaneko, T., Itoh, T., and Ozaki, Y. (2009) *J. Phys. Chem. C* **113**, 11689.

47. Han, X. X., Kitahama, Y., Tanaka, Y., Guo, J., Xu, W. Q., Zhao, B., and Ozaki, Y. (2008) *Anal. Chem.* **80**, 6567.

48. Lakowicz, J. R., Geddes, C. D., Gryczynski, I., Malicka, J., Gryczynski, Z., Aslan, K., and Lukomska, J. (2004) *J. Fluoresc.* **14**, 425.

49. Geddes, C. D. and Lakowicz, J. R. (2002) *J. Fluoresc.* **12**, 121.

50. Kneipp, K., Kneipp, H., Itzkan, I., Dasari, R. R., and Feld, M. S. (2002) *J. Phys. Condens. Matter* **14**, R597.

51. Eliasson, C., Loren, A., Engelbrektsson, J., Josefson, M., Abrahamsson, J., and Abrahamsson, K. (2005) *Spectrochim. Acta A* **61**, 755.

52. Kneipp, K., Kneipp, H., Itzkan, I., Dasari, R. R., and Feld, M. S. (1999) *Chem. Rev.* **99**, 2957.

53. Morjani, H., Riou, J. F., Nabiev, I., Lavelle, F., and Manfait, M. M. (1993) *Cancer Res.* **53**, 4784.

54. Manfait, M., Morjani, H., and Nabiev, I. (1992) *J. Cell. Pharmacol.* **3**, 120.

55. Nabiev, I., Morjani, H., and Manfait, M. (1991) *Eur. Biophys. J.* **19**, 311.

56. Cao, Y. C., Jin, R., and Mirkin, C. A. (2002) *Science* **297**, 1536.

57. Rejman, J., Oberle, V., Zuhorn, I. S., and Hoekstra, D. (2004) *Biochem. J.* **377**, 159.

58. Arlein, W. J., Shearer, J. D., and Caldwell, M. D. (1998) *Am. J. Physiol.* **44**, R1041.

59. Tkachenko, A. G., Xie, H., Liu, Y. L., Coleman, D., Ryan, J., Glomm, W. R., Shipton, M. K., Franzen, S., and Feldheim, D. L. (2004) *Bioconjugate Chem.* **15**, 482.

60. Kneipp, J., Kneipp, H., McLaughlin, M., Brown, D., and Kneipp, K. (2006) *Nano Lett.* **6**, 2225.

61. Shamsaie, A., Heim, J., Yanik, A. A., and Irudayaraj, J. (2008) *Chem. Phys. Lett.* **461**, 131.

62. Talley, C. E., Huser, T. R., Hollars, C. W., Jusinski, L., Laurence, T., and Lane, S. M. (2005) *Nanoparticle Based Surface-Enhanced Raman Spectroscopy*, UCRL-PROC-208863, NATO Advanced Study Institute, Biophotonics Ottawa, Canada.

63. Talley, C. E., Jusinski, L., Hollars, C. W., Lane, S. M., and Huser, T. (2004) *Anal. Chem.* **76**, 7064.

64. Alberts, B., Bray, D., Lewis, D. J., Raff, M., Roberts, K., and Watson, J. D. (1994) *Molecular Biology of the Cell*, Garland Publishing, New York.

65. Fukasawa, M., Sekine, F., Miura, M., Nishijima, M., and Hanada, K. (1997) *Exp. Cell Res.* **230**, 154.

66. Chourpa, I., Lei, F. H., Dubois, P., Manfaita, M., and Sockalingum, G. D. (2008) *Chem. Soc. Rev.* **37**, 993.

67. Kapteyn, J. C., Ende, H. V. D., and Klis, F. M. (1999) *Biochim. Biophys. Acta* **1426**, 373.

68. Dijkstra, R. J., Scheenen, W. J. J. M., Dama, N., Roubos, E. W., and ter Meulen, J. J. (2007) *J. Neurosci. Meth.* **159**, 43.

69. Lee, S., Kim, S., Choo, J., Shin, S. Y., Lee, Y. H., Choi, H. Y., Ha, S., Kang, K., and Oh, C. H. (2007) *Anal. Chem.* **79**, 916.

70. Kneipp, J., Kneipp, H., Rice, W. L., and Kneipp, K. (2005) *Anal. Chem.* **77**, 2381.

71. Sujith, A., Itoh, T., Abe, H., Anas, A., Yoshida, K., Biju, V., and Ishikawa, M. (2008) *Appl. Phys. Lett.* **92**, 103901.

72. Ahern, A. M. and Garrell, R. L. (1991) *Langmuir* **7**, 254.

73. Lipke, P. N. and Ovalle, R. (1998) *J. Bacteriol.* **180**, 3735.

74. Stewart, S. and Fredericks, P. M. (1999) *Spectrochim. Acta A* **55**, 1615.

75. Heme, T. M., Ahern, A. M., and Garrell, R. L. (1991) *Anal. Chim. Acta* **246**, 75.

76. Chaffin, W. L., Pez-Ribot, J. L., Casanova, M., Gozalbo, D., and Martinez, J. P. (1998) *Microbiol. Mol. Biol. Rev.* **62**, 130.

77. Tamarat, Ph. Maali, A., Lounis, B., and Orrit, M. (2000) *J. Phys. Chem. A*, **104**, 1.

78. Kitahama, Y., Tanaka, Y., Itoh., T., and Ozaki, Y. (2009) *Chem. Lett.* **38**, 54.

79. Shegai, T. O. and Haran, G. (2006) *J. Phys. Chem. B* **110**, 2459.

80. Takazawa, K., Kitahama, Y., Kimura, Y., and Kido, G. (2005) *Nano Lett.* **5**, 1293.

81. Weiss, A. and Haran, G. (2001) *J. Phys. Chem. B* **105**, 12348.

82. Emony, S. R., Jensen, R. A., Wenda, T., Han, M., and Nie, S. (2006) *Faraday Discuss.* **132**, 249.

83. Bizzarri, A. R. and Cannistraro, S. (2005) *Phys. Rev. Lett.* **94**, 068303.

16

LINEAR AND NONLINEAR RAMAN MICROSPECTROSCOPY: FROM A MOLECULE TO SINGLE LIVING CELLS

HIDEAKI KANO,[1,2] YU-SAN HUANG,[1] YASUAKI NAITO,[3] RINTARO SHIMADA,[1] AND HIRO-O HAMAGUCHI[1]

[1]*Department of Chemistry, School of Science, The University of Tokyo, Tokyo, Japan*

[2]*PRESTO (Precursory Research for Embryonic Science and Technology), Japan Science and Technology Agency, Saitama, Japan*

[3]*Department of Chemistry, Gakushuin University, Tokyo, Japan*

16.1 INTRODUCTION

"Spectra are letters from the molecule." This romantic phrase states impressively how molecules send us messages about themselves in the form of spectra. In particular, vibrational spectra such as Raman and infrared embody many characteristic features that are specific to a molecule. By analyzing vibrational spectra, we can identify chemical species and elucidate in details their structure and dynamics. Thus, vibrational spectra are often called "molecular fingerprints." In the mid-infrared "fingerprint" region, we observe many vibrational bands that are attributable to skeletal modes characteristic to a molecule. Since biological systems are made up of molecules, vibrational spectroscopy should be useful in life sciences as much as it is in material sciences. Owing to its noninvasive and nondestructive nature, Raman spectroscopy is more suitable for biological applications than infrared. Thanks to recent technical developments, we can now investigate a living cell *in vivo* under a microscope. Thus, quite a number of Raman microspectroscopic studies have already been reported on living cells [1–11], though it

was very difficult to confirm that the cells were really living. We recently found a strong Raman band in mitochondria of a living fission yeast cell, which sharply reflects the metabolic activity of mitochondria [8, 9]. We call this band the "Raman spectroscopic signature of life." By monitoring this signature, we can indeed confirm that the cell is living. It means that we can visualize not only the distributions of molecular species but also the cell activity of the growing and dying yeast cells. In this chapter, we review our recent studies on the structure, transformation, and bioactivity of single living yeast cells by linear and nonlinear Raman microspectroscopy. Last but not least, hyper-Raman (HR) microspectroscopy will be introduced, by which we can investigate infrared active vibrational modes with submicrometer spatial resolution under a microscope. Combination of Raman and hyper-Raman opens up a new scope for high spatial resolution vibrational microspectroscopy that is not restricted by the selection rule. We note that this chapter is a compilation of several of our previous papers cited in the reference and therefore contains some overlaps of descriptions.

Raman, Infrared, and Near-Infrared Chemical Imaging   Edited by Slobodan Šašić and Yukihiro Ozaki

16.2   *IN VIVO* REAL-TIME PURSUIT OF THE CELL ACTIVITY OF SINGLE LIVING FISSION YEAST CELLS BY TIME- AND SPACE-RESOLVED RAMAN MICROSPECTROSCOPY

16.2.1   Experimental

We used a confocal Raman microspectrometer (Nanofinder, Tokyo Instrument, Inc.). The 632.8 nm line of a He–Ne laser (Melles Griot 05-LHP-991) was used with a power of 1–4 mW at the sample. The spatial resolutions were 0.3 and 1.7 μm for the lateral and the axial directions, respectively. Yeast cells (*Schizosaccharomyces pombe*), whose nucleus was labeled by green fluorescent protein (GFP), were studied.

16.2.2   Space-Resolved Raman Spectra

Figure 16.1 shows the space-resolved Raman spectra of a single living fission yeast cell in the G1/S phase. These spectra were obtained under a low nutrition condition, so that the cell cycle was slowed down to allow a long exposure time of 300 s. Figure 16.1a, b, and d corresponds to the spectra for nucleus, mitochondria, and septum, respectively. The positions from where Raman spectra are measured are indicated by letters a, b, and d in the inset. Using a cell whose mitochondria are tagged by GFP, we have confirmed that the spectrum in Figure 16.1b comes from mitochondria. The spectra from nuclei are dominated by known protein

Raman bands. In particular, the bands in Figure 16.1a show the amide I mode of the main chain (1655–1660 cm^{-1}), the C–H bend of the aliphatic chain (1450 and 1340 cm^{-1}), the amide III mode of the main chain (1250–1300 cm^{-1}), and the breathing mode of the phenylalanine residue in proteins (1003 cm^{-1}). It is well known that the frequencies of the amide I and III bands are sensitive markers of the secondary structure of the protein main chain. In the present study, the amide I band is observed in the range of 1654–1659 cm^{-1}, which indicates the domination of α-helix structures [12]. Concerning the secondary structure, we need to investigate further in details because this band is broad and thus we cannot neglect the contribution from other secondary structures. New insight into the secondary structure of proteins in a living cell is highly important in connection with the presence of natively unfolded proteins [12, 13], which has been discussed intensively in the past few years. In addition to the protein bands, weak bands are observed at 781 and 1576 cm^{-1}, which can be assigned to nucleic acids. According to the result of a component analysis of isolated nuclei, the DNA/RNA/protein chemical composition ratio in a *S. pombe* nucleus is 1/9.4/115 [14]. This ratio means that proteins are about 10 times more abundant in the isolated nucleus than nucleic acids. This result is consistent with the result of the *in vivo* Raman spectra observed in the present study. The intensity ratio of the band at 853 cm^{-1} to that at 825 cm^{-1} is known to be an indicator of the H-bonding strength of the phenolic hydroxyl group [15].

The Raman spectrum from mitochondria (Figure 16.1b) is similar to that of phosphatidylcholine (Figure 16.1c) except for an intense band at 1602 cm^{-1}. Apart from this 1602 cm^{-1} band, all prominent bands in Figure 16.1b are ascribed to the known phospholipid vibrational modes with reference to the assignments of the spectrum of phosphatidylcholine [16, 17]. The skeletal C–C stretch modes in the region of 1000–1150 cm^{-1} are known to be sensitive to the conformation of the hydrocarbon chains [18]. The bands at 1062 and 1122 cm^{-1} are assigned to the out-of-phase and in-phase modes of the all-*trans* chain. On the other hand, the band at 1082 cm^{-1} is attributed to the *gauche* conformation. The ratio of the intensity of the *gauche* band to that of the *trans* band is larger in the Raman spectrum from mitochondria (Figure 16.1b) than in that from phosphatidylcholine (Figure 16.1c). This finding indicates that the hydrocarbon chains of the mitochondrial membrane are conformationally less ordered than that in pure phosphatidylcholine.

The Raman spectrum from septum is shown in Figure 16.1d. The bands are mostly assigned to polysaccharides. We found the change of the Raman spectrum of the septum in the course of the cell division process. Based on the normal mode analysis of disaccharides [19, 20], it is considered that this change reflects the gradual polymerization of the saccharide molecules.

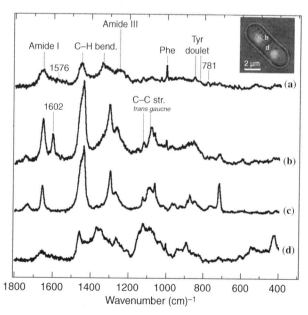

FIGURE 16.1   Spatial-resolved Raman spectra of a single living fission yeast cell in the G1/S phase; (a) nucleus, (b) mitochondria, (c) phosphatidylcholine (model compound of lipid bilayer), and (d) septum.

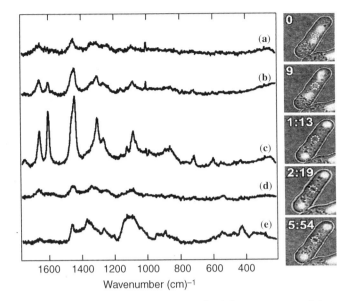

FIGURE 16.2 Time- and space-resolved Raman spectra of the central part of a living yeast cell. The laser beam spot is denoted with a white broken circle.

16.2.3 Time- and Space-Resolved Raman Spectra of a Dividing Fission Yeast

As the cell division proceeds, the Raman spectrum is expected to change drastically, reflecting the changes in molecular composition of the organelles. Figure 16.2 shows the time- and space-resolved Raman spectra of a yeast cell dispersed in YE broth. We start the Raman measurement from the early M phase (a), in which a dividing nucleus is observed at the center of the cell. At 9 min (b), the two nuclei are put apart symmetrically toward the perimeter of the cell. At 1 h, 13 min (G1/S phase), the nuclei are completely separated and located at the two ends of the cell. In the following G1/S stage, a septum starts to form from the plasma membrane, as shown in (d). Finally, the septum becomes mature at 5 h, 54 min (e). In the course of the mitosis process, the Raman spectrum changes significantly. The Raman bands at 0 min (a) are assigned to the proteins in the nucleus. The spectrum at 9 min is a superposition of those of the mitochondrion and cytoplasm. It means that the mitochondria started to be generated at the central part of the cell. At 1 h, 13 min (c), the phospholipid bands due to mitochondria are observed dominantly in the Raman spectrum. It should also be noticed that the intense band is found at the Raman shift of 1602 cm^{-1}. The intensity of this band relative to that of the 1654 cm^{-1} band is clearly higher than those observed in the space-resolved experiment in Figure 16.1. This result is intriguing with regard to the relevance of the 1602 cm^{-1} band to the metabolic activity of mitochondria. The time- and space-resolved Raman spectra were obtained from the yeast cells dispersed in YE broth, while those for the space-resolved Raman spectra (Figure 16.1) were measured under

a low nutrition condition. It means that the band intensity at 1602 cm^{-1} depends on the nutrient condition. The stronger band at 1602 cm^{-1} in Figure 16.2c is indicative of higher metabolic activities in a yeast cell in YE broth.

16.2.4 Discovery of the "Raman Spectroscopic Signature of Life"

To investigate in further details the relationship between the band intensity at 1602 cm^{-1} and the metabolic activity of a mitochondrion, the following experiment has been carried out. We added a KCN aqueous solution to the yeast cell sample in order to look at the effect of a respiration inhibitor on the intensity of the 1602 cm^{-1} band. The time- and space-resolved Raman spectra of a KCN-treated yeast cell are shown in Figure 16.3. The temporal resolution is 100 s. Five minutes before the addition of KCN, the Raman spectrum shows a strong band at 1602 cm^{-1} and the well-known phospholipid bands at 1655, 1446, and 1300 cm^{-1}. Three minutes after the addition of KCN (b), the intensity of the band at 1602 cm^{-1} decreases considerably, while the other phospholipid bands remain unchanged. As time goes on, the 1602 cm^{-1} band becomes weaker ((c) and (d)), and finally disappears at 36 min (e). Concomitantly, the phospholipid bands gradually change from well-resolved peaks to diffuse broad bands. The protein band at 1003 cm^{-1} does not change, and no additional peaks appear throughout the time course of the experiment. We consider that the addition of KCN affects a mitochondrion of a living yeast cell in the following

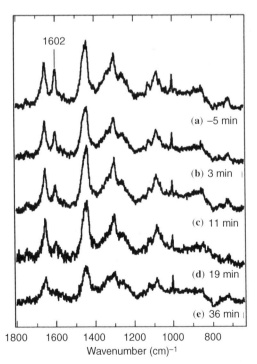

FIGURE 16.3 Time- and space-resolved Raman spectra of a KCN-treated yeast cell.

two steps. First, the cellular respiration is inhibited by the action of CN⁻, and the metabolic activity of the mitochondrion is lowered. This process has been monitored by the drastic decrease in the intensity of the $1602\,\text{cm}^{-1}$ band. Second, the double-membrane structure of the mitochondrion is degraded by the lowered metabolic activities, and it is eventually destroyed. This process has been probed by the changes in the phospholipid bands. It is highly likely that the $1602\,\text{cm}^{-1}$ band probes the primary dying process of the KCN-treated yeast cell through the metabolic activity of a mitochondrion. Therefore, we call this band the "Raman spectroscopic signature of life."

16.3 *IN VIVO* TIME-RESOLVED RAMAN IMAGING OF A SPONTANEOUS DEATH PROCESS OF A SINGLE BUDDING YEAST CELL

16.3.1 Vacuole and Dancing Body in a Budding Yeast Cell

Vacuole is one of the biggest organelles in a yeast cell. The function of vacuole includes amino acid storage and detoxification. In a budding yeast (*S. cerevisiae*) vacuole, a particle called dancing body occasionally appears and moves actively for a while. Although a previous fluorescence study suggested that the main molecular component of the dancing body was polyphosphates [21], it was not clear. One of the difficulties in studying dancing body is that we cannot separate it from the vacuole. We have investigated the molecular component of a dancing body by time- and space-resolved Raman spectroscopy (data not shown), and have proved that the dancing body consists of crystal-like polyphosphate [10].

16.3.2 Spontaneous Death Process Following the Appearance of a Dancing Body

Figure 16.4 shows the time-resolved Raman images (a) and the optical microscope images (b) in a timespan of about 20 h of a single living budding yeast *S. cerevisiae*. From the microscope images in Figure 16.4b, it is seen that a dancing body forms in a vacuole between 5 h, 50 min and 6 h. Then, the vacuole disappears between 8 h, 41 min and 9 h, 31 min. Finally, the cell becomes totally destructed between 9 h, 31 min and 19 h, 37 min. It is obvious that these changes in the microscope images reflect a spontaneous cell death process of an *S. cerevisiae* cell. We confirmed that this spontaneous death process observed in Figure 16.4 also occurred for cells without laser irradiation. We have examined 642 cells to find that once a dancing body was formed in a vacuole, the cell eventually died without any exception. In Figure 16.4b, we trace these changes at the molecular level by the Raman images at 1602, 1440, 1160, and

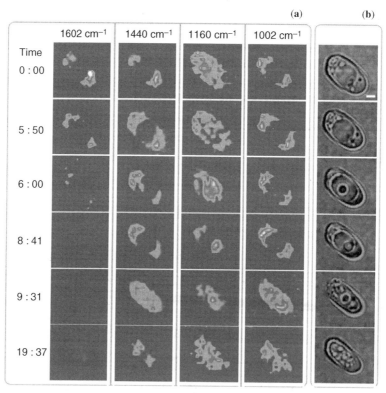

FIGURE 16.4   (a) Time-resolved Raman images and (b) corresponding optical microscope images of a dying *S. cerevisiae* cell. (See the color version of this figure in Color Plate section.)

$1002 \, cm^{-1}$. The $1602 \, cm^{-1}$ band, the "Raman spectroscopic signature of life," reflects the metabolic activity of mitochondria. Therefore, it shows the distribution of active mitochondria in a cell. The band at $1445 \, cm^{-1}$ is due to the C–H bending modes of phospholipids. This band gives an image of phospholipid distribution, which corresponds to the location of mitochondria that contain high concentration of phospholipids. The polyphosphate band at $1160 \, cm^{-1}$ provides an image of phosphate distribution including the location of a dancing body. The band at $1002 \, cm^{-1}$ is assigned to the breathing mode of phenylalanine and shows the protein distribution in a cell. At 0 min, the cell has active mitochondria as shown by the $1602 \, cm^{-1}$ image. The phospholipids ($1440 \, cm^{-1}$) and proteins ($1002 \, cm^{-1}$) are located only outside of a vacuole, but some amount of polyphosphates do exist inside the vacuole as well as outside. At 6 h, when a dancing body becomes suddenly visible, the metabolic activity in mitochondria is markedly lowered as seen from the $1602 \, cm^{-1}$ image, though the mitochondrial distribution ($1440 \, cm^{-1}$) does not change appreciably. The $1160 \, cm^{-1}$ band gives an image covering a large part of the vacuole, indicating that a dancing body is trapped by the laser field and that it moves with the scanning laser spot within the vacuole. There is no change in the protein distribution. At 8 h, 41 min, the dancing body stops moving and stays at the lower part of the vacuole. At this stage, the metabolic activity of mitochondria is completely lost, while the mitochondrial and protein distributions do not change much. The $1160 \, cm^{-1}$ band image corresponds to the remaining of the dancing body that does not move any more. At 9 h, 31 min, the vacuole is lost, while the remaining of the dancing body still exists at the center of the cell. The Raman images show that the molecular distributions at this stage are highly randomized. It indicates the loss of structures in the cell. At 19 h, 37 min, the molecular distributions become totally random, showing that the cell is not alive any more. Thus, a spontaneous cell death process is visualized by *in vivo* time-resolved Raman imaging at the molecular level.

16.4 NONLINEAR RAMAN MICROSPECTROSCOPY AND IMAGING OF SINGLE LIVING CELLS

16.4.1 Ultrabroadband Multiplex Coherent Anti-Stokes Raman Scattering Process

As shown above, spontaneous Raman microspectroscopy is powerful for elucidating intracellular structure *in vivo* with three-dimensional sectioning capability. However, it may not be suitable to trace a detailed dynamical behavior inside the cell, because of its relatively low efficiency. Spontaneous Raman process often requires several minutes to obtain one spectrum. This low efficiency originates from the small scattering cross section of the spontaneous Raman process. An alternative approach to obtain vibrational images with high speed is coherent anti-Stokes Raman scattering (CARS) [5, 22–27]. In particular, multiplex CARS microspectroscopy is promising because of its capability to obtain vibrational spectra efficiently [24, 28–31]. Figure 16.5 shows an energy diagram for the multiplex CARS process. The multiplex CARS process requires two laser sources, namely, a narrow-band pump laser (ω_1) and a broadband Stokes laser (ω_2). The multiple vibrational coherences are created because of the wide spectral range of the frequency difference, $\omega_1 - \omega_2$. If we can prepare ultrashort laser pulses, an impulsive Raman excitation and a subsequent narrow-band probe can also generate a multiplex CARS spectrum [32, 33]. One of the most prominent features of multiplex CARS microspectroscopy lies in the fact that it can easily distinguish the concentration change of a particular molecule from the structural change through the spectral analysis [11]. It should be emphasized that a single-wavenumber CARS detection, which is widely adopted in CARS microscopy, cannot discriminate these two phenomena. Although there were several restrictions on the spectral coverage of multiplex CARS microspectroscopy mainly due to the bandwidth of the laser emission [24, 28–31], the spectral coverage has been significantly broadened using the supercontinuum light source

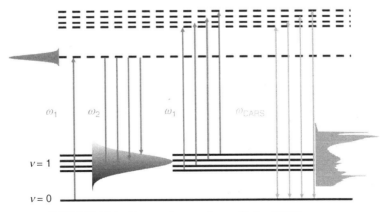

FIGURE 16.5 Energy diagram for multiplex CARS process.

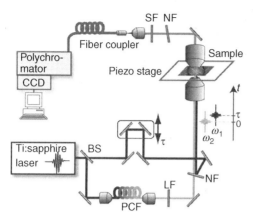

FIGURE 16.6 Experimental setup for multiplex CARS microspectroscopy; BS: beam splitter; NBF: narrow bandpass filter; LF: long-wavelength pass filter; NF: notch filter; SF: short-wavelength pass filter.

generated from a photonic crystal fiber [11, 34–36] or a tapered fiber [37]. Recently, the spectral coverage of the multiplex CARS microspectroscopy has been extended to be more than 3500 cm^{-1}.

16.4.2 Experimental Setup for Ultrabroadband Multiplex CARS Microspectroscopy

Figure 16.6 shows a schematic of the multiplex CARS microspectrometer. An unamplified mode-locked Ti:sapphire laser (Coherent, Vitesse-800) provides wavelength fixed radiation at 800 nm, which is used as the pump laser. Typical duration, pulse energy, and repetition rate are 100 fs, 12 nJ, and 80 MHz, respectively. A portion of the output from the oscillator is used as a seed for generating a supercontinuum in the PCF (crystal fiber, NL-PM-750). The input pulse energy for the supercontinuum generation is less than 2.3 nJ. As shown in Figure 16.6, the fundamentals of the Ti:sapphire laser and the supercontinuum are used for the pump (ω_1) and Stokes (ω_2) lasers, respectively. To obtain Raman spectrum

with high frequency resolution, the pump laser pulses are spectrally filtered using a narrow bandpass filter. The bandwidth was measured to be about 20 cm^{-1}. Because the pump laser is in the near-infrared (NIR) region, the Stokes laser must also be in the NIR. The visible component in the supercontinuum is thus blocked by a long-wavelength pass filter. Thanks to the NIR excitation, we can expect several advantages such as low photodamage, suppression of the nonresonant background signal, and a deep penetration depth for opaque samples. Two laser pulses are superimposed collinearly using an 800 nm notch filter, and then tightly focused onto the sample with a 40× 0.9 NA microscope objective. Under the tight focusing condition, the phase-matching conditions are relaxed due to the large angular dispersion and the small interaction volume [38, 39]. The relaxation of the phase-matching condition is important especially for CARS spectroscopy, because a wide range of vibrational resonances can be accomplished simultaneously using the ultrabroadband Stokes laser. A 40× 0.6 NA microscope objective is used to collect the forward-propagating CARS signal. Finally, the CARS signal is guided to a polychromator (Acton, SpectraPro-300i), and is detected by a CCD camera (Roper Scientific, Spec-10:400BR/XTE). The multiplex CARS images are measured by a point-by-point acquisition of the CARS spectrum. The sample is moved using piezo-driven *xyz* translators (MadCity, Nano-LP-100). We used fission yeast *S. pombe* as a sample [7–9]. The nuclei of yeast cells were labeled by GFP.

16.4.3 CARS Imaging of Single Living Cells

Figure 16.7a shows a typical spectral profile of the CARS signal of a living yeast cell. As clearly shown, a strong signal is observed at the Raman shift of 2840 cm^{-1}. This band originates from C−H stretch vibrational mode, which shows a slightly dispersive lineshape due to an interference with a nonresonant background. On the basis of our previous spontaneous Raman [7–9] and CARS [11] studies, the signal at the

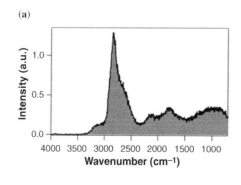

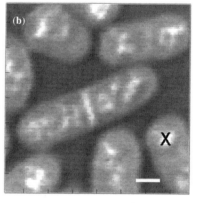

FIGURE 16.7   (a) Typical spectral profile of the CARS signal from a living yeast cell; (b) CARS image of living yeast cells at the Raman shift of C−H stretching vibrational mode. The short bar measures 2 μm.

(a)   (b)

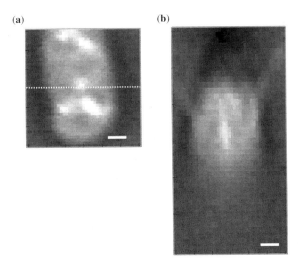

FIGURE 16.8 Lateral (a) and axial (b) CARS images of a yeast cell at C–H stretching vibrational mode. The short bars correspond to 1 μm.

Raman shift of 2840 cm^{-1} is found particularly strong in mitochondria, because mitochondrion is an organelle containing a high concentration of phospholipids. Figure 16.7b shows the vibrationally resonant CARS image at C–H stretching mode using a differentiation method [11]. The CARS spectrum of a yeast cell in Figure 16.7a is obtained at $(x, y) = (5.65 \, \mu m, -3.05 \, \mu m)$, which is indicated as a black cross in Figure 16.7b. In Figure 16.7b, yeast cells at various cell-cycle stages are clearly imaged. Especially, a septum is visualized in the yeast cell around the center of Figure 16.7b. The septum is composed of carbohydrates such as polysaccharide, which is also rich in C–H bonds.

Thanks to the three-dimensional sectioning capability, CARS microscopy enables us to obtain not only a lateral but also an axial slice of a living yeast cell. Figure 16.8a shows a lateral CARS image of a yeast cell. Figure 16.8b corresponds to the vertical slice of the yeast cell at the position of $y = 0$. The CARS signal is weaker at the top part than at the bottom part. This variation is due to the imperfect focusing of the two laser beams because of the spatially heterogeneous refractive index inside the cell.

16.4.4 Multi-Nonlinear Optical Image of Single Living Cells

The supercontinuum light source can also be used as an excitation light source for the two-photon excitation fluorescence (TPEF) [40–42]. Owing to the broadband spectral profile of the supercontinuum, the two-photon allowed electronic state can be excited efficiently in comparison with conventional TPEF microscopy using a Ti:sapphire oscillator. Figure 16.9a shows the CARS and the TPEF spectra obtained in a 100 ms exposure time. The nuclei of the yeast cells in the present study are labeled by GFP. A broad but

distinguishable peak is observed around 506 nm. Taking into account the spectral profile of this signal, it is assigned to the TPEF signal due to GFP. Figure 16.9b and c shows CARS and TPEF images. It should be emphasized that both images of multiplex CARS and TPEF signals are obtained simultaneously in one single measurement. Since there is no overlap between the CARS and TPEF signals in the spectral domain, it can be easily differentiated using a spectrometer. Although dual imaging of the CARS and the TPEF signals has been reported using two synchronized Ti:sapphire oscillators [43], full spectral information is obtained for the first time in the present study.

In conclusion, we can obtain Raman spectrum and Raman image with high speed using ultrabroadband multiplex CARS microspectroscopy. At the moment, the exposure time is 30 ms, which is about 1/30 in comparison with that of spontaneous Raman microscopy.

16.4.5 *In Vivo* Measurement of a Cell Division Process

CARS technique enables us to obtain vibrational images with a high speed. We have applied it to the cell division process [44]. Figure 16.10a and b shows a CARS at the C–H stretch mode and TPEF images, respectively. The sample is a living fission yeast cell whose nucleus is labeled by GFP. The number at the upper left shows the time of observation. Exposure time at each spatial point is 50 ms. Each image is constructed from 61 × 61 CARS spectra obtained by scanning the sample from the bottom to the top. It took 3.8 min for obtaining one image, which determined the temporal resolution in the present experiment. Two living cells at the G1 phase are observed at the beginning. Both cells have septa at around the center of the cell. A strong CARS signal due to membranous organelles is also observed at each compartment. First, we focus on the yeast cell at the lower side of the images. The CARS signal from the septum decreases in the cell division (from (a-1) to (a-4)). The cell finally splits into two daughter cells (a-4). One of the daughter cells moves almost out of the field of view at (a-8). For the other daughter cell, the CARS signal inside the cell does not show significant distribution change from (a-10) to (a-21) in comparison with that in the dividing process (from (a-1) to (a-4)). Second, the yeast cell at the upper side is discussed. The CARS signal from the septum gradually increases from (a-1) to (a-10). Next, it decreases slightly at around (a-13), and the cell splits into two daughter cells at (a-15). After the cell division, the daughter cell in the field of view still shows dynamic distribution change of the CARS intensity inside the cell, which is in contrast to the cell at the lower part. It could be explained by the migration of organelles in the axial direction. As shown in Figure 16.10b, the relative positions of nuclei inside the cell do not change significantly in the course of the cell cycle during the observation. It is also noted that the TPEF signal intensity becomes weaker and weaker in the

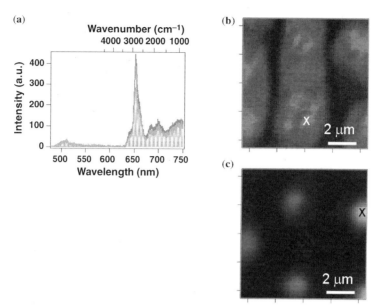

FIGURE 16.9   (a) Spectral profile of the CARS and TPEF signals of a living yeast cell; (b) CARS lateral images of living yeast cells for C–H stretching mode; (c) TPEF lateral images of the same system at 506 nm. The red and green spectra in (a) are obtained at the white and the black crosses in (b) and (c), respectively. (See the color version of this figure in Color Plate section.)

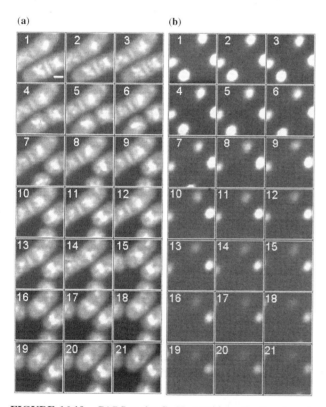

FIGURE 16.10   CARS at the C–H stretching vibrational mode (a) and TPEF (b) images, respectively. The sample is a living yeast cell, whose nucleus is labeled by GFP. The scale bar corresponds to 2 μm. The number at the upper left shows the time course of the observation. Exposure time at each spatial point is 50 ms. Lateral (XY) images consist of 61 × 61 pixels, and are measured in 3.8 min per one image.

course of the cell division. It is probably due to a photobleaching effect by laser irradiation. On the other hand, the CARS signal intensity does not deteriorate. This result manifests another advantage of the CARS imaging, which does not suffer inherently from the photobleaching effect.

16.5   HYPER-RAMAN MICROSPECTROSCOPY

Using spontaneous Raman and/or CARS microspectroscopy, we can investigate the structure and dynamics of chemical and biological systems, such as living cells *in vivo*. However, Raman and CARS allow us to observe only the Raman active vibrational modes because of the selection rule. Thus, Raman or CARS microspectroscopy alone is not sufficient to obtain full vibrational information on molecules under investigation. Imagine a situation in which we have the Raman spectrum of an unknown species in a biological system under a microscope and we cannot extract it for analysis with X-ray crystallography and/or NMR. Then, we would like to have the infrared (IR) spectrum measured under the same microscope. However, the spatial resolution of infrared microscopy is very limited. It is on the order of several micrometers due to the diffraction limit of infrared light. Although a near-field technique has been introduced to IR microscopy, it is still in a developing stage [45, 46]. To overcome this difficulty, we have developed hyper-Raman microspectroscopy that allows investigation of Raman inactive but infrared active vibrational modes with a spatial resolution as high as that of Raman microspectroscopy. This development is a new approach to achieve complete vibrational microspectroscopy that is not restricted by the selection rule and that provides full vibrational information on molecules detected only under a microscope [47].

Hyper-Raman scattering, one of the nonlinear Raman effects, was first observed in 1965 [48]. Since then, there have been many researches, both theoretically and experimentally [48–54] focused on this phenomenon. According to the selection rule [49–52], any IR active vibrational mode is HR active. Therefore, with the use of a visible or near-IR laser source, HR microscopy can obtain IR equivalent vibrational information with spatial resolution as high as that of Raman microscopy. Moreover, the nonlinear property of the process provides some advantages over conventional Raman microscopy, such as inherent three-dimensional sectioning capability and the absence of interference from one-photon fluorescence.

The sample we have chosen for the present study is all-*trans*-β-carotene, which is a typical carotenoid widely found in nature. There are two reasons for the choice. One is to show that a natural substance can be a good HR probe under a microscope. The other is to demonstrate the selection rule under the mutual exclusion principle. Note that all-*trans*-β-carotene has the inversion symmetry.

The light source is a mode-locked Ti:sapphire laser (Coherent, Vittesse-800), which is the same light source for multiplex CARS microspectroscopy. The output from the femtosecond laser is too broad to be used for HR excitation. Thus, it is spectrally filtered using a narrow bandpass filter or a grating and slit pair. A variable neutral density filter is used for adjusting the excitation power. The filtered laser output is introduced to an inverted microscope (Nikon, TE2000-S) and then focused onto a sample with an objective (40× NA 0.9). HR scattering is collected with the same objective, passed through a dichroic mirror and a couple of short-wavelength pass filters (Asahi Bunko), and is coupled into an optical fiber to be guided into a polychromator (Acton, SpectraPro-300i). Finally, HR spectra are recorded on a charge-coupled device (CCD, Roper Scientific, Spec-10:400BR). A HR image is acquired by scanning the sample with a piezo-driven *xyz* translator (MadCity, Nano-LP-100). It is noted that this setup is part of the multiplex CARS system. By blocking the Stokes supercontinuum, the system can be changed from CARS to HR microspectroscopy.

Spontaneous Raman spectra were obtained with a confocal Raman microspectrometer (Nanofinder, Tokyo Instruments, Inc.). The light source was a cw He–Ne laser. Infrared absorption spectra in KBr disks were recorded on a JASCO FT/IR-670 spectrometer.

All-*trans*-β-carotene and benzene were purchased from WAKO Pure Chemical Industries and from Nacalai Tesque, Inc., respectively. All reagents were used without further purification. Microcrystals of β-carotene were obtained by recrystallization from benzene solutions. A typical size of the microcrystals used in the present study

was 20 μm in length. All sample preparation was made under deep red light.

HR, Raman, and IR spectra of crystalline all-*trans*-β-carotene are shown in Figure 16.11. In the HR spectrum, an intense band is observed at 1564 cm^{-1}. This wavenumber is different from that in the Raman spectrum at 1523 cm^{-1} but is close to that in the IR spectrum at 1561 cm^{-1}. The wavenumber is also consistent with the normal coordinate analysis of all-*trans*-β-carotene [55]. Assuming that all-*trans*-β-carotene belongs to the point group C_{2h}, HR active vibrational modes should be infrared active. Therefore, the HR signal at 1564 cm^{-1} is safely ascribed to a C=C and C–C stretch vibration of the conjugated chain [55]. Comparing HR and IR spectra in Figure 16.11, we notice that the relative intensities of HR and IR bands are different. In fact, some IR active bands are not observed in the HR spectrum. It can be explained by considering the origins of HR and IR signal intensities. The HR signals are generated through the first derivatives of hyperpolarizability with respect to the vibrational normal coordinates, while IR signals are through such derivatives of the dipole moment. The electronic resonance effect may also contribute to the HR signal. We also measured all-*trans*-β-carotene in cyclohexane dilute solutions. The overall spectral profile in solution was similar to that in the microcrystals.

Figure 16.12a shows an overall spectral profile of an all-*trans*-β-carotene microcrystal measured in 1 s. An intense

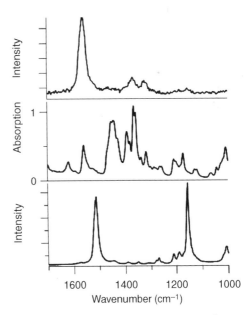

FIGURE 16.11 Vibrational spectra of all-*trans*-β-carotene microcrystals; HR spectrum (top), Raman spectrum (middle), infrared absorption spectrum (bottom) measured in a KBr disk. A vertical solid line indicates 1564 cm^{-1}. HR spectrum was obtained with the pulse energy of 3 mW at the sample point. The exposure time was 5 min. The HR spectrum has been corrected for the two-photon fluorescence background.

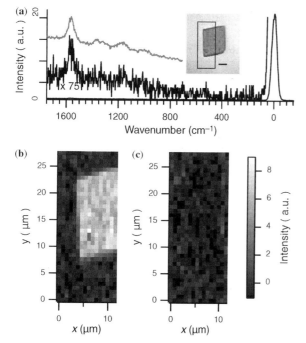

FIGURE 16.12 (a) HR spectrum of an all-*trans*-β-carotene microcrystal measured in 1 s. The gray line is obtained by averaging the 220 spatially resolved spectra of the microcrystal; the HR images of the microcrystal at the Raman shift of (b) 1564 cm^{-1} and (c) 1944 cm^{-1}. An inset shows a microscopic image of the same crystal. A black bar in the inset indicates 5 μm. The pulse energy of the incident laser was 8 mW at the sample point. The exposure time was 1 s for each pixel. The whole image was acquired in 10 min.

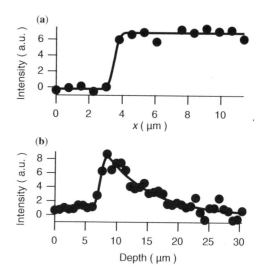

FIGURE 16.13 (a) Lateral and (b) axial intensity profiles of the HR signal (filled circles) of (a) a microcrystal at 1564 cm^{-1} and (b) all-*trans*-β-carotene in cyclohexane at 1574 cm^{-1}. Fitted curves are indicated by solid lines.

signal is observed at 0 cm^{-1}, which corresponds to the hyper-Rayleigh scattering or the second harmonic of the excitation laser field. The broad background is probably due to the two-photon fluorescence of the sample. Figure 16.12b and c shows two HR lateral images of the microcrystal at the Raman shifts of 1564 and 1944 cm^{-1}, respectively. As clearly shown in Figure 16.12b, HR image is successfully obtained at 1564 cm^{-1}. On the other hand, no vibrational contrast is observed at 1944 cm^{-1} (Figure 16.12c). The high vibrational contrast in Figure 16.12b with the exposure time as short as 1 s for a pixel demonstrates the feasibility of HR microspectroscopy as a new tool for vibrational imaging.

As shown in the following, HR microspectroscopy has a spatial resolution much higher than that of conventional IR microspectroscopy. First, the lateral spatial resolution is evaluated. The HR intensity profile at the edge of the crystal shown in Figure 16.12 is shown in Figure 16.13a. By fitting this profile with a step function convoluted with a Gaussian, the full-width at half-maximum is determined to be 0.6 ± 0.2 μm. Although the value 0.6 μm is considered to be the largest possible spatial resolution, it corresponds well to the theoretical value [3] $0.61\lambda/\sqrt{2}NA = 0.38$ μm from the laser beam spot size. It is emphasized that we have successfully obtained a vibrational image of an infrared active vibrational

mode with submicrometer resolution. The spatial resolution of a microscope is grossly determined by the spot size of the focused laser beam. The spot size is proportional to the wavelength of incident light if it is tightly focused to the diffraction limit. Therefore, it is difficult to achieve high spatial resolution with IR microscopy, which uses longer wavelength light. IR microscopy with submicrometer resolution has been demonstrated by IR near-field scanning optical microscopy (IR-NSOM) with tip-enhanced infrared absorption technique [7] or with specially designed aperture probes [46]. However, the application of IR-NSOM to biological sample is intrinsically limited, because IR-NSOM cannot provide three-dimensional sectioning capability. A strong absorption by water is also a drawback for IR microscopy. The HR microspectroscopy, which has been developed by us, can be regarded as a unique alternative, because it enables us to obtain a vibrational image with an IR active mode with a submicrometer spatial resolution. Furthermore, HR microspectroscopy can also be employed to measure IR and Raman inactive vibrational modes.

Second, we discuss the depth resolution. Utilizing the second order nonlinear optical process, HR microspectroscopy has an intrinsic high axial resolution. Figure 16.13b shows the depth dependence of the HR signal of β-carotene in a cyclohexane solution at 1574 cm^{-1}, which was measured across the interface between the cover glass and the solution. The intensity profile is fitted well with a single exponential function convoluted by a Gaussian. The exponential decay is most probably due to the reabsorption of the scattered light by the solution. From the rise of the signal in glass/solution interface, the axial spatial resolution is estimated to be 1.4 ± 0.4 μm. The observed axial spatial resolution is comparable to the theoretical estimation [3], $2\lambda/\sqrt{2}NA^2 = 1.4\,\mu m$.

HR microspectroscopy is shown to provide vibrational information under a microscope on Raman inactive but infrared active modes. With a careful choice of the exciting laser wavelength and the use of electronic resonances, we believe that this technique can be applied to a wide variety of organic materials. The lateral and axial spatial resolution is evaluated to be less than 0.6 and 1.4 µm, respectively. The lateral resolution of HR microspectroscopy is much better than that of conventional infrared microscopy. Thus, combination of Raman and HR methods accomplishes high spatial resolution vibrational microspectroscopy that is not restricted by the selection rule.

16.6 CONCLUSIONS

Owing to its inherent high molecular specificity, linear and nonlinear Raman spectroscopies provide rich information on molecular composition, structure, and dynamics in a living cell. In addition, the *"Raman spectroscopic signature of life"* enables us to visualize metabolically active mitochondria without any pretreatment like dye labeling. In the near future, we will be able to discuss the life and death of a single living cell quantitatively at the molecular level, using time- and space-resolved linear and nonlinear Raman spectroscopy.

ACKNOWLEDGMENTS

It is a pleasure to compile our recent studies on Raman spectroscopy and imaging of living cells carried out in our laboratory at the University of Tokyo. The authors are grateful to all the members of the laboratory for their cooperation and support. They also thank Dr. T. Karashima, Prof. M. Yamamoto, and Prof. A. Toh-e for their collaboration.

REFERENCES

1. Puppels, G. J., De Mul, F. F. M., Otto, C., Greve, J., Robert-Nicoud, M., Arndt-Jovin, D. J., and Jovin, T. M. (1990) *Nature* **347**, 301.

2. Maquelin, K., Choo-Smith, L. P., van Vreeswijk, T., Endtz, H. P., Smith, B., Bennett, R., Bruining, H. A., and Puppels, G. J. (2000) *Anal. Chem.* **72**, 12.

3. Schuster, K. C., Urlaub, E., and Gapes, J. R. (2000) *J. Microbiol. Methods* **42**, 29.

4. Mohacek-Grosev, V., Bozac, R., and Puppels, G. J. (2001) *Spectrochim. Acta* **57A**, 2815.

5. Cheng, J.-X., Jia, Y. K., Zheng, G., and Xie, X. S. (2002) *Biophys. J.* **83**, 502.

6. Xie, C. and Li, Y.-Q. (2003) *J. Appl. Phys.* **94**, 6138.

7. Huang, Y.-S., Karashima, T., Yamamoto, M., and Hamaguchi, H. (2003) *J. Raman Spectrosc.* **34**, 1.

8. Huang, Y.-S., Karashima, T., Yamamoto, M., Ogura, T., and Hamaguchi, H. (2004) *J. Raman Spectrosc.* **35**, 525.

9. Huang, Y.-S., Karashima, T., Yamamoto, M., and Hamaguchi, H. (2005) *Biochemistry* **44**, 10009.

10. Naito, Y., Toh-e, A., and Hamaguchi, H. (2005) *J. Raman Spectrosc.* **36**, 837.

11. Kano, H. and Hamaguchi, H. (2005) *Opt. Express* **13**, 1322.

12. Maiti, N. C., Apetri, M. M., Zagorski, M. G., Carey, P. R., and Anderson, V. E. (2004) *J. Am. Chem. Soc.* **126**, 2399.

13. Jeong, H., Mason, S. P., Barabasi, A. L., and Oltvai, Z. N. (2001) *Nature* **411**, 41.

14. Duffus, J. H. (1975) In: Prescott, D. M. (Ed.), *Methods in Cell Biology*, Academic Press, New York.

15. Siamwiza, M. N., Lord, R. C., Chen, M. C., Takamatsu, T., Harada, I., Matsuura, H., and Shimanouchi, T. (1975) *Biochemistry* **14**, 4870.

16. Gaber, B. P. and Peticolas, W. L. (1977) *Biochim. Biophys. Acta* **465**, 260.

17. Takai, Y., Masuko, T., and Takeuchi, H. (1977) *Biochim. Biophys. Acta* **465**, 260.

18. Lippert, J. L. and Peticolas, W. L. (1971) *Proc. Natl. Acad. Sci. USA* **68**, 1572.

19. Dauchez, M., Derreumaux, P., Lagant, P., Vergoten, G., Sekkal, M., and Legrand, P. (1994) *Spectrochim. Acta* **50A**, 87.

20. Dauchez, M., Lagant, P., Derreumaux, P., Vergoten, G., Sekkal, M., Legrand, P., and Sombret, B. (1994) *Spectrochim. Acta* **50A**, 105.

21. Allan, R. A. and Miller, J. J. (1980) *Can. J. Microbiol.* **26**, 912.

22. Hashimoto, M., Araki, T., and Kawata, S. (2000) *Opt. Lett.* **25**, 1768.

23. Zumbusch, A., Holtom, G. R., and Xie, X. S. (1999) *Phys. Rev. Lett.* **82**, 4142.

24. Wurpel, G. W. H., Schins, J. M., and Mueller, M. (2002) *Opt. Lett.* **27**, 1093.

25. Paulsen, H. N., Hilligsoe, K. M., Thogersen, J., Keiding, S. R., and Larsen, J. J. (2003) *Opt. Lett.* **28**, 1123.

26. Schaller, R. D., Ziegelbauer, J., Lee, L. F., Haber, L. H., and Saykally, R. J. (2002) *J. Phys. Chem. B* **106**, 8489.

27. Ichimura, T., Hayazawa, N., Hashimoto, M., Inouye, Y., and Kawata, S. (2004) *Phys. Rev. Lett.* **92**, 220801/1.

28. Otto, C., Voroshilov, A., Kruglik, S. G., and Greve, J. (2001) *J. Raman Spectrosc.* **32**, 495.

29. Cheng, J.-X., Volkmer, A., Book, L. D., and Xie, X. S. (2002) *J. Phys. Chem. B* **106**, 8493.

30. Oron, D., Dudovich, N., and Silberberg, Y. (2002) *Phys. Rev. Lett.* **89**, 273001.

31. Oron, D., Dudovich, N., Yelin, D., and Silberberg, Y. (2002) *Phys. Rev. Lett.* **88**, 063004/1.

32. Lim, S.-H., Caster, A. G., and Leone, S. R. (2005) *Phys. Rev. A* **30**, 2805.

33. Kano, H. and Hamaguchi, H. (2006) *J. Raman Spectrosc.* **37**, 411.

34. Konorov, S. O., Akimov, D. A., Serebryannikov, E. E., Ivanov, A. A., Alfimov, M. V., and Zheltikov, A. M. (2004) *Phys. Rev. E* **70**, 057601.

35. Kano, H. and Hamaguchi, H. (2005) *Appl. Phys. Lett.* **86**, 121113/1.

36. Petrov, I. G. and Yakovlev, V. V. (2005) *Opt. Express* **13**, 1299.

37. Kee, T. W. and Cicerone, M. T. (2004) *Opt. Lett.* **29**, 2701.

38. Toleutaev, B. N., Tahara, T., and Hamaguchi, H. (1994) *Appl. Phys. B* **59**, 369.

39. Cheng, J.-X., Volkmer, A., and Xie, X. S. (2002) *J. Opt. Soc. Am. B* **19**, 1363.

40. McConnell, C. and Riis, E. (2004) *Phys. Med. Biol.* **49**, 4757.

41. Isobe, K., Watanabe, W., Matsunaga, S., Higashi, T., Fukui, K., and Itoh, K. (2005) *Jpn. J. Appl. Phys. Part 2*, **44**, L167.

42. Palero, J. A., Boer, V. O., Vijverberg, J. C., Gerritsen, H. C., and Sterenborg, H. J. C. M. (2005) *Opt. Express* **13**, 5363.

43. Wang, H., Fu, Y., Zickmund, P., Shi, R., and Cheng, J.-X. (2005) *Biophys. J.* **89**, 581.

44. Kano, H. and Hamaguchi, H. (2007) *Anal. Chem.* **79**, 8967.

45. Knoll, B. and Keilmann, F. (1999) *Nature (London)* **399**, 134.

46. Masaki, T., Inouye, Y., and Kawata, S. (2004) *Rev. Sci. Instrum.* **75**, 3284.

47. Shimada, R., Kano, H., and Hamaguchi, H. (2006) *Opt. Lett.* **31**, 320.

48. Terhune, R. W., Maker, P. D., and Savage, C. M. (1965) *Phys. Rev. Lett.* **14**, 681.

49. Long, D. A. and Stanton, L. (1970) *Proc. Roy. Soc. Lond. A* **318**, 441.

50. Andrews, D. L. and Thirunamachandran, T. (1978) *J. Chem. Phys.*

51. Ziegler, L. D. (1990) *J. Raman Spectrosc.* **21**, 769.

52. Bonang, C. C. and Cameron, S. M. (1992) *Chem. Phys. Lett.* **192**, 303.

53. Mizuno, M., Hiroo, H., and Tahara, T. (2002) *J. Phys. Chem. A* **106**, 3599.

54. Kelley, A. M., Leng, W., and Blanchard-Desce, M. (2003) *J. Am. Chem. Soc.* **125**, 10520.

55. Saito, S. and Tasumi, M. (1983) *J. Raman Spectrosc.* **14**, 310.

INDEX